STUDY GUIDE AND SELECTED SOLUTIONS MANUAL

Karen C. Timberlake
Los Angeles Valley College

Mark Quirie
Algonquin College

BASIC CHEMISTRY

FOURTH EDITION

PEARSON

Boston Columbus Indianapolis New York San Francisco Upper Saddle River
Amsterdam Cape Town Dubai London Madrid Milan Munich Paris Montréal Toronto
Delhi Mexico City São Paulo Sydney Hong Kong Seoul Singapore Taipei Tokyo

Editor in Chief: Adam Jaworski
Senior Marketing Manager: Jonathan Cottrell
Project Editor: Jessica Moro
Assistant Editor: Coleen Morrison
Managing Editor, Chemistry and Geosciences: Gina M. Cheselka
Senior Project Manager: Beth Sweeten
Composition/Production Management: PreMediaGlobal
Cover Design: Seventeenth Street Studios
Cover Credit: Kenneth Libbrecht / Photo Researchers, Inc.

1 2 3 4 5 6 7 8 9 10—EBM—16 15 14 13 12 11

PEARSON

www.pearsonhighered.com

ISBN 10: 0-321-83443-7
ISBN 13: 978-0-321-83443-0

Contents

Answers to Selected Problems

Preface

This Study Guide is intended to accompany *Basic Chemistry, Fourth Edition.* The purpose of this Study Guide is to provide you with additional learning resources to increase your understanding of the key concepts in the text. Each chapter in the Study Guide correlates with a chapter in the text. For every Section in each chapter, there are Learning Exercises and Answers that focus on problem solving, which in turn promote an understanding of the chemical principles. Following the Learning Exercises, a Check List of Learning Goals, and a multiple-choice Practice Test provide a review of the chapter content. The Answers and Solutions for all the odd-numbered problems in each chapter of the textbook are included.

I hope that this Study Guide will help in the learning of chemistry. If you wish to make comments or corrections, or ask questions, you can send me an e-mail message at khemist@aol.com.

Karen Timberlake

To the Student

One must learn by doing the thing;
though you think you know it, you
have no certainty until you try.
—*Sophocles*

Here you are in a chemistry class with your textbook in front of you. Perhaps you have already been assigned some reading or some problems to do in the book. Looking through the chapter, you may see words, terms, and pictures that are new to you. This may very well be your first experience with a science class like chemistry. At this point you may have some questions about what you can do to learn chemistry. This *Study Guide* is written with those considerations in mind.

Learning chemistry is similar to learning something new such as tennis or skiing or driving. If I asked you how you learn to play tennis or ski or drive a car, you would probably tell me that you would need to practice often. It is the same with learning chemistry. Learning the chemical concepts and learning to work the problems depends on the time and effort you invest in it. If you practice every day, you will find that learning chemistry is an exciting experience and a way to understand the current issues of the environment, health, and medicine.

Manage Your Study Time

I often recommend a study system to students in which you read one section of the text and immediately practice the questions and problems that are at the end of that section. In this way, you concentrate on a small amount of information and actively use what you learned to answer questions. This helps you to organize and review the information without being overwhelmed by the entire chapter. It is important to understand each section, because they build like steps. Information presented in each chapter proceeds from the basic to the more complex. Perhaps you can only study three or four sections of the chapter. As long as you also practice doing some problems at the same time, the information will stay with you.

Form a Study Group

I highly recommend that you form a study group in the first week of your chemistry class. Working with your peers will help you use the language of chemistry. Scheduling a time to meet each week will help you study and prepare to discuss problems. You will be able to teach some things to other students in the group, and sometimes they will help you understand a topic that puzzles you. You won't always understand a concept right away. Your group will help you see your way through it. Most of all, a study group creates a strong support system when students help each other successfully complete the class.

Go to Office Hours

Try to go to your tutor's and/or professor's office hours. Your professor wants you to understand and enjoy learning this material. Often a tutor is assigned to a class or there are tutors available at your college. Don't be intimidated. Going to see a tutor or your professor is one of the best ways to clarify what you need to learn in chemistry.

Using This Study Guide

Now you are ready to sit down and study chemistry. Let's go over some methods that can help you learn chemistry. This *Study Guide* is written specifically to help you understand and practice the chemical concepts that are presented in your class and in your text. The following features are part of this *Study Guide:*

1. Learning Goals

The Learning Goals give you an overview of what each Section in the chapter is about and what you can expect to accomplish when you complete your study and learning of that Section.

2. Key Terms

Key Terms appear throughout each Chapter in the Study Guide. As you complete the description of the Key Terms, you will have an overview of the topics you have studied in that chapter. Because many of the Key Terms may be new to you, this is an opportunity to review their meaning.

3. Chapter Sections

Each section begins with a list of the important ideas to guide you through each of the learning activities. When you are ready to begin your study, read the matching section in the textbook and review the *Sample Problems* in the text. Included are the Guides to Problem Solving with step-by-step directions for working out a problem as well as the icons for *Key Math Skills and Core Chemistry Skills* that are found in the textbook.

4. Learning Exercises

The Learning Exercises give you an opportunity to practice problem solving related to the chemical principles in the chapter. Each set of exercises reviews one chemical principle. The answers are found immediately following each exercise. Check your answers right away. If they don't match the answer in the study guide, go back to the textbook and review the material again. It is important to make corrections before you go on. Chemistry, involves a layering of skills such that each one must be understood before the next one can be learned.

At various times, you will notice some essay questions that illustrate one of the concepts. I believe that writing out your ideas is a very important way of learning content. If you can put your problem-solving techniques into words, then you understand the patterns of your thinking and you will find that you have to memorize less.

5. Checklist

Use the Checklist to check your understanding of the Learning Goals. This gives you an overview of the major topics in the section. If something does not sound familiar, go back and review. One aspect of being a strong problem-solver is the ability to check your knowledge and understanding as you go along.

6. Practice Test

A Practice Test is found at the end of each chapter. When you have learned the material in a chapter, you can apply your understanding to the Practice Test. If the results of this test indicate that you know the material, you are ready to proceed to the next chapter. The Practice Test questions are keyed to the Learning Goals, which allows you to identify the Sections you may need to review. Answers are found at the end of the Practice Test.

7. Answers and Solutions to Selected Problems in the Text

The complete solutions to all of the odd-numbered problems for each chapter can be found at the end of this Study Guide. As you work the odd-numbered questions and problems in the text, you can check your answers and see the solutions.

1
Chemistry in Our Lives

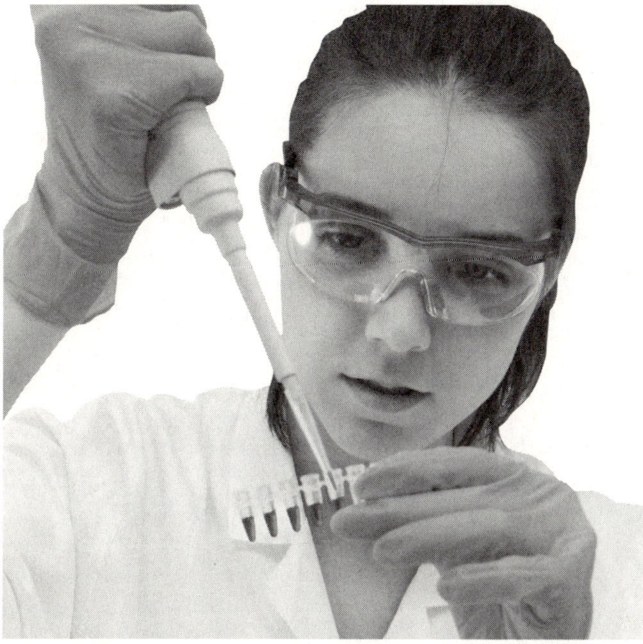

When a female victim is found dead in her home, samples of her blood and stomach contents are sent to Sarah, who works as a forensic scientist. After using a variety of chemical tests, Sarah concludes that the victim was poisoned when she ingested ethylene glycol. Since the initial symptoms of ethylene glycol poisoning are similar to being intoxicated, the victim was unaware of the poisoning. When ethylene glycol is oxidized, the products can cause renal failure and may be toxic to the body. Two weeks later, police arrested the victim's husband after finding a bottle of ethylene glycol in the kitchen of the couple's home. How does this result show the use of the scientific method?

1.1 Chemistry and Chemicals

Learning Goal: Define the term chemistry and identify substances as chemicals.

- Chemistry is the study of the composition, structure, properties, and reactions of matter.
- A chemical is a substance containing one type of material that has the same composition and properties wherever it is found.

♦ Learning Exercise 1.1

Indicate if each of the following is a chemical or not:

a. _____ aluminum **b.** _____ heat

c. _____ sodium fluoride in toothpaste **d.** _____ ammonium nitrate in fertilizer

e. _____ time

Answers **a.** yes **b.** no **c.** yes **d.** yes **e.** no

1.2 Scientific Method: Thinking Like a Scientist

Learning Goal: Describe the activities that are part of the scientific method.

- The scientific method is a process of making observations, writing a hypothesis, and testing the hypothesis with experiments.
- A theory develops when experiments that validate a hypothesis are repeated by many scientists with consistent results.

Key Terms for Sections 1.1–1.2

Match each of the following key terms with the correct definition:

a. scientific method　**b.** experiment　**c.** hypothesis
d. theory　**e.** chemistry

1. _____ an explanation of nature validated by many experiments
2. _____ the study of substances and how they interact
3. _____ a possible explanation of a natural phenomenon
4. _____ the process of making observations, writing a hypothesis, and testing with experiments
5. _____ a procedure used to test a hypothesis

Answers　**1.** d　**2.** e　**3.** c　**4.** a　**5.** b

♦ Learning Exercise 1.2

Identify each of the following as an observation (O), a hypothesis (H), or an experiment (E).

Students make observations in the chemistry laboratory.

a. _____ Sunlight is necessary for the growth of plants.

b. _____ Plants in the shade were shorter than plants in the sun.

c. _____ Plant leaves are covered with aluminum foil and their growth measured.

d. _____ Fertilizer is added to plants.

e. _____ Ozone slows plant growth by interfering with photosynthesis.

f. _____ Ozone causes brown spots on plant leaves.

Answers　**a.** H　**b.** O　**c.** E　**d.** E　**e.** H　**f.** O

1.3 Learning Chemistry: A Study Plan

Learning Goal: Develop a study plan for learning chemistry.

- Components of the text that promote learning include: *Looking Ahead, Learning Goals, Chapter Readiness, Key Math Skills, Core Chemical Skills, Concept Checks, Sample Problems, Study Checks, Guides to Problem Solving (GPS), Chemistry Link to Health, Chemistry Link to the Environment, Chemistry Link to History, Chemistry Link to Industry, Questions and Problems, Concept Maps, Chapter Reviews, Key Terms, Understanding the Concepts, Additional Questions and Problems, Challenge Problems, Glossary/Index, Answers to Selected Questions and Problems, and Combining Ideas.*
- An active learner continually interacts with chemical concepts while reading the text and attending lecture.
- Working with a study group clarifies ideas and illustrates problem solving.

♦ Learning Exercise 1.3

Which of the following activities would be included in a successful study plan for learning chemistry?

a. _____ attending lecture once in a while

b. _____ working problems with friends from class

c. _____ attending review sessions

d. _____ planning a regular study time

e. _____ not doing the assigned problems

f. _____ going to the instructor's office hours

Answers **a.** no **b.** yes **c.** yes **d.** yes **e.** no **f.** yes

1.4 Learning Chemistry: Key Math Skills

Learning Goal: Review math concepts used in chemistry: place values, positive and negative numbers, percentages, solving equations, and interpreting line graphs.

- Basic math skills are important in learning chemistry.
- In each number, we identify the place value of each digit.
- A positive number is *greater* than zero and has a positive sign (+); a negative number is *less* than zero and has a negative sign (−).
- A percentage is calculated as the parts divided by the whole, then multiplied by 100%.
- To solve a linear equation, it is rearranged to place the unknown value of x on one side.
- A line graph represents the relationship between two variables, which are plotted on perpendicular axes.

♦ Learning Exercise 1.4A

KEY MATH SKILL
Identifying Place Values

Identify the place value for each of the digits in the number 825.10.

Digit	Place Value
8	
2	
5	
1	
0	

Answer

Digit	Place Value
8	hundreds
2	tens
5	ones
1	tenths
0	hundredths

♦ Learning Exercise 1.4B

KEY MATH SKILL
Using Positive and Negative Numbers in Calculations

Solve the following:

$$\frac{-14 + 22}{-4} = ?$$

Answer

$$\frac{-14 + 22}{-4} = -2$$

♦ Learning Exercise 1.4C

KEY MATH SKILL
Calculating a Percentage

A set of pencils contains 4 red pencils, 6 green pencils, and 10 blue pencils. What is the percentage of red pencils in the set?

Answer

$$\frac{\text{4 red pencils}}{\text{20 pencils in the set}} \times 100\% = 20\% \text{ red pencils}$$

♦ Learning Exercise 1.4D

KEY MATH SKILL
Solving Equations

Solve the following equation for y:

$$3y - 8 = 22$$

Answer

$$3y - 8 = 22$$
$$3y - 8\,(+8) = 22\,(+8)$$
$$3y = 30$$
$$\frac{3y}{3} = \frac{30}{3}$$
$$y = 10$$

♦ Learning Exercise 1.4E

KEY MATH SKILL

Interpreting a Line Graph

Use the following graph to answer questions **a** and **b**.

Time for Cooling of Coffee versus Temperature

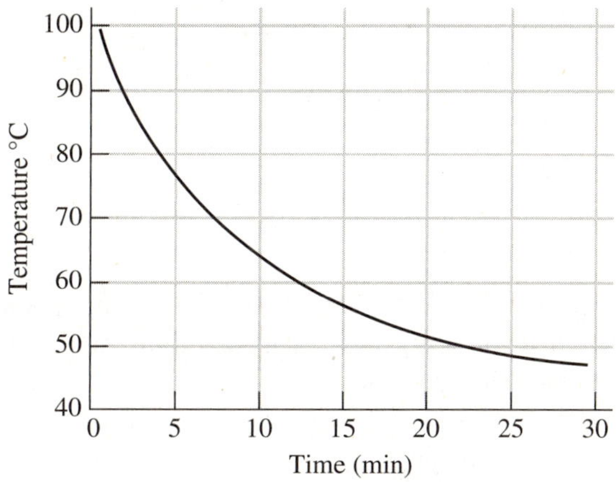

a. How many minutes elapse when coffee cools from 100 °C to 65 °C?
b. What is the temperature after the coffee has cooled for 15 min?

Answer

a. Draw a horizontal line from 65 °C on the temperature axis to intersect the curved line. Then drop a vertical line to the time axis and read the time where it crosses the x axis, which is estimated to be 9 min.

Time for Cooling of Coffee versus Temperature

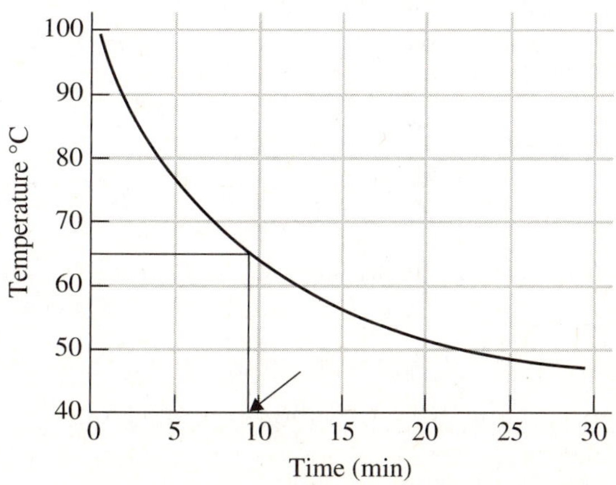

b. Draw a vertical line from 15 min on the time axis to intersect the curved line. Draw a horizontal line from the intersect to the temperature axis and read the temperature where it crosses the *y* axis, which is estimated to be 56 °C.

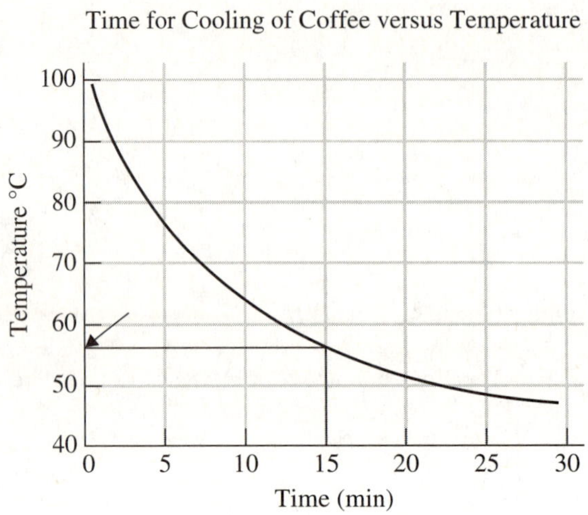

Checklist for Chapter 1

You are ready to take the Practice Test for Chapter 1. Be sure you have accomplished the following learning goals for this chapter. If not, review the section listed at the end of the goal. Then, apply your new skills and understanding to the Practice Test.

After studying Chapter 1, I can successfully:

_____ Describe a substance as a chemical (1.1).

_____ Identify the steps of the scientific method (1.2).

_____ Design a study plan for successfully learning chemistry (1.3).

_____ Use math skills needed for chemistry: place values, positive and negative numbers, percentages, solving equations, and interpreting line graphs (1.4).

Practice Test for Chapter 1

1. Which of the following would be described as a chemical? (1.1)
   A. sleeping **B.** salt **C.** singing
   D. listening to a concert **E.** energy

2. Which of the following is not a chemical? (1.1)
   A. wool **B.** sugar **C.** feeling cold
   D. salt **E.** vanilla

For questions 3 through 7, identify each statement as an observation (O), a hypothesis (H), or an experiment (E): (1.2)

3. _____ More sugar dissolves in 50 mL of hot water than in 50 mL of cold water.

4. _____ Samples containing 20 g of sugar each are placed separately in a glass of cold water and a glass of hot water.

5. _____ Sugar consists of white crystals.

6. _____ Water flows downhill.

7. _____ Drinking ten glasses of water a day will help me lose weight.

For questions 8 through 12, answer yes or no. (1.3)

To learn chemistry, I will:

8. _____ work the problems in the chapter and check answers

9. _____ attend some lectures, but not all

10. _____ form a study group

11. _____ set up a regular study time

12. _____ wait until the night before the exam to start studying

13. Solve for the value of a in the equation $5a + 11 = 36$. (1.4)
 A. 1 B. 5 C. 15 D. 25 E. 30

14. Give the answer for the problem $\frac{22 + 23}{+1 - 4}$. (1.4)
 A. 5 B. −5 C. −7 D. −15 E. 15

15. At the zoo, there are 8 lions, 6 tigers, and 2 panthers. What percentage of the animals at the zoo are lions? (1.4)
 A. 50% B. 30% C. 20% D. 10% E. 5%

16. Use the following graph to determine the solubility of carbon dioxide in water, in g CO_2/100 g water, at 40 °C. (1.4)

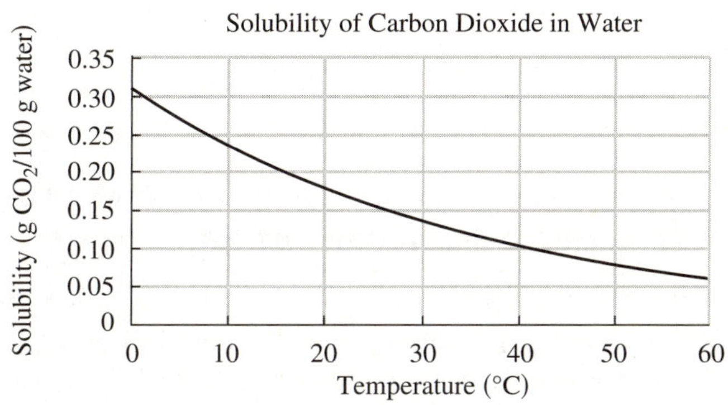

 A. 0.10 B. 0.15 C. 0.20 D. 0.25 E. 0.30

Answers to the Practice Test

1. B 2. C 3. O 4. E 5. O
6. O 7. H 8. yes 9. no 10. yes
11. yes 12. no 13. B 14. D 15. A
16. A

2
Measurements

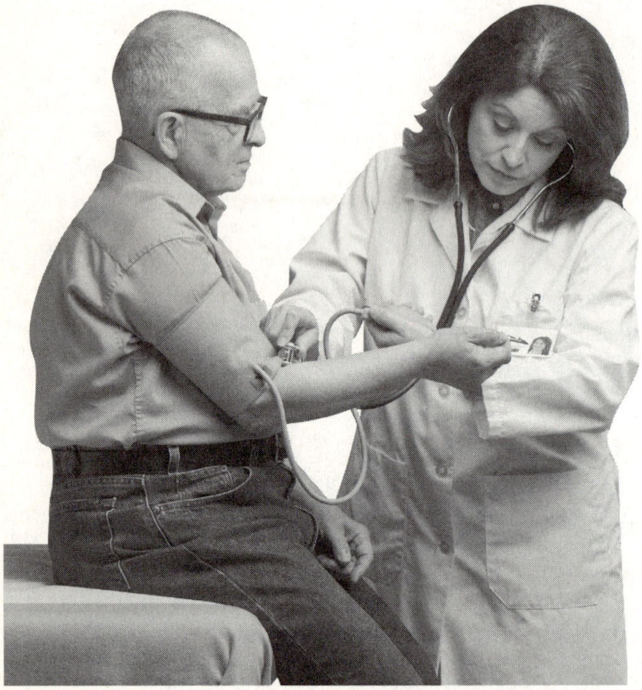

Greg has been on Inderal for high blood pressure. Every two months, he has his blood pressure taken by a registered nurse. During his last visit, his blood pressure was lower, 130/85. Thus, the dosage of Inderal was reduced to 45 mg to be taken twice daily. If one tablet contains 15 mg of Inderal, how many tablets does Greg need to take in one day?

2.1 Units of Measurement

Learning Goal: Write the names and abbreviations for the metric or SI units used in measurements of length, volume, mass, temperature, and time.

- In science, physical quantities are described in units of the metric or International System of Units (SI).
- Some important units of measurement are meter (m) for length, liter (L) for volume, gram (g) and kilogram (kg) for mass, degree Celsius (°C) and kelvin (K) for temperature, and second (s) for time.

Key Terms for Section 2.1

Match each of the following key terms with the correct definition:

a. International System of Units
b. volume
c. gram
d. second
e. meter
f. liter
g. mass

1. _____ the metric unit for volume that is slightly larger than a quart

2. _____ the metric unit used in measurements of mass

3. _____ the amount of space occupied by a substance

4. ____ the metric and SI unit for time
5. ____ a measure of the quantity of material in an object
6. ____ the official system of measurement used by scientists and used in most countries of the world
7. ____ the metric unit for length that is slightly longer than a yard

Answers **1.** f **2.** c **3.** b **4.** d **5.** g **6.** a **7.** e

♦ Learning Exercise 2.1

Match the measurement with the quantity in each of the following:

On an electronic balance, the digital readout gives the mass of a nickel, which is 5.01 g.

1. length **2.** mass **3.** volume **4.** temperature **5.** time

a. ____ 45 g **b.** ____ 8.2 m **c.** ____ 215 °C **d.** ____ 45 L
e. ____ 825 K **f.** ____ 8.8 s **g.** ____ 140 g **h.** ____ 50 s

Answers **a.** 2 **b.** 1 **c.** 4 **d.** 3 **e.** 4 **f.** 5 **g.** 2 **h.** 5

2.2 Scientific Notation

Learning Goal: Write a number in scientific notation.

- Large and small numbers can be written using scientific notation in which the decimal point is moved to give a coefficient of at least 1 but *less* than 10 and in which the number of spaces moved is shown as a power of 10.
- A value written in scientific notation has three parts: a number that is at least 1 but *less* than 10 called a coefficient, a power of 10, and a unit of measurement.
- A large number will have a positive power of 10, while a small number will have a negative power of 10.
- For numbers *greater* than 10, the decimal point is moved to the left to give a positive power of 10.
- For numbers *less* than 1, the decimal point is moved to the right to give a negative power of 10.

Study Note	
1. For a number *greater* than 10, the decimal point is moved to the left to give a number that is at least 1 but *less* than 10 and a positive power of 10. For a number *less* than 1, the decimal point is moved to the right to give a number that is at least 1 but *less* than 10 and a negative power of 10.	
2. To express 2500 cm in scientific scientific notation, we can write it as 2.5 (1000) cm. Because 1000 is the same as 10^3, we can express 2500 cm in scientific notation as 2.5×10^3 cm.	
3. To express 0.082 L in scientific scientific notation, we move the decimal point 2 places to the right and write 8.2 (0.01) L. Because 0.01 is the same as 10^{-2}, we can express 0.082 L in scientific notation as 8.2×10^{-2} L.	
Guide to Writing a Number in Scientific Notation	
Step 1	Move the decimal point to obtain a coefficient that is at least 1 but *less* than 10.
Step 2	Express the number of places moved as a power of 10.
Step 3	Write the product of the coefficient multiplied by the power of 10 with the unit.

KEY MATH SKILL

Using Scientific Notation

♦ Learning Exercise 2.2A

Write each of the following measurements in scientific notation:

a. 240 000 cm ______________ **b.** 825 m ______________

c. 230 000 kg ______________ **d.** 53 000 y ______________

e. 0.002 m ______________ **f.** 0.000 001 5 g ______________

g. 0.08 kg ______________ **h.** 0.000 24 s ______________

Answers **a.** 2.4×10^5 cm **b.** 8.25×10^2 m **c.** 2.3×10^5 kg **d.** 5.3×10^4 y
e. 2×10^{-3} m **f.** 1.5×10^{-6} g **g.** 8×10^{-2} kg **h.** 2.4×10^{-4} s

♦ Learning Exercise 2.2B

Indicate the larger number in each pair.

a. 2500 or 2.5×10^2 ______________ **b.** 0.04 or 4×10^{-3} ______________

c. 65 000 or 6.5×10^5 ______________ **d.** 0.000 35 or 3.5×10^{-3} ______________

e. 300 000 or 3×10^6 ______________ **f.** 0.002 or 2×10^{-4} ______________

Answers **a.** 2500 **b.** 0.04 **c.** 6.5×10^5
d. 3.5×10^{-3} **e.** 3×10^6 **f.** 0.002

♦ Learning Exercise 2.2C

Write each of the following in standard form:

Examples: 2×10^2 m = 200 m and 3.2×10^{-4} g = 0.000 32 g

a. 4×10^3 m ________ **b.** 5.2×10^4 g ________

c. 1.8×10^5 g ________ **d.** 8×10^{-3} L ________

e. 6×10^{-2} kg ________ **f.** 3.1×10^{-5} g ________

Answers **a.** 4000 m **b.** 52 000 g **c.** 180 000 g
d. 0.008 L **e.** 0.06 kg **f.** 0.000 031 g

2.3 Measured Numbers and Significant Figures

Learning Goal: Identify a number as measured or exact; determine the number of significant figures in a measured number.

- A *measured number* is obtained when you use a measuring device.
- An *exact number* is obtained by counting items or from a definition within the same measuring system.
- There is uncertainty in every measured number, but not in exact numbers.
- Significant figures in a measured number are all the digits, including the estimated digit.
- A number is a significant figure if it is
  a. not a zero
  b. a zero between nonzero digits
  c. a zero at the end of a decimal number
- Zeros in front of a decimal or at the end of a nondecimal number are not significant.

Key Terms for Sections 2.2–2.4

Match each of the following key terms with the correct definition:

a. measured number **b.** significant figures (SFs)
c. exact number **d.** scientific notation

1. _____ a form of writing large and small numbers using a coefficient that is at least 1 but *less* than 10, followed by a power of 10

2. _____ a number obtained by counting

3. _____ the numbers recorded in a measurement

4. _____ a number obtained when a quantity is determined by using a measuring device

Answers **1.** d **2.** c **3.** b **4.** a

♦ Learning Exercise 2.3A

Are the numbers in each of the following statements measured (M) or exact (E)?

a. ____ There are 7 days in 1 week. **b.** ____ A concert lasts for 73 min.

c. ____ There are 1000 g in 1 kg. **d.** ____ The potatoes have a mass of 2.5 kg.

e. ____ A student has 26 DVDs. **f.** ____ A snake is 1.2 m long.

Answers **a.** E (counted) **b.** M (use a watch) **c.** E (metric definition)
d. M (use a balance) **e.** E (counted) **f.** M (use a metric ruler)

Study Note

Significant figures (abbreviated SFs) are all the numbers reported in a measurement including the estimated digit. Zeros are significant unless they are placeholders appearing at the beginning of a decimal number or in a large number without a decimal point.

4.255 g (Four SFs) 0.0042 m (Two SFs) 46 500 L (Three SFs)

♦ Learning Exercise 2.3B

State the number of significant figures in each of the following measured numbers:

CORE CHEMISTRY SKILL
Counting Significant Figures

a. 35.24 g ____ **b.** 0.000 080 m ____

c. 55 000 m ____ **d.** 805 mL ____

e. 5.025 L ____ **f.** 0.006 kg ____

g. 268 200 mm ____ **h.** 25.0 °C ____

Answers **a.** 4 **b.** 2 **c.** 2 **d.** 3 **e.** 4 **f.** 1 **g.** 4 **h.** 3

2.4 Significant Figures in Calculations

Learning Goal: Adjust calculated answers to give the correct number of significant figures.

- In multiplication or division, the final answer is written so that it has the same number of significant figures as in the measurement with the fewest significant figures.
- In addition or subtraction, the final answer is written so that it has the same number of decimal places as the measurement with the fewest decimal places.
- To evaluate a calculator answer, count the significant figures in the measurements, and round off the calculator answer properly.
- Answers for chemical calculations rarely use all the numbers that appear in the calculator display. Exact numbers are not included in the determination of the number of significant figures (SFs).

Study Note

1. To round off a number when the first digit to be dropped is *less* than 5, keep the digits you need and drop all the digits that follow.

   Round off 42.8254 to 3 SFs → 42.8 (drop 254)

2. To round off a number when the first digit to be dropped is 5 or *greater*, keep the proper number of digits and increase the last retained digit by 1.

   Round off 8.4882 to 2 SFs → 8.5 (drop 882, increase final digit by 1)

3. When rounding off large numbers without decimal points, maintain the value of the answer by adding nonsignificant zeros as placeholders.

   Round off 356 835 to 3 SFs → 357 000 (drop 835; increase final digit by 1; add placeholder zeros)

♦ Learning Exercise 2.4A

CORE CHEMISTRY SKILL
Rounding off

Round off each of the following to give two significant figures:

a. 88.75 m ________ **b.** 0.002 923 g ________

c. 50.525 g ________ **d.** 1.672 m ________

e. 0.001 055 8 kg ________ **f.** 82 080 mL ________

Answers **a.** 89 m **b.** 0.0029 g **c.** 51 g
d. 1.7 m **e.** 0.0011 kg **f.** 82 000 mL

Study Note

1. An answer obtained from multiplying and dividing has the same number of significant figures as the measurement with the fewest significant figures.

   1.5 (Two SFs) × 32.546 (Five SFs) = 48.819 → 49 *Answer rounded off to two SFs*

2. An answer obtained from adding or subtracting has the same number of decimal places as the measurement with the fewest decimal places.

   82.223 (Three decimal places) + 4.1 (One decimal place) = 86.323 → 86.3 *Answer rounded off to one decimal place*

♦ Learning Exercise 2.4B

CORE CHEMISTRY SKILL
Using Significant Figures in Calculations

Solve each problem and give the answer with the correct number of significant figures or decimal places:

a. $1.3 \times 71.5 =$ **b.** $\frac{8.00}{4.00} =$

c. $\frac{0.082 \times 25.4}{0.116 \times 3.4} =$ **d.** $\frac{3.05 \times 1.86}{118.5} =$

e. $\frac{376}{0.0073} =$ **f.** $38.520 - 11.4 =$

g. $4.2 + 8.15 =$ **h.** $102.56 + 8.325 - 0.8825 =$

Answers	**a.** 93	**b.** 2.00	**c.** 5.3	**d.** 0.0479
	e. 52 000 (5.2×10^4)	**f.** 27.1	**g.** 12.4	**h.** 110.00

2.5 Prefixes and Equalities

Learning Goal: Use the numerical values of prefixes to write a metric equality.

- In the metric system, larger and smaller units use prefixes to change the size of the unit by factors of 10. For example, a prefix such as centi or milli preceding the unit meter gives a smaller length than a meter. A prefix such as kilo added to gram gives a unit that measures a mass that is 1000 times *greater* than a gram.
- An equality contains two units that measure the *same quantity* such as length, volume, mass, or time.
- Some common metric equalities are: 1 m = 100 cm; 1 L = 1000 mL; 1 kg = 1000 g.
- Some useful metric–U.S. equalities are: 2.54 cm = 1 in.; 1 kg = 2.205 lb; 1 L = 1.057 qt
- Some metric (SI) prefixes are:

Prefix	Symbol	Numerical Value	Scientific Notation	Equality
Prefixes That Increase the Size of the Unit				
peta	P	1 000 000 000 000 000	10^{15}	1 Pg = 1×10^{15} g 1 g = 1×10^{-15} Pg
tera	T	1 000 000 000 000	10^{12}	1 Tg = 1×10^{12} g 1 g = 1×10^{-12} Tg
giga	G	1 000 000 000	10^{9}	1 Gm = 1×10^{9} m 1 m = 1×10^{-9} Gm
mega	M	1 000 000	10^{6}	1 Mg = 1×10^{6} g 1 g = 1×10^{-6} Mg
kilo	k	1 000	10^{3}	1 km = 1×10^{3} m 1 m = 1×10^{-3} km
Prefixes That Decrease the Size of the Unit				
deci	d	0.1	10^{-1}	1 dL = 1×10^{-1} L 1 L = 10 dL
centi	c	0.01	10^{-2}	1 cm = 1×10^{-2} m 1 m = 100 cm
milli	m	0.001	10^{-3}	1 ms = 1×10^{-3} s 1 s = 1×10^{3} ms
micro	μ	0.000 001	10^{-6}	1 μg = 1×10^{-6} g 1 g = 1×10^{6} μg
nano	n	0.000 000 001	10^{-9}	1 nm = 1×10^{-9} m 1 m = 1×10^{9} nm
pico	p	0.000 000 000 001	10^{-12}	1 ps = 1×10^{-12} s 1 s = 1×10^{12} ps
femto	f	0.000 000 000 000 001	10^{-15}	1 fs = 1×10^{-15} s 1 s = 1×10^{15} fs

CORE CHEMISTRY SKILL
Using Prefixes

♦ Learning Exercise 2.5A

Match the items in column A with those from column B.

	A		B
1.	____ megameter	a.	nanometer
2.	____ 1000 meters	b.	decimeter
3.	____ 0.1 m	c.	10^{-6} m
4.	____ millimeter	d.	kilometer
5.	____ centimeter	e.	0.01 m
6.	____ 10^{-9} m	f.	1000 m
7.	____ micrometer	g.	10^{-3} m
8.	____ kilometer	h.	10^{6} m

Answers **1.** h **2.** d **3.** b **4.** g **5.** e **6.** a **7.** c **8.** f

♦ Learning Exercise 2.5B

Place the following units in order from smallest to largest.

a. kilogram milligram gram ____________

b. centimeter kilometer millimeter ____________

c. dL mL L ____________

d. kg pg mg ____________

Answers **a.** milligram, gram, kilogram **b.** millimeter, centimeter, kilometer
c. mL, dL, L **d.** pg, mg, kg

♦ Learning Exercise 2.5C

Complete each of the following metric relationships:

a. 1 L = ________ mL **b.** 1 L = ________ dL

c. 1 m = ________ cm **d.** 1 s = ________ ms

e. 1 kg = ________ g **f.** 1 cm = ________ mm

g. 1 mg = ________ μg **h.** 1 dL = ________ L

i. 1 m = ________ mm **j.** 1 cm = ________ m

Answers **a.** 1000 **b.** 10 **c.** 100 **d.** 1000 **e.** 1000
f. 10 **g.** 1000 **h.** 0.1 **i.** 1000 **j.** 0.01

2.6 Writing Conversion Factors

Learning Goal: Write a conversion factor for two units that describe the same quantity.

Conversion factors are equalities in the form of a fraction that are used to change one unit to another.

- Each conversion factor represents an equality expressed in the form of a fraction.
- Two conversion factors can be written for any relationship between equal quantities.

  For example, the metric—U.S. equality 2.54 cm = 1 in. can be written as follows:

$$\frac{2.54\text{ cm}}{1\text{ in.}} \quad \text{and} \quad \frac{1\text{ in.}}{2.54\text{ cm}}$$

- A percentage is written as a conversion factor by expressing matching units in the relationship as the parts of a specific substance in 100 parts of the whole.
- Extremely small conversion factors can be written as parts per million (ppm) or parts per billion (ppb).

Study Note

Metric conversion factors are obtained from metric prefixes. For example, the metric equality 1 m = 100 cm is written as two factors:

$$\frac{1\text{ m}}{100\text{ cm}} \quad \text{and} \quad \frac{100\text{ cm}}{1\text{ m}}$$

When a unit is squared or cubed, the value of the conversion factor is also squared or cubed. For example, when the metric equality 1 m = 100 cm is squared, the relationship becomes

$$(1\text{ m})^2 = (100\text{ cm})^2$$

This equality gives the factors

$$\frac{(1\text{ m})^2}{(100\text{ cm})^2} \quad \text{and} \quad \frac{(100\text{ cm})^2}{(1\text{ m})^2}$$

♦ Learning Exercise 2.6A

CORE CHEMISTRY SKILL

Writing Conversion Factors from Equalities

Write two conversion factors for each of the following pairs of units:

a. millimeters and meters

b. kilogram and grams

c. kilograms and pounds

d. square inches and square centimeters

e. centimeters and meters

f. liters and quarts

g. deciliters and liters

h. cubic millimeters and cubic centimeters

Answers

a. $\dfrac{1000\text{ mm}}{1\text{ m}}$ and $\dfrac{1\text{ m}}{1000\text{ mm}}$

b. $\dfrac{1000\text{ g}}{1\text{ kg}}$ and $\dfrac{1\text{ kg}}{1000\text{ g}}$

c. $\dfrac{2.205\text{ lb}}{1\text{ kg}}$ and $\dfrac{1\text{ kg}}{2.205\text{ lb}}$

d. $\dfrac{(2.54\text{ cm})^2}{(1\text{ in.})^2}$ and $\dfrac{(1\text{ in.})^2}{(2.54\text{ cm})^2}$

e. $\dfrac{100\text{ cm}}{1\text{ m}}$ and $\dfrac{1\text{ m}}{100\text{ cm}}$

f. $\dfrac{1\text{ L}}{1.057\text{ qt}}$ and $\dfrac{1.057\text{ qt}}{1\text{ L}}$

g. $\dfrac{10\text{ dL}}{1\text{ L}}$ and $\dfrac{1\text{ L}}{10\text{ dL}}$

h. $\dfrac{(1\text{ cm})^3}{(10\text{ mm})^3}$ and $\dfrac{(10\text{ mm})^3}{(1\text{ cm})^3}$

Study Note

1. Sometimes, a statement within a problem gives an equality that is only true for that problem. Then conversion factors can be written that are true only for that problem. For example, a problem states that there are 50 mg of vitamin B in a tablet. The conversion factors are

$$\frac{1\text{ tablet}}{50\text{ mg vitamin B}} \quad\text{and}\quad \frac{50\text{ mg vitamin B}}{1\text{ tablet}}$$

2. If a problem gives a percentage (%), it can be stated as parts per 100 parts. For example, a candy bar contains 45% by mass chocolate. This percentage (%) equality can be written with factors using the same mass unit.

$$\frac{45\text{ g chocolate}}{100\text{ g candy bar}} \quad\text{and}\quad \frac{100\text{ g candy bar}}{45\text{ g chocolate}}$$

3. When a problem gives ppm or ppb, it can be stated as parts per million (ppm), which is mg/kg or parts per billion (ppb), which is μg/kg. For example, the level of nitrate in the Los Angeles water supply is 2.5 ppm. This ppm can be written as conversion factors using mg/kg.

$$\frac{2.5\text{ mg of nitrate}}{1\text{ kg of water}} \quad\text{and}\quad \frac{1\text{ kg of water}}{2.5\text{ mg of nitrate}}$$

The maximum level of bromate allowed in water is 10 ppb. This ppb can be written as conversion factors.

$$\frac{10\ \mu\text{g bromate}}{1\text{ kg water}} \quad\text{and}\quad \frac{1\text{ kg water}}{10\ \mu\text{g bromate}}$$

♦ Learning Exercise 2.6B

Write two conversion factors for each of the following statements:

a. A cheese contains 55% fat by mass.

b. In the city, a car gets 14 miles to the gallon.

c. A 125-g steak contains 45 g of protein.

d. Pink gold contains 25% copper by mass.

e. A cadmium level of 1.8 ppm in food causes liver damage in rats.

f. A water sample contains 5.4 ppb of arsenic.

Answers

a. $\dfrac{55\text{ g fat}}{100\text{ g cheese}}$ and $\dfrac{100\text{ g cheese}}{55\text{ g fat}}$

b. $\dfrac{14\text{ mi}}{1\text{ gal}}$ and $\dfrac{1\text{ gal}}{14\text{ mi}}$

c. $\dfrac{45\text{ g protein}}{125\text{ g steak}}$ and $\dfrac{125\text{ g steak}}{45\text{ g protein}}$

d. $\dfrac{25\text{ g copper}}{100\text{ g pink gold}}$ and $\dfrac{100\text{ g pink gold}}{25\text{ g copper}}$

e. $\dfrac{1.8\text{ mg cadmium}}{1\text{ kg food}}$ and $\dfrac{1\text{ kg food}}{1.8\text{ mg cadmium}}$

f. $\dfrac{5.4\ \mu\text{g arsenic}}{1\text{ kg water}}$ and $\dfrac{1\text{ kg water}}{5.4\ \mu\text{g arsenic}}$

2.7 Problem Solving Using Unit Conversion

Learning Goal: Use conversion factors to change from one unit to another.

- Conversion factors can be used to change a quantity expressed in one unit to a quantity expressed in another unit.
- The process of solving a problem requires the change of the initial unit to one or more units until the final unit of the answer is obtained.

Guide to Problem Solving Using Conversion Factors	
Step 1	State the given and needed quantities.
Step 2	Write a plan to convert the given unit to the needed unit.
Step 3	State the equalities and conversion factors.
Step 4	Set up the problem to cancel units and calculate the answer.

Example: How many liters is 2850 mL?

Step 1: State the given unit and the needed quantities.

Analyze the Problem	Given	Need
	2850 mL	liters

Step 2: Write a plan to convert the given unit to the final unit.

$$\text{milliliters} \xrightarrow{\text{Metric factor}} \text{liters}$$

Step 3: State the equalities and conversion factors.

$$1\text{ L} = 1000\text{ mL}$$

$$\frac{1\text{ L}}{1000\text{ mL}} \quad \text{and} \quad \frac{1000\text{ mL}}{1\text{ L}}$$

Step 4: Set up the problem to cancel units and calculate the answer.

$$2850\ \cancel{\text{mL}} \times \frac{1\text{ L}}{1000\ \cancel{\text{mL}}} = 2.85\text{ L}$$

♦ Learning Exercise 2.7A

CORE CHEMISTRY SKILL
Using Conversion Factors

Use metric conversion factors to solve each of the following problems:

a. 189 mL = ________ L **b.** 2.7 cm = ________ mm

c. 0.274 m^2 = ________ cm^2 **d.** 76 mg = ________ g

e. How many meters tall is a person whose height is 175 cm?

Analyze the Problem	**Given**	**Need**

f. There are 285 mL in a cup of tea. How many liters is that?

Analyze the Problem	**Given**	**Need**

g. If a ring contains 13 500 mg of gold, how many grams of gold are in the ring?

Analyze the Problem	**Given**	**Need**

h. You walked a distance of 1.5 km on the treadmill at the gym. How many meters did you walk?

Analyze the Problem	**Given**	**Need**

Answers

a. 0.189 L **b.** 27 mm

c. 2740 cm^2 **d.** 0.076 g

e.

Analyze the Problem	**Given**	**Need**	**Answer**
	175 cm	meters (m)	1.75 m

f.

Analyze the Problem	**Given**	**Need**	**Answer**
	285 mL	liters (L)	0.285 L

g.

Analyze the Problem	**Given**	**Need**	**Answer**
	13 500 mg	grams (g)	13.5 g

h.

Analyze the Problem	**Given**	**Need**	**Answer**
	1.5 km	meters (m)	1500 m

♦ Learning Exercise 2.7B

Use metric–U.S. conversion factors to solve each of the following problems:

a. 18 in. = ________ cm   **b.** 4.0 qt = ________ L

c. 275 mL = ________ qt   **d.** 1300 mg = ________ lb

e. 150 lb = ________ kg   **f.** 840 g = ________ lb

g. 15 ft = ________ cm   **h.** 8.50 oz = ________ g

Answers

a. 46 cm **b.** 3.8 L **c.** 0.291 qt **d.** 0.0029 lb
e. 68 kg **f.** 1.9 lb **g.** 460 cm **h.** 241 g

♦ Learning Exercise 2.7C

Use conversion factors to solve each of the following problems:

a. A piece of plastic tubing measures 120 mm. What is the length of the tubing, in inches?

b. A statue weighs 245 lb. What is the mass of the statue, in kilograms?

c. Your friend has a height of 6 ft 3 in. What is your friend's height, in meters?

d. In a triple-bypass surgery, a patient requires 3.00 pt of whole blood. How many milliliters of blood were given?

e. Calculate the number of square meters in a rug with an area of 48 000 cm^2.

f. A mouthwash contains 62% alcohol by volume. How many milliliters of alcohol are in a 1.18-L bottle of mouthwash?

Answers

a. 4.7 in. **b.** 111 kg **c.** 1.9 m **d.** 1420 mL
e. 4.8 m^2 **f.** 730 mL

2.8 Density

Learning Goal: Calculate the density of a substance; use the density to calculate the mass or volume of a substance.

- The density of a substance is the ratio of its mass to its volume, usually in units of g/mL or g/cm^3. For example, the density of sugar is 1.59 g/mL and silver is 10.5 g/mL.

$$\text{Density} = \frac{\text{mass of substance}}{\text{volume of substance}}$$

- The volume of 1 mL is equal to the volume of 1 cm^3.

Key Terms for Sections 2.5–2.8

Match each of the following key terms with the correct definition:

a. equality　　**b.** conversion factor
c. density　　**d.** prefix

1. _____ The relationship of the mass of an object to its volume expressed as grams per cubic centimeter (g/cm^3) or grams per milliliter (g/mL).

2. _____ The part of the name of a metric unit that precedes the base unit and specifies the size of the measurement.

3. _____ A ratio in which the numerator and denominator are quantities from an equality or given relationship.

4. _____ A relationship between two units that measure the same quantity.

Answers **1.** c **2.** d **3.** b **4.** a

Guide to Calculating Density	
Step 1	State the given and needed quantities.
Step 2	Write the density expression.
Step 3	Express mass in grams and volume in milliliters (mL) or cm^3.
Step 4	Substitute mass and volume into the density expression and calculate the density.

♦ Learning Exercise 2.8A

Calculate the density for each of the following:

a. What is the density (g/mL) of glycerol if a 200.0-mL sample has a mass of 252 g?

b. A small solid has a mass of 5.5 oz. When placed in a graduated cylinder with a water level of 25.2 mL, the object causes the water level to rise to 43.8 mL. What is the density of the object, in g/mL?

Answers **a.** 1.26 g/mL **b.** 8.4 g/mL

Guide to Using Density	
Step 1	State the given and needed quantities.
Step 2	Write a plan to calculate the needed quantity.
Step 3	Write equalities and their conversion factors, including density.
Step 4	Set up the problem to calculate the needed quantity.

Example: Density can be used as a factor to convert between the mass (g) and volume (mL) of a substance. The density of silver is 10.5 g/mL. What is the mass, in grams, of 6.0 mL of silver?

Step 1: State the given and needed quantities.

Given 6.0 mL of silver; density of silver 10.5 g/mL

Need grams of silver

Step 2: Write a plan to calculate the needed quantity.

$$\text{milliliters} \xrightarrow[\text{factor}]{\text{density}} \text{grams}$$

Step 3: 1 mL = 10.5 g

$$\frac{1\text{ mL}}{10.5\text{ g}} \quad \text{and} \quad \frac{10.5\text{ g}}{1\text{ mL}}$$

Step 4:

$$0.60\ \cancel{\text{ml Ag}} \times \underbrace{\frac{10.5\text{ g Ag}}{1\ \cancel{\text{ml Ag}}}}_{\textit{Density factor}} = 63\text{ g of Ag}$$

CORE CHEMISTRY SKILL

Using Density as a Conversion Factor

♦ Learning Exercise 2.8B

Using density as a convertion factor to solve each of the following:

a. A sugar solution has a density of 1.20 g/mL. What is the mass, in grams, of 0.225 L of the solution?

b. A piece of pure gold weighs 0.26 lb. If gold has a density of 19.3 mL, what is the volume, in milliliters, of the piece of gold?

c. A salt solution has a density of 1.15 g/mL and a volume of 425 mL. What is the mass, in grams, of the solution?

d. A 50.0-g sample of a glucose solution has a density of 1.28 g/mL. What is the volume, in liters, of the sample?

Answers **a.** 270. g **b.** 6.1 mL **c.** 489 g **d.** 0.0391 L

Checklist for Chapter 2

You are ready to take the Practice Test for Chapter 2. Be sure that you have accomplished the following learning goals for this chapter. If not, review the section listed at the end of the goal. Then apply your new skills and understanding to the Practice Test.

After studying Chapter 2, I can successfully:

_____ Write the names and abbreviations for the metric (SI) units of measurement (2.1).

_____ Write large or small numbers using scientific notation (2.2).

_____ Identify a number as a measured number or an exact number (2.3).

_____ Count the number of significant figures in measured numbers (2.3).

_____ Report a calculated answer with the correct number of significant figures (2.4).

_____ Write a metric equality from the numerical values of metric prefixes (2.5).

_____ Write two conversion factors for an equality (2.6).

_____ Use conversion factors to change from one unit to another unit (2.7).

_____ Calculate the density of a substance, or use density to calculate the mass or volume (2.8).

Practice Test for Chapter 2

1. Which of the following is a metric measurement of volume? (2.1)
A. kilogram **B.** kilowatt **C.** kiloliter **D.** kilometer **E.** kiloquart

2. The measurement 24 000 g written in scientific notation is (2.2)
A. 24 g **B.** 24×10^3 g **C.** 2.4×10^3 g **D.** 2.4×10^{-3} g **E.** 2.4×10^4 g

3. The measurement 0.005 m written in scientific notation is (2.2)
A. 5 m **B.** 5×10^{-3} m **C.** 5×10^{-2} m **D.** 0.5×10^{-4} m **E.** 5×10^3 m

4. The measured number in the following is (2.3)
A. 1 book **B.** 2 cars **C.** 4 flowers **D.** 5 rings **E.** 45 g

5. The number of significant figures in 105.4 m is (2.3)
A. 1 **B.** 2 **C.** 3 **D.** 4 **E.** 5

6. The number of significant figures in 0.000 82 g is (2.3)
A. 1 **B.** 2 **C.** 3 **D.** 4 **E.** 5

7. The calculator answer 5.7805 rounded to two significant figures is (2.3)
A. 5 **B.** 5.7 **C.** 5.8 **D.** 5.78 **E.** 6.0

8. The calculator answer 3486.512 rounded to three significant figures is (2.3)
A. 4000 **B.** 3500 **C.** 349 **D.** 3487 **E.** 3490

9. The reported answer for $16.0 \div 8.0$ is (2.4)
A. 2 **B.** 2.0 **C.** 2.00 **D.** 0.2 **E.** 5.0

10. The reported answer for $58.5 + 9.158$ is (2.4)
A. 67 **B.** 67.6 **C.** 67.7 **D.** 67.66 **E.** 67.658

11. The reported answer for $\dfrac{2.5 \times 3.12}{4.6}$ is (2.4)
A. 0.54 **B.** 7.8 **C.** 0.85 **D.** 1.7 **E.** 1.69

12. Which of these prefixes has the largest value? (2.5)
A. centi **B.** deci **C.** milli **D.** kilo **E.** micro

13. What is the decimal equivalent of the prefix *centi*? (2.5)
A. 0.001 **B.** 0.01 **C.** 0.1 **D.** 10 **E.** 100

14. Which of the following is the smallest unit of mass measurement? (2.5)
A. gram **B.** milligram **C.** kilogram **D.** decigram **E.** centigram

15. Which volume is the largest? (2.5)
A. mL **B.** dL **C.** cm^3 **D.** L **E.** kL

16. Which of the following is a conversion factor? (2.6)
A. 12 in. **B.** 3 ft **C.** 20 m **D.** $\frac{1000\text{ g}}{1\text{ kg}}$ **E.** 2 cm^3

17. Which is the correct conversion factor that relates milliliters to liters? (2.6)
A. $\frac{1000\text{ mL}}{1\text{ L}}$ **B.** $\frac{100\text{ mL}}{1\text{ L}}$ **C.** $\frac{10\text{ mL}}{1\text{ L}}$ **D.** $\frac{0.01\text{ mL}}{1\text{ L}}$ **E.** $\frac{0.001\text{ mL}}{1\text{ L}}$

18. Which is the conversion factor for millimeters and centimeters? (2.6)
A. $\frac{1\text{ mm}}{1\text{ cm}}$ **B.** $\frac{10\text{ mm}}{1\text{ cm}}$ **C.** $\frac{100\text{ cm}}{1\text{ mm}}$ **D.** $\frac{100\text{ mm}}{1\text{ cm}}$ **E.** $\frac{10\text{ cm}}{1\text{ mm}}$

19. Which is the conversion factor for ppm? (2.6)
A. $\frac{1\text{ g}}{1\text{ kg}}$ **B.** $\frac{1\text{ dg}}{1\text{ kg}}$ **C.** $\frac{1\ \mu\text{g}}{1\text{ g}}$ **D.** $\frac{1\text{ mg}}{1\text{ kg}}$ **E.** $\frac{1\ \mu\text{g}}{1\text{ kg}}$

20. 294 mm is equal to (2.7)
A. 2940 m **B.** 29.4 m **C.** 2.94 m **D.** 0.294 m **E.** 0.0294 m

21. The area of the face of a tennis racket measures 115 $in.^2$. What is that area, in square centimeters? (2.7)
A. 742 cm^2 **B.** 17.8 cm^2 **C.** 45.3 cm^2 **D.** 292 cm^2 **E.** 115 cm^2

22. What is the volume of 65 mL, in liters? (2.7)
A. 650 L **B.** 65 L **C.** 6.5 L **D.** 0.65 L **E.** 0.065 L

23. What is the mass, in kilograms, of a 22-lb turkey? (2.7)
A. 10. kg **B.** 48 kg **C.** 10 000 kg **D.** 0.048 kg **E.** 22 000 kg

24. The number of milliliters in 2 dL of a liquid is (2.7)
A. 20 mL **B.** 200 mL **C.** 2000 mL **D.** 20 000 mL **E.** 500 000 mL

25. What is the height, in meters, of a person who is 5 ft 4 in. tall? (2.7)
A. 64 m **B.** 25 m **C.** 14 m **D.** 1.6 m **E.** 1.3 m

26. How many ounces are in 1500 grams? (1 lb = 16 oz) (2.7)
A. 94 oz **B.** 53 oz **C.** 24 000 oz **D.** 33 oz **E.** 3.3 oz

27. How many quarts of orange juice are in 255 mL of juice? (2.7)
A. 0.255 qt **B.** 270 qt **C.** 236 qt **D.** 0.270 qt **E.** 0.400 qt

28. Your doctor places you on a 2200-Cal diet, with 19% of the Calories from fat. How many Calories are you allowed from fat? (2.7)
A. 19 Cal **B.** 2200 Cal **C.** 1900 Cal **D.** 420 Cal **E.** 4200 Cal

29. How many milliliters of a salt solution with a density of 1.8 g/mL are needed to provide 450 g of salt solution? (2.8)
A. 250 mL **B.** 25 mL **C.** 720 mL **D.** 450 mL **E.** 45 mL

30. Three liquids have densities of 1.15 g/mL, 0.79 g/mL, and 0.95 g/mL. When the liquids, which do not mix, are poured into a graduated cylinder, the liquid at the top is the one with a density, in g/mL, of (2.8)
A. 1.15 g/mL **B.** 1.00 g/mL **C.** 0.95 g/mL **D.** 0.79 g/mL **E.** 0.16 g/mL

31. A sample of oil has a mass of 65 g and a volume of 80.0 mL. What is the density, in g/mL, of the oil? (2.8)
A. 1.5 g/mL **B.** 1.4 g/mL **C.** 1.2 g/mL **D.** 0.90 g/mL **E.** 0.81 g/mL

32. What is the mass, in grams, of a 10.0-mL sample of liquid with a density of 1.04 g/mL? (2.8)
A. 104 g **B.** 10.4 g **C.** 1.04 g **D.** 1.40 g **E.** 9.62 g

33. Ethyl alcohol has a density of 0.785 g/mL. What is the mass, in grams, of 0.250 L of the alcohol? (2.8)
A. 196 g **B.** 158 g **C.** 3.95 g **D.** 0.253 g **E.** 0.160 g

34. 1.5 ft is the same length, in centimeters, as (2.7)
A. 46 cm **B.** 7.1 cm **C.** 18 cm **D.** 3.8 cm **E.** 0.59 cm

35. The level of lead in water was 4.6 ppb. What mass of lead is in 2.0 kg of water? (2.7)
A. 9.2 μg **B.** 4.6 μg **C.** 2.3 μg **D.** 9.2 mg **E.** 4.6 mg

Answers to the Practice Test

1. C	**2.** E	**3.** B	**4.** E	**5.** D
6. B	**7.** C	**8.** E	**9.** B	**10.** C
11. D	**12.** D	**13.** B	**14.** B	**15.** E
16. D	**17.** A	**18.** B	**19.** D	**20.** D
21. A	**22.** E	**23.** A	**24.** B	**25.** D
26. B	**27.** D	**28.** D	**29.** A	**30.** D
31. E	**32.** B	**33.** A	**34.** A	**35.** A

3
Matter and Energy

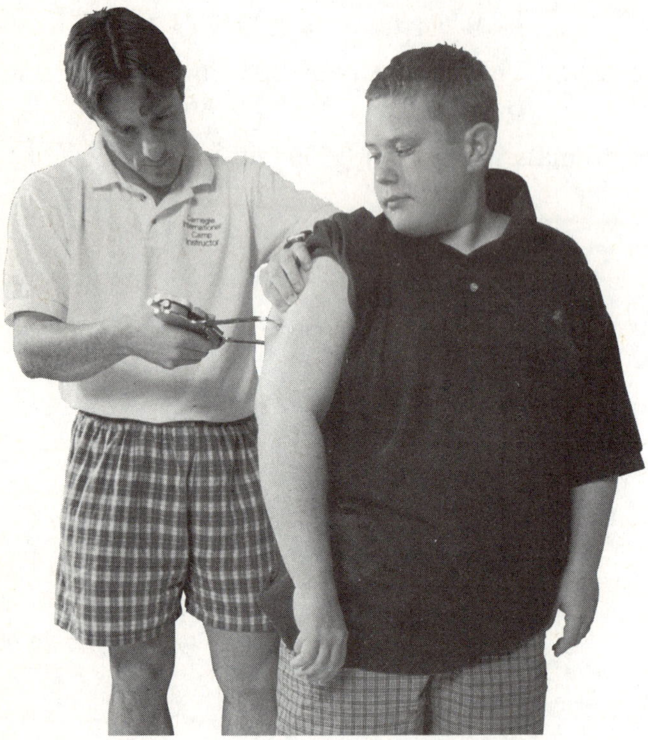

Since Charles had a visit with Daniel, a dietitian, he and his mother have been preparing healthier foods with smaller serving sizes. Charles has lost 6 lb. Yesterday he had a blood glucose test that showed his blood glucose level was normal. One of the meals that Charles likes consists of rice, chicken, and carrots. If Charles consumes 6 g of carbohydrate, 4 g of fat, and 12 g of protein, what is the energy content of the meal, in kilojoules and kilocalories?

3.1 Classification of Matter

Learning Goal: Classify examples of matter as pure substances or mixtures.

- Matter is anything that has mass and occupies space.
- A pure substance, whether it is an element or compound, has a definite composition.
- Elements are the simplest type of matter; compounds consist of a combination of two or more elements.
- Mixtures contain two or more substances that are physically, not chemically, combined.
- Mixtures are classified as homogeneous or heterogeneous.

Key Terms for Section 3.1

Match the following terms with the definitions:

a. element **b.** mixture **c.** pure substance
d. matter **e.** compound

1. _____ anything that has mass and occupies space

2. _____ an element or a compound that has a definite composition

3. ____ the physical combination of two or more substances that does not change the identities of the substances

4. ____ a pure substance consisting of two or more elements with a definite composition that can be broken down into simpler substances only by chemical methods

5. ____ a pure substance containing only one type of matter that cannot be broken down by chemical methods

Answers **1.** d **2.** c **3.** b **4.** e **5.** a

♦ Learning Exercise 3.1A

CORE CHEMISTRY SKILL
Classifying Matter

Identify each of the following as an element (E) or a compound (C):

1. ____ iron	**2.** ____ carbon dioxide
3. ____ potassium iodide	**4.** ____ gold
5. ____ aluminum	**6.** ____ pure table salt (sodium chloride)

Answers **1.** E **2.** C **3.** C **4.** E **5.** E **6.** C

♦ Learning Exercise 3.1B

Identify each of the following as a pure substance (P) or mixture (M):

1. ____ bananas and milk	**2.** ____ sulfur
3. ____ silver	**4.** ____ a bag of raisins and nuts
5. ____ water	**6.** ____ sand and water

Answers **1.** M **2.** P **3.** P **4.** M **5.** P **6.** M

♦ Learning Exercise 3.1C

Identify each of the following mixtures as homogeneous (Ho) or heterogeneous (He):

1. ____ chocolate milk	**2.** ____ sand and water
3. ____ lemon drink	**4.** ____ a bag of raisins and nuts
5. ____ air	**6.** ____ vinegar

Answers **1.** Ho **2.** He **3.** Ho **4.** He **5.** Ho **6.** Ho

3.2 States and Properties of Matter

Learning Goal: Identify the states and the physical and chemical properties of matter.

- The states of matter are solid, liquid, and gas.
- Physical properties are those characteristics of a substance that can change without affecting the identity of the substance.
- A substance undergoes a physical change when its shape, size, or state changes, but the type of substance itself does not change.

- Chemical properties are those characteristics of a substance that change when a new substance is produced.
- Melting, freezing, boiling, and condensing are changes of state.

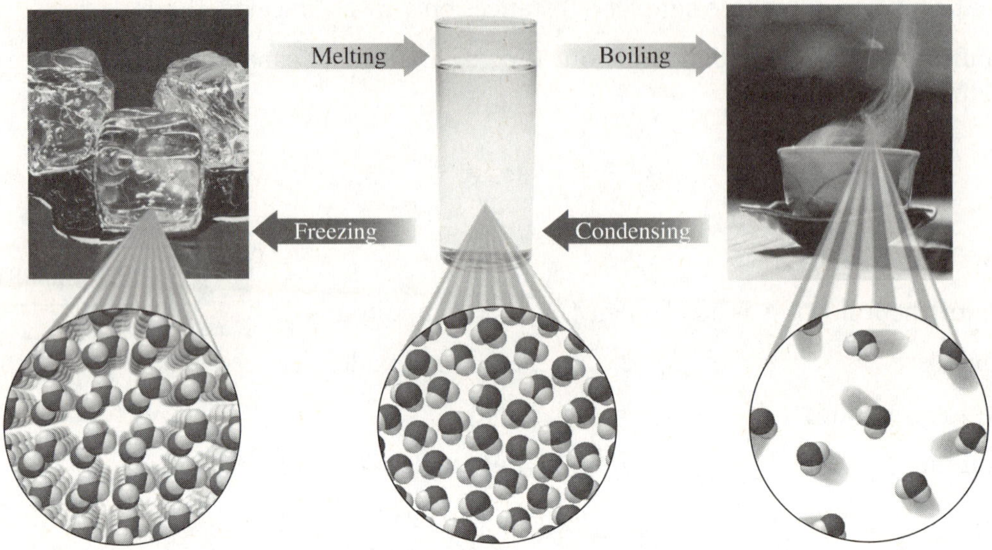

♦ Learning Exercise 3.2A

State whether the following statements describe a gas (G), a liquid (L), or a solid (S):

1. _____ There are no attractions among the particles.
2. _____ The particles are held close together in a definite pattern.
3. _____ This substance has a definite volume, but no definite shape.
4. _____ The particles are moving extremely fast.
5. _____ This substance has no definite shape and no definite volume.
6. _____ This substance has a definite volume, but takes the shape of its container.
7. _____ The particles in this substance are vibrating slowly in fixed positions.
8. _____ This substance has a definite volume and a definite shape.

Answers **1.** G **2.** S **3.** L **4.** G **5.** G **6.** L **7.** S **8.** S

♦ Learning Exercise 3.2B

Classify each of the following as a physical (P) or chemical (C) property:

a. _____ Silver is shiny.
b. _____ Water is a liquid at 25 °C.
c. _____ Wood burns.
d. _____ Mercury is a very dense liquid.
e. _____ Helium is not reactive.
f. _____ Ice cubes float in water.

Answers **a.** P **b.** P **c.** C **d.** P **e.** C **f.** P

♦ Learning Exercise 3.2C

CORE CHEMISTRY SKILL
Identifying Physical and Chemical Changes

Classify each of the following as a physical (P) or chemical change (C):

a. ____ Sodium melts at 98 °C.
b. ____ Iron forms rust in air and water.
c. ____ Water condenses on a cold window.
d. ____ Fireworks explode when ignited.
e. ____ Gasoline burns in a car engine.
f. ____ Paper is cut to make confetti.

Answers **a.** P **b.** C **c.** P **d.** C **e.** C **f.** P

3.3 Temperature

Learning Goal: Given a temperature, calculate a corresponding value on another temperature scale.

- Temperature is measured in degrees Celsius (°C) or kelvins (K). In the United States, the Fahrenheit scale (°F) is still in use.
- The equation $T_F = 1.8T_C + 32$ is used to convert a Celsius temperature to a Fahrenheit temperature.
- When rearranged for T_C, the equation is used to convert from T_F to T_C.

$$T_C = \frac{(T_F - 32)}{1.8}$$

- The temperature on the Celsius scale is related to the Kelvin scale: $T_K = T_C + 273$.

Guide to Calculating Temperature	
Step 1	State the given and needed quantities.
Step 2	Write a temperature equation.
Step 3	Substitute in the known values and calculate the new temperature.

♦ Learning Exercise 3.3

CORE CHEMISTRY SKILL
Converting between Temperature Scales

Calculate the temperature in each of the following problems:

a. To prepare yogurt, milk is warmed to 185 °F. What Celsius temperature is needed to prepare the yogurt?

b. On a chilly day in Alaska, the temperature drops to −12 °C. What is that temperature on a Fahrenheit thermometer?

c. A patient has a temperature of 39.5 °C. What is that temperature in °F?

d. On a hot summer day, the temperature is 95 °F. What is the temperature on the Celsius scale?

e. A pizza is cooked at a temperature of 425 °F. What is the °C temperature?

f. A research experiment requires the use of liquid nitrogen to cool the reaction flask to −45 °C. What temperature will this be on the Kelvin scale?

Answers **a.** 85.0 °C **b.** 10 °F **c.** 103.1 °F **d.** 35 °C **e.** 218 °C **f.** 228 K

3.4 Energy

Learning Goal: Identify energy as potential or kinetic; convert between units of energy.

- Energy is the ability to do work.
- Potential energy is stored energy; that is determined by position or composition; kinetic energy is the energy of motion.
- The SI unit of energy is the joule (J), the metric unit is the calorie (cal). One calorie (cal) is equal to 4.184 joules exactly (J).

$$\frac{4.184\ \text{J}}{1\ \text{cal}} \quad \text{and} \quad \frac{1\ \text{cal}}{4.184\ \text{J}}$$

Key Terms for Section 3.4

Match the following terms with the statements:

a. potential energy **b.** joule **c.** kinetic energy
d. calorie **e.** heat **f.** energy

1. ____ the SI unit of energy

2. ____ the energy of motion

3. ____ the amount of heat energy that raises the temperature of exactly 1 g of water exactly 1 °C

4. ____ the ability to do work

5. ____ a type of energy that is stored for future use

6. ____ the energy associated with the motion of particles

Answers **1.** b **2.** c **3.** d **4.** f **5.** a **6.** e

♦ Learning Exercise 3.4A

State whether each of the following statements describes potential (P) or kinetic (K) energy:

1. ____ a potted plant sitting on a ledge **2.** ____ your breakfast cereal

3. ____ logs sitting in a fireplace **4.** ____ a piece of candy

5. ____ an arrow shot from a bow **6.** ____ a ski jumper standing at the top of the ski jump

7. _____ a jogger running

8. _____ a skydiver waiting to jump

9. _____ water flowing down a stream

10. _____ a bowling ball striking the pins

Answers 1. P 2. P 3. P 4. P 5. K
6. P 7. K 8. P 9. K 10. K

♦ Learning Exercise 3.4B

Match the words in column A with the descriptions in column B.

A	B
1. _____ calorie	a. the SI unit of work and energy
2. _____ kilocalorie	b. the heat needed to raise 1 g of water by 1 °C
3. _____ joule	c. 1000 cal

Answers: 1. b 2. c 3. a

♦ Learning Exercise 3.4C

CORE CHEMISTRY SKILL
Using Energy Units

Calculate an answer for each of the following conversions:

a. 58 000 cal to kcal

b. 3450 J to cal

c. 2.8 kJ to cal

d. 15 200 cal to kJ

Answers a. 58 kcal b. 825 cal c. 670 cal d. 63.6 kJ

3.5 Specific Heat

Learning Goal: Use specific heat to calculate heat loss or gain, temperature change, or mass of a sample.

- Specific heat is the amount of energy required to raise the temperature of 1 g of a substance by 1 °C.
- The specific heat for liquid water is 4.184 J/g °C or 1.00 cal/g °C.
- The heat lost or gained can be calculated using the mass of the substance, temperature difference, and its specific heat: Heat $(q) = \text{mass} \times \Delta T \times SH$

Guide to Calculating Specific Heat	
Step 1	State the given and needed quantities.
Step 2	Write the relationship for specific heat.
Step 3	Set up the problem to calculate the specific heat.

♦ Learning Exercise 3.5A

CORE CHEMISTRY SKILL
Calculating Specific Heat

Calculate the specific heat, in J/g °C, for each of the following:

a. a 15.2-g sample of a metal that absorbs 231 J when its temperature rises from 84.5 °C to 125.3 °C

b. a 31.8-g sample of a metal that absorbs 816 J when its temperature rises from 23.7 °C to 56.2 °C

c. a 38.2-g sample of a metal that absorbs 125 J when its temperature rises from 62.1 °C to 68.4 °C

Answers **a.** 0.372 J/g °C **b.** 0.790 J/g °C **c.** 0.52 J/g °C

Guide to Calculations Using Specific Heat	
Step 1	List the given and the needed data.
Step 2	Calculate the temperature change (ΔT).
Step 3	Write the heat equation.
Step 4	Substitute in the given values and solve, making sure units cancel.

♦ Learning Exercise 3.5B

CORE CHEMISTRY SKILL
Using the Heat Equation

Calculate the kilojoules (kJ) and kilocalories (kcal) gained or released during the following:

1. heating 20.0 g of water from 22 °C to 77 °C

2. heating 10.0 g of water from 12.4 °C to 67.5 °C

3. cooling 0.450 kg of water from 80.0 °C to 35.0 °C

4. cooling 125 g of water from 72.0 °C to 45.0 °C

Answers
1. 4.6 kJ; 1.1 kcal (gained) **2.** 2.31 kJ; 0.551 kcal (gained)
3. 84.7 kJ; 20.3 kcal (released) **4.** 14.1 kJ; 3.38 kcal (released)

Study Note

The mass of a substance is calculated by rearranging the heat equation:

$$\text{Mass} = \frac{\text{heat } (q)}{\Delta T \times SH}$$

The temperature change (ΔT) for a substance is calculated by rearranging the heat equation:

$$\Delta T = \frac{\text{heat } (q)}{\text{mass} \times SH}$$

♦ Learning Exercise 3.5C

a. Copper has a specific heat of 0.385 J/g °C. When 1250 J are added to a copper sample, its temperature rises from 24.6 °C to 61.3 °C. What is the mass, in grams, of the copper sample?

b. A sample of aluminum has a specific heat of 0.897 J/g °C. When 785 J are added to an aluminum sample, its temperature rises from 14 °C to 106 °C. What is the mass of the aluminum sample?

c. Silver has a specific heat of 0.235 J/g °C. What is the temperature change, in degrees Celsius, for a 15.2-g piece of silver at 10.0 °C when 426 J is added?

d. What is the final temperature, in degrees Celsius, of 15.0 g of water initially at 5.5 °C when 1140 J is added? (*Hint:* Determine the ΔT first.)

e. A 25.6-g sample of a metal heated to 100.0 °C is placed in 48.2 g of water initially at 14.5 °C. If the final temperature of the water (and metal) is 18.6 °C, what is the specific heat, in J/g °C, of the metal?

Answers

a. 88.5 g
b. 9.5 g
c. 119 °C
d. ΔT = 18.2 °C; T_f = 23.7 °C
e. 0.398 J/g °C

3.6 Energy and Nutrition

Learning Goal: Use the energy values to calculate the kilojoules (kJ) or kilocalories (kcal) in a food.

- The nutritional calorie (Cal) is the same amount of energy as 1 kcal, or 1000 calories.
- When a substance is burned in a calorimeter, the water that surrounds the reaction chamber absorbs the heat given off. The heat absorbed by the water is calculated, and the energy value for the substance (energy per gram) is determined.
- The energy content of a food is the sum of kilojoules or kilocalories from carbohydrate, fat, and protein.

Key Terms for Sections 3.5–3.6

Match the following terms with the definitions:

a. energy value **b.** specific heat **c.** Calorie (Cal) **d.** kilojoule

1. _____ a nutritional unit of energy equal to 1000 cal, or 1 kcal

2. _____ a quantity of heat that changes the temperature of exactly 1 g of a substance by exactly 1 °C

3. _____ the number of kilojoules or kilocalories obtained per gram of carbohydrate, fat, or protein

4. _____ a unit of energy equal to 1000 J

Answers **1.** c **2.** b **3.** a **4.** d

Guide to Calculating the Energy from a Food	
Step 1	State the given and needed quantities.
Step 2	Use the energy value for each food type and calculate the kJ or kcal rounded off to the tens place.
Step 3	Add the energy for each food type to give the total energy from the food.

♦ Learning Exercise 3.6A

Calculate the kilojoules (kJ) and kilocalories (kcal) for each food using the following data (round off the kilocalories for each food type to the tens place):

Food	Carbohydrate	Fat	Protein	kJ	kcal
a. green peas, cooked, 1 cup	19 g	1 g	9 g	_____	_____
b. potato chips, 10 chips	10 g	8 g	1 g	_____	_____
c. cream cheese, 8 oz	5 g	86 g	18 g	_____	_____
d. lean hamburger, 3 oz	0	10 g	23 g	_____	_____

e. banana, 1	26 g	0	1 g	______	______

Answers **a.** 510 kJ; 120 kcal **b.** 490 kJ; 120 kcal **c.** 3660 kJ; 870 kcal
d. 770 kJ; 180 kcal **e.** 460 kJ; 110 kcal

♦ Learning Exercise 3.6B

Use energy values to solve each of the following:

1. How many kcal are in a single serving of pudding that contains 5 g of protein, 31 g of carbohydrate, and 5 g of fat (round off the kilocalories for each food type to the tens place)?

2. One serving of peanut butter (2 tbsp) has an energy value of 190 kcal. If there are 8 g of protein and 10 g of carbohydrate, how many grams of fat are in one serving of peanut butter (round off the kilocalories for each food type to the tens place)?

3. A serving of breakfast cereal provides 220 kcal. In this serving, there are 8 g of protein and 6 g of fat. How many grams of carbohydrate are in the cereal (round off the kilocalories for each food type to the tens place)?

4. Complete the following table listing ingredients for a peanut butter sandwich (round off the kilocalories for each food type to the tens place):

	Protein	**Fat**	**Carbohydrate**	**kcal**
2 slices of bread	5 g	0 g	30 g	______
2 tbsp of peanut butter	8 g	13 g	10 g	______
2 tsp of jelly	0 g	0 g	10 g	______
1 tsp of margarine	0 g	7 g	0 g	______

Total kcal in sandwich ______

Answers

1. protein = 20 kcal; carbohydrate = 120 kcal; fat = 50 kcal; Total = 190 kcal
2. protein = 30 kcal; carbohydrate = 40 kcal; 190 kcal − 30 kcal − 40 kcal = 120 kcal of fat

$$120\ \cancel{\text{kcal}} \times \frac{1\ \text{g fat}}{9\ \cancel{\text{kcal}}} = 13\ \text{g of fat}$$

3. protein = 30 kcal; fat = 50 kcal; 30 kcal + 50 kcal = 80 kcal from protein and fat
   220 kcal − 80 kcal = 140 kcal from carbohydrate

$$140\ \cancel{\text{kcal}} \times \frac{1\ \text{g carbohydrate}}{4\ \cancel{\text{kcal}}} = 35\ \text{g of carbohydrate}$$

4. bread = 140 kcal; peanut butter = 190 kcal; jelly = 40 kcal; margarine = 60 kcal
   total kcal in sandwich = 430 kcal

Checklist for Chapter 3

You are ready to take the Practice Test for Chapter 3. Be sure you have accomplished the following learning goals for this chapter. If not, review the section listed at the end of the goal. Then apply your new skills and understanding to the Practice Test.

After studying Chapter 3, I can successfully:

____ Identify matter as a pure substance or mixture (3.1).

____ Identify a mixture as homogeneous or heterogeneous (3.1).

____ Identify the physical state of a substance as a solid, liquid, or gas (3.2).

____ Calculate a temperature in degrees Fahrenheit, degrees Celsius, or kelvins (3.3).

____ Describe some forms of energy (3.4).

____ Change a quantity in one energy unit to another energy unit (3.4).

____ Calculate the specific heat of a substance (3.5).

____ Given the mass of a sample, specific heat, and the temperature change, calculate the heat lost or gained (3.5).

____ Using the energy values, calculate the energy, in kilocalories (Cal) or kilojoules (kJ), for a food sample (3.6).

Practice Test for Chapter 3

For questions 1 through 4, classify each of the following as a pure substance (P) or a mixture (M): (3.1)

1. toothpaste ____
2. platinum ____
3. chromium ____
4. mouthwash ____

For questions 5 through 8, classify each of the following mixtures as homogeneous (Ho) or heterogeneous (He): (3.1)

5. noodle soup ____
6. mineral water ____
7. chocolate chip cookie ____
8. mouthwash ____

9. Which of the following is a chemical property? (3.1)
A. dynamite explodes
B. a shiny metal
C. a melting point of 110 °C
D. rain on a cool day
E. breaking up cement

For questions 10 through 12, answer with solid (S), liquid (L), or gas (G): (3.2)

10. _____ has a definite volume, but takes the shape of a container

11. _____ does not have a definite shape or definite volume

12. _____ has a definite shape and a definite volume

13. Which of the following is a chemical property of silver? (3.2)
A. density of 10.5 g/mL
B. shiny
C. melts at 961 °C
D. good conductor of heat
E. reacts to form tarnish

14. Which of the following is a physical property of silicon? (3.2)
A. burns in chlorine
B. has a black to gray color
C. reacts with nitric acid
D. reacts with oxygen to form sand
E. used to form silicone

For problems 15 through 19, answer as physical change (P) or chemical change (C): (3.2)

15. _____ Butter melts in a hot pan.

16. _____ Iron forms rust with oxygen.

17. _____ Baking powder forms bubbles (CO_2) as a cake is baking.

18. _____ Water boils.

19. _____ Propane burns in a camp stove.

20. 105 °F = _______ °C (3.3)
A. 73 °C **B.** 41 °C **C.** 58 °C **D.** 90 °C **E.** 189 °C

21. The melting point of gold is 1064 °C. The Fahrenheit temperature needed to melt gold would be (3.3)
A. 129 °F **B.** 623 °F **C.** 1031 °F **D.** 1913 °F **E.** 1947 °F

22. The average daytime temperature on the planet Mercury is 683 K. What is this temperature on the Celsius scale? (3.3)
A. 956 °C **B.** 715 °C **C.** 680. °C **D.** 410. °C **E.** 303 °C

23. Which of the following would be described as potential energy? (3.4)
A. a car going around a racetrack
B. a rabbit hopping
C. oil in an oil well
D. a moving merry-go-round
E. a bouncing ball

24. Which of the following would be described as kinetic energy? (3.4)
A. a car battery
B. a can of tennis balls
C. gasoline in a car fuel tank
D. a box of matches
E. a tennis ball crossing over the net

25. When 34 J are added to a 10.0-g piece of iron, the temperature rises by 7.6 °C. (3.4)
What is the specific heat of iron?
A. 2.2 J/g °C
B. 0.22 J/g °C
C. 0.45 J/g °C
D. 0.39 J/g °C
E. 2.6 J/g °C

26. The number of joules needed to raise the temperature of 5.0 g of water from 25 °C to 55 °C is (3.5)
A. 5.0 J **B.** 36 J **C.** 30.0 J **D.** 335 J **E.** 630 J

27. The number of kilojoules (kJ) released when 15 g of water cools from 58 °C to 22 °C is (3.5)
A. 0.13 kJ **B.** 0.54 kJ **C.** 2.3 kJ **D.** 63 kJ **E.** 150 kJ

For questions 28 to 30, consider a cup of milk that is 3.5% butterfat with an energy value of 690 kJ. In the cup of milk, there are 9 g of fat, 12 g of carbohydrate, and some protein: (3.6)

28. The number of kcal provided by the carbohydrate is
A. 4 kcal **B.** 9 kcal **C.** 36 kcal **D.** 48 kcal **E.** 81 kcal

29. The number of kJ provided by the fat is
A. 9.0 kJ **B.** 38 kJ **C.** 150 kJ **D.** 200 kJ **E.** 340 kJ

30. The number of kcal provided by the protein is
A. 4 kcal **B.** 9 kcal **C.** 36 kcal **D.** 48 kcal **E.** 81 kcal

Answers to the Practice Test

1. M	**2.** P	**3.** P	**4.** M	**5.** He
6. Ho	**7.** He	**8.** Ho	**9.** A	**10.** L
11. G	**12.** S	**13.** E	**14.** B	**15.** P
16. C	**17.** C	**18.** P	**19.** C	**20.** B
21. E	**22.** D	**23.** C	**24.** E	**25.** C
26. E	**27.** C	**28.** D	**29.** E	**30.** C

4

Atoms and Elements

John is ready to prepare his fields to plant a new crop. Although the element nitrogen in the soil is important, he will also check the soil levels of the elements phosphorus and potassium, which are also important for a good yield. Tests show that the level of phosphorus in the soil is below 15 ppm, which is low. Tests also show that the potassium level in the soil is in the optimal range from 150 to 200 ppm. Thus, John will apply a fertilizer that contains phosphorus but no potassium. What are the symbols of the elements potassium, phosphorus, and nitrogen found in fertilizers?

4.1 Elements and Symbols

Learning Goal: Given the name of an element, write its correct symbol; from the symbol, write the correct name.

- Chemical symbols are one- or two-letter abbreviations for the names of the elements; only the first letter of a chemical symbol is a capital letter.
- Element names are derived from planets, mythology, and names of geographical features and famous scientists.

♦ Learning Exercise 4.1A

Write the symbol for each of the following elements:

1. carbon ____	**2.** iron ____	**3.** sodium ____
4. phosphorus ____	**5.** oxygen ____	**6.** nitrogen ____
7. iodine ____	**8.** sulfur ____	**9.** potassium ____
10. lead ____	**11.** calcium ____	**12.** gold ____

Answers						
	1. C	**2.** Fe	**3.** Na	**4.** P	**5.** O	**6.** N
	7. I	**8.** S	**9.** K	**10.** Pb	**11.** Ca	**12.** Au

♦ Learning Exercise 4.1B

Write the name of the element represented by each of the following symbols:

1. Mg ______________	**2.** H ______________
3. Ag ______________	**4.** F ______________
5. Cr ______________	**6.** Be ______________
7. Si ______________	**8.** Br ______________
9. Zn ______________	**10.** Al ______________

Answers

1. magnesium	**2.** hydrogen	**3.** silver	**4.** fluorine	**5.** chromium
6. beryllium	**7.** silicon	**8.** bromine	**9.** zinc	**10.** aluminum

4.2 The Periodic Table

Learning Goal: Use the periodic table to identify the group and the period of an element; identify the element as a metal, a nonmetal, or a metalloid.

- The periodic table is an arrangement of the elements into vertical columns and horizontal rows.
- Each vertical column is a group of elements which have similar properties.
- A horizontal row of elements is called a period.
- On the periodic table, the metals are located on the left of the heavy zigzag line, the nonmetals are to the right, and metalloids, in dark gray boxes, are found along the dark zigzag line.
- Main group or representative elements are found in Groups 1A (1), 2A (2) and 3A (13) to 8A (18). Transition elements are B-group elements (3–12).

Periodic Table of Elements

Representative elements: Groups 1–2 and 13–18; Alkali metals: Group 1; Alkaline earth metals: Group 2; Transition elements: Groups 3–12; Halogens: Group 17; Noble gases: Group 18

Period number	1 Group 1A	2 Group 2A	3 3B	4 4B	5 5B	6 6B	7 7B	8 8B	9 8B	10 8B	11 1B	12 2B	13 Group 3A	14 Group 4A	15 Group 5A	16 Group 6A	17 Group 7A	18 Group 8A
1	1 H																	2 He
2	3 Li	4 Be											5 B	6 C	7 N	8 O	9 F	10 Ne
3	11 Na	12 Mg											13 Al	14 Si	15 P	16 S	17 Cl	18 Ar
4	19 K	20 Ca	21 Sc	22 Ti	23 V	24 Cr	25 Mn	26 Fe	27 Co	28 Ni	29 Cu	30 Zn	31 Ga	32 Ge	33 As	34 Se	35 Br	36 Kr
5	37 Rb	38 Sr	39 Y	40 Zr	41 Nb	42 Mo	43 Tc	44 Ru	45 Rh	46 Pd	47 Ag	48 Cd	49 In	50 Sn	51 Sb	52 Te	53 I	54 Xe
6	55 Cs	56 Ba	57* La	72 Hf	73 Ta	74 W	75 Re	76 Os	77 Ir	78 Pt	79 Au	80 Hg	81 Tl	82 Pb	83 Bi	84 Po	85 At	86 Rn
7	87 Fr	88 Ra	89† Ac	104 Rf	105 Db	106 Sg	107 Bh	108 Hs	109 Mt	110 Ds	111 Rg	112 Cn	113 —	114 Fl	115 —	116 Lv	117 —	118 —

*Lanthanides	58 Ce	59 Pr	60 Nd	61 Pm	62 Sm	63 Eu	64 Gd	65 Tb	66 Dy	67 Ho	68 Er	69 Tm	70 Yb	71 Lu
†Actinides	90 Th	91 Pa	92 U	93 Np	94 Pu	95 Am	96 Cm	97 Bk	98 Cf	99 Es	100 Fm	101 Md	102 No	103 Lr

Metals | Metalloids | Nonmetals

On the periodic table, groups are the elements arranged as vertical columns, and periods are the elements in each horizontal row.

Study Note

1. The periodic table consists of horizontal rows called *periods* and vertical columns called *groups*, which are also called *families*.
2. Group 1A (1) contains the *alkali metals,* Group 2A (2) contains the *alkaline earth metals,* Group 7A (17) contains the *halogens,* and Group 8A (18) contains the *noble gases.*

Key Terms for Sections 4.1–4.2

Match each of the following key terms with the correct definition.

a. nonmetal	**b.** period	**c.** chemical symbol	**d.** metal
e. group number	**f.** halogen	**g.** periodic table	**h.** transition element

1. _____ a horizontal row of elements on the periodic table

2. _____ an element with little or no luster that is a poor conductor of heat and electricity

3. _____ a number that appears at the top of each vertical column (group) in the periodic table

4. _____ an element that is shiny, malleable, ductile, and a good conductor of heat and electricity

5. _____ an element in the center of the periodic table that has a group number of 3 through 12

6. _____ an arrangement of elements by increasing atomic number

7. _____ a one- or two-letter abbreviation that represents the name of an element

8. _____ an element in Group 7A (17)

Answers	**1.** b	**2.** a	**3.** e	**4.** d
	5. h	**6.** g	**7.** c	**8.** f

♦ Learning Exercise 4.2A

Indicate whether the following elements are part of the same group (G), the same period (P), or neither (N):

a. Li, C, and O _________	**b.** Br, Cl, and F _________
c. Al, Si, and Cl _________	**d.** C, N, and O _________
e. Mg, Ca, and Ba _________	**f.** C, S, and Br _________
g. Li, Na, and K _________	**h.** K, Ca, and Br _________

Answers	**a.** P	**b.** G	**c.** P	**d.** P
	e. G	**f.** N	**g.** G	**h.** P

♦ Learning Exercise 4.2B

Complete the list of elements and symbols, group numbers, and period numbers in the following table:

Element and Symbol	Group Number	Period Number
	2A (2)	3
Silicon, Si		
	5A (15)	2
Aluminum, Al		
	4A (14)	5
	1A (1)	6

Answers

Element and Symbol	Group Number	Period Number
Magnesium, Mg	2A (2)	3
Silicon, Si	4A (14)	3
Nitrogen, N	5A (15)	2
Aluminum, Al	3A (13)	3
Tin, Sn	4A (14)	5
Cesium, Cs	1A (1)	6

♦ Learning Exercise 4.2C

Identify each of the following elements as a metal (M), a nonmetal (NM), or a metalloid (ML):

1. Cl ____ **2.** N ____ **3.** Fe ____ **4.** Si ____ **5.** Al ____

6. C ____ **7.** Ca ____ **8.** Zn ____ **9.** Sb ____ **10.** Mg ____

Answers **1.** NM **2.** NM **3.** M **4.** ML **5.** M
6. NM **7.** M **8.** M **9.** ML **10.** M

♦ Learning Exercise 4.2D

Match the name of each of the following chemical groups with the elements K, Cl, He, Fe, Mg, Ne, Li, Cu, and Br:

1. halogen ____________________

2. noble gas ____________________

3. alkali metal ____________________

4. alkaline earth metal ____________________

5. transition element ____________________

Answers **1.** Cl, Br **2.** He, Ne **3.** K, Li **4.** Mg **5.** Fe, Cu

4.3 The Atom

Learning Goal: Describe the electrical charge and location in an atom for a proton, a neutron, and an electron.

- An atom is the smallest particle that retains the characteristics of an element.
- Atoms are composed of three subatomic particles. Protons have a positive charge (+), electrons carry a negative charge (−), and neutrons are electrically neutral.
- The protons and neutrons, with masses of about 1 amu, are found in the tiny, dense nucleus.
- The electrons are located outside the nucleus and have masses about 1/2000 that of a neutron or proton.

Dalton's Atomic Theory

1. All matter is made up of tiny particles called atoms.
2. All atoms of a given element are identical to one another and different from atoms of other elements.
3. Atoms of two or more different elements combine to form compounds. A particular compound is always made up of the same kinds of atoms and the same number of each kind of atom.
4. A chemical reaction involves the rearrangement, separation, or combination of atoms. Atoms are never created or destroyed during a chemical reaction.

♦ Learning Exercise 4.3A

True (T) or false (F): Each of the following statements is consistent with current atomic theory:

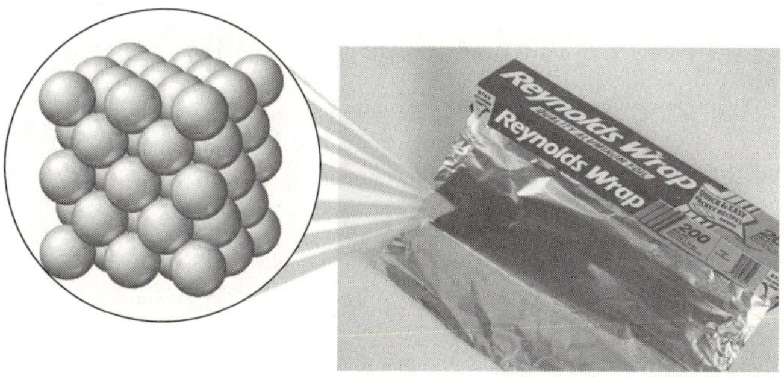

Aluminum foil consists of atoms of aluminum

1. All matter is composed of atoms.
2. All atoms of an element are identical.
3. Atoms can combine to form compounds.
4. Most of the mass of the atom is in the nucleus.

Answers **1.** T **2.** F **3.** T **4.** T

♦ Learning Exercise 4.3B

Match the following terms with the correct statements:

a. proton **b.** neutron **c.** electron **d.** nucleus

1. _____ is found in the nucleus of an atom

2. _____ has a 1− charge

3. _____ is found outside the nucleus

4. _____ has a mass of about 1 amu

5. _____ is the small, dense center of the atom

6. _____ is neutral

Answers **1.** a and b **2.** c **3.** c **4.** a and b **5.** d **6.** b

4.4 Atomic Number and Mass Number

Learning Goal: Given the atomic number and the mass number of an atom, state the number of protons, neutrons, and electrons.

- The atomic number, which can be found on the periodic table, is the number of protons in a given atom of an element.
- In a neutral atom, the number of protons is equal to the number of electrons.
- The mass number is the total number of protons and neutrons in an atom.

CORE CHEMISTRY SKILL
Counting Protons and Neutrons

♦ Learning Exercise 4.4A

Give the number of protons in each of the following neutral atoms:

a. an atom of carbon _____

b. an atom of the element with atomic number 15 _____

c. an atom with a mass number of 40 and atomic number 19 _____

d. an atom with 9 neutrons and a mass number of 19 _____

e. a neutral atom that has 18 electrons _____

Answers **a.** 6 **b.** 15 **c.** 19 **d.** 10 **e.** 18

Study Note

1. The *atomic number* is the number of protons in every atom of an element. In neutral atoms, the number of electrons equals the number of protons.
2. The *mass number* is the total number of neutrons and protons in the nucleus of an atom.
3. The number of neutrons is *mass number − atomic number*.

Example: How many protons and neutrons are in the nucleus of a strontium atom with a mass number of 89?

Solution: The atomic number of Sr is 38. Thus, an atom of Sr has 38 protons.
Mass number − atomic number = number of neutrons: 89 − 38 = 51 neutrons

♦ Learning Exercise 4.4B

Determine the number of neutrons in each of the following atoms:

a. has a mass number of 42 and atomic number 20 ________

b. has a mass number of 10 and 5 protons ________

c. is a silicon atom with a mass number of 30 ________

d. has a mass number of 9 and an atomic number of 4 ________

e. has a mass number of 22 and 10 protons ________

f. is a zinc atom with a mass number of 66 ________

Answers **a.** 22 **b.** 5 **c.** 16 **d.** 5 **e.** 12 **f.** 36

4.5 Isotopes and Atomic Mass

Learning Goal: Give the number of protons, electrons, and neutrons in one or more of the isotopes of an element; calculate the atomic mass of an element using the percent abundance and mass of its naturally occurring isotopes.

- *Isotopes* are atoms of the same element that have the same number of protons, but different numbers of neutrons.
- The atomic mass of an element is the weighted average mass of all isotopes in a naturally occurring sample of that element.

♦ Key Terms for Sections 4.3–4.5

Match each of the following key terms with the correct definition.

a. electron **b.** atom **c.** atomic mass **d.** mass number **e.** neutron
f. isotope **g.** atomic number **h.** nucleus **i.** proton

1. ____ the total number of neutrons and protons in the nucleus of an atom

2. ____ the smallest particle of an element that retains the characteristics of the element

3. ____ a negatively charged subatomic particle having a minute mass

4. ____ a neutral subatomic particle found in the nucleus of an atom

5. ____ a number that is equal to the number of protons in an atom

6. ____ the compact, extremely dense center of an atom where the protons and neutrons are located

7. ____ a positively charged subatomic particle found in the nucleus of an atom

8. ____ the weighted average mass of all the naturally occurring isotopes of an element

9. ____ An atom that differs in mass number from another atom of the same element

Answers **1.** d **2.** b **3.** a **4.** e **5.** g
6. h **7.** i **8.** c **9.** f

Study Note

In the atomic symbol for a particular atom, the mass number appears in the upper left corner and the atomic number in the lower left corner of the atomic symbol. Mass number → $^{32}_{16}S$ ← Atomic number

♦ Learning Exercise 4.5A

CORE CHEMISTRY SKILL

Writing Atomic Symbols for Isotopes

Complete the following table for neutral atoms:

Atomic Symbol	Atomic Number	Mass Number	Number of Protons	Number of Neutrons	Number of Electrons
	12			12	
			20	22	
		55		29	
	35			45	
		35	17		
$^{120}_{50}Sn$					

Answers

Atomic Symbol	Atomic Number	Mass Number	Number of Protons	Number of Neutrons	Number of Electrons
$^{24}_{12}Mg$	12	24	12	12	12
$^{42}_{20}Ca$	20	42	20	22	20
$^{55}_{26}Fe$	26	55	26	29	26
$^{80}_{35}Br$	35	80	35	45	35
$^{35}_{17}Cl$	17	35	17	18	17
$^{120}_{50}Sn$	50	120	50	70	50

♦ Learning Exercise 4.5B

Identify the sets of atoms that are isotopes.

A. $^{20}_{10}X$ **B.** $^{20}_{11}X$ **C.** $^{21}_{11}X$ **D.** $^{19}_{10}X$ **E.** $^{19}_{9}X$

Answer Atoms A and D are isotopes (at. no. 10); atoms B and C are isotopes (atomic number 11).

♦ Learning Exercise 4.5C

Essay Copper has two naturally occurring isotopes, $^{63}_{29}Cu$ and $^{65}_{29}Cu$. If that is the case, why is the atomic mass of copper listed as 63.55 on the periodic table?

Answer Copper in nature consists of two isotopes that have different atomic masses. The atomic mass is the weighted average of the individual masses of the two isotopes, which takes into consideration their percent abundances in the sample. The atomic mass does not represent the mass of any individual atom.

♦ Learning Exercise 4.5D

Rubidium is the element with atomic number 37.

1. What is the symbol for rubidium? ______________
2. What is the group number for rubidium? ______________
3. What is the name of the family that includes rubidium? ______________
4. Rubidium has two naturally occurring isotopes, $^{85}_{37}Rb$ and $^{87}_{37}Rb$. How many protons, neutrons, and electrons are in each isotope?

5. In naturally occurring rubidium, 72.17% is Rb-85 with a mass of 84.91 amu, whereas 27.83% is Rb-87 with a mass of 86.91 amu. What is the atomic mass of rubidium?

Answers

1. Rb
2. Rubidium is in Group 1A (1).
3. Rubidium is in the family called the alkali metals.
4. The isotope Rb-85 has 37 protons, 48 neutrons, and 37 electrons. The isotope Rb-87 has 37 protons, 50 neutrons, and 37 electrons.
5. 84.91 amu (72.17/100) + 86.91 amu (27.83/100) = 61.28 amu + 24.19 amu = 85.47 amu for the atomic mass of Rb.

Checklist for Chapter 4

You are ready to take the Practice Test for Chapter 4. Be sure you have accomplished the following learning goals for this chapter. If not, review the section listed at the end of the goal. Then apply your new skills and understanding to the Practice Test.

After studying Chapter 4, I can successfully:

_____ Write the correct symbol or name for an element (4.1).

_____ Use the periodic table to identify the group and period of an element, and describe it as a metal, nonmetal, or metalloid and a representative or transition element (4.2).

_____ Use the periodic table to identify an element as a representative or transition element (4.2).

_____ State the electrical charge, mass, and location of the protons, neutrons, and electrons in an atom (4.3).

_____ Given the atomic number and mass number of an atom, state the number of protons, neutrons, and electrons (4.4).

_____ Identify an isotope (4.5).

_____ Calculate the atomic mass of an element from the masses of the naturally occurring isotopes and percent abundances (4.5).

Practice Test for Chapter 4

For questions 1 through 5, write the correct symbol for each of the elements listed: (4.1)

1. potassium ____ **2.** phosphorus ____ **3.** calcium ____

4. carbon ____ **5.** sodium ____

For questions 6 through 10, write the correct element name for each of the symbols listed: (4.1)

6. Fe ________________ **7.** Cu ________________

8. Cl ________________ **9.** Pb ________________

10. Ag ________________

11. The elements C, N, and O are part of a (4.2)
A. period **B.** group **C.** neither

12. The elements Li, Na, and K are part of a (4.2)
A. period **B.** group **C.** neither

13. What is the classification of an atom with 15 protons and 17 neutrons? (4.2)
A. metal **B.** nonmetal **C.** transition element
D. noble gas **E.** halogen

14. What is the group number of the element with atomic number 3? (4.2)
A. 1A (1) **B.** 2A (2) **C.** 3B (3) **D.** 3A (13) **E.** 5A (15)

For questions 15 through 17, indicate whether the statement is True or False: (4.3)

15. ____ The electrons of an atom are located in the nucleus.

16. ____ Most of the mass of an atom is due to the protons and neutrons.

17. ____ All matter is composed of atoms.

For questions 18 through 21, consider an atom with 12 protons and 13 neutrons: (4.4)

18. This atom has an atomic number of
A. 12 **B.** 13 **C.** 23 **D.** 24 **E.** 25

19. This atom has a mass number of
A. 12 **B.** 13 **C.** 23 **D.** 24 **E.** 25

20. This is an atom of
A. carbon **B.** sodium **C.** magnesium **D.** aluminum **E.** manganese

21. The number of electrons in this atom is
A. 12 **B.** 13 **C.** 23 **D.** 24 **E.** 25

For questions 22 through 25, consider an atom of calcium with a mass number of 42: (4.4)

22. This atom of calcium has an atomic number of
A. 20 **B.** 22 **C.** 40 **D.** 41 **E.** 42

23. The number of protons in this atom of calcium is
A. 20 **B.** 22 **C.** 40 **D.** 41 **E.** 42

24. The number of neutrons in this atom of calcium is
A. 20 **B.** 22 **C.** 40 **D.** 41 **E.** 42

25. The number of electrons in this atom of calcium is
A. 20 **B.** 22 **C.** 40 **D.** 41 **E.** 42

26. A platinum atom, $^{195}_{78}Pt$, has (4.4)
A. 78 p^+, 78 e^-, 78 n^0 **B.** 195 p^+, 195 e^-, 195 n^0 **C.** 78 p^+, 78 e^-, 195 n^0
D. 78 p^+, 78 e^-, 117 n^0 **E.** 78 p^+, 117 e^-, 117 n^0

For questions 27 and 28, use the following list of atoms: (4.5)

$^{14}_{7}G$ $^{16}_{8}J$ $^{19}_{9}M$ $^{16}_{7}Q$ $^{18}_{8}R$

27. Which atom(s) is (are) isotopes of an atom with 8 protons and 9 neutrons?
A. J **B.** J, R **C.** M, Q **D.** M **E.** Q

28. Which atom(s) are isotopes of an atom with 7 protons and 8 neutrons?
A. G **B.** J **C.** G, Q **D.** J, R **E.** none

29. Which element would you expect to have properties most like oxygen? (4.2)
A. nitrogen **B.** carbon **C.** chlorine **D.** argon **E.** sulfur

30. Which of the following is an isotope of nitrogen? (4.5)
A. $^{14}_{8}N$ **B.** $^{7}_{3}N$ **C.** $^{10}_{5}N$ **D.** $^{4}_{2}He$ **E.** $^{15}_{7}N$

Answers for the Practice Test

1. K	**2.** P	**3.** Ca	**4.** C	**5.** Na
6. Iron	**7.** Copper	**8.** Chlorine	**9.** Lead	**10.** Silver
11. A	**12.** B	**13.** B	**14.** A	**15.** False
16. True	**17.** True	**18.** A	**19.** E	**20.** C
21. A	**22.** A	**23.** A	**24.** B	**25.** A
26. D	**27.** B	**28.** C	**29.** E	**30.** E

5
Electronic Structure and Periodic Trends

Robert and Jennifer are materials engineers who develop and test materials used in the manufacture of consumer goods. In their research, the two work with some of the elements in Groups 3A (13), 4A (14), and 5A (15) of the periodic table. These elements, such as silicon, have properties that make them good semiconductors.

A microchip requires growing a single crystal of a semiconductor, such as pure silicon. When small amounts of impurities are added to the crystalline structure, holes form through which electrons can travel with little obstruction. Microchips are manufactured for use in computers, cell phones, satellites, televisions, calculators, GPSs, and many other devices. What elements are found in Group 4A (14) on the periodic table?

5.1 Electromagnetic Radiation

Learning Goal: Compare the wavelength of radiation with its frequency.

- Electromagnetic radiation is energy that travels as waves in space at the speed of light.
- The wavelength (λ) is the distance between the crests of adjacent waves.
- The frequency (ν) is the number of waves that pass a certain point in 1 second.
- Frequency is measured in hertz (Hz), which is equal to cycles/s or s^{-1}.
- The electromagnetic spectrum is all forms of electromagnetic radiation, which can be arranged in order of decreasing wavelength.
- The speed of light (c) 3.00×10^8 m/s = wavelength (λ) $\times$ frequency (ν)
- The wave equation, $c = \lambda\nu$, expresses the relationship of the speed of light (m/s) to wavelength (m) and frequency (s^{-1})

♦ Learning Exercise 5.1A

What is the wavelength, in meters, for a police radar monitor that operates at 2.2×10^{10} Hz?

Answer

$2.2 \times 10^{10}\text{ Hz} = 2.2 \times 10^{10}\text{ s}^{-1}$

$$\lambda = \frac{3.00 \times 10^8\text{ m }\cancel{\text{s}^{-1}}}{2.2 \times 10^{10}\ \cancel{\text{s}^{-1}}} = 1.4 \times 10^{-2}\text{ m}$$

♦ Learning Exercise 5.1B

Using the electromagnetic spectrum, identify the type of radiation in each of the following pairs that would have the shorter wavelength:

a. microwave or infrared ____________

b. radio waves or X-rays ____________

c. gamma rays or radio waves ____________

d. ultraviolet or microwave ____________

Answers **a.** infrared **b.** X-rays **c.** gamma rays **d.** ultraviolet

5.2 Atomic Spectra and Energy Levels

Learning Goal: Explain how atomic spectra correlate with the energy levels in atoms.

- An atomic spectrum is a series of colored lines each of which corresponds to a photon of a specific energy emitted by a heated element.
- A photon is a particle of electromagnetic radiation of a specified amount of energy called a quantum.
- In an atom, the energy levels indicated by the principal quantum number, *n*, contain electrons of similar energies.
- The principal quantum number, *n*, increases as the energy of the electrons increase.
- When electrons change energy levels, photons of specific energies are absorbed or emitted.

♦ Learning Exercise 5.2

Identify the photon in each of the following pairs with the greater energy:

a. _____ yellow light or green light **b.** _____ $\lambda = 10^9$ nm or $\lambda = 10^5$ nm

c. _____ purple light or red light **d.** _____ $\lambda = 10^{-3}$ nm or $\lambda = 10^2$ nm

Answers **a.** green light **b.** $\lambda = 10^5$ nm **c.** purple light **d.** $\lambda = 10^{-3}$ nm

5.3 Sublevels and Orbitals

Learning Goal: Describe the sublevels and orbitals in atoms.

- Within each energy level, electrons with identical energy are grouped in sublevels.
- An *s* sublevel can accommodate 2 electrons; a *p* sublevel can accommodate 6 electrons; a *d* sublevel can accommodate 10 electrons; and an *f* sublevel can accommodate 14 electrons.
- The energies of the sublevels increase in order: $s < p < d < f$.
- The number of sublevels in each energy level is equal to its principal quantum number, *n*. For example, $n = 1$ has one sublevel (s), $n = 2$ has two sublevels (s, p), $n = 3$ has three sublevels (s, p, d), and $n = 4$ has four sublevels (s, p, d, f).
- An orbital is a region in an atom where there is the greatest probability of finding an electron of certain energy.
- An *s* orbital is spherical, and *p* orbitals have two lobes along an axis. The *d* and *f* orbitals have more complex shapes.

- Each sublevel consists of a set of orbitals: an *s* sublevel consists of one orbital; a *p* sublevel consists of three orbitals; a *d* sublevel consists of five orbitals; and an *f* sublevel consists of seven orbitals.
- An orbital can contain a maximum of two electrons, which have opposite spins.

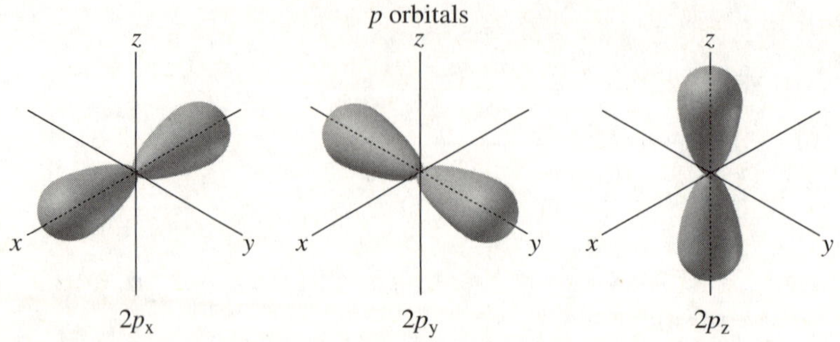

Each *p* orbital, which has two regions of high probability, is aligned along a different axis.

Key Terms for Sections 5.1–5.3

Match each the following key terms with the correct definition:

a. photon **b.** wavelength **c.** electromagnetic radiation **d.** orbital

1. _____ the distance between the peaks of two adjacent waves

2. _____ energy that travels as waves at the speed of light

3. _____ the smallest particle of light

4. _____ the region within an atom where an electron of a certain energy is most likely to be found

Answers **1.** b **2.** c **3.** a **4.** d

♦ Learning Exercise 5.3A

What is similar about the following orbitals?

a. 4*s* and 4*d*

b. 3*p* and 5*p*

Answers

a. The 4*s* and 4*d* orbitals are found in the same energy level with the same principal quantum number.
b. The 3*p* and 5*p* orbitals are the same type, which have the same general shape.

♦ Learning Exercise 5.3B

State the maximum number of electrons for each of the following:

a. 3*p* sublevel _____ **b.** 3*d* sublevel _____

c. 2*s* orbital _____ **d.** energy level $n = 4$ _____

e. 1*s* sublevel _____ **f.** 4*p* orbital _____

g. 5*p* sublevel _____ **h.** 4*f* sublevel _____

Answers **a.** 6 **b.** 10 **c.** 2 **d.** 32 **e.** 2 **f.** 2 **g.** 6 **h.** 14

5.4 Orbital Diagrams and Electron Configurations

Learning Goal: Draw the orbital diagram and write the electron configuration for an element.

- An orbital diagram shows the orbitals in an atom with squares that represent the orbitals and arrows that represent the electrons. In this example, the orbital diagram of helium consists of a square with two arrows.

- The electron configuration shows the number of electrons in each sublevel in order of increasing energy. For example, the electron configuration for Mg (atomic number 12) is $1s^22s^22p^63s^2$.
- In the abbreviated electron configuration, the electron configuration of the preceding noble gas is replaced by its symbol in square brackets. For example, the abbreviated electron configuration for Mg is [Ne] $3s^2$.

Guide to Drawing Orbital Diagrams	
Step 1	Draw boxes to represent the occupied orbitals.
Step 2	Place a pair of electrons with opposite spins in each filled orbitals.
Step 3	Place remaining electrons in the last occupied sublevel in separate orbitals.

♦ Learning Exercise 5.4A

Draw the orbital diagram for each of the following elements:

a. beryllium ______________ **b.** carbon ______________

c. sodium ______________ **d.** nitrogen ______________

e. fluorine ______________ **f.** magnesium ______________

Answers

a. 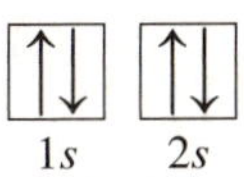

b. 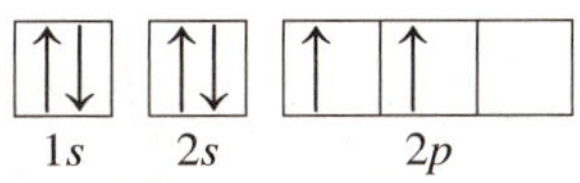

c. 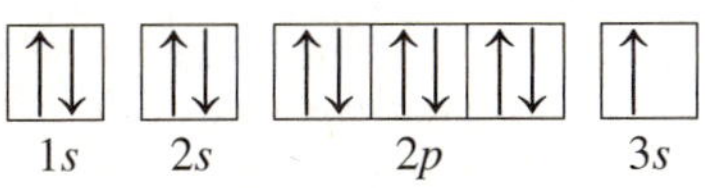

d. 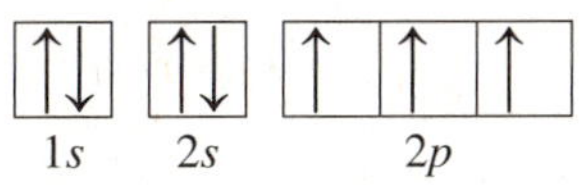

e. 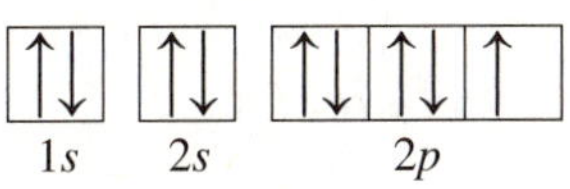

f. 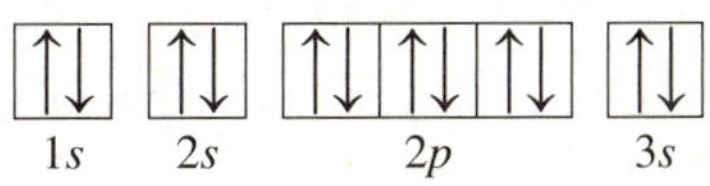

CORE CHEMISTRY SKILL
Writing Electron Configurations

♦ Learning Exercise 5.4B

Write the electron configuration for each of the following elements:

a. carbon ______________________________

b. magnesium ______________________________

c. iron ______________________________

d. silicon ______________________________

e. chlorine ______________________

f. phosphorus ______________________

Answers

a. $1s^22s^22p^2$ **b.** $1s^22s^22p^63s^2$ **c.** $1s^22s^22p^63s^23p^64s^23d^6$
d. $1s^22s^22p^63s^23p^2$ **e.** $1s^22s^22p^63s^23p^5$ **f.** $1s^22s^22p^63s^23p^3$

♦ Learning Exercise 5.4C

Write the abbreviated electron configuration for each of the following elements:

a. carbon ______________________

b. magnesium ______________________

c. iron ______________________

d. silicon ______________________

e. chlorine ______________________

f. phosphorus ______________________

Answers

a. [He]$2s^22p^2$ **b.** [Ne]$3s^2$ **c.** [Ar]$4s^23d^6$
d. [Ne]$3s^23p^2$ **e.** [Ne]$3s^23p^5$ **f.** [Ne]$3s^23p^3$

♦ Learning Exercise 5.4D

Name the element with an electron configuration ending with each of the following notations:

a. $3p^5$ __________ **b.** $2s^1$ __________

c. $3d^8$ __________ **d.** $4p^1$ __________

e. $5p^5$ __________ **f.** $3p^2$ __________

g. $1s^1$ __________ **h.** $6s^2$ __________

Answers

a. chlorine **b.** lithium **c.** nickel **d.** gallium
e. iodine **f.** silicon **g.** hydrogen **h.** barium

5.5 Electron Configurations and the Periodic Table

Learning Goal: Write the electron configuration for an atom using the sublevel blocks on the periodic table.

- The electron sublevels are organized as blocks on the periodic table. The *s* block corresponds to Groups 1A (1) and 2A (2), the *p* block extends from Group 3A (13) to 8A (18), the *d* block contains the transition elements from 3B (3) to 2B (12), and the *f* block contains the lanthanides and actinides.

Guide to Writing Electron Configurations Using Sublevel Blocks	
Step 1	Locate the element on the periodic table.
Step 2	Write the filled sublevels in order, going across each period.
Step 3	Complete the configuration by counting the number of electrons in the last occupied sublevel block.

♦ Learning Exercise 5.5A

Use the sublevel blocks on the periodic table to write the electron configuration for each of the following:

a. fluorine ______________________

b. sulfur ______________________

c. cadmium ______________________

d. strontium ______________________

Answers
a. $1s^22s^22p^5$
b. $1s^22s^22p^63s^23p^4$
c. $1s^22s^22p^63s^23p^64s^23d^{10}4p^65s^24d^{10}$
d. $1s^22s^22p^63s^23p^64s^23d^{10}4p^65s^2$

♦ Learning Exercise 5.5B

CORE CHEMISTRY SKILL
Using the Periodic Table to Write Electron Configurations

Use the sublevel blocks on the periodic table to write the abbreviated electron configuration for the each of the following elements:

a. iodine __________ b. barium __________

c. zinc __________ d. tin __________

e. cesium __________ f. bromine __________

Answers
a. $[Kr]5s^24d^{10}5p^5$
b. $[Xe]6s^2$
c. $[Ar]4s^23d^{10}$
d. $[Kr]5s^24d^{10}5p^2$
e. $[Xe]6s^1$
f. $[Ar]4s^23d^{10}4p^5$

♦ Learning Exercise 5.5C

Give the symbol of the element that has each of the following:

a. six 3*d* electrons ____

b. first to completely fill four *s* orbitals ____

c. $[Kr]5s^1$ ____

d. $[Ar]4s^23d^{10}4p^5$ ____

e. two 6*p* electrons ____

f. first to have nine completely filled *p* orbitals ____

Answers **a.** Fe **b.** Ca **c.** Rb **d.** Br **e.** Pb **f.** Kr

5.6 Trends in Periodic Properties

Learning Goal: Use the electron configurations of elements to explain the trends in periodic properties.

- The physical and chemical properties of elements change in a periodic manner going across each period and are repeated in each successive period.
- Representative elements in a group have similar behavior.
- The group number of a representative element gives the number of valence electrons (outermost electrons).
- An electron-dot symbol is a convenient way to represent the valence electrons, which are shown as dots on the sides, top, or bottom of the symbol for the element. For beryllium, the electron-dot symbol is:

$$\dot{\text{Be}}\cdot$$

- The atomic radius of representative elements generally increases going down a group and decreases going from left to right across a period.
- The ionization energy generally decreases going down a group and increases going from left to right across a period.
- The metallic character increases going down a group and decreases going left to right across a period.

Key Terms for Sections 5.5–5.6

Match each of the following key terms with the correct definition:

a. *f* block	**b.** *s* block	**c.** electron-dot symbol
d. *d* block	**e.** ionization energy	**f.** valence electrons

1. ____ the block of 14 elements in the rows at the bottom of the periodic table

2. ____ the elements in Groups 1A (1) and 2A (2) in which electrons fill the *s* orbitals

3. ____ the block of ten elements from Groups 3B (3) to 2B (12)

4. ____ the representation of an atom that shows valence electrons around the symbol of the element

5. ____ a measure of how easily an element loses a valence electron

6. ____ the electrons in the highest energy level of an atom

Answers **1.** a **2.** b **3.** d **4.** c **5.** e **6.** f

♦ Learning Exercise 5.6A

CORE CHEMISTRY SKILL
Identifying Trends in Periodic Properties

State the number of valence electrons, the group number, and the electron-dot symbol for each of the following:

Element	Valence Electrons	Group Number	Electron-Dot Symbol
a. nitrogen			
b. oxygen			
c. magnesium			
d. hydrogen			
e. fluorine			
f. aluminum			

Answers

Element	Valence Electrons	Group Number	Electron-Dot Symbol
a. Nitrogen	5 e^-	Group 5A (15)	·N̈· (with one dot below)
b. Oxygen	6 e^-	Group 6A (16)	·Ö: (with one dot below)
c. Magnesium	2 e^-	Group 2A (2)	Ṁg·
d. Hydrogen	1 e^-	Group 1A (1)	H·
e. Fluorine	7 e^-	Group 7A (17)	·F̤̈:
f. Aluminum	3 e^-	Group 3A (13)	·Ȧl·

♦ Learning Exercise 5.6B

Indicate the element in each of the following pairs that has the larger atomic radius:

a. ____________ Mg or Ca b. ____________ Si or Cl

c. ____________ Sr or Rb d. ____________ Br or Cl

e. ____________ Li or Cs f. ____________ Li or N

g. ____________ N or P h. ____________ As or Ca

Answers **a.** Ca **b.** Si **c.** Rb **d.** Br
e. Cs **f.** Li **g.** P **h.** Ca

♦ Learning Exercise 5.6C

Indicate the element in each pair that has the lower ionization energy.

a. ____________ Mg or Na b. ____________ P or Cl

c. ____________ K or Rb d. ____________ Br or F

e. ____________ Li or O f. ____________ Sb or N

g. ____________ K or Br h. ____________ S or Na

Answers **a.** Na **b.** P **c.** Rb **d.** Br
e. Li **f.** Sb **g.** K **h.** Na

♦ Learning Exercise 5.6D

Indicate the element in each of the following pairs that has more metallic character:

1. ____________ K or Na 2. ____________ O or Se

3. ____________ Ca or Br 4. ____________ I or F

5. ____________ Li or N 6. ____________ Pb or Ba

Answers **1.** K **2.** Se **3.** Ca **4.** I **5.** Li **6.** Ba

Checklist for Chapter 5

You are ready to take the Practice Test for Chapter 5. Be sure you have accomplished the following learning goals for this chapter. If not, review the section listed at the end of the goal. Then apply your new skills and understanding to the Practice Test.

After studying Chapter 5, I can successfully:

_____ Compare the wavelength of radiation with its energy (5.1).

_____ Compare frequency, wavelength, and energy of radiation on the electromagnetic spectrum (5.1).

_____ Explain how atomic spectra correlate with the energy levels in atoms (5.2).

_____ Describe the sublevels and orbitals in atoms (5.3).

_____ Write the orbital diagrams and electron configurations for hydrogen to argon (5.4).

_____ Use the sublevel blocks on the periodic table to write electron configurations (5.5).

_____ Explain the trends in valence electrons, group numbers, atomic size, and ionization energy going down a group and across a period (5.6).

_____ Determine which of two elements is the more metallic. (5.6)

Practice Test for Chapter 5

1. Which of the following has the longest wavelength? (5.1)
A. microwave **B.** ultraviolet **C.** radio waves **D.** infrared **E.** visible

2. Which of the following has the highest frequency? (5.1)
A. microwave **B.** ultraviolet **C.** radio waves **D.** infrared **E.** visible

3. What color of visible light has the highest frequency? (5.1)
A. violet **B.** blue **C.** green **D.** yellow **E.** red

4. What color of visible light has the longest wavelength? (5.1)
A. violet **B.** blue **C.** green **D.** yellow **E.** red

5. Arrange the following in order of increasing energy: infrared, ultraviolet, radio waves, X-rays. (5.1)
A. X-rays, infrared, ultraviolet, radio waves
B. radio waves, ultraviolet, infrared, X-rays
C. ultraviolet, infrared, radio waves, X-rays
D. radio waves, X-rays, infrared, ultraviolet
E. radio waves, infrared, ultraviolet, X-rays

6. The wavelength is (5.1)
A. the height to the crest of a wave
B. the length of a wave
C. the distance between two adjacent crests of a wave
D. the number of waves that pass a point in 1 s
E. related directly to the frequency of a wave of electromagnetic radiation

7. A microwave has a wavelength of 3.0×10^{-2} m. This radiation has a frequency, in s^{-1}, of (5.1)
A. $1.0 \times 10^{10}\ s^{-1}$ **B.** $1.0 \times 10^{-10}\ s^{-1}$ **C.** $1.0 \times 10^{-6}\ s^{-1}$
D. $1.0 \times 10^{6}\ s^{-1}$ **E.** $9.0 \times 10^{6}\ s^{-1}$

8. Atomic spectra (5.2)
A. consist of lines of different colors
B. are different for each of the elements
C. occur when energy changes produce photons of certain wavelengths
D. indicate that electrons in atoms have specific energy levels
E. all of these

9. The maximum number of electrons that can be accommodated in the principal energy level $n = 4$ is (5.3)
A. 2 **B.** 4 **C.** 8 **D.** 18 **E.** 32

10. The sublevels that make up the principal energy level $n = 3$ are (5.3)
A. $1s2s3s$ **B.** $3s3p$ **C.** $3s3p3d$ **D.** $3s3p3d3f$ **E.** $3p3d3f$

11. The spherical volume within an atom in which an electron is likely to be found is called a(n) (5.3)
A. *d* orbital **B.** *p* orbital **C.** *s* orbital **D.** *f* orbital **E.** *g* orbital

12. The number of orbitals in a $4p$ sublevel is (5.3)
A. one **B.** two **C.** three **D.** four **E.** five

13. The principal energy level that consists of one s orbital and three p orbitals has principal quantum number (5.3)
A. $n = 1$ **B.** $n = 2$ **C.** $n = 3$ **D.** $n = 4$ **E.** $n = 5$

14. The maximum number of electrons that the $5d$ sublevel can accommodate is (5.3)
A. 2 **B.** 3 **C.** 6 **D.** 10 **E.** 14

15. In the orbital diagram of nitrogen, the $2p$ sublevel would be shown as (5.4)

A. [↑↓][ ][↑] **B.** [↑↑][↑][ ] **C.** [↑↓][↑↓][↑]
D. [↑][↑][↑] **E.** [↑][ ][↑↓]

16. The electron configuraton for silicon is (5.4)
A. $1s^22s^22p^63s^4$ **B.** $1s^22s^22p^63s^23p^2$ **C.** $1s^22s^22p^63s^23d^2$
D. $1s^22s^22p^63s^23p^4$ **E.** $1s^22s^22p^63p^4$

17. The electron configuration for oxygen is (5.4)
A. $2s^22p^4$ **B.** $1s^22s^42p^4$ **C.** $1s^22s^6$
D. $1s^22s^22p^23s^2$ **E.** $1s^22s^22p^4$

18. The electron configuration for aluminum is (5.4)
A. $1s^22s^22p^9$ **B.** $1s^22s^22p^63p^5$ **C.** $1s^22s^22p^63s^23p^1$
D. $1s^22s^22p^83p^1$ **E.** $1s^22s^22p^63p^3$

For questions 19 through 22, match the final notation in the electron configuration with the following elements: (5.4)
A. As **B.** Rb **C.** Na **D.** N **E.** Xe

19. $4p^3$ ____ **20.** $5s^1$ ____

21. $3s^1$ ____ **22.** $5p^6$ ____

23. The element that has an electron configuration that ends with $5p^2$ is (5.4)
A. strontium **B.** germanium **C.** titanium **D.** tin **E.** lead

24. The element that has an abbreviated electron configuration of $[Ar]4s^23d^6$ is (5.4)
A. chromium **B.** iron **C.** krypton **D.** calcium **E.** zinc

25. The element that begins filling the $6s$ sublevel is (5.5)
A. lithium **B.** rubidium **C.** cesium **D.** scandium **E.** barium

26. The element that has 5 electrons in its principal energy level $n = 4$ is (5.5)
A. boron **B.** phosphorus **C.** vanadium **D.** bromine **E.** arsenic

27. The number of electrons in the $5p$ sublevel of iodine is (5.5)
A. 2 **B.** 3 **C.** 5 **D.** 7 **E.** 15

28. The number of valence electrons in gallium is (5.5)
A. 1 **B.** 2 **C.** 3 **D.** 13 **E.** 31

29. The atomic radius of oxygen is larger than that of (5.6)
A. lithium **B.** sulfur **C.** fluorine **D.** boron **E.** argon

30. Of C, Si, Ge, Sn, and Pb, the element with the greatest ionization energy is (5.6)
A. C **B.** Si **C.** Ge **D.** Sn **E.** Pb

31. The electron-dot symbol ·X· would be correct for an element in (5.6)
A. Group 1A (1) B. Group 2A (2) C. Group 4A (14) D. Group 6A (16) E. Group 7A (17)

32. The electron-dot symbol X· would be correct for (5.6)
A. magnesium B. chlorine C. sulfur D. cesium E. nitrogen

33. Of Li, Na, K, Rb, and Cs, the element with the most metallic character is (5.6)
A. Li B. Na C. K D. Rb E. Cs

Answers for the Practice Test

1. C	2. B	3. A	4. E	5. E
6. C	7. A	8. E	9. E	10. C
11. C	12. C	13. B	14. D	15. D
16. B	17. E	18. C	19. A	20. B
21. C	22. E	23. D	24. B	25. C
26. E	27. C	28. C	29. C	30. A
31. B	32. D	33. E		

6
Ionic and Molecular Compounds

Sarah, a pharmacist, talked with her customer regarding questions about stomach upset from the low-dose aspirin, $C_9H_8O_4$, he was taking to prevent a heart attack or stroke. Sarah explained that when he takes a low-dose aspirin, he needs to drink a glass of water. If his stomach is still upset, he can take the aspirin with food or milk each time. A few days later, Sarah talked with her customer again. He told her he followed her suggestions and that he no longer had any upset stomach from the low-dose aspirin. Is aspirin an ionic or a molecular compound?

6.1 Ions: Transfer of Electrons

Learning Goal: Write the symbols for the simple ions of the representative elements.

- The stability of the noble gases is associated with a stable electron configuration of eight electrons (s^2p^6), an octet, in their outer energy level. Helium is stable with two electrons in the outer energy level.
- Atoms of elements other than the noble gases achieve stability by losing, gaining, or sharing valence electrons with other atoms in the formation of compounds.
- Metals of the representative elements in Groups 1A (1), 2A (2) and 3A (13) achieve a noble gas electron configuration by losing their valence electron(s) to form positively charged cations with a charge of 1+, 2+, or 3+, respectively.
- Nonmetals in Groups 5A (15), 6A (16) and 7A (17) gain valence electrons to achieve an octet forming negatively charged anions with a charge of 3−, 2−, or 1−, respectively.
- *Ionic bonds* occur when the valence electrons of atoms of a metal are transferred to atoms of nonmetals. For example, sodium atoms lose electrons and chlorine atoms gain electrons to form the ionic compound NaCl.
- *Covalent bonds* form when atoms of nonmetals share valence electrons. In the molecular compounds H_2O and C_3H_8, atoms share electrons.

Study Note

When an atom loses or gains electrons, it acquires the electron configuration of the nearest noble gas. For example, sodium loses one electron, which gives the Na^{+} ion the electron configuration of neon. Oxygen gains two electrons to give an oxide ion, O^{2-}, the electron configuration of neon.

♦ Learning Exercise 6.1A

CORE CHEMISTRY SKILL
Writing Positive and Negative Ions

The following elements lose electrons when they form ions. Indicate the group number, the number of electrons lost, and the ion (symbol and charge) for each of the following:

Element	Group Number	Electrons Lost	Ion Formed
Magnesium			
Sodium			
Calcium			
Potassium			
Aluminum			

Answers

Element	Group Number	Electrons Lost	Ion Formed
Magnesium	2A (2)	2	Mg^{2+}
Sodium	1A (1)	1	Na^{+}
Calcium	2A (2)	2	Ca^{2+}
Potassium	1A (1)	1	K^{+}
Aluminum	3A (13)	3	Al^{3+}

♦ Learning Exercise 6.1B

The following elements gain electrons when they form ions. Indicate the group number, the number of electrons gained, and the ion (symbol and charge) for each of the following:

Element	Group Number	Electrons Gained	Ion Formed
Chlorine			
Oxygen			
Nitrogen			
Fluorine			
Sulfur			

Answers

Element	Group Number	Electrons Gained	Ion Formed
Chlorine	7A (17)	1	Cl^-
Oxygen	6A (16)	2	O^{2-}
Nitrogen	5A (15)	3	N^{3-}
Fluorine	7A (17)	1	F^-
Sulfur	6A (16)	2	S^{2-}

6.2 Writing Formulas for Ionic Compounds

Learning Goal: Using charge balance, write the correct formula for an ionic compound.

- In the formulas of ionic compounds, the total positive charge is equal to the total negative charge. For example, the compound magnesium chloride, $MgCl_2$, contains Mg^{2+} and two Cl^-. The sum of the charges is zero: $(2+) + 2(1-) = 0$.
- When two or more ions are needed for charge balance, that number is indicated by subscripts in the formula.

For the following exercises, you may want to cut pieces of paper that represent typical positive and negative ions, as shown. To determine an ionic formula, place the positive and negative pieces together to complete a geometric shape. Write the number of positive ions and negative ions as the subscripts for the formula.

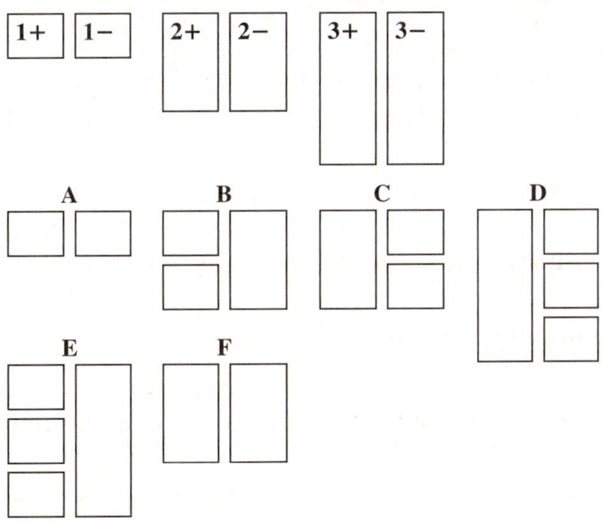

♦ Learning Exercise 6.2A

Give the letter (A, B, C, etc.) that matches the arrangement of ions in the following compounds:

Compound	Combination	Compound	Combination
1. $MgCl_2$	______	**2.** Na_2S	______
3. LiCl	______	**4.** CaO	______
5. K_3N	______	**6.** $AlBr_3$	______
7. MgS	______	**8.** $BaCl_2$	______

Answers **1.** C **2.** B **3.** A **4.** F **5.** E **6.** D **7.** F **8.** C

Study Note

You can check that the ionic formula you write is electrically neutral by multiplying each of the ionic charges by their subscripts. When added together, their sum should equal zero. For example, the formula Na_2O gives $2(1+) + 1(2-) = (2+) + (2-) = 0$.

♦ Learning Exercise 6.2B

Write the correct formula of the ionic compound formed from each of the following pairs of ions:

1. Na^+ and Cl^- ____________
2. K^+ and S^{2-} ____________
3. Al^{3+} and O^{2-} ____________
4. Mg^{2+} and Cl^- ____________
5. Ca^{2+} and S^{2-} ____________
6. Al^{3+} and Cl^- ____________
7. Li^+ and N^{3-} ____________
8. Ba^{2+} and P^{3-} ____________

Answers
1. NaCl 2. K_2S 3. Al_2O_3 4. $MgCl_2$
5. CaS 6. $AlCl_3$ 7. Li_3N 8. Ba_3P_2

6.3 Naming Ionic Compounds

Learning Goal: Given the formula of an ionic compound, write the correct name; given the name of an ionic compound, write the correct formula.

- In naming ionic compounds, the positive ion is named first, followed by the name of the negative ion. The name of a representative metal ion in Group 1A (1), 2A (2), or 3A (13) is the same as its elemental name. The name of a nonmetal ion is obtained by replacing the end of its element name with *ide*.
- Most transition elements form cations with two or more ionic charges. Then the ionic charge must be written as a Roman numeral in parentheses after the name of the metal. For example, the cations of iron, Fe^{2+} and Fe^{3+}, are named iron(II) and iron(III). The ions of copper are Cu^+, copper(I), and Cu^{2+}, copper(II).
- The only transition elements with fixed charges are zinc, Zn^{2+}; silver, Ag^+; and cadmium, Cd^{2+}.

Guide to Naming Ionic Compounds with Metals that Form a Single Ion	
Step 1	Identify the cation and anion.
Step 2	Name the cation by its element name.
Step 3	Name the anion by using the first syllable of its element name followed by *ide*.
Step 4	Write the name of the cation first and the anion second.

♦ Learning Exercise 6.3A

Write the ions and the correct formula for each of the following ionic compounds:

Formula	**Ions**		**Name**
1. Cs_2O	________	________	____________________
2. $BaBr_2$	________	________	____________________

3. Mg_3P_2 ________ ________ ______________________

4. Na_2S ________ ________ ______________________

Answers
1. Cs^+, O^{2-}, cesium oxide
2. Ba^{2+}, Br^-, barium bromide
3. Mg^{2+}, P^{3-}, magnesium phosphide
4. Na^+, S^{2-}, sodium sulfide

♦ Learning Exercise 6.3B

Write the name of each of the following ions:

1. Cl^- ________
2. Fe^{2+} ________
3. Cu^+ ________
4. Ag^+ ________
5. O^{2-} ________
6. Ca^{2+} ________
7. S^{2-} ________
8. Al^{3+} ________
9. Fe^{3+} ________
10. Ba^{2+} ________
11. Cu^{2+} ________
12. N^{3-} ________

Answers
1. chloride
2. iron(II)
3. copper(I)
4. silver
5. oxide
6. calcium
7. sulfide
8. aluminum
9. iron(III)
10. barium
11. copper(II)
12. nitride

♦ Learning Exercise 6.3C

Write the symbol of each of the following ions:

Name	Symbol	Name	Symbol
1. chromium(III) ion	________	2. cobalt(II) ion	________
3. zinc ion	________	4. lead(IV) ion	________
5. gold(I) ion	________	6. silver ion	________
7. potassium ion	________	8. nickel(III) ion	________

Answers
1. Cr^{3+}
2. Co^{2+}
3. Zn^{2+}
4. Pb^{4+}
5. Au^+
6. Ag^+
7. K^+
8. Ni^{3+}

Guide to Naming Ionic Compounds with Variable Charge Metals	
Step 1	Determine the charge of the cation from the anion.
Step 2	Name the cation by its element name and use a Roman numeral in parentheses for the charge.
Step 3	Name the anion by using the first syllable of its element name followed by *ide*.
Step 4	Write the name of the cation first and the name of the anion second.

♦ Learning Exercise 6.3D

CORE CHEMISTRY SKILL
Naming Ionic Compounds

Write the ions and a correct name for each of the following ionic compounds:

Formula	Ions		Name
1. $CrCl_2$	________	________	____________________________
2. $SnBr_4$	________	________	____________________________
3. Na_3P	________	________	____________________________
4. Ni_2O_3	________	________	____________________________
5. CuO	________	________	____________________________
6. Mg_3N_2	________	________	____________________________

Answers

1. Cr^{2+}, Cl^-, chromium(II) chloride
2. Sn^{4+}, Br^-, tin(IV) bromide
3. Na^+, P^{3-}, sodium phosphide
4. Ni^{3+}, O^{2-}, nickel(III) oxide
5. Cu^{2+}, O^{2-}, copper(II) oxide
6. Mg^{2+}, N^{3-}, magnesium nitride

Guide to Writing Formulas from the Name of An Ionic Compound	
Step 1	Identify the cation and anion.
Step 2	Balance the charges.
Step 3	Write the formula, cation first, using subscripts from charge balance.

♦ Learning Exercise 6.3E

CORE CHEMISTRY SKILL
Writing Ionic Formulas

Write the ions and the correct ionic formula for each of the following ionic compounds:

Compound	Positive Ion	Negative Ion	Formula of Compound
aluminum sulfide			
copper(II) chloride			
magnesium oxide			
gold(III) bromide			
silver oxide			

Answers

Compound	Positive Ion	Negative Ion	Formula of Compound
aluminum sulfide	Al^{3+}	S^{2-}	Al_2S_3
copper(II) chloride	Cu^{2+}	Cl^-	$CuCl_2$
magnesium oxide	Mg^{2+}	O^{2-}	MgO
gold(III) bromide	Au^{3+}	Br^-	$AuBr_3$
silver oxide	Ag^+	O^{2-}	Ag_2O

6.4 Polyatomic Ions

Learning Goal: Write the name and formula for an ionic compound containing a polyatomic ion.

- A polyatomic ion is a group of nonmetal atoms that carries an electrical charge, usually negative, 1−, 2−, or 3−. The polyatomic ion NH_4^+ has a positive charge.
- Polyatomic ions cannot exist alone, but are combined with an ion of the opposite charge.
- The names of ionic compounds containing polyatomic anions often end with *ate* or *ite*.
- When there is more than one polyatomic ion in the formula for a compound, the entire polyatomic ion formula is enclosed in parentheses and the subscript written outside the parentheses.

Key Terms for Sections 6.1–6.4

Match each of the following key terms with the correct definition:

a. ion **b.** cation **c.** anion
d. polyatomic ion **e.** ionic bond

1. _____ a positively charged ion

2. _____ a negatively charged ion

3. _____ a group of covalently bonded nonmetal atoms that has an overall charge

4. _____ an atom or group of atoms having an electrical charge

5. _____ the attraction between positively charged and negatively charged ions

Answers **1.** b **2.** c **3.** d **4.** a **5.** e

Study Note

By learning the most common polyatomic ions such as nitrate NO_3^-, carbonate CO_3^{2-}, sulfate SO_4^{2-}, and phosphate PO_4^{3-}, you can derive their related polyatomic ions. For example, the nitrite ion, NO_2^-, has one oxygen atom less than the nitrate ion: NO_3^-, nitrate, and NO_2^-, nitrite.

♦ Learning Exercise 6.4A

Write the polyatomic ion (symbol and charge) for each of the following:

1. sulfate ion _____ **2.** hydroxide ion _____

3. carbonate ion _____ **4.** sulfite ion _____

5. ammonium ion _____ **6.** phosphate ion _____

7. nitrate ion _____ **8.** nitrite ion _____

Answers **1.** SO_4^{2-} **2.** OH^- **3.** CO_3^{2-} **4.** SO_3^{2-}
5. NH_4^+ **6.** PO_4^{3-} **7.** NO_3^- **8.** NO_2^-

Guide to Writing Formulas with Polyatomic Ions	
Step 1	Identify the cation and polyatomic ion (anion).
Step 2	Balance the charges.
Step 3	Write the formula, cation first, using the subscripts from charge balance.

♦ Learning Exercise 6.4B

Write the formula of the ions, and the correct formula for each of the following compounds:

Compound	Positive Ion	Negative Ion	Formula
sodium phosphate			
iron(II) hydroxide			
ammonium carbonate			
silver bicarbonate			
chromium(III) sulfate			
lead(II) nitrate			
potassium sulfite			
barium phosphate			

Answers

Compound	Positive Ion	Negative Ion	Formula
sodium phosphate	Na^+	PO_4^{3-}	Na_3PO_4
iron(II) hydroxide	Fe^{2+}	OH^-	$Fe(OH)_2$
ammonium carbonate	NH_4^+	CO_3^{2-}	$(NH_4)_2CO_3$
silver bicarbonate	Ag^+	HCO_3^-	$AgHCO_3$
chromium(III) sulfate	Cr^{3+}	SO_4^{2-}	$Cr_2(SO_4)_3$
lead(II) nitrate	Pb^{2+}	NO_3^-	$Pb(NO_3)_2$
potassium sulfite	K^+	SO_3^{2-}	K_2SO_3
barium phosphate	Ba^{2+}	PO_4^{3-}	$Ba_3(PO_4)_2$

Guide to Naming Ionic Compounds with Polyatomic Ions	
Step 1	Identify the cation and polyatomic ion (anion).
Step 2	Name the cation, using a Roman numeral, if needed.
Step 3	Name the polyatomic ion. Write the name for the compound, cation first and the polyatomic ion second.

♦ Learning Exercise 6.4C

Write the ions and a correct name for each of the following ionic compounds:

Formula	Ions		Name
1. $Ba(NO_3)_2$	______	______	____________
2. $Fe_2(SO_4)_3$	______	______	____________
3. Na_3PO_3	______	______	____________
4. $Al(ClO_3)_3$	______	______	____________
5. $(NH_4)_2CO_3$	______	______	____________
6. $Cr(OH)_2$	______	______	____________

Answers
1. Ba^{2+}, NO_3^-, barium nitrate
2. Fe^{3+}, SO_4^{2-}, iron(III) sulfate
3. Na^+, PO_3^{3-}, sodium phosphite
4. Al^{3+}, ClO_3^-, aluminum chlorate
5. NH_4^+, CO_3^{2-}, ammonium carbonate
6. Cr^{2+}, OH^-, chromium(II) hydroxide

6.5 Molecular Compounds: Sharing Electrons

Learning Goal: Given the formula of a molecular compound, write its correct name; given the name of a molecular compound, write its formula.

- In a covalent bond, atoms of nonmetals share electrons to achieve an octet. For example, oxygen with six valence electrons shares electrons with two hydrogen atoms to form the covalent compound water (H_2O). In water, oxygen has two bonding pairs of shared electrons and two lone pairs of nonbonding electrons.

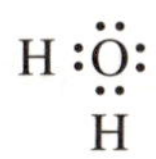

- Covalent compounds are composed of nonmetals bonded together to give discrete units called molecules.
- The formula of a covalent compound is written using the order of the symbols of the elements in the name, followed by subscripts indicated by prefixes.

Guide to Naming Molecular Compounds	
Step 1	Name the first nonmetal by its element name.
Step 2	Name the second nonmetal by using the first syllable of its element name followed by *ide*.
Step 3	Add prefixes to indicate the number of atoms (subscripts).

CORE CHEMISTRY SKILL
Writing the Names and Formulas for Molecular Compounds

♦ Learning Exercise 6.5A

Use the appropriate prefixes to name the following covalent compounds:

1. CS_2 ____________
2. CCl_4 ____________
3. CO ____________
4. SO_3 ____________
5. N_2O_4 ____________
6. PCl_3 ____________
7. P_4S_6 ____________
8. IF_7 ____________
9. ClO_2 ____________
10. S_2O ____________

Answers

1. carbon disulfide
2. carbon tetrachloride
3. carbon monoxide
4. sulfur trioxide
5. dinitrogen tetroxide
6. phosphorus trichloride
7. tetraphosphorus hexasulfide
8. iodine heptafluoride
9. chlorine dioxide
10. disulfur oxide

Guide to Writing Formulas for Molecular Compounds	
Step 1	Write the symbols in the order of the elements in the name.
Step 2	Write any prefixes as subscripts.

♦ Learning Exercise 6.5B

Write the formula of each of the following covalent compounds:

1. dinitrogen oxide ____________
2. silicon tetrabromide ____________
3. nitrogen trichloride ____________
4. carbon dioxide ____________
5. sulfur hexafluoride ____________
6. oxygen difluoride ____________
7. phosphorus trifluoride ____________
8. phosphorus trihydride ____________
9. iodine trifluoride ____________
10. sulfur dioxide ____________

Answers

1. N_2O
2. $SiBr_4$
3. NCl_3
4. CO_2
5. SF_6
6. OF_2
7. PF_3
8. PH_3
9. IF_3
10. SO_2

6.6 Organic Compounds: Extended Topic

Learning Goal: Identify the properties characteristic of organic and inorganic compounds.

- Organic compounds are compounds of carbon that typically have covalent bonds, low melting and boiling points, burn vigorously, are nonelectrolytes, and are soluble in nonpolar solvents.
- Alkanes are hydrocarbons that have only single bonds, C—C and C—H.
- Each carbon in an alkane has four bonds arranged so that the bonded atoms are in the corners of a tetrahedron.

♦ Learning Exercise 6.6A

Identify the following as typical of organic (O) or inorganic (I) compounds:

1. _____ have covalent bonds
2. _____ have low boiling points
3. _____ burn in air
4. _____ are soluble in water
5. _____ have high melting points
6. _____ are soluble in nonpolar solvents
7. _____ have ionic bonds
8. _____ form long chains
9. _____ contain carbon and hydrogen
10. _____ do not burn in air

Answers

1. O
2. O
3. O
4. I
5. I
6. O
7. I
8. O
9. O
10. I

6.7 Names and Formulas of Alkanes: Extended Topic

Learning Goal: Write the IUPAC name and draw the condensed structural formula for an alkane.

- An expanded structural formula shows the bonds between all the atoms.
- A condensed structural formula depicts each carbon atom and its attached hydrogen atoms as a group.

Expanded Structural Formula	Condensed Structural Formula	Skeletal Formula	Molecular Formula
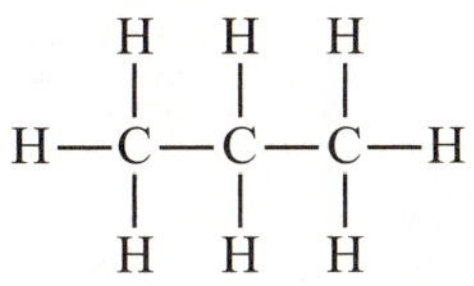	$CH_3—CH_2—CH_3$		C_3H_8

- In the IUPAC system used to name organic compounds, the alkanes containing one, two, three, or four carbon atoms connected in a row or a *continuous* chain are named: methane, ethane, propane, and butane, respectively, as shown:

Name	Number of Carbon Atoms	Condensed Structural Formula
Methane	1	CH_4
Ethane	2	$CH_3—CH_3$
Propane	3	$CH_3—CH_2—CH_3$
Butane	4	$CH_3—CH_2—CH_2—CH_3$

- Alkanes with five or more carbon atoms in a chain are named using the prefix *pent* (5), *hex* (6), *hept* (7), *oct* (8), *non* (9), or *dec* (10) and the suffix *ane*, which indicates single bonds (alkane).

Key Terms for Sections 6.5–6.7

Match each the following key terms with the correct definition:

a. covalent bond **b.** alkane **c.** octet **d.** hydrocarbon

1. _____ an arrangement of eight valence electrons

2. _____ a hydrocarbon that contains only carbon–carbon single bonds

3. _____ sharing of valence electrons by two atoms

4. _____ an organic compound that consist of only carbon and hydrogen atoms

Answers **1.** c **2.** b **3.** a **4.** d

♦ Learning Exercise 6.7A

CORE CHEMISTRY SKILL
Naming and Drawing Alkanes

Write the IUPAC name for each of the following:

1. $CH_3—CH_2—CH_3$ ______________
2. $CH_3—CH_3$ ______________
3. $CH_3—CH_2—CH_2—CH_3$ ______________
4. $CH_3—CH_2—CH_2—CH_2—CH_2—CH_3$ ______________

Answers **1.** propane **2.** ethane **3.** butane **4.** hexane

♦ Learning Exercise 6.7B

Draw the condensed structural formula for each of the following compounds:

1. pentane ______________________________
2. methane ______________________________
3. octane ______________________________
4. heptane ______________________________

Answers
1. $CH_3—CH_2—CH_2—CH_2—CH_3$
2. CH_4
3. $CH_3—CH_2—CH_2—CH_2—CH_2—CH_2—CH_2—CH_3$
4. $CH_3—CH_2—CH_2—CH_2—CH_2—CH_2—CH_3$

Checklist for Chapter 6

You are ready to take the Practice Test for Chapter 6. Be sure you have accomplished the following learning goals for this chapter. If not, review the section listed at the end of the goal. Then apply your new skills and understanding to the Practice Test.

After studying Chapter 6, I can successfully:

____ Illustrate the octet rule for the formation of ions (6.1).

____ Write the formulas of ionic compounds containing the ions of metals and nonmetals of representative elements (6.2).

____ Use charge balance to write an ionic formula (6.2).

____ Write the name of an ionic compound (6.3).

____ Write the formula of a compound containing a polyatomic ion (6.4).

____ Write the names and formulas of molecular compounds (6.5).

____ Identify properties as characteristic of organic or inorganic compounds (6.6).

____ Use the IUPAC system to name alkanes with one to ten carbon atoms (6.7).

____ Draw the condensed structural formulas of alkanes with one to ten carbon atoms (6.7).

Practice Test for Chapter 6

For questions 1 through 4, consider an atom of phosphorus: (6.1)

1. Phosphorus is in group
   A. 2A (2) B. 3A (13) C. 5A (15) D. 7A (17) E. 8A (18)
2. How many valence electrons does an atom of phosphorus have?
   A. 2 B. 3 C. 5 D. 8 E. 15
3. To achieve an octet, a phosphorus atom in an ionic compound will
   A. lose 1 electron B. lose 2 electrons C. lose 5 electrons
   D. gain 2 electrons E. gain 3 electrons
4. A phosphide ion has charge of
   A. 1+ B. 2+ C. 5+ D. 2− E. 3−
5. To achieve an octet, a calcium atom (6.1)
   A. loses 1 electron B. loses 2 electrons C. loses 3 electrons
   D. gains 1 electron E. gains 2 electrons
6. To achieve an octet, a chlorine atom (6.1)
   A. loses 1 electron B. loses 2 electrons C. loses 3 electrons
   D. gains 1 electron E. gains 2 electrons
7. Another name for a positive ion is (6.1)
   A. anion B. cation C. proton D. positron E. sodium
8. The correct ionic charge for a calcium ion is (6.1)
   A. 1+ B. 2+ C. 1− D. 2− E. 3−
9. The silver ion has a charge of (6.1)
   A. 1+ B. 2+ C. 1− D. 2− E. 3−
10. The correct ionic charge for a phosphate ion is (6.1)
    A. 1+ B. 2+ C. 1− D. 2− E. 3−
11. The correct ionic charge for fluoride is (6.1)
    A. 1+ B. 2+ C. 1− D. 2− E. 3−
12. The correct ionic charge for a sulfate ion is (6.1)
    A. 1+ B. 2+ C. 1− D. 2− E. 3−
13. When the elements magnesium and sulfur react, (6.2)
    A. an ionic compound forms
    B. a covalent compound forms
    C. no reaction occurs
    D. the two repel each other and will not combine
    E. none of the above
14. An ionic bond typically occurs between (6.2)
    A. two different nonmetals
    B. two of the same type of nonmetals
    C. two noble gases
    D. two different metals
    E. a metal and a nonmetal
15. The formula for a compound between sodium and sulfur is (6.2)
    A. SoS B. NaS C. Na_2S D. NaS_2 E. Na_2SO_4

16. The formula for a compound between aluminum and oxygen is (6.2)
A. AlO B. Al_2O C. AlO_3 D. Al_2O_3 E. Al_3O_2

17. The formula for a compound between barium and sulfur is (6.2)
A. BaS B. Ba_2S C. BaS_2 D. Ba_2S_2 E. $BaSO_4$

18. The correct formula for iron(III) chloride is (6.2, 6.3)
A. FeCl B. $FeCl_2$ C. Fe_2Cl D. Fe_3Cl E. $FeCl_3$

19. The correct formula for ammonium sulfate is (6.2, 6.4)
A. AmS B. $AmSO_4$ C. $(NH_4)_2S$ D. NH_4SO_4 E. $(NH_4)_2SO_4$

20. The correct formula for copper(II) chloride is (6.2, 6.3)
A. CoCl B. CuCl C. $CoCl_2$ D. $CuCl_2$ E. Cu_2Cl

21. The correct formula for lithium phosphate is (6.2, 6.4)
A. $LiPO_4$ B. Li_2PO_4 C. Li_3PO_4 D. $Li_2(PO_4)_3$ E. $Li_3(PO_4)_2$

22. The correct formula for silver oxide is (6.2, 6.3)
A. AgO B. Ag_2O C. AgO_2 D. Ag_3O_2 E. Ag_3O

23. The correct formula for magnesium carbonate is (6.2, 6.4)
A. $MgCO_3$ B. Mg_2CO_3 C. $Mg(CO_3)_2$ D. MgCO E. $Mg_2(CO_3)_3$

24. The correct formula for copper(I) sulfate is (6.3, 6.4)
A. $CuSO_3$ B. $CuSO_4$ C. Cu_2SO_3 D. $Cu(SO_4)_2$ E. Cu_2SO_4

25. The name of $Al_2(HPO_4)_3$ is (6.3, 6.4)
A. aluminum hydrogen phosphite
B. aluminum hydrogen phosphate
C. aluminum hydrogen phosphorus
D. aluminum hydrogen phosphorus oxide
E. trialuminum diphosphate

26. The name of CoS is (6.2, 6.3)
A. copper sulfide B. cobalt(II) sulfate C. cobalt(I) sulfide
D. cobalt sulfide E. cobalt(II) sulfide

27. The name of $MnCl_2$ is (6.2, 6.3)
A. magnesium chloride B. manganese(II) chlorine C. manganese(II) chloride
D. manganese chlorine E. manganese(III) chloride

28. The name of $ZnCO_3$ is (6.3, 6.4)
A. zinc(III) carbonate B. zinc(II) carbonate C. zinc bicarbonate
D. zinc carbon trioxide E. zinc carbonate

29. The name of Al_2O_3 is (6.2)
A. aluminum oxide B. aluminum(II) oxide C. aluminum trioxide
D. dialuminum trioxide E. aluminum oxygenate

30. The name of $Cr_2(SO_3)_3$ is (6.3, 6.4)
A. chromium sulfite B. dichromium trisulfite C. chromium(III) sulfite
D. chromium(III) sulfate E. chromium sulfate

31. The name of PH_3 is (6.5)
A. potassium trihydrogen B. phosphorus trihydride C. phosphorus trihydrogen
D. potassium trihydride E. phosphorus trihydrogenide

32. The name of NCl_3 is (6.5)
A. nitrogen chloride B. nitrogen trichloride C. trinitrogen chloride
D. nitrogen chlorine three E. nitrogen trichlorine

33. The name of CO is (6.5)
A. carbon monoxide **B.** carbonic oxide **C.** carbon oxide
D. carbonious oxide **E.** carboxide

34. The formula for a compound between carbon and chlorine is (6.5)
A. CCl **B.** CCl_2 **C.** C_4Cl **D.** CCl_4 **E.** C_4Cl_2

35. 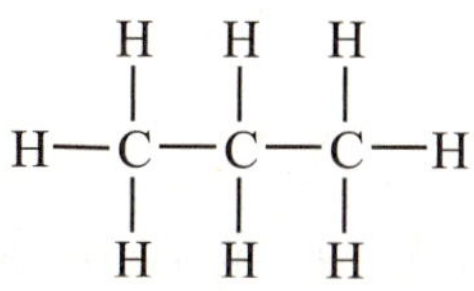

This is an example of a(n) (6.7)
A. ball-and-stick model **B.** condensed structural formula **C.** space-filling model
D. tetrahedral model **E.** expanded structural formula

For questions 36 through 41, indicate whether the following characteristics are typical of organic (O) compounds or inorganic (I) compounds: (6.6)

36. _____ higher melting points

37. _____ covalent bonds

38. _____ soluble in water

39. _____ ionic bonds

40. _____ burn easily in air

41. _____ soluble in nonpolar solvents

For questions 42 and 43, match the name of the hydrocarbon with each of the following: (6.7)
A. ethane **B.** butane **C.** hexane **D.** heptane **E.** octane

42. $CH_3—CH_3$

43. $CH_3—CH_2—CH_2—CH_2—CH_2—CH_2—CH_2—CH_3$

44. $CH_3—CH_2—CH_2—CH_2—CH_2—CH_3$

45. $CH_3—CH_2—CH_2—CH_3$

46. $CH_3—CH_2—CH_2—CH_2—CH_2—CH_2—CH_3$

Answers to the Practice Test

1. C **2.** C **3.** E **4.** E **5.** B
6. D **7.** B **8.** B **9.** A **10.** E
11. C **12.** D **13.** A **14.** E **15.** C
16. D **17.** A **18.** E **19.** E **20.** D
21. C **22.** B **23.** A **24.** E **25.** B
26. E **27.** C **28.** E **29.** A **30.** C
31. B **32.** B **33.** A **34.** D **35.** E
36. I **37.** O **38.** I **39.** I **40.** O
41. O **42.** A **43.** E **44.** C **45.** B
46. D

7
Chemical Quantities

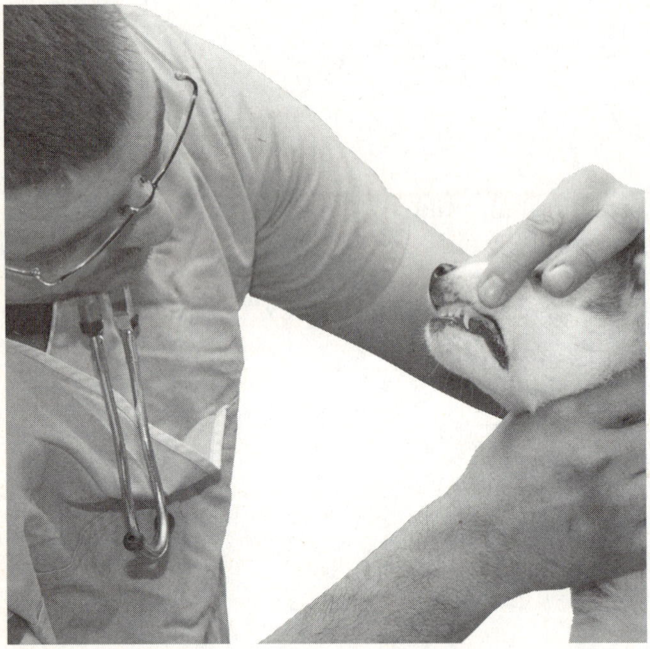

Max, a six-year-old dog, was taken to his veterinarian two months ago for an examination. After a sample for a blood chemistry profile determined that Max had an elevated white blood count, his veterinarian treated him with 375-mg tablets of Clavamox, a broad-spectrum antibiotic that combines amoxicillin and clavulanic acid. A few days later, the lab tests were repeated and indicated a normal white blood count, revealing that the infection had cleared. Max's electrolytes, which are also indicators of good health, were all in the normal ranges. The electrolyte chloride, Cl^-, was in the normal range of 0.106–0.118 mol/L, and the electrolyte sodium, Na^+, was in the normal range of 0.144–0.160 mol/L. If amoxicillin has a formula of $C_{16}H_{19}N_3O_5S$ and a molar mass of 365 g/mol, how many moles of amoxicillin are in 250 mg of amoxicillin?

7.1 The Mole

Learning Goal: Use Avogadro's number to determine the number of particles in a given number of moles.

- A mole of any compound contains Avogadro's number, 6.022×10^{23}, of particles.
  1 mol = 6.022×10^{23} particles (atoms, molecules, ions, formula units)
- Avogadro's number provides conversion factors between moles and the number of particles:

$$\frac{1 \text{ mol}}{6.022 \times 10^{23} \text{ particles}} \quad \text{and} \quad \frac{6.022 \times 10^{23} \text{ particles}}{1 \text{ mol}}$$

Guide to Calculating the Atoms or Molecules of a Substance	
Step 1	State the given and needed quantities.
Step 2	Write a plan to convert moles to atoms or molecules.
Step 3	Use Avogadro's number to write conversion factors.
Step 4	Set up the problem to calculate the number of particles.

♦ Learning Exercise 7.1A

CORE CHEMISTRY SKILL
Converting Particles to Moles

Use Avogadro's number to calculate each of the following:

a. number of Ca atoms in 3.00 mol of Ca

b. number of Zn atoms in 0.250 mol of Zn

c. number of SO_2 molecules in 0.118 mol of SO_2

d. number of moles of Ag in 4.88×10^{23} atoms of Ag

e. number of moles of NH_3 (ammonia) in 7.52×10^{23} molecules of NH_3

Answers
a. 1.81×10^{24} atoms of Ca
b. 1.51×10^{23} atoms of Zn
c. 7.11×10^{22} molecules of SO_2
d. 0.810 mol of Ag
e. 1.25 mol of NH_3

Study Note

The subscripts in the formula of a compound indicate the number of moles of each element in one mol of that compound. For example, consider the formula Mg_3N_2:

$$1 \text{ mol of } Mg_3N_2 = 3 \text{ mol of Mg atoms and } 2 \text{ mol of N atoms}$$

For this formula, the conversion factors for the moles of elements can be written as

$$\frac{3 \text{ mol Mg atoms}}{1 \text{ mol } Mg_3N_2} \text{ and } \frac{1 \text{ mol } Mg_3N_2}{3 \text{ mol Mg atoms}} \qquad \frac{2 \text{ mol N atoms}}{1 \text{ mol } Mg_3N_2} \text{ and } \frac{1 \text{ mol } Mg_3N_2}{2 \text{ mol N atoms}}$$

Guide to Calculating Moles	
Step 1	State the given and needed quantities.
Step 2	Write a plan to convert moles of a compound to moles of an element.
Step 3	Write equalities and conversion factors using the subscripts.
Step 4	Set up the problem to calculate the moles of an element.

♦ Learning Exercise 7.1B

Vitamin C (ascorbic acid) has the formula $C_6H_8O_6$.

a. How many moles of carbon are in 2.0 mol of vitamin C?

b. How many moles of hydrogen are in 5.0 mol of vitamin C?

c. How many moles of oxygen are in 1.5 mol of vitamin C?

Answers

a. 12 mol of carbon

b. 40. mol of hydrogen

c. $6 \times 1.5 = 9.0$ mol of oxygen

♦ Learning Exercise 7.1C

For the compound ibuprofen ($C_{13}H_{18}O_2$) used in Advil and Motrin, determine each of the following:

a. moles of carbon atoms in 2.20 mol of ibuprofen

b. moles of hydrogen in 4.40×10^{24} molecules of ibuprofen

c. atoms of oxygen in 0.750 mol of ibuprofen

d. molecules of ibuprofen that contain 15 mol of hydrogen

e. molecules of ibuprofen that contain 2.84×10^{23} atoms of hydrogen

Answers

a. 28.6 mol of C

b. 132 mol of H

c. 9.03×10^{23} atoms of O

d. 5.0×10^{23} molecules of ibuprofen

e. 1.58×10^{22} molecules of ibuprofen

7.2 Molar Mass

Learning Goal: Given the chemical formula of a substance, calculate its molar mass.

- The molar mass (g/mol) of an element is numerically equal to its atomic mass in grams.
- The molar mass (g/mol) of a compound is the sum of the mass (grams) for each element as indicated by the subscripts in the formula.

Guide to Calculating Molar Mass	
Step 1	Obtain the molar mass of each element.
Step 2	Multiply each molar mass by the number of moles (subscript) in the formula.
Step 3	Calculate the molar mass by adding the masses of the elements.

Example: What is the molar mass of silver nitrate, $AgNO_3$?

Step 1: Obtain the molar mass of each element.

$$\frac{107.9 \text{ g Ag}}{1 \text{ mol Ag}} \qquad \frac{14.01 \text{ g N}}{1 \text{ mol N}} \qquad \frac{16.00 \text{ g O}}{1 \text{ mol O}}$$

Step 2: Multiply each molar mass by the number of moles (subscript) in the formula.

$$1 \cancel{\text{mol Ag}} \times \frac{107.9 \text{ g Ag}}{1 \cancel{\text{mol Ag}}} = 107.9 \text{ g of Ag}$$

$$1 \cancel{\text{mol N}} \times \frac{14.01 \text{ g N}}{1 \cancel{\text{mol N}}} = 14.01 \text{ g of N}$$

$$3 \cancel{\text{mol O}} \times \frac{16.00 \text{ g O}}{1 \cancel{\text{mol O}}} = 48.00 \text{ g of O}$$

Step 3: Calculate the molar mass by adding the masses of the elements.

$$\text{Molar mass of } AgNO_3 = (107.9 \text{ g} + 14.01 \text{ g} + 48.00 \text{ g}) = 169.9 \text{ g}$$

♦ Learning Exercise 7.2

CORE CHEMISTRY SKILL

Calculating Molar Mass

Determine the molar mass for each of the following:

a. K_2O

b. $AlCl_3$

c. $C_{13}H_{18}O_2$

d. C_4H_{10}

e. $Ca(NO_3)_2$

f. Mg_3N_2

g. $FeCO_3$

h. $(NH_4)_3PO_4$

Answers **a.** 94.20 g **b.** 133.33 g **c.** 206.3 g **d.** 58.12 g **e.** 164.10 g **f.** 100.95 g **g.** 115.86 g **h.** 149.10 g

7.3 Calculations Using Molar Mass

Learning Goal: Given the number of moles of a substance, calculate the mass in grams; given the mass, calculate the number of moles.

- The molar mass is useful as a conversion factor to change a given quantity in moles to grams or grams to moles.
- The two conversion factors for the molar mass of NaOH (40.01 g/mol) are:

$$\frac{40.01 \text{ g NaOH}}{1 \text{ mol NaOH}} \quad \text{and} \quad \frac{1 \text{ mol NaOH}}{40.01 \text{ g NaOH}}$$

Guide to Calculating the Moles (or Grams) of a Substance from Grams (or Moles)	
Step 1	State the given and needed quantities.
Step 2	Write a plan to convert moles to grams (or grams to moles).
Step 3	Determine the molar mass and write conversion factors.
Step 4	Set up problem to convert moles to grams (or grams to moles).

Example: What is the mass in grams of 0.254 mol of Na_2CO_3?

Step 1: State the given and needed quantities.

Given 0.254 mol of Na_2CO_3 **Need** mass of grams of Na_2CO_3

Step 2: Write a plan to convert moles to grams. moles $\xrightarrow{\text{Molar mass}}$ grams

Step 3: Determine the molar mass, and write conversion factors.

1 mol of Na_2CO_3 = 106.0 g of Na_2CO_3

$$\frac{106.0 \text{ g } Na_2CO_3}{1 \text{ mol } Na_2CO_3} \quad \text{and} \quad \frac{1 \text{ mol } Na_2CO_3}{106.0 \text{ g } Na_2CO_3}$$

Step 4: Set up the problem to convert moles to grams.

$$0.254 \cancel{\text{mol } Na_2CO_3} \times \frac{106.0 \text{ g } Na_2CO_3}{1 \cancel{\text{mol } Na_2CO_3}} = 26.9 \text{ g of } Na_2CO_3$$

Molar mass (cancels mol Na_2CO_3)

♦ Learning Exercise 7.3A

CORE CHEMISTRY SKILL

Using Molar Mass as a Conversion Factor

Calculate the number of grams in each of the following:

a. 0.75 mol of S

b. 3.18 mol of K_2SO_4

c. 2.50 mol of NH_4Cl

d. 4.08 mol of $FeCl_3$

e. 2.28 mol of PCl_3

f. 0.815 mol of $Mg(NO_3)_2$

Answers **a.** 24 g **b.** 554 g **c.** 134 g **d.** 662 g **e.** 313 g **f.** 121 g

Study Note

When the grams of a substance are given, the molar mass is used to calculate the number of moles of substance present.

$$\cancel{\text{grams substance}} \times \underset{\textit{Molar mass}}{\frac{1 \text{ mol substance}}{\cancel{\text{grams substance}}}} = \text{mol of substance}$$

Example: How many moles of NaOH are in 4.00 g of NaOH?

$$4.00 \text{ g } \cancel{\text{NaOH}} \times \underset{\textit{Molar Mass (inverted, cancels g NaOH)}}{\frac{1 \text{ mol NaOH}}{40.01 \text{ g } \cancel{\text{NaOH}}}} = 0.100 \text{ mol of NaOH}$$

♦ Learning Exercise 7.3B

Calculate the number of moles in each of the following quantities:

a. 108 g of CH_4

b. 6.12 g of CO_2

c. 38.7 g of $CaBr_2$

d. 236 g of Cl_2

e. 128 g of $Mg(OH)_2$

f. 172 g of Al_2O_3

g. The methane used to heat a home has a formula of CH_4. If 725 g of methane are used in one month, how many moles of methane were used during this time?

h. A vitamin tablet contains 18 mg of iron. If there are 100 tablets in a bottle, how many moles of iron are contained in the vitamins in the bottle?

Answers

a. 6.73 mol **b.** 0.139 mol **c.** 0.194 mol **d.** 3.33 mol
e. 2.19 mol **f.** 1.69 mol **g.** 45.2 mol **h.** 0.032 mol

Guide to Calculating the Particles of a Substance from Grams	
Step 1	State the given and needed quantities.
Step 2	Write a plan to convert grams to particles.
Step 3	Write the equalities and conversion factors for molar mass and Avogadro's number.
Step 4	Set up the problem to convert grams to particles.

Example: How many molecules of CH_4 are in 5.25 g of CH_4?

Step 1: State the given and needed quantities.

Analyze the Problem	**Given** 5.25 g of CH_4	**Need** molecules of CH_4

Step 2: Write a plan to convert grams to particles.

$$\text{grams of } CH_4 \xrightarrow{\text{Molar mass}} \text{moles of } CH_4 \xrightarrow{\text{Avogadro's number}} \text{molecules of } CH_4$$

Step 3: Write the equalities and conversion factors for molar mass and Avogadro's number.

$$\text{Molar Mass of } CH_4 = \left(1\ \cancel{\text{mol C}} \times \frac{12.01\text{ g C}}{1\ \cancel{\text{mol C}}}\right) + \left(4\ \cancel{\text{mol H}} \times \frac{1.008\text{ g H}}{1\ \cancel{\text{mol H}}}\right) = 16.04\text{ g } CH_4$$

$$1\text{ mol of } CH_4 = 16.04\text{ g/mol}$$

$$\frac{16.04\text{ g } CH_4}{1\text{ mol } CH_4} \quad \text{and} \quad \frac{1\text{ mol } CH_4}{16.04\text{ g } CH_4}$$

$$1\text{ mol of } CH_4 = 6.022 \times 10^{23}\text{ molecules of } CH_4$$

$$\frac{6.022 \times 10^{23}\text{ molecules } CH_4}{1\text{ mol } CH_4} \quad \text{and} \quad \frac{1\text{ mol } CH_4}{6.022 \times 10^{23}\text{ molecules } CH_4}$$

Step 4: Set up the problem to convert grams to particles.

$$5.25\ \cancel{\text{g } CH_4} \times \underset{(\textit{cancels g } CH_4)}{\frac{1\ \cancel{\text{mol } CH_4}}{16.04\ \cancel{\text{g } CH_4}}} \times \underset{(\textit{cancels mol } CH_4)}{\frac{6.022 \times 10^{23}\text{ molecules } CH_4}{1\ \cancel{\text{mol } CH_4}}}$$

$$= 1.97 \times 10^{23}\text{ molecules of } CH_4$$

♦ Learning Exercise 7.3C

Calculate the number of particles in each of the following quantities by using the subscripts in the formula to write conversion factors.

a. molecules of C_3H_8 in 20.8 g of C_3H_8

b. molecules of CO_2 in 6.12 g of CO_2

c. bromine atoms in 38.7 g of PBr_3

d. chlorine atoms in 236 g of Cl_2

e. hydroxide ions in 128 g of $Mg(OH)_2$

f. oxide ions in 172 g of Al_2O_3

Answers

a. 2.84×10^{23} molecules of C_3H_8
b. 8.37×10^{22} molecules of CO_2
c. 2.58×10^{23} atoms of bromine
d. 4.01×10^{24} atoms of chlorine
e. 2.64×10^{24} hydroxide ions
f. 3.05×10^{24} oxide ions

7.4 Mass Percent Composition and Empirical Formulas

Learning Goal: Given the formula of a compound, calculate the mass percent composition; from the mass percent composition, determine the empirical formula of a compound.

- Percent composition is the percent by mass of each element in a compound.
- An empirical (simplest) formula gives the lowest whole-number ratio of atoms in a compound.

Guide to Calculating Mass Percent Composition	
Step 1	Determine the total mass of each element in the molar mass of a formula.
Step 2	Divide the total mass of each element by the molar mass and multiply by 100%.

Example: What is the percent Na in NaOH?

Step 1: Determine the total mass of each element in the molar mass of a formula.

$$1\ \cancel{\text{mol Na}} \times \frac{22.99\text{ g Na}}{1\ \cancel{\text{mol Na}}} = 22.99\text{ g of Na}$$

$$1\ \cancel{\text{mol O}} \times \frac{16.00\text{ g O}}{1\ \cancel{\text{mol O}}} = 16.00\text{ g of O}$$

$$1\ \cancel{\text{mol H}} \times \frac{1.008\text{ g H}}{1\ \cancel{\text{mol H}}} = 1.008\text{ g of H}$$

$$\text{Molar mass of NaOH} = 40.01\text{ g of NaOH}$$

Step 2: Divide the total mass of each element by the molar mass and multiply by 100%.

$$\frac{22.99\text{ g Na}}{40.01\text{ g NaOH}} \times 100\% = 57.46\%\text{ Na}$$

♦ Learning Exercise 7.4A

CORE CHEMISTRY SKILL

Calculating Mass Percent Composition

Calculate the mass percent composition of each element for each of the following compounds:

a. C_3H_6O

b. $BaSO_4$

c. $Ca(NO_3)_2$

d. NaOCl

e. $(NH_4)_3PO_4$

Answers

a. C 62.04%, H 10.41%, O 27.55%
b. Ba 58.83%, S 13.74%, O 27.42%
c. Ca 24.42%, N 17.07%, O 58.50%
d. Na 30.88%, O 21.49%, Cl 47.62%
e. N 28.19%, H 8.11%, P 20.77%, O 42.92%

Guide to Calculating Empirical Formula	
Step 1	Calculate the moles of each element.
Step 2	Divide by the smallest number of moles.
Step 3	Use the lowest whole-number ratio of moles as subscripts.

Example: What is the empirical formula of a compound containing 27.3% C and 72.7% O by mass? If we assume we have *exactly* 100 g of the compound, there would be 27.3 g of C and 72.7 g of O.

Step 1: Calculate the moles of each element.

$$27.3 \text{ g C} \times \frac{1 \text{ mol C}}{12.01 \text{ g C}} = 2.27 \text{ mol of C} \qquad 72.7 \text{ g O} \times \frac{1 \text{ mol O}}{16.00 \text{ g O}} = 4.54 \text{ mol of O}$$

Step 2: Divide by the smallest number of moles.

$$\frac{2.27 \text{ mol C}}{2.27} = 1.00 \text{ mol of C} \qquad \frac{4.54 \text{ mol O}}{2.27} = 2.00 \text{ mol of O}$$

Step 3: Use the lowest whole-number ratio of moles as subscripts.

$$\text{Empirical formula} = C_{1.00}O_{2.00} \longrightarrow CO_2$$

Study Note

When the number of moles is a decimal that is within 0.1 of a whole number, round off to the whole number. For example, 1.98 mol is rounded off to 2 mol.

However, if a decimal is obtained that is *greater* than 0.1 or *less* than 0.9, it is not rounded off. Instead, multiply by a small integer to obtain a whole number. Examples of using small integers are given in Table 7.1. Then all other subscripts must be increased by multiplying by the same small integer.

Table 7.1 Examples of Small Integers That Convert Decimal Values to Whole Numbers

Decimal	Small Integer	Example		Whole Number
0.20	5	1.20 × 5	=	6
0.25	4	2.25 × 4	=	9
0.33	3	1.33 × 3	=	4
0.50	2	2.50 × 2	=	5

♦ Learning Exercise 7.4B

CORE CHEMISTRY SKILL
Calculating an Empirical Formula

Determine the empirical formula of each of the following compounds:

a. a compound that is 75.7% by mass tin (Sn) and 24.3% by mass fluorine (F)

b. a compound that is 20.2% by mass magnesium (Mg), 26.7% sulfur (S), and 53.2% oxygen (O)

c. a 4.58-g sample of a compound that contains 2.00 g of phosphorus (P) and the rest is oxygen (O)

d. a compound that contains 1.00 g of C, 0.168 g of H, and 0.890 g of O

Answers **a.** SnF_2 **b.** $MgSO_4$ **c.** P_2O_5 **d.** $C_3H_6O_2$

7.5 Molecular Formulas

Learning Goal: Determine the molecular formula of a substance from the empirical formula and molar mass.

- A molecular formula gives the actual number of atoms of each element in the compound.
- A molecular formula is related to the empirical formula by a small integer such as 1, 2, or 3.
- The molecular formula is obtained using the empirical formula mass and molar mass of the compound.

Key Terms for Sections 7.1–7.5

Match the following terms with the statements below:

a. mole **b.** Avogadro's number **c.** empirical formula
d. molar mass **e.** molecular formula

1. _____ the formula that gives the actual number of atoms of each element in a compound
2. _____ the number of items in 1 mol is equal to 6.022×10^{23}
3. _____ the formula that gives the smallest whole-number ratio of atoms in a formula
4. _____ the amount of a substance that contains 6.022×10^{23} particles of that substance
5. _____ the mass in grams of an element or compound that is equal numerically to its atomic mass or sum of atomic masses as indicated by the formula

Answers **1.** e **2.** b **3.** c **4.** a **5.** d

Guide to Calculating a Molecular Formula from an Empirical Formula	
Step 1	Obtain the empirical formula and calculate the empirical formula mass.
Step 2	Divide the molar mass by the empirical formula mass to obtain a small integer.
Step 3	Multiply the empirical formula by the small integer to obtain the molecular formula.

Example: Lactic acid has the empirical formula CH_2O. If the molar mass of lactic acid is about 90 g, what is the molecular formula of lactic acid?

Step 1: Obtain the empirical formula and calculate the empirical formula mass.

$$\text{Mass of } CH_2O = 12.01 \text{ g of C} + 2(1.008) \text{ g of H} + 16.00 \text{ g of O} = 30.03 \text{ g}$$

Step 2: Divide the molar mass by the empirical formula mass to obtain a small integer.

$$\frac{\text{Molar mass}}{\text{Empirical formula mass}} = \frac{90 \text{ g}}{30.03 \text{ g}} = 3 \text{ (small integer)}$$

Step 3: Multiply the empirical formula by the small integer to obtain the molecular formula.

$$\text{Molecular formula} = 3 \times (CH_2O) = C_3H_6O_3 \text{ Formula of lactic acid}$$

♦ Learning Exercise 7.5A

CORE CHEMISTRY SKILL
Calculating a Molecular Formula

Determine the molecular formula of each compound from its empirical formula and approximate molar mass:

a. NS, 184 g/mol

b. KCO_2, 166 g/mol

c. HIO_3, 176 g/mol

d. $NaPO_3$, 306 g/mol

e. $C_7H_8O_3$, 420 g/mol

f. C_3H_5O, 228 g/mol

Answers **a.** N_4S_4 **b.** $K_2C_2O_4$ **c.** HIO_3 **d.** $Na_3P_3O_9$ **e.** $C_{21}H_{24}O_9$ **f.** $C_{12}H_{20}O_4$

♦ Learning Exercise 7.5B

Determine the molecular formula for each of the following:

a. pyrogallol with a composition of C 57.14%, H 4.80%, and O 38.06% and a molar mass of about 126 g

b. triazaborane with a composition of B 40.31%, H 7.51%, and N 52.18% and a molar mass of about 80 g

c. a substance used to tan leather with a composition of K 26.58%, Cr 35.36%, and O 38.07% and a molar mass of about 294 g

d. pyrocatechol, a topical anesthetic with a composition of C 65.44%, H 5.49%, and O 29.06% and a molar mass of about 110 g

Answers **a.** $C_6H_6O_3$ **b.** $B_3H_6N_3$ **c.** $K_2Cr_2O_7$ **d.** $C_6H_6O_2$

Checklist for Chapter 7

You are ready to take the Practice Test for Chapter 7. Be sure you have accomplished the following learning goals for this chapter. If not, review the section listed at the end of the goal. Then apply your new skills and understanding to the Practice Test.

After studying Chapter 7, I can successfully:

_____ Determine the number of particles in a mole of a substance (7.1).

_____ Calculate the molar mass given the formula of a substance (7.2).

_____ Convert the grams of a substance to moles, and moles to grams (7.3).

_____ Convert the number of grams of a substance to the number of particles, and number of particles to the number of grams (7.3).

_____ Calculate the percent composition of a compound (7.4).

_____ Determine the empirical formula for a compound (7.4).

_____ Determine the molecular formula for a compound (7.5).

Practice Test for Chapter 7

1. The number of carbon atoms in 2.0 mol of C is (7.1)
   A. 6.0 **B.** 12 **C.** 6.0×10^{23} **D.** 3.0×10^{23} **E.** 1.2×10^{24}
2. The number of molecules in 0.25 mol of NH_3 is (7.1)
   A. 0.25 **B.** 2.5×10^{23} **C.** 4.2×10^{-25} **D.** 1.5×10^{23} **E.** 2.43×10^{24}
3. The number of moles in 8.8×10^{24} molecules of SO_3 is (7.1)
   A. 0.068 **B.** 2.6 **C.** 1.5 **D.** 15 **E.** 53
4. The moles of oxygen (O) in 2.0 mol of $Al(OH)_3$ is (7.1)
   A. 1 **B.** 2 **C.** 3 **D.** 4 **E.** 6
5. The number of atoms of phosphorus in 1.00 mol of Be_3P_2 is (7.1)
   A. 3.01×10^{23} **B.** 6.02×10^{23} **C.** 1.20×10^{23} **D.** 9.03×10^{23} **E.** 1.20×10^{24}
6. What is the molar mass of Li_2SO_4? (7.2)
   A. 55.01 g **B.** 62.10 g **C.** 103.01 g **D.** 109.95 g **E.** 103.11 g
7. What is the molar mass of $NaNO_3$? (7.2)
   A. 34.00 g **B.** 37.00 g **C.** 53.00 g **D.** 75.00 g **E.** 85.00 g
8. The number of grams in 0.600 mol of Cl_2 is (7.3)
   A. 71.0 g **B.** 21.3 g **C.** 42.5 g **D.** 84.5 g **E.** 4.30 g
9. How many grams are in 4.00 mol of NH_3? (7.3)
   A. 4.00 g **B.** 17.0 g **C.** 34.0 g **D.** 68.1 g **E.** 0.240 g
10. How many grams are in 4.50 mol of N_2? (7.3)
    A. 6.22 g **B.** 28.0 g **C.** 56.0 g **D.** 112 g **E.** 126 g
11. How many moles is 8.00 g of NaOH? (7.3)
    A. 0.100 mol **B.** 0.200 mol **C.** 0.400 mol **D.** 2.00 mol **E.** 4.00 mol
12. The number of moles of aluminum in 54.0 g of Al is (7.3)
    A. 0.500 mol **B.** 1.00 mol **C.** 2.00 mol **D.** 3.00 mol **E.** 4.00 mol

13. The number of moles of water in 3.60 g of H_2O is (7.3)
A. 0.0500 mol **B.** 0.100 mol **C.** 0.200 mol **D.** 0.300 mol **E.** 0.400 mol

14. What is the number of moles in 2.20 g of CO_2? (7.3)
A. 2.00 mol **B.** 1.00 mol **C.** 0.200 mol **D.** 0.0500 mol **E.** 0.0100 mol

15. 0.200 g of H_2 = ____ mol of H_2. (7.3)
A. 0.100 mol **B.** 0.200 mol **C.** 0.400 mol **D.** 0.0400 mol **E.** 0.0100 mol

16. The mass percent of carbon in C_3H_8 is (7.4)
A. 18.29% **B.** 27.32% **C.** 59.84% **D.** 81.71% **E.** 92.24%

17. What is the empirical formula of a compound that is 27.3% C and 72.7% O? (7.4)
A. CO **B.** CO_2 **C.** C_2O **D.** C_2O_3 **E.** CO_3

18. What is the empirical formula of a compound that has 3.88 g of Cl and 6.12 g of O? (7.4)
A. ClO **B.** Cl_2O_7 **C.** Cl_3O **D.** Cl_2O_3 **E.** ClO_4

19. What is the molecular formula of a compound that has an empirical formula of CH_2 and molar mass of about 70 g? (7.5)
A. CH_2 **B.** C_2H_4 **C.** C_3H_6 **D.** C_4H_8 **E.** C_5H_{10}

20. What is the molecular formula of a compound that has an empirical formula of P_2O_3 and molar mass of about 220 g? (7.5)
A. P_2O_3 **B.** P_2O_5 **C.** P_4O_6 **D.** P_4O_8 **E.** P_4O_{10}

Answers to the Practice Test

1. E	2. D	3. D	4. E	5. E
6. D	7. E	8. C	9. D	10. E
11. B	12. C	13. C	14. D	15. A
16. D	17. B	18. B	19. E	20. C

8 Chemical Reactions

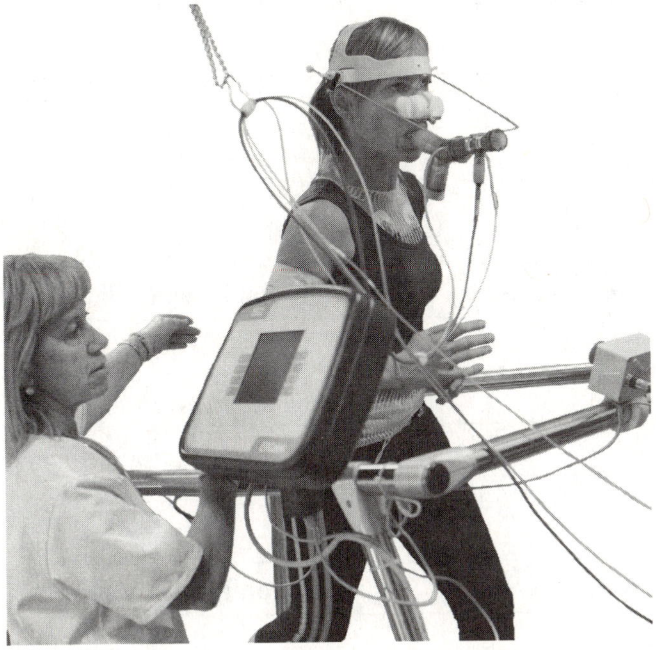

Since Natalie's diagnosis of mild emphysema due to secondhand cigarette smoke, she has been working with Angela, an exercise physiologist. Natalie's workout started with low-intensity exercises that use smaller muscles. Now that Natalie has increased the oxygen concentration in her blood, Angela has added some exercises that use larger muscles, which need more O_2. Today Natalie had a stress test and an EKG to assess and evaluate her progress. Her results show that her heart is functioning more efficiently and that she needs less oxygen. Her doctor told her that she has begun to reverse the progression of emphysema. Why is oxygen required during aerobic exercise?

8.1 Equations for Chemical Reactions

Learning Goal: Identify a balanced chemical equation and determine the number of atoms in the reactants and products.

- A chemical change may be occurs when a substance is converted into one or more new substances.
- Chemical change may be indicated by a formation of a gas (bubbles), change in color, formation of a solid, and/or heat produced or heat absorbed.
- When new substances form, a chemical reaction has taken place.
- A chemical equation shows the formulas of the reactants on the left side of the arrow and the formulas of the products on the right side.
- In a balanced equation, numbers called *coefficients*, which appear in front of the symbols or formulas, provide the same number of atoms for each kind of element on the reactant and product sides.
- Each formula in an equation is followed by an abbreviation, in parentheses, that gives the physical state of the substance: solid (s), liquid (l), or gas (g), and, if dissolved in water, an aqueous solution (aq).
- The Greek letter delta (Δ) over the arrow in an equation represents the application of heat to the reaction.

♦ Learning Exercise 8.1

State the number of atoms of each element on the reactant side and on the product side for each of the following balanced equations:

a. $CaCO_3(s) \longrightarrow CaO(g) + CO_2(g)$

Element	Atoms on Reactant Side	Atoms on Product Side
Ca		
C		
O		

b. $2Na(s) + 2H_2O(l) \longrightarrow 2NaOH(aq) + H_2(g)$

Element	Atoms on Reactant Side	Atoms on Product Side
Na		
H		
O		

c. $C_5H_{12}(g) + 8O_2(g) \longrightarrow 5CO_2(g) + 6H_2O(g)$

Element	Atoms on Reactant Side	Atoms on Product Side
C		
H		
O		

d. $2AgNO_3(aq) + K_2S(aq) \longrightarrow 2KNO_3(aq) + Ag_2S(s)$

Element	Atoms on Reactant Side	Atoms on Product Side
Ag		
N		
O		
K		
S		

e. $2Al(OH)_3(aq) + 3H_2SO_4(aq) \longrightarrow Al_2(SO_4)_3(s) + 6H_2O(l)$

Element	Atoms on Reactant Side	Atoms on Product Side
Al		
O		
H		
S		

Answers

a. $CaCO_3(s) \longrightarrow CaO(g) + CO_2(g)$

Element	Atoms on Reactant Side	Atoms on Product Side
Ca	1	1
C	1	1
O	3	3

b. $2Na(s) + 2H_2O(l) \longrightarrow 2NaOH(aq) + H_2(g)$

Element	Atoms on Reactant Side	Atoms on Product Side
Na	2	2
H	4	4
O	2	2

c. $C_5H_{12}(g) + 8O_2(g) \longrightarrow 5CO_2(g) + 6H_2O(g)$

Element	Atoms on Reactant Side	Atoms on Product Side
C	5	5
H	12	12
O	16	16

d. $2AgNO_3(aq) + K_2S(aq) \longrightarrow 2KNO_3(aq) + Ag_2S(s)$

Element	Atoms on Reactant Side	Atoms on Product Side
Ag	2	2
N	2	2
O	6	6
K	2	2
S	1	1

e. $2Al(OH)_3(aq) + 3H_2SO_4(aq) \longrightarrow Al_2(SO_4)_3(s) + 6H_2O(l)$

Element	Atoms on Reactant Side	Atoms on Product Side
Al	2	2
O	18	18
H	12	12
S	3	3

8.2 Balancing a Chemical Equation

Learning Goal: Write a balanced chemical equation from the formulas of the reactants and products for a chemical reaction.

- A chemical equation is balanced by placing numbers called coefficients in front of the symbols or formulas in the equation.

Guide to Balancing a Chemical Equation	
Step 1	Write an equation using the correct formulas of the reactants and products.
Step 2	Count the atoms of each element in reactants and products.
Step 3	Use coefficients to balance each element.
Step 4	Check the final equation to confirm it is balanced.

Example: Balance the following equation:

$$N_2(g) + H_2(g) \longrightarrow NH_3(g)$$

1. Write an equation using the correct formulas of the reactants and products.

$$N_2(g) + H_2(g) \longrightarrow NH_3(g)$$

2. Count the atoms of N and H on the reactant side and on the product side.

$$N_2(g) + H_2(g) \longrightarrow NH_3(g)$$

2N, 2H 1N, 3H

3. Balance the N atoms by placing a coefficient of 2 in front of NH_3. (This increases the number of H atoms, too.) Recheck the number of N atoms and the number of H atoms.

$$N_2(g) + H_2(g) \longrightarrow 2NH_3(g)$$

2N, 2H 2N, 6H

Balance the H atoms by placing a coefficient of 3 in front of H_2.

$$N_2(g) + 3H_2(g) \longrightarrow 2NH_3(g)$$

4. Recheck the number of N atoms and the number of H atoms.

$$N_2(g) + 3H_2(g) \longrightarrow 2NH_3(g)$$

2N, 6H 2N, 6H *The equation is balanced.*

♦ Learning Exercise 8.2

Balance each of the following equations by placing appropriate coefficients in front of the formulas as needed:

a. ____ $MgO(s) \longrightarrow$ ____ $Mg(s) +$ ____ $O_2(g)$

b. ____ $Zn(s) +$ ____ $HCl(aq) \longrightarrow$ ____ $ZnCl_2(aq) +$ ____ $H_2(g)$

c. ____ $Al(s) +$ ____ $CuSO_4(aq) \longrightarrow$ ____ $Cu(s) +$ ____ $Al_2(SO_4)_3(aq)$

d. ____ $Al_2S_3(s) +$ ____ $HCl(aq) \longrightarrow$ ____ $AlCl_3(aq) +$ ____ $H_2S(aq)$

e. ____ $BaCl_2(aq) +$ ____ $Na_2SO_4(aq) \longrightarrow$ ____ $BaSO_4(s) +$ ____ $NaCl(aq)$

f. ____ $CO(g) +$ ____ $Fe_2O_3(s) \longrightarrow$ ____ $Fe(s) +$ ____ $CO_2(g)$

g. ____ $K(s) +$ ____ $H_2O(l) \longrightarrow$ ____ $KOH(aq) +$ ____ $H_2(g)$

h. ____ $Fe(OH)_3(s) \xrightarrow{\Delta}$ ____ $Fe_2O_3(s) +$ ____ $H_2O(g)$

Answers

a. $2MgO(s) \longrightarrow 2Mg(s) + O_2(g)$
b. $Zn(s) + 2HCl(aq) \longrightarrow ZnCl_2(aq) + H_2(g)$
c. $2Al(s) + 3CuSO_4(aq) \longrightarrow 3Cu(s) + Al_2(SO_4)_3(aq)$
d. $Al_2S_3(s) + 6HCl(aq) \longrightarrow 2AlCl_3(aq) + 3H_2S(aq)$
e. $BaCl_2(aq) + Na_2SO_4(aq) \longrightarrow BaSO_4(s) + 2NaCl(aq)$
f. $3CO(g) + Fe_2O_3(s) \longrightarrow 2Fe(s) + 3CO_2(g)$
g. $2K(s) + 2H_2O(l) \longrightarrow 2KOH(aq) + H_2(g)$
h. $2Fe(OH)_3(s) \xrightarrow{\Delta} Fe_2O_3(s) + 3H_2O(g)$

8.3 Types of Reactions

Learning Goal: Identify a reaction as a combination, decomposition, single replacement, double replacement, or combustion.

- Reactions can be classified as: combination, decomposition, single replacement, double replacement, or combustion.
- In a *combination* reaction, two or more reactants form one product: $A + B \longrightarrow AB$
- In a *decomposition* reaction, a reactant splits into two or more simpler products: $AB \longrightarrow A + B$
- In a *single replacement* reaction, an uncombined element takes the place of an element in a compound:

$$A + BC \longrightarrow AC + B$$

- In a *double replacement* reaction, the positive ions in the reactants switch places:

$$AB + CD \longrightarrow AD + CB$$

- In a *combustion* reaction, a compound of carbon and hydrogen reacts with oxygen to form CO_2, H_2O, and energy.

$$C_xH_y(g) + ZO_2(g) \xrightarrow{\Delta} XCO_2(g) + Y/2H_2O(g) + \text{energy}$$

CORE CHEMISTRY SKILL
Classifying Types of Chemical Reactions

♦ Learning Exercise 8.3A

Match each of the following reactions with the type of reaction:

a. combination **b.** decomposition **c.** single replacement
d. double replacement **e.** combustion

1. ____ $N_2(g) + 3H_2(g) \longrightarrow 2NH_3(g)$

2. ____ $BaCl_2(aq) + K_2CO_3(aq) \longrightarrow BaCO_3(s) + 2KCl(aq)$

3. ____ $2H_2O_2(aq) \longrightarrow 2H_2O(l) + O_2(g)$

4. ____ $CuO(s) + H_2(g) \longrightarrow Cu(s) + H_2O(l)$

5. ____ $N_2(g) + 2O_2(g) \longrightarrow 2NO_2(g)$

6. ____ $C_2H_4(g) + 3O_2(g) \longrightarrow 2CO_2(g) + 2H_2O(g) = \text{energy}$

7. ____ $PbCO_3(s) \longrightarrow PbO(s) + CO_2(g)$

8. ____ $2Al(s) + Fe_2O_3(s) \longrightarrow 2Fe(s) + Al_2O_3(s)$

Answers **1.** a **2.** d **3.** b **4.** c
5. a **6.** e **7.** b **8.** c

♦ Learning Exercise 8.3B

1. Tarnish, Ag_2S, is removed from silver objects in a reaction with aluminum.

$$Al(s) + Ag_2S(s) \longrightarrow Ag(s) + Al_2S_3(s)$$

a. Write the balanced equation for this reaction.

b. What type of reaction takes place?

CORE CHEMISTRY SKILL
Balancing Combustion Reactions

2. Heptane, C_7H_{16}, a component of gasoline, reacts with O_2.

$$C_7H_{16}(l) + O_2(g) \xrightarrow{\Delta} CO_2(g) + H_2O(g)$$

a. Write the balanced equation for this reaction.

b. What type of reaction takes place?

Answers
1 a. $2Al(s) + 3Ag_2S(s) \longrightarrow 6Ag(s) + Al_2S_3(s)$
b. single replacement reaction
2 a. $C_7H_{16}(l) + 11O_2(g) \xrightarrow{\Delta} 7CO_2(g) + 8H_2O(g)$
b. combustion reaction

8.4 Functional Groups and Reactions of Organic Compounds: Extended Topic

Learning Goal: Classify organic molecules according to their functional groups; draw the condensed structural formulas for the products of reactions of organic compounds.

- Organic compounds are classified by *functional groups*, which are atoms or groups of atoms that undergo specific chemical reactions.
- Alkenes are hydrocarbons that contain one or more double bonds (C=C); alkynes contain a triple bond (C≡C).
- Hydrogenation adds hydrogen atoms to the double bond of an alkene to yield an alkane.
- Alcohols contain a hydroxyl (—OH) group bonded to a hydrocarbon chain.
- Aldehydes contain a carbonyl group (C=O) bonded to at least one H atom; ketones contain the carbonyl group bonded to two other carbon groups.
- Carboxylic acids contain a carboxyl group (—COOH), which is a combination of a carbonyl group (C=O) and a hydroxyl group (—OH). Esters contain the carboxyl groups, in which an oxygen is attached to another carbon atom.
- Amines are derived from ammonia, NH_3, in which carbon groups replace one or more of the H atoms. Amides replace the —OH group of a carboxylic acid with a nitrogen group.

Key Terms for Sections 8.1–8.4

Match the following terms with the following statements:

a. chemical change	**b.** chemical equation	**c.** combination reaction
d. single replacement reaction	**e.** decomposition	**f.** functional group
g. reactant	**h.** alkene	**i.** ester

1. ______ the type of reaction in which an element replaces a different element in a reacting compound

2. ______ a change that alters the composition of a substance, producing a new substance with new properties

3. ______ the type of reaction in which a reactant breaks up into two or more products

4. ______ the type of reaction in which reactants combine to form a single product

5. ______ an atom or group of atoms that influences the chemical reactions of an organic compound

6. ______ a shorthand method of writing a chemical reaction with the formulas of the reactants written on the left side of an arrow and the formulas of the products on the right side

7. ______ an organic compound that contains a carboxyl group (—COO—) with an oxygen atom bonded to a carbon atom

8. ______ an initial substance that undergoes change in a chemical reaction

9. ______ a type of hydrocarbon that contains carbon to carbon double bonds (C=C).

Answers	**1.** d	**2.** a	**3.** e	**4.** c	**5.** f
	6. b	**7.** i	**8.** g	**9.** h	

♦ Learning Exercise 8.4A

CORE CHEMISTRY SKILL
Identifying Functional Groups

Classify the organic compounds shown according to the following functional groups:

a. alkane **b.** alkene **c.** alcohol **d.** aldehyde **e.** carboxylic acid

1. ______ $CH_3-CH_2-CH=CH_2$

2. ______ $CH_3-CH_2-CH_3$

3. ______ $CH_3-CH_2-\overset{\overset{\large O}{\|}}{C}-H$

4. ______ $CH_3-CH_2-CH_2-OH$

5. ______ $CH_3-CH_2-CH_2-CH_3$

6. ______ $CH_3-CH_2-CH_2-COOH$

Answers	**1.** b	**2.** a	**3.** d	**4.** c	**5.** a	**6.** e

♦ Learning Exercise 8.4B

CORE CHEMISTRY SKILL
Identifying Functional Groups

Classify the following compounds according to their functional groups:

a. alcohol **b.** aldehyde **c.** ketone **d.** amide **e.** amine

1. ____ $CH_3-CH_2-CH_2-\overset{O}{\overset{\|}{C}}-H$

2. ____ $CH_3-CH_2-CH_2-NH_2$

3. ____ $CH_3-CH_2-\overset{O}{\overset{\|}{C}}-CH_2-CH_3$

4. ____ $CH_3-CH_2-\overset{O}{\overset{\|}{C}}-NH_2$

5. ____ $CH_3-\overset{O}{\overset{\|}{C}}-CH_2-CH_3$

6. ____ $CH_3-\overset{O}{\overset{\|}{C}}-H$

7. ____ $CH_3-CH_2-\overset{NH_2}{\overset{|}{CH}}-CH_3$

8. ____ $CH_3-CH_2-\overset{OH}{\overset{|}{CH}}-CH_3$

Answers **1.** b **2.** e **3.** c **4.** d **5.** c **6.** b **7.** e **8.** a

♦ Learning Exercise 8.4C

Write the products for each of the following reactions:

1. $CH_3-CH=CH_2 + H_2 \xrightarrow{Pt}$
2. $CH_3-CH=CH-CH_2-CH_3 + H_2 \xrightarrow{Pt}$

Answers
1. $CH_3-CH_2-CH_3$
2. $CH_3-CH_2-CH_2-CH_2-CH_3$

8.5 Biochemical Compounds: Extended Topic

Learning Goal: Identify the type of a biochemical compound from its structure and name its functional groups.

- The carbohydrates are a group of biochemical compounds consisting of monosaccharides, which contain many alcohol groups and an aldehyde or ketone group, disaccharides, and polysaccharides.
- Carbohydrates are classified as monosaccharides (simple sugars), disaccharides (two monosaccharide units), and polysaccharides (many monosaccharide units).
- Lipids are nonpolar compounds that are not soluble in water but are soluble in organic solvents.
- Classes of lipids include fatty acids, waxes, triacylglycerols, and steroids.
- A group of 20 amino acids provides the molecular building blocks of proteins.
- Peptides form when an amide bond forms between the carboxylic acid group of one amino acid and the amino group of the second amino acid.
- Proteins are biologically active molecules containing amino acids connected by peptide bonds.

- Amino acids contain two functional groups, an amine group and a carboxyl group, attached to a central carbon atom.

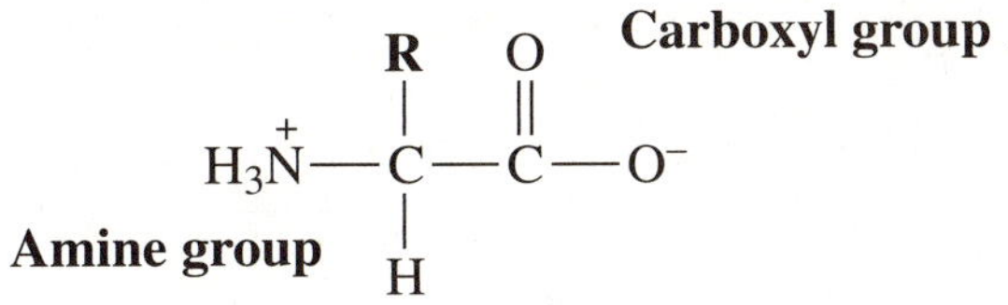

Ionized form in biological systems

- A peptide bond is an amide bond between the carboxyl group of one amino acid and the amino group of the second amino acid.

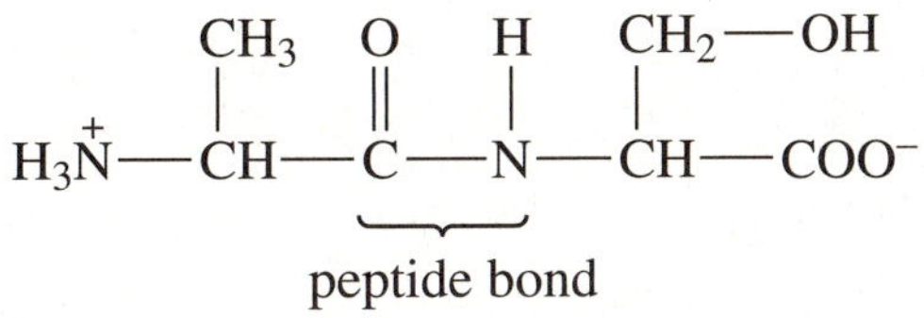

Key Terms for Section 8.5

Match the following terms with the correct definition below:

a. lipid	**b.** carbohydrate	**c.** glucose	**d.** disaccharide
e. fatty acid	**f.** triacylglycerol	**g.** protein	**h.** peptide

1. _____ a simple or complex sugar composed of a carbon chain with an aldehyde or ketone group and several hydroxyl groups

2. _____ two or more amino acids linked by amide bonds

3. _____ a type of compound that is not soluble in water but is soluble in nonpolar solvents

4. _____ a lipid consisting of glycerol bonded to three fatty acids

5. _____ an aldohexose that is the most prevalent monosaccharide in the diet

6. _____ a biologically active polymer of amino acids

7. _____ a carbohydrate that contains two monosaccharides linked by a glycosidic bond

8. _____ a long-chain carboxylic acid found in triacylglycerols

Answers				
	1. b	**2.** h	**3.** a	**4.** f
	5. c	**6.** g	**7.** d	**8.** e

♦ Learning Exercise 8.5A

Identify the functional groups in each of the following monosaccharides:

1. 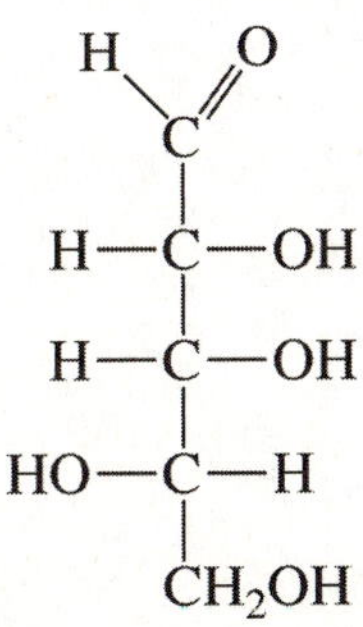

2. 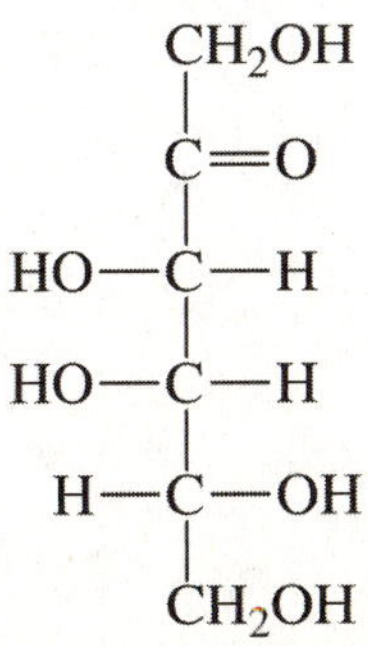

3. 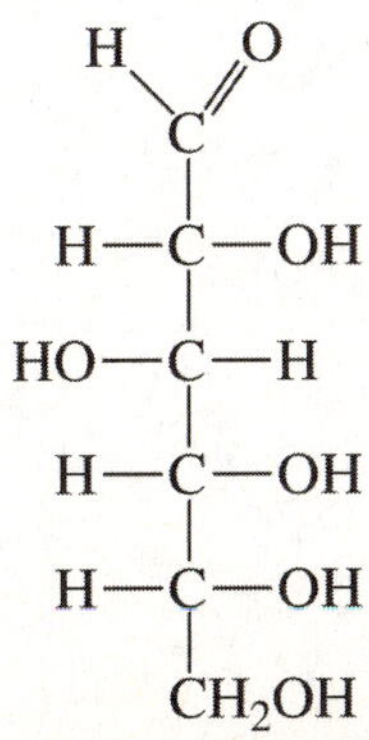

4. 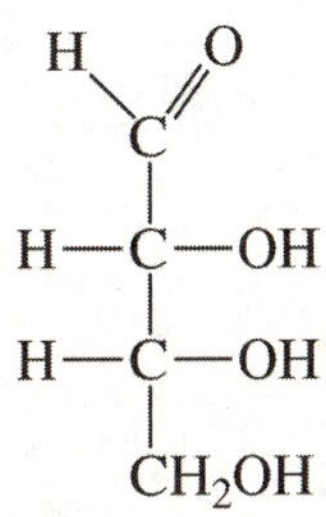

1. ______________________________

2. ______________________________

3. ______________________________

4. ______________________________

Answers **1.** aldehyde group, hydroxyl groups **2.** ketone group, hydroxyl groups **3.** aldehyde group, hydroxyl groups **4.** aldehyde group, hydroxyl groups

♦ Learning Exercise 8.5B

Match the correct class of lipids with the compositions below:

a. wax **b.** triacylglycerol **c.** fatty acid **d.** steroid

1. _____ a fused structure of four cycloalkanes

2. _____ a long-chain alcohol and a fatty acid

3. _____ glycerol and three fatty acids

4. _____ a long-chain carboxylic acid

Answers **1.** d **2.** a **3.** b **4.** c

♦ Learning Exercise 8.5C

Draw the condensed structural formula of each of the following fatty acids:

1. linoleic acid

2. stearic acid

Answers

1. Linoleic acid $CH_3-(CH_2)_4-CH=CH-CH_2-CH=CH-(CH_2)_7-\overset{\overset{\displaystyle O}{\|}}{C}-OH$

2. Stearic acid $CH_3-(CH_2)_{16}-\overset{\overset{\displaystyle O}{\|}}{C}-OH$

♦ Learning Exercise 8.5D

Using the appropriate side chain, complete the condensed structural formula of each of the following amino acids:

Glycine

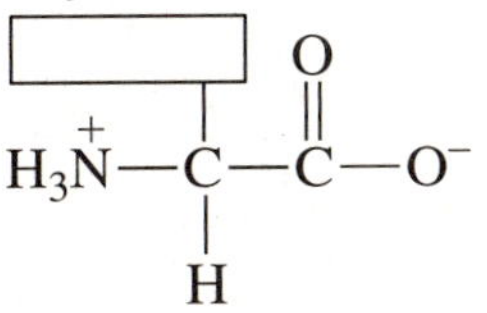

Alanine

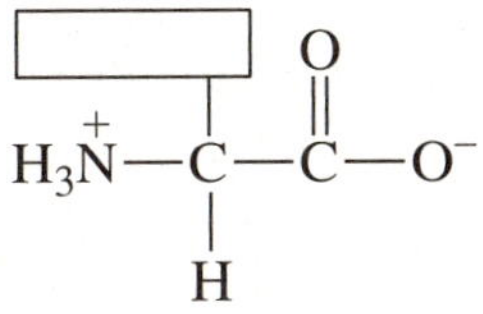

Serine

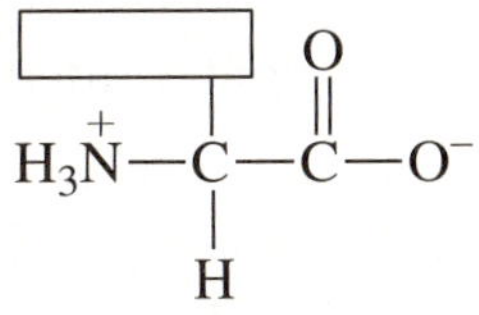

Answers

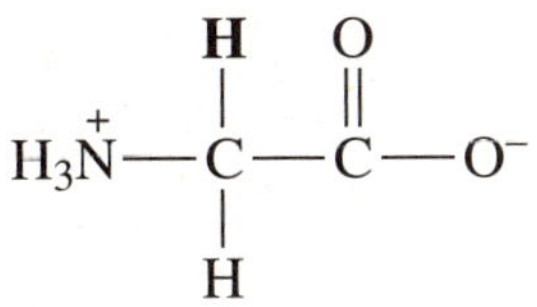

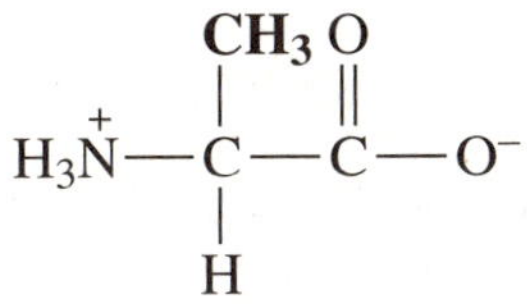

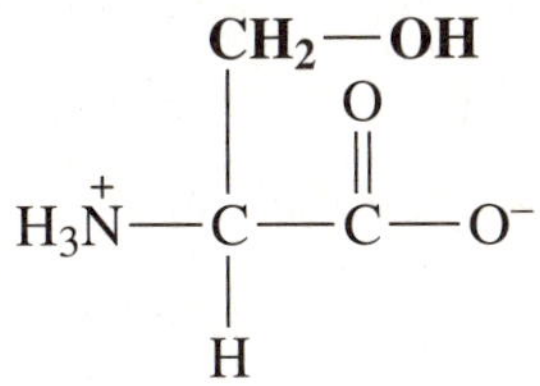

♦ Learning Exercise 8.5E

Draw the structural formulas of the following dipeptides:

1. Ser—Gly

2. Cys—Val

Answers

1. 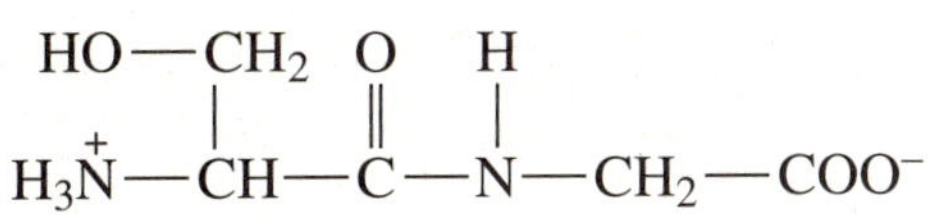

2. 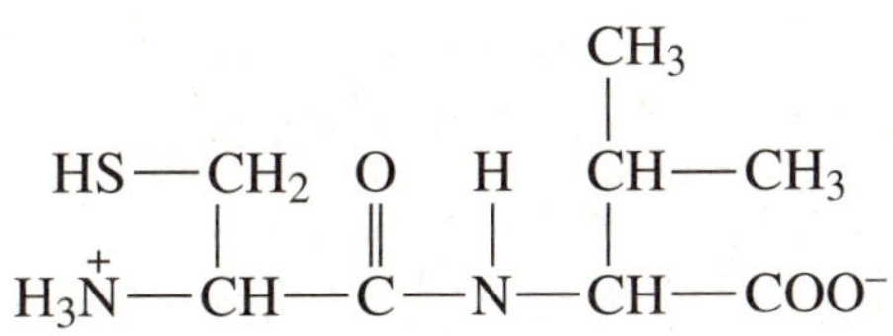

Checklist for Chapter 8

You are ready to take the Practice Test for Chapter 8. Be sure you have accomplished the following learning goals for this chapter. If not, review the section listed at the end of the goal. Then apply your new skills and understanding to the Practice Test.

After studying Chapter 8, I can successfully:

_____ State a chemical equation in words and calculate the total atoms of each element in the reactants and products (8.1).

_____ Write a balanced equation for a chemical reaction from the formulas of the reactants and products (8.2).

_____ Identify a reaction as a combination, decomposition, single replacement, double replacement, or combustion (8.3).

_____ Identify the functional groups in organic compounds (8.4).

_____ Balance equations for the combustion reaction of organic compounds (8.4).

_____ Draw the condensed structural formula for the product of a hydrogenation reaction of an alkene (8.4).

_____ Classify carbohydrates as monosaccharides, disaccharides, and polysaccharides (8.5).

_____ Identify a fatty acid as saturated or unsaturated (8.5).

_____ Write the condensed structural formula of a wax or triacylglycerol produced by the reaction of a fatty acid and an alcohol or glycerol (8.5).

_____ Describe the functions of proteins in the body (8.5).

_____ Draw the ionized structural formula for an amino acid (8.5).

_____ Draw the structure for an amino acid (8.5).

_____ Describe a peptide bond and draw the structure for a peptide (8.5).

Practice Test for Chapter 8

For questions 1 through 5, balance and select the correct coefficient for the component in the equation written in boldface type (8.1, 8.2).

A. 1 **B.** 2 **C.** 3 **D.** 4 **E.** 5

1. _____ $Sn(s) + \mathbf{Cl_2}(g) \longrightarrow SnCl_4(s)$
2. _____ $Al(s) + H_2O(l) \longrightarrow Al_2O_3(s) + \mathbf{H_2}(g)$
3. _____ $C_3H_8(g) + \mathbf{O_2}(g) \longrightarrow CO_2(g) + H_2O(g)$
4. _____ $\mathbf{NH_3}(g) + O_2(g) \longrightarrow N_2(g) + H_2O(g)$
5. _____ $N_2O(g) \longrightarrow N_2(g) + \mathbf{O_2}(g)$

For questions 6 through 11, classify each reaction as one of the following: (8.3)

A. combination **B.** decomposition **C.** single replacement
D. double replacement **E.** combustion

6. _____ $S(s) + O_2(g) \longrightarrow SO_2(g)$
7. _____ $Fe_2O_3(s) + 3C(s) \longrightarrow 2Fe(s) + 3CO(g)$
8. _____ $CaCO_3(s) \longrightarrow CaO(s) + CO_2(g)$
9. _____ $C_4H_8(g) + 6O_2(g) \xrightarrow{\Delta} 4CO_2(g) + 4H_2O(g)$
10. _____ $Na_2S(aq) + Pb(NO_3)_2(aq) \longrightarrow PbS(s) + 2NaNO_3(aq)$
11. _____ $2KClO_3(s) \longrightarrow 2KCl(s) + 3O_2(g)$

For questions 12 through 16, classify each of the following compounds according to its functional group: (8.4)

A. alkane **B.** alkene **C.** alcohol **D.** aldehyde
E. ester **F.** amine

12. ____ $CH_3-CH_2-C(=O)-O-CH_3$

13. ____ $CH_3-CH_2-CH_2-OH$

14. ____ $CH_3-CH=C(CH_3)-CH_3$

15. ____ $CH_3-CH_2-NH-CH_3$

16. ____ $CH_3-C(=O)-H$

In questions 17 through 20, classify each of the following compounds according to its functional group: (8.4)

A. alcohol **B.** aldehyde **C.** carboxylic acid **D.** ester **E.** ketone

17. ____ $CH_3-CH_2-C(=O)-O-CH_3$

18. ____ $CH_3-CH_2-C(=O)-H$

19. ____ $CH_3-C(=O)-CH_2-CH_3$

20. ____ $CH_3-CH_2-C(=O)-OH$

21. The correctly balanced equation for the complete combustion of ethane is: (8.2)

A. $C_2H_6(g) + O_2(g) \xrightarrow{\Delta} 2CO(g) + 3H_2O(g) + \text{heat}$
B. $C_2H_6(g) + O_2(g) \xrightarrow{\Delta} CO_2(g) + H_2O(g) + \text{heat}$
C. $C_2H_6(g) + 2O_2(g) \xrightarrow{\Delta} 2CO_2(g) + 3H_2O(g) + \text{heat}$
D. $2C_2H_6(g) + 7O_2(g) \xrightarrow{\Delta} 4CO_2(g) + 6H_2O(g) + \text{heat}$
E. $2C_2H_6(g) + 4O_2(g) \xrightarrow{\Delta} 4CO_2(g) + 6H_2O(g) + \text{heat}$

22. Hydrogenation of $CH_3-CH=CH_2$ gives: (8.4)

A. $3CO_2(g) + 6H_2(g)$
B. $CH_3-CH_2-CH_3$
C. $CH_2=CH-CH_3$
D. $CH_3-CH_2-CH_2-CH_3$
E. no reaction

23. Hydrogenation of $CH_3—CH_2—CH=CH_2$ gives: (8.4)
A. $3CO_2(g) + 6H_2(g)$
B. $CH_3—CH_2—CH_3$
C. $CH_2=CH—CH_3$
D. $CH_3—CH_2—CH_2—CH_3$
E. no reaction

24. What functional groups are in the open-chain forms of monosaccharides? (8.4, 8.5)
A. hydroxyl groups **B.** aldehyde groups **C.** ketone groups
D. hydroxyl and aldehyde or ketone groups **E.** hydroxyl and ether groups

25. An ester of a fatty acid is a (an) (8.5)
A. carbohydrate **B.** lipid **C.** protein
D. oxyacid **E.** amino acid

26. The following formula represents a (an) (8.5)

$$CH_3—(CH_2)_{16}—COOH$$

A. unsaturated fatty acid **B.** saturated fatty acid **C.** amino acid
D. triacylglycerol **E.** protein

For questions 27 through 30, classify each as a (an) (8.5)

A. amino acid **B.** triacylglycerol **C.** monosaccharide **D.** fatty acid

27.

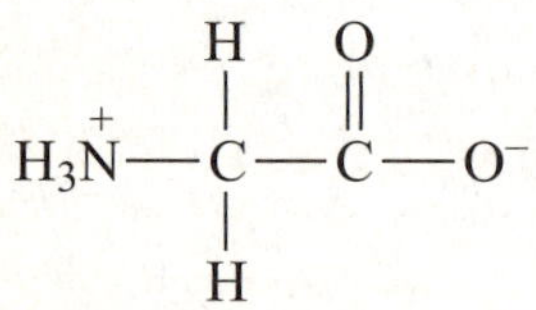

28. $CH_3—(CH_2)_{16}—COOH$

29.

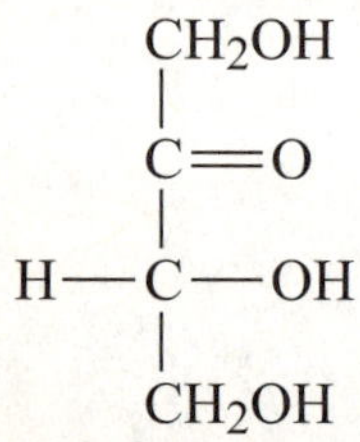

30. an ester of glycerol with three palmitic acid molecules

Answers to the Practice Test

1. B	2. C	3. E	4. D	5. A
6. A	7. C	8. B	9. E	10. D
11. B	12. E	13. C	14. B	15. F
16. D	17. D	18. B	19. E	20. C
21. D	22. B	23. D	24. D	25. B
26. B	27. A	28. D	29. C	30. B

9
Chemical Quantities in Reactions

Lance, an environmental health practitioner, has returned to a farm where he found small amounts of fenbendazole, a sheep dewormer, in the soil. His recent tests indicate that the fenbendazole is now within acceptable range in both the soil and the water in the lake near the farm. When he is working on-site, he wears protective clothing while he obtains samples of soil and water.

Lance specializes in measuring levels of contaminants in soil and water caused by manufacturing processes or farming. In his work, Lance also acts as an advisor, consultant, and safety officer to ensure that soil and water on the farm are safe.

Fenbendazole has a molecular formula $C_{15}H_{13}N_3O_2S$ with a molar mass of 299 g/mol. If a dose for deworming sheep is 8.6 g of fenbendazole, how many moles of fenbendazole is adminstered?

9.1 Mole Relationships in Chemical Equations

Learning Goal: Given a quantity in moles of reactant or product, use a mole-mole factor from the balanced equation to calculate the number of moles of another substance in the reaction.

- The law of conservation of mass states that mass is neither lost nor gained during a chemical reaction.
- The coefficients in a balanced chemical equation describe the moles of reactants and products in the reactions.
- Using the coefficients, mole–mole conversion factors can be written to relate any two substances in the equation.
- For the reaction of oxygen to form ozone, $3O_2(g) \longrightarrow 2O_3(g)$, the mole–mole conversion factors are the following:

$$\frac{3 \text{ mol } O_2}{2 \text{ mol } O_3} \quad \text{and} \quad \frac{2 \text{ mol } O_3}{3 \text{ mol } O_2}$$

♦ Learning Exercise 9.1A

Write all of the possible conversion factors for the following equation:

$$N_2(g) + O_2(g) \longrightarrow 2NO(g)$$

Answers

For N_2 and O_2: $\frac{1 \text{ mol } N_2}{1 \text{ mol } O_2}$ and $\frac{1 \text{ mol } O_2}{1 \text{ mol } N_2}$

For N_2 and NO: $\frac{1 \text{ mol } N_2}{2 \text{ mol NO}}$ and $\frac{2 \text{ mol NO}}{1 \text{ mol } N_2}$

For NO and O_2: $\frac{1 \text{ mol } O_2}{2 \text{ mol NO}}$ and $\frac{2 \text{ mol NO}}{1 \text{ mol } O_2}$

Guide to Calculating the Quantities of Reactants and Products in a Chemical Reaction	
Step 1	State the given and needed quantities (moles or grams).
Step 2	Write a plan to convert the given to the needed quantity (moles or grams).
Step 3	Use coefficients to write mole–mole factors; write molar mass factors if needed.
Step 4	Set up the problem to give the needed quantity (moles or grams).

Example: For the equation $N_2(g) + O_2(g) \rightarrow 2NO(g)$, calculate the number of moles of NO from 3.0 mole of N_2.

Step 1: State the given and needed quantities.

Given 3.0 mole of N_2 **Need** moles of NO

Step 2: Write a plan to convert the given to the needed moles.

moles of N_2 $\rightarrow$ moles of NO

Step 3: Use coefficients to write relationships and mole–mole factors; write molar mass factors if needed.

Mole–Mole Factors: $\frac{2 \text{ mol NO}}{1 \text{ mol } N_2}$ and $\frac{1 \text{ mol } N_2}{2 \text{ mol NO}}$

Step 4: Set up the problem to give the needed quantity (moles or grams).

$$3.0\ \cancel{\text{mol } N_2} \times \frac{2 \text{ mol NO}}{1\ \cancel{\text{mol } N_2}} = 6.0 \text{ mol NO}$$

♦ Learning Exercise

CORE CHEMISTRY SKILL

Using Mole–Mole Factors

Use this equation to answer the following questions:

$$C_3H_8(g) + 5O_2(g) \xrightarrow{\Delta} 3CO_2(g) + 4H_2O(g)$$

a. How many moles of O_2 are needed to react with 2.00 mol of C_3H_8?

b. How many moles of CO_2 are produced when 4.00 mol of O_2 reacts?

c. How many moles of C_3H_8 react with 3.00 mol of O_2?

d. How many moles of H_2O are produced from 0.500 mol of C_3H_8?

Answers

a. 10.0 mol of O_2 **b.** 2.40 mol of CO_2
c. 0.600 mol of C_3H_8 **d.** 2.00 mol of H_2O

9.2 Mass Calculations for Reactions

Learning Goal: Given the mass in grams of a substance in a reaction, calculate the mass in grams of another substance in the reaction.

- The number of grams or moles of one substance in an equation is converted to grams or moles of another substance in an equation using appropriate molar masses and mole–mole factors.

Substance A — Substance B

grams of A → Molar mass factor A → moles of A → Mole–mole factor B/A → moles of B → Molar mass factor B → grams of B

Guide to Calculating the Quantities of Reactants and Products in a Chemical Reaction	
Step 1	State the given and needed quantitites.
Step 2	Write a plan to convert the given to the needed quantity (moles or grams).
Step 3	Use coefficients to write mole–mole factors; write molar mass factors if needed.
Step 4	Set up the problem to give the needed quantity (moles or grams).

♦ Learning Exercise 9.2

CORE CHEMISTRY SKILL
Converting Grams to Grams

Consider the equation for the following questions:

$$2C_2H_6(g) + 7O_2(g) \longrightarrow 4CO_2(g) + 6H_2O(g)$$

a. How many grams of oxygen (O_2) are needed to react with 120. g of C_2H_6?

b. How many grams of C_2H_6 are needed to react with 115 g of O_2?

c. How many grams of C_2H_6 react if 2.00 g of CO_2 is produced?

d. How many grams of CO_2 are produced when 60.0 g of C_2H_6 reacts with sufficient oxygen?

e. How many grams of water are produced when 82.5 g of O_2 reacts with sufficient C_2H_6?

Answers

a. 448 g of O_2 **b.** 30.9 g of C_2H_6 **c.** 0.683 g of C_2H_6
d. 176 g of CO_2 **e.** 39.8 g of H_2O

9.3 Limiting Reactants

Learning Goal: Identify a limiting reactant when given the quantities of two reactants; calculate the amount of product formed from the limiting reactant.

- In a limiting reactant problem, the availability of one of the reactants limits the amount of product.
- The reactant that is used up is the limiting reactant; the reactant that remains is the excess reactant.
- The limiting reactant produces the smaller number of moles or grams of product.

Guide to Calculating the Grams of Product from a Limiting Reactant	
Step 1	State the given and needed grams.
Step 2	Write a plan to convert the grams of each reactant to grams of product.
Step 3	Write the molar mass factors and the mole-mole factors from the equation.
Step 4	Calculate the grams of product from each reactant and select the smaller number of grams as the amount of product from the limiting reactant.

Example: In the reaction $S(s) + 3F_2(g) \longrightarrow SF_6(g)$, how many grams of SF_6 are possible when 16.1 g of S is mixed with 45.6 g of F_2?

Solution:

Step 1: State the given and needed grams.

Given 16.1 g of S and 45.6 g of F_2 **Need** grams of SF_6

Step 2: Write a plan to convert the grams of each reactant to grams of product.

$$\text{g of S (or F}_2) \xrightarrow{\text{Molar mass}} \text{mol of S (or F}_2) \xrightarrow{\text{Mole}-\text{mole factor}} \text{moles of CO} \xrightarrow{\text{Molar mass}} \text{g of CO}$$

Step 3: Write the molar mass factors and the mole-mole factors from the equation.

$$1 \text{ mol of S} = 1 \text{ mol of SF}_6 \qquad 3 \text{ mol of F}_2 = 1 \text{ mol of SF}_6$$

$$\frac{1 \text{ mol S}}{1 \text{ mol SF}_6} \text{ and } \frac{1 \text{ mol SF}_6}{1 \text{ mol S}} \qquad \frac{3 \text{ mol F}_2}{1 \text{ mol SF}_6} \text{ and } \frac{1 \text{ mol SF}_6}{3 \text{ mol F}_2}$$

$$1 \text{ mol of S} = 32.1 \text{ g of S} \qquad 1 \text{ mol of SF}_6 = 146.1 \text{ g of SF}_6$$

$$\frac{32.1 \text{ g S}}{1 \text{ mol S}} \text{ and } \frac{1 \text{ mol S}}{32.1 \text{ g S}} \qquad \frac{146.1 \text{ g SF}_6}{1 \text{ mol SF}_6} \text{ and } \frac{1 \text{ mol SF}_6}{146.1 \text{ g SF}_6}$$

Step 4: Calculate the grams of product from each reactant and select the smaller number of grams as the amount of product from the limiting reactant.

$$16.1\ \cancel{g\ S} \times \frac{1\ \cancel{mol\ S}}{32.1\ \cancel{g\ S}} \times \frac{1\ \cancel{mol\ SF_6}}{1\ \cancel{mol\ S}} \times \frac{146.1\ g\ SF_6}{1\ \cancel{mol\ SF_6}} = 73.3\ g\ SF_6$$

$$45.6\ \cancel{g\ F_2} \times \frac{1\ \cancel{mol\ F_2}}{38.0\ \cancel{g\ F_2}} \times \frac{1\ \cancel{mol\ SF_6}}{3\ \cancel{mol\ F_2}} \times \frac{146.1\ g\ SF_6}{1\ \cancel{mol\ SF_6}}$$

$= 58.4\ g\ SF_6$ Smaller amount, therefore F_2 is the limiting reactant

♦ **Learning Exercise 9.3**

CORE CHEMISTRY SKILL

Calculating Quantity of Product from a limiting Reactant

a. How many grams of Co_2S_3 can be produced from the reaction of 130. g of Co and 116. g of S?

$$2Co(s) + 3S(s) \longrightarrow Co_2S_3(s)$$

b. How many grams of NO_2 can be produced from the reaction of 32.0 g of NO and 24.0 g of O_2?

$$2NO(g) + O_2(g) \longrightarrow 2NO_2(g)$$

Answers

a. 236 g of Co_2S_3 (Co is the limiting reactant)
b. 49.2 g of NO_2 (NO is the limiting reactant)

9.4 Percent Yield

Learning Goal: Given the actual quantity of product, determine the percent yield for a reaction.

- In most reactions, some product is lost due to incomplete reactions or side reactions. The amount of product collected and measured is called the actual yield.
- Theoretical yield is the maximum amount of product calculated for a given amount of a reactant.
- Percent yield is the ratio of the actual amount (yield) of product obtained to the theoretical yield.

$$\text{Percent yield} = \frac{\text{actual yield}}{\text{theoretical yield}} \times 100\%$$

Study Note

The percent yield is the ratio of the actual yield obtained to the theoretical yield which is calculated for a given amount of starting reactant. If we calculate that the reaction of 35.5 g of N_2 can theoretically produce 43.1 g of NH_3, but the actual yield is 26.0 g of NH_3, the percent yield is:

$$\frac{26.0\ g\ NH_3\ (\text{actual})}{43.1\ g\ NH_3\ (\text{theoretical})} \times 100\% = 60.3\% \qquad \text{Percent yield}$$

Guide to Calculations for Percent Yield	
Step 1	Write the given and needed quantities.
Step 2	Write a plan to calculate the theoretical yield and the percent yield.
Step 3	Write the molar mass factors and the mole–mole factor from the balanced equation.
Step 4	Calculate the percent yield by dividing the actual yield (given) by the theoretical yield and multiplying the result by 100%.

CORE CHEMISTRY SKILL
Calculating Percent Yield

♦ Learning Exercise 9.4

Consider the following reaction: $2H_2S(g) + 3O_2(g) \longrightarrow 2SO_2(g) + 2H_2O(g)$

a. If 60.0 g of H_2S reacts with oxygen to produce 45.5 g of SO_2, what is the percent yield of SO_2?

b. If 25.0 g of O_2 reacts with H_2S to produce 18.6 g of SO_2, what is the percent yield of SO_2?

Consider the reaction $2C_2H_6(g) + 7O_2(g) \longrightarrow 4CO_2(g) + 6H_2O(g)$
Ethane

c. If 125 g of C_2H_6 reacting with sufficient oxygen produces 175 g of CO_2, what is the percent yield of CO_2?

d. When 35.0 g of O_2 reacts with sufficient ethane to produce 12.5 g of H_2O, what is the percent yield of water?

Answers **a.** 40.3% **b.** 55.7% **c.** 47.8% **d.** 74.0%

9.5 Energy in Chemical Reactions

Learning Goal: Given the heat of reaction (enthalpy change), calculate the loss or gain of heat for an exothermic or endothermic reaction.

- The heat of reaction is the energy released or absorbed when the reactions occur.
- The heat of reaction or enthalpy change, ΔH, is the energy difference between the reactants and the products.
- In exothermic reactions, the energy of the reactants is greater than that of the products, which means that energy is released. The heat of reaction for an exothermic reaction is designated by a $-\Delta H$, by a negative value for the heat, or by writing the heat of reaction as a product.

- In endothermic reactions, the energy of the reactants is less than that of the products, which means that energy is absorbed. The heat of reaction for an endothermic reaction is designated by a $+\Delta H$, by a positive value for the heat, or by writing the heat of reaction as a reactant.

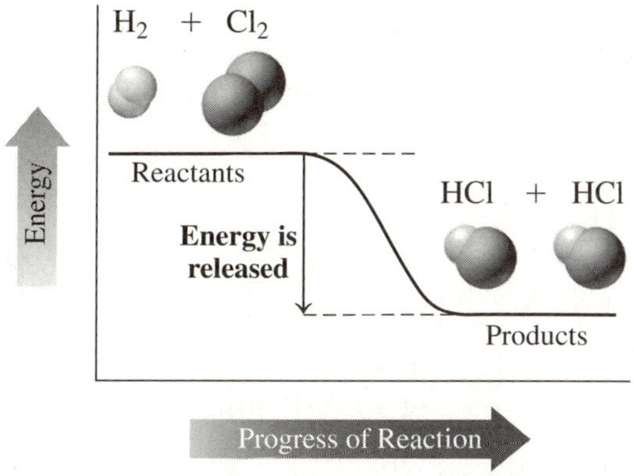

- The value of ΔH refers to the heat change for the number of moles of each substance in the balanced chemical equation.
- Heat conversion factors can be written for each substance: heat/mol of substance and mol of substance/heat.

Key Terms for Sections 9.1–9.5

Match each of the following terms with the correct definition below:

a. actual yield **b.** exothermic reaction **c.** endothermic reaction
d. theoretical yield **e.** mole–mole factor

1. _____ a reaction in which the energy of the reactants is less than that of the products

2. _____ the maximum amount of product that a reaction can produce from a given amounts of reactants

3. _____ a reaction in which the energy of the reactants is greater than that of the products

4. _____ a conversion factor that relates the number of moles of two compounds derived from the coefficients in a balanced equation

5. _____ the amount of product produced by a reaction

Answers **1.** c **2.** d **3.** b **4.** e **5.** a

♦ Learning Exercise 9.5A

Indicate whether each of the following is an endothermic or exothermic reaction:

1. $2H_2(g) + O_2(g) \longrightarrow 2H_2O(g) + 582\text{ kJ}$ _______________

2. $C_2H_4(g) \longrightarrow H_2(g) + C_2H_2(g) \quad \Delta H = +176\text{ kJ}$ _______________

3. $2C(s) + O_2(g) \longrightarrow 2CO(g) \quad \Delta H = -220\text{ kJ}$ _______________

4. $C_6H_{12}O_6(s) + 6O_2(g) \longrightarrow 6CO_2(g) + 6H_2O(l) + 1350\text{ kcal}$ _______________

5. $C_2H_5OH(l) + 21\text{ kcal} \longrightarrow C_2H_4(g) + H_2O(g)$ _______________

Answers **1.** exothermic **2.** endothermic **3.** exothermic
4. exothermic **5.** endothermic

Guide to Calculations Using the Heat of Reaction (ΔH)	
Step 1	State the given and needed quantities.
Step 2	Write a plan using the heat of reaction and any molar mass needed.
Step 3	Write the conversion factors including heat of reaction.
Step 4	Set up the problem to calculate the heat.

CORE CHEMISTRY SKILL

Using the Heat of Reaction

♦ Learning Exercise 9.5B

Consider the following reaction:

$$C_6H_{12}O_6(s) + 6O_2(g) \longrightarrow 6CO_2(g) + 6H_2O(l) \qquad \Delta H = -1350 \text{ kcal}$$

How much heat, in kcal and kJ, is produced when 50.0 g of CO_2 is formed?

Answer 256 kcal; −1070 J

♦ Learning Exercise 9.5C

Calculate the number of kilojoules in each of the following reactions and specify if energy is absorbed or released:

1. $C_2H_4(g) \longrightarrow H_2(g) + C_2H_2(g) \qquad \Delta H = +176 \text{ kJ}$

a. 3.50 mol of C_2H_2 is produced

b. 75.0 g of C_2H_4 reacts

2. $2H_2(g) + O_2(g) \longrightarrow 2H_2O(g) \qquad \Delta H = -582 \text{ kJ}$

a. 0.820 mol of H_2 reacts

b. 2.25 g of H_2O is produced

Answers **1 a.** 616 kJ is absorbed **b.** 471 kJ is absorbed
2 a. −239 kJ is released **b.** −36.3 kJ is released

Checklist for Chapter 9

You are ready to take the Practice Test for Chapter 9. Be sure you have accomplished the following learning goals for this chapter. If not, review the section listed at the end of the goal. Then apply your new skills and understanding to the Practice Test.

After studying Chapter 9, I can successfully:

_____ Use mole–mole factors for the mole relationships in an equation to calculate the moles of another substance in an equation for a chemical reaction (9.1).

_____ Calculate the mass in grams of a substance in an equation using mole factors and molar masses (9.2).

_____ Given the mass of reactants, find the limiting reactant and calculate the amount of product formed (9.3).

_____ Calculate the percent yield for a reaction given the actual yield of a product (9.4).

_____ Given the heat of reaction, describe a reaction as exothermic or endothermic (9.5).

_____ Given the mass of reactants or products, calculate the heat, in kcal or kilojoules, released or absorbed by a chemical reaction (9.5).

Practice Test for Chapter 9

For questions 1 to 7, use the reaction: $\underset{\text{Ethanol}}{C_2H_5OH(g)} + 3O_2(g) \longrightarrow 2CO_2(g) + 3H_2O(g)$ (9.1, 9.2)

1. How many moles of oxygen are needed to react with 1.00 mol of ethanol?
   A. 1.00 mol **B.** 2.00 mol **C.** 3.00 mol **D.** 4.00 mol **E.** 5.00 mol

2. How many moles of water are produced when 12.0 mol of oxygen reacts?
   A. 3.00 mol **B.** 6.00 mol **C.** 8.00 mol **D.** 12.0 mol **E.** 36.0 mol

3. How many grams of carbon dioxide are produced when 92.0 g of ethanol reacts?
   A. 22.0 g **B.** 44.0 g **C.** 88.0 g **D.** 92.0 g **E.** 176 g

4. How many moles of oxygen would be needed to produce 1.00 mol of CO_2?
   A. 0.670 mol **B.** 1.00 mol **C.** 1.50 mol **D.** 2.00 mol **E.** 3.00 mol

5. How many grams of water will be produced if 23.0 g of ethanol reacts?
   A. 54.0 g **B.** 27.0 g **C.** 18.0 g **D.** 9.00 g **E.** 6.00 g

6. How many grams of water will be produced if 138 g of ethanol and 160. g of O_2 react?
   A. 3.60 g **B.** 18.0 g **C.** 54.1 g **D.** 90.1 g **E.** 162 g

7. How many grams of water will be produced if 230. g of ethanol and 384 g of O_2 react?
   A. 270 g **B.** 216 g **C.** 54.1 g **D.** 36.0 g **E.** 18.0 g

For questions 8 and 9, use the reaction: $C_3H_8(g) + 5O_2(g) \rightarrow 3CO_2(g) + 4H_2O(g)$ (9.3)

8. If 50.0 g of C_3H_8 and 150 g of O_2 react, the limiting reactant is:
   A. C_3H_8 **B.** O_2 **C.** both **D.** neither **E.** CO_2

9. If 50.0 g of C_3H_8 and 150 g of O_2 react, the mass of CO_2 formed is:
   A. 150 g **B.** 206 g **C.** 200 g **D.** 124 g **E.** 132 g

10. In an experiment using the equation below, 36.0 g of NO is produced from 20.0 g of N_2. What is the percent yield?

$N_2(g) + O_2(g) \rightarrow 2NO(g)$ (9.4)

A. 42.9% **B.** 55.6% **C.** 71.4% **D.** 84.1% **E.** 93.4%

For questions 11 and 12, use the following equation: $N_2(g) + 3H_2(g) \longrightarrow 2NH_3(g)$ (9.3, 9.4)

11. When 25.0 g of N_2 reacts, what is the theoretical yield of NH_3?
A. 7.60 g **B.** 15.2 g **C.** 28.02 g **D.** 30.4 g **E.** 60.8 g

12. When 25.0 g of N_2 reacts, the actual yield of NH_3 is 20.8 g. What is the percent yield?
A. 68.4% **B.** 20.8% **C.** 25.0% **D.** 82.5% **E.** 100%

For questions 13 through 16, use the following equation and heat of reaction: (9.5)
$C_2H_6O(l) + 3O_2(g) \longrightarrow 2CO_2(g) + 3H_2O(l)\ \Delta H = -1420\text{ kJ}$

13. The reaction is
A. thermic **B.** endothermic **C.** exothermic **D.** nonthermic **E.** subthermic

14. How much heat, in kilojoules, is released when 0.500 mol of C_2H_6O reacts?
A. −710 kJ **B.** −1420 kJ **C.** −2130 kJ **D.** −2840 kJ **E.** −4260 kJ

15. How many grams of CO_2 are produced if −355 kJ of heat are released?
A. 11.0 g **B.** 22.0 g **C.** 33.0 g **D.** 44.0 g **E.** 88.0 g

16. How much heat, in kilojoules, is released when 16.0 g of O_2 reacts?
A. −4260 kJ **B.** −1420 kJ **C.** −710 kJ **D.** −237 kJ **E.** −118 kJ

Answers to the Practice Test

1. C	**2.** D	**3.** E	**4.** C	**5.** B
6. D	**7.** B	**8.** B	**9.** D	**10.** D
11. D	**12.** A	**13.** C	**14.** A	**15.** B
16. D				

10
Structures of Solids and Liquids

Lisa, a histologist, prepares thin tissue samples and mounts them on slides for review by a pathologist. She then adds dyes that stain the structures in the tissue and make them visible. The types of tissue include epithelial muscle, connective, and nerve. Recently, a dermatologist excised a lesion from the calf of a 37-year-old woman. When Lisa received the sample, she froze it in liquid nitrogen, which has a heat of vaporization of 198 J/g at the boiling point of nitrogen, which is −196 °C. Then she cut thin slices of the frozen tissue and mounted them on slides. The pathologist reported that the lesion contained a variety of abnormal growth patterns that led to a diagnosis of Rosai-Dorfman disease, which is sometimes associated with herpes virus 6. Because this disease is benign, there is minimal risk that it will progress and will instead probably regress. Eosin, which is used to stain cytoplasm and muscle fibers, is prepared from fluorescein.

How many kilojoules are required to convert 15.0 g of nitrogen from liquid to gas?

10.1 Electron-Dot Formulas

Learning Goal: Draw the electron-dot formulas for molecular compounds or polyatomic ions with multiple bonds and show resonance structures.

- In a covalent bond, atoms of nonmetals share electrons to achieve an octet.
- For example, oxygen with six valence electrons shares electrons with two hydrogen atoms to form the covalent compound water (H_2O).

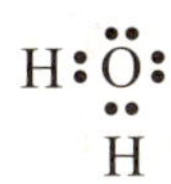

- In a double bond, two pairs of electrons are shared between the same two atoms to complete octets.
- In a triple bond, three pairs of electrons are shared to complete octets.
- Bonding pairs of electrons are shared between atoms. Nonbonding electron pairs are also called lone pairs.
- When two or more electron-dot formulas can be written for the same compound, the resulting formulas are called *resonance structures*.

♦ Learning Exercise 10.1A

List the number of bonds typically formed by the following atoms in covalent compounds:

a. N ________ **b.** S ________ **c.** P ________ **d.** C ________

e. Cl ________ **f.** O ________ **g.** H ________ **h.** F ________

Answers **a.** 3 **b.** 2 **c.** 3 **d.** 4
e. 1 **f.** 2 **g.** 1 **h.** 1

Guide to Drawing Electron-Dot Formulas	
Step 1	Determine the arrangement of atoms.
Step 2	Determine the total number of valence electrons.
Step 3	Attach each bonded atom to the central atom with a pair of electrons.
Step 4	Place the remaining electrons using single or multiple bonds to complete the octets.

Example: An electron-dot formula for SCl_2 can be drawn as follows:

Step 1: Determine the arrangement of atoms.

Cl S Cl

Step 2: Determine the total number of valence electrons.

$1 \text{ S and } 2 \text{ Cl} = 1(6\,e^-) + 2(7\,e^-) = 20$ valence electrons

Step 3: Attach each bonded atom to the central atom with a pair of electrons.

Cl—S—Cl $20\,e^- - 4\,e^- = 16$ remaining valence electrons (8 pairs)

Step 4: Place the remaining electrons using single or multiple bonds to complete the octets.

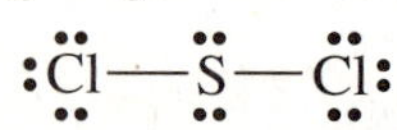

CORE CHEMISTRY SKILL
Drawing Electron-Dot Formulas

♦ Learning Exercise 10.1B

Draw the electron-dot formulas for each of the following covalent compounds:

H_2 NCl_3 HCl

Cl_2 H_2S CCl_4

Answers H:H    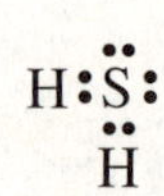 

Study Note

An electron-dot formula for CO_2 can be drawn with double bonds as follows:

Step 1 Determine the arrangement of atoms.

O C O

Step 2 Determine the total number of valence electrons.

$$1\text{ C and }2\text{ O} = 1(4\,e^-) + 2(6\,e^-) = 16\text{ valence electrons}$$

Step 3 Attach each bonded atom to the central atom with a pair of electrons. Each bonding pair can be represented by a bond line.

O—C—O $16\,e^- - 4\,e^- = 12$ remaining valence electrons (6 pairs)

Step 4 Place the remaining electrons using single or multiple bonds to complete octets.

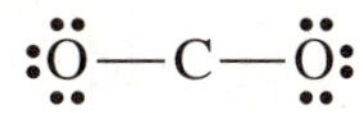

When octets cannot be completed with available valence electrons, one or more lone pairs are shared between the central atom and the bonded atoms to form double or triple bonds.

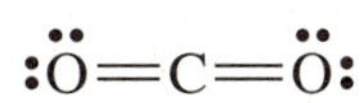

♦ Learning Exercise 10.1C

Each of the following compounds contains a multiple bond. Draw the electron-dot formulas.

a. CS_2

b. HCN

c. H_2CCH_2

d. HONO

Answers

a. 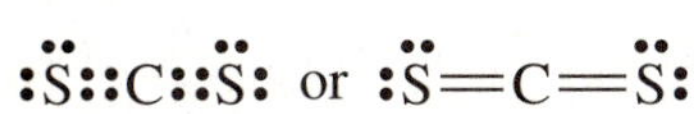

b. H:C:::N: or H—C≡N:

c. 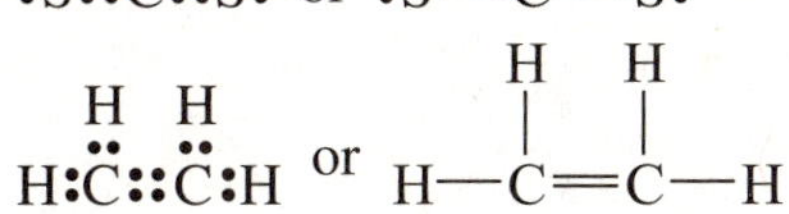

d. 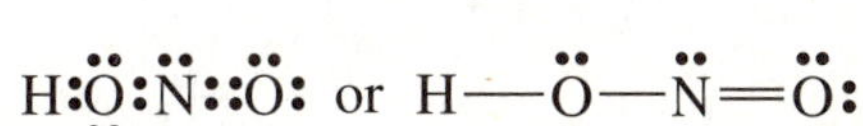

Study Note

When a molecule or polyatomic ion contains multiple bonds, it may be possible to draw more than one electron-dot formula. Draw two resonance structures for selenium dioxide, SeO_2.

Solution

Step 1 Determine the arrangement of atoms. In SeO_2, the Se atom is the central atom.

O Se O

Step 2 Determine the total number of valence electrons. We can use the group numbers to determine the valence electrons.

1 Se atom × 6 e^-	= 6 valence e^-
2 O atoms × 6 e^-	= 12 valence e^-
Total valence electrons	= 18 valence e^-

Step 3 Attach each bonded atom to the central atom with a pair of electrons. Four electrons are used to form two single bonds to connect the Se atom and each O atom. We will use a single line to represent a pair of bonding electrons.

O—Se—O

Step 4 Place the remaining electrons using single or multiple bonds to complete the octets. The remaining 14 electrons are drawn as lone pairs to complete the octets of the O atoms, but not the Se atom.

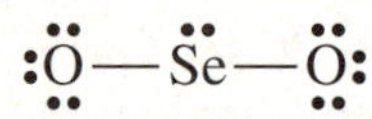

To complete the octet for Se, a lone pair from one of the O atoms is shared with Se to form a double bond. Because the lone pair that is shared can come from either O atom, two resonance structures can be drawn with a double-headed arrow shown between them.

♦ Learning Exercise 10.1D

CORE CHEMISTRY SKILL
Drawing Resonance Structures

When a polyatomic ion has a negative charge, electron(s) are added to the total number of valence electrons. Draw two or more resonance structures for each of the following:

a. SO_2

b. CO_3^{2-}

c. NCO^-

d. NO_3^-

Answers

a. 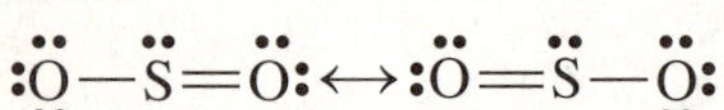

b. 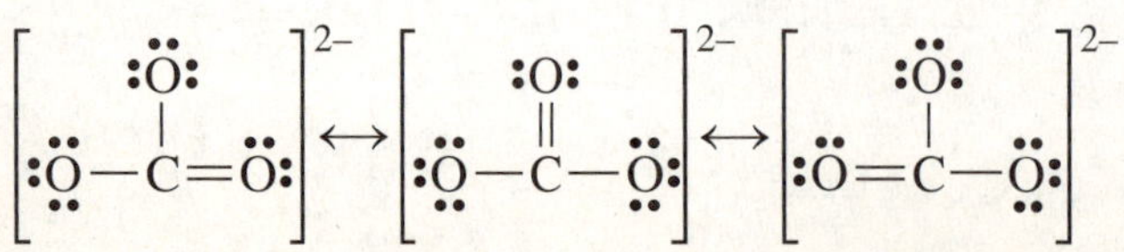

Resonance structures with charges are drawn in brackets with the charge written at the top right outside the bracket.

c. 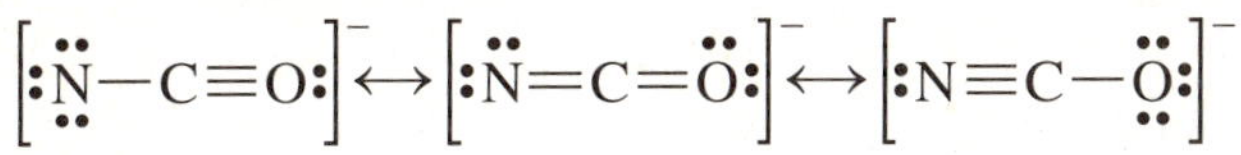

d. 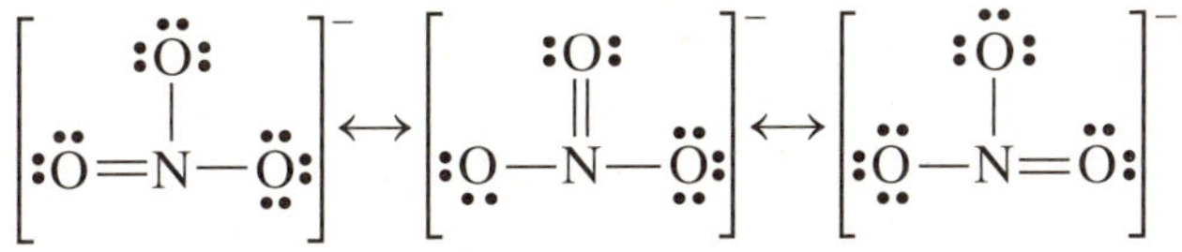

10.2 Shapes of Molecules and Ions (VSEPR Theory)

Learning Goal: Predict the three-dimensional structure of a molecule or polyatomic ion and classify it as polar or nonpolar.

- The VSEPR theory predicts the geometry of a molecule by placing the electron groups around a central atom as far apart as possible.
- One electron group may consist of the electrons in a lone pair, single, double, or triple bond.
- The atoms bonded to the central atom determine the three-dimensional geometry of a molecule or a polyatomic ion.
- A central atom with two electron groups has a linear geometry (180°); three electron groups give a trigonal planar geometry (120°); and four electron pairs give a tetrahedral geometry (109°).
- A linear molecule has a central atom bonded to two atoms and no lone pairs.

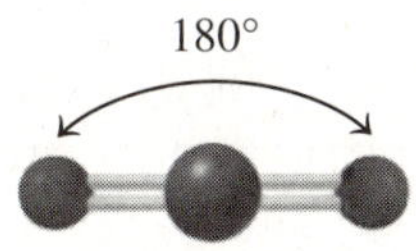

Linear shape

- A trigonal planar molecule has a central atom bonded to three atoms and no lone pairs. A bent molecule with a bond angle of 120° has a central atom bonded to two atoms and one lone pair.

Trigonal planar shape

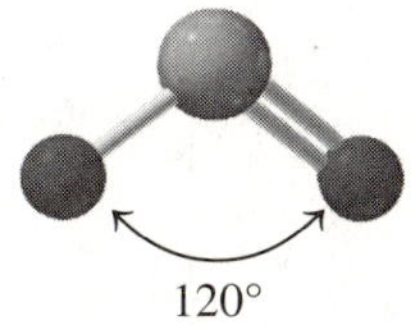

Bent shape

- A tetrahedral molecule has a central atom bonded to four atoms and no lone pairs. In a trigonal pyramidal molecule, a central atom is bonded to three atoms and one lone pair. In a bent molecule with a bond angle of 109°, a central atom is bonded to two atoms and two lone pairs.

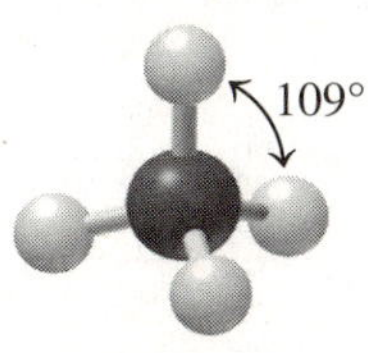

Tetrahedral shape

Trigonal pyramidal shape

Bent shape

♦ Learning Exercise 10.2A

Match the shape of a molecule with the following descriptions of the electron groups around the central atoms and the number of bonded atoms:

a. linear	**b.** trigonal planar	**c.** tetrahedral
d. trigonal pyramidal	**e.** bent (120°)	**f.** bent (109°)

1. three electron groups with three bonded atoms ________
2. two electron groups with two bonded atoms ________
3. four electron groups with three bonded atoms ________
4. three electron groups with two bonded atoms ________
5. four electron groups with four bonded atoms ________
6. four electron groups with two bonded atoms ________

Answers **1.** b **2.** a **3.** d **4.** e **5.** c **6.** f

Guide to Predicting Molecular Shape (VSEPR Theory)	
Step 1	Draw the electron-dot formula.
Step 2	Arrange the electron groups around the central atom to minimize repulsion.
Step 3	Use the atoms bonded to the central atom to determine the shape.

♦ Learning Exercise 10.2B

CORE CHEMISTRY SKILL
Predicting Shape

For each of the following, draw the electron-dot formula, state the number of electron groups and bonded atoms, and predict the shape and bond angles of each of the following molecules or ions:

Molecule or Ion	Electron-dot Formula	Number of Electron Groups	Number of Bonded Atoms	Shape and Bond Angle
CH_4				
PCl_3				
SO_3				
H_2S				
NO_2^-				

Answers

Molecule or Ion	Electron-dot Formula	Number of Electron Groups	Number of Bonded Atoms	Shape and Bond Angle
CH_4	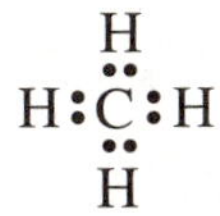	4	4	Tetrahedral, 109°
PCl_3	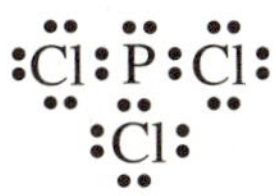	4	3	Trigonal pyramidal, 109°
SO_3	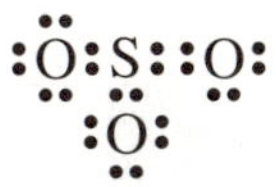	3	3	Trigonal planar, 120°
H_2S		4	2	Bent, 109°
NO_2^-		3	2	Bent, 120°

10.3 Electronegativity and Polarity

Learning Goal: Use electronegativity to determine the polarity of a bond or a molecule.

- Electronegativity values indicate the ability of an atom to attract electrons in a chemical bond. In general, metals have low electronegativity values and nonmetals have high values.

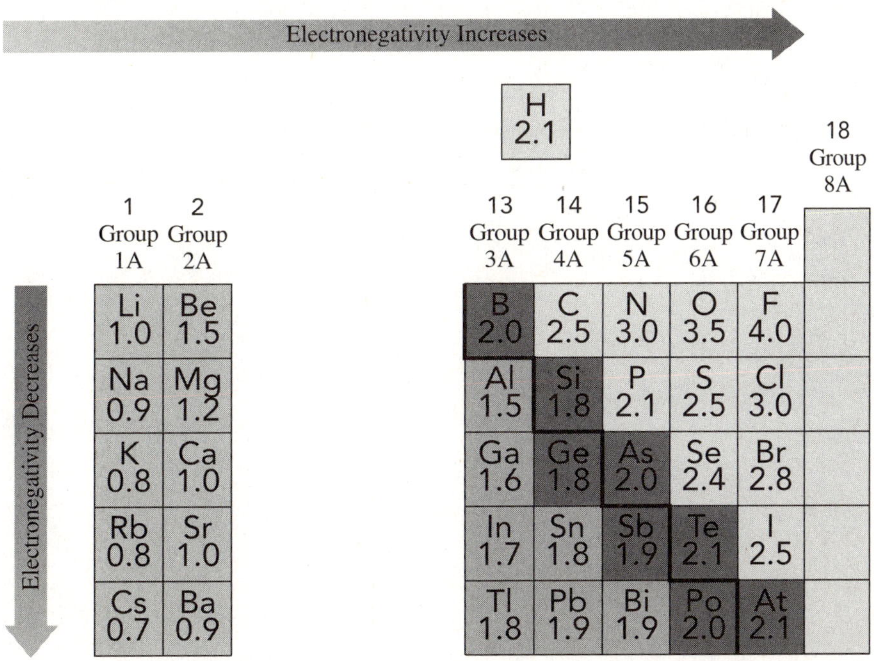

- When atoms sharing electrons have the same or similar electronegativity values, electrons are shared equally and the bond is *nonpolar covalent*.

- Electrons are shared unequally in *polar covalent* bonds because they are attracted to the more electronegative atom.
- An electronegativity difference of 0 to 0.4 indicates a nonpolar covalent bond, whereas a difference of 0.5 to 1.8 indicates a polar covalent bond.
- An electronegativity difference greater than 1.8 indicates a bond that is ionic.
- A polar bond has a charge separation, which is called a dipole; the positive end of the dipole is marked as δ^+ and the negative end as δ^-.
- Nonpolar molecules can have polar bonds when the dipoles are in a symmetric arrangement and the bond dipoles cancel each other.
- In polar molecules, the bond dipoles do not cancel each other.

♦ Learning Exercise 10.3A

CORE CHEMISTRY SKILL
Using Electronegativity

Using the electronegativity values, determine the following:

1. the electronegativity difference for each pair
2. the type of bonding as (I) ionic, (PC) polar covalent, or (NC) nonpolar covalent

Elements	Electronegativity Difference	Bonding	Elements	Electronegativity Difference	Bonding
a. H and O	______	______	**b.** N and S	______	______
c. Al and O	______	______	**d.** Li and F	______	______
e. H and Cl	______	______	**f.** Cl and Cl	______	______
g. S and F	______	______	**h.** H and C	______	______

Answers

a. 1.4, PC	**b.** 0.5, PC	**c.** 2.0, I	**d.** 3.0, I
e. 0.9, PC	**f.** 0, NP	**g.** 1.5, PC	**h.** 0.4, NP

♦ Learning Exercise 10.3B

Write the symbols δ^+ and δ^- over the appropriate atoms in polar bonds.

1. H—O **2.** N—H **3.** C—Cl

4. O—F **5.** N—F **6.** P—Cl

Answers

1. $\overset{\delta^+}{\mathrm{H}}$—$\overset{\delta^-}{\mathrm{O}}$ **2.** $\overset{\delta^-}{\mathrm{N}}$—$\overset{\delta^+}{\mathrm{H}}$ **3.** $\overset{\delta^+}{\mathrm{C}}$—$\overset{\delta^-}{\mathrm{Cl}}$

4. $\overset{\delta^+}{\mathrm{O}}$—$\overset{\delta^-}{\mathrm{F}}$ **5.** $\overset{\delta^+}{\mathrm{N}}$—$\overset{\delta^-}{\mathrm{F}}$ **6.** $\overset{\delta^+}{\mathrm{P}}$—$\overset{\delta^-}{\mathrm{Cl}}$

Guide to Determination of Polarity of a Molecule	
Step 1	Determine if the bonds are polar covalent or nonpolar covalent.
Step 2	If the bonds are polar covalent, draw the electron-dot formula and determine if the dipoles cancel.

CORE CHEMISTRY SKILL
Identifying Polarity of Molecules

♦ Learning Exercise 10.4C

Using the dipole arrows, draw the bond dipoles in each of the following molecules and determine whether the molecule is polar or nonpolar:

1. CF_4
2. HCl
3. NH_3
4. OF_2

Answers

1. CF_4

```
   F
   ↑
F←C→F
   ↓
   F
```

bond dipoles cancel, nonpolar

2. HCl $\overset{\delta^+}{H}\rightarrow\overset{\delta^-}{Cl}$ ⟼ bond dipole does not cancel, polar

3. NH_3

```
   ••
H→N←H
   ↑
   H
```

(net dipole pointing up) bond dipoles do not cancel, polar

4. OF_2

```
 ••
:O→F
 ↓
 F
```

(net dipole pointing down and to the right) bond dipoles do not cancel, polar

10.4 Attractive Forces In Compounds

Learning Goal: Describe the attractive forces between ions, polar covalent molecules, and nonpolar covalent molecules.

- The attractive forces between particles in solids and liquids determine their melting and boiling points.
- Ionic solids have high melting points due to strong ionic interactions between positive and negative ions.
- In polar substances, dipole–dipole attractions occur between the positive end of one molecule and the negative end of another.

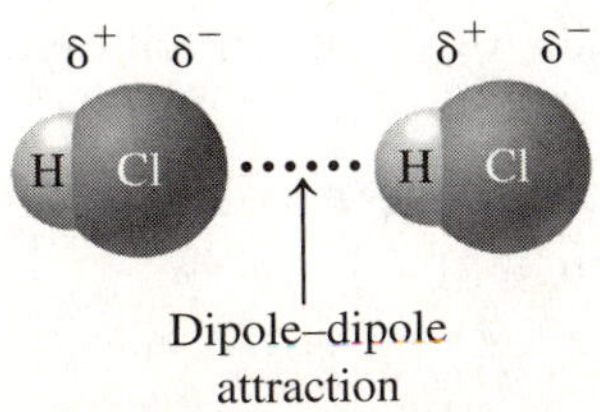

- Hydrogen bonding, a particularly strong type of dipole–dipole attraction, occurs between partially positive hydrogen atoms and the strongly electronegative atoms F, O, or N.

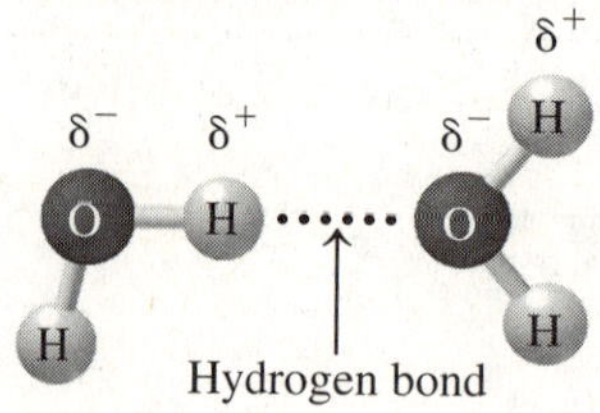

- Dispersion forces occur when temporary dipoles form within the nonpolar molecules, causing weak attractions to other nonpolar molecules. Dispersion forces are the only attractive force that occur in nonpolar molecules. Dispersion forces occur in polar molecules, but they are not the strongest type of attractive force.

♦ Learning Exercise 10.4A

CORE CHEMISTRY SKILL
Identifying Attractive Forces

Indicate the major type of attractive force that occurs in each of the following substances:

a. ionic **b.** dipole–dipole attractions **c.** hydrogen bond **d.** dispersion forces

1. ________ KCl **2.** ________ NF_3 **3.** ________ SBr_2 **4.** ________ Cl_2

5. ________ HF **6.** ________ H_2O **7.** ________ C_4H_{10} **8.** ________ Na_2O

Answers **1.** a **2.** b **3.** b **4.** d
5. c **6.** c **7.** d **8.** a

♦ Learning Exercise 10.4B

Identify the substance that would have the higher boiling point in each pair and give the major type of attractive force:

1. NaCl or HCl ________ **2.** Br_2 or HBr ________ **3.** H_2O or H_2S ________

4. C_2H_6 or C_8H_{18} ________ **5.** $MgCl_2$ or OCl_2 ________ **6.** NH_3 or PH_3 ________

Answers
1. NaCl (ionic)
2. HBr (dipole–dipole) attractions
3. H_2O (hydrogen bonding)
4. C_8H_{18} (dispersion forces are stronger in the larger molecule)
5. $MgCl_2$ (ionic)
6. NH_3 (hydrogen bonding)

10.5 Changes of State

Learning Goal: Describe the changes of state between solids, liquids, and gases; calculate the energy involved.

- The three states of matter are solid, liquid, and gas.
- Physical properties such as shape, state, or color can change without affecting the identity of a substance.
- A substance melts and freezes at its melting (freezing) point. During the process of melting or freezing, the temperature remains constant.

- The *heat of fusion* is the energy required to change 1 g of solid to liquid at the melting point. For ice to melt at 0 °C, 334 J/g is required. This quantity is called the heat of fusion. This is also the amount of heat lost when 1 g of water freezes at 0 °C.
- A substance boils and condenses at its boiling point. During the process of boiling or condensing, the temperature remains constant.
- The *heat of vaporization* is the energy required to change 1 g of liquid to 1 g of gas at the boiling point.
- When water boils at 100 °C, 2260 J is required to change 1 g of liquid to 1 g of gas (steam); it is also the amount of heat released when 1 g of water vapor condenses at 100 °C.
- Evaporation is a surface phenomenon while boiling occurs throughout the liquid.
- Sublimation is the change of state from a solid directly to a gas. Deposition is the opposite process.
- A heating or cooling curve illustrates the changes in temperature and state as heat is added to or removed from a substance. An example of a cooling curve for water is shown below.

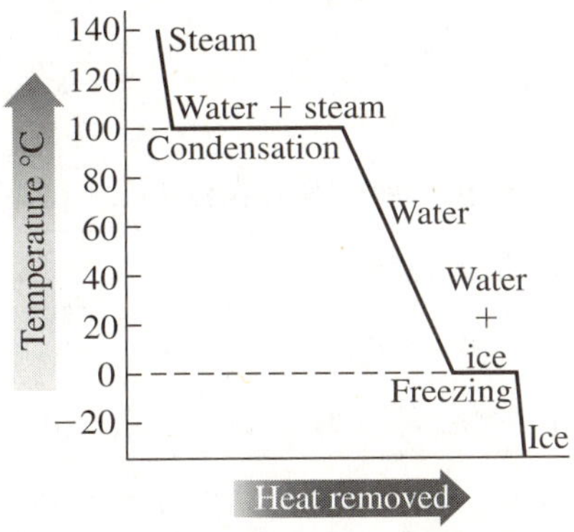

A cooling curve for water illustrates the change in temperature and changes of state as heat is removed.

- When a substance is heated or cooled, the total energy gained or released is the total energy involved in temperature changes as well as the energy involved in changes of state.

Key Terms for Sections 10.1–10.5

Match each the following key terms with the correct definition:

a. polar covalent bond **b.** multiple bond **c.** VSEPR theory
d. nonpolar covalent bond **e.** resonance **f.** dipole–dipole attraction
g. deposition **h.** dispersion forces

1. _____ equal sharing of valence electrons by two atoms

2. _____ occurs when two or more electron-dot formulas can be written for the same compound

3. _____ the unequal attraction for shared electrons in a covalent bond

4. _____ the atoms in a molecule are arranged to minimize repulsion between electrons

5. _____ the sharing of two or three pairs of electrons by two atoms

6. _____ interaction between the positive end of a polar molecule and the negative end of another

7. _____ results from momentary polarization of nonpolar molecules

8. _____ change of a gas directly to a solid

Answers **1.** d **2.** e **3.** a **4.** c **5.** b
6. f **7.** i **8.** g

♦ Learning Exercise 10.5A

Identify each of the following as

1. melting **2.** freezing **3.** sublimation **4.** condensation

a. ________ A liquid changes to a solid.

b. ________ Ice forms on the surface of a lake in winter.

c. ________ Dry ice in an ice cream cart changes to a gas.

d. ________ Butter in a hot pan turns to liquid.

e. ________ A gas changes to a liquid.

Answers **a.** 2 **b.** 2 **c.** 3 **d.** 1 **e.** 4

Guide to Calculations Using Heat of Fusion	
Step 1	State the given and needed quantities.
Step 2	Write a plan to convert the given quantity to the needed quantity.
Step 3	Write the heat of fusion factor and any metric factor needed.
Step 4	Set up the problem and calculate the needed quantity.

Study Note

The amount of heat needed or released during melting or freezing can be calculated using the heat of fusion: Heat = mass × heat of fusion

♦ Learning Exercise 10.5B

CORE CHEMISTRY SKILL
Calculating Heat or Change of State

Calculate each of the following when a substance melts or freezes:

a. How many joules of heat are needed to melt 24.0 g of ice at 0 °C?

b. How much heat, in kilojoules, is released when 325 g of water freezes at 0 °C?

c. How many joules of heat are required to melt 125 g of ice at 0 °C?

d. How many grams of ice would melt when 1680 J of heat were absorbed?

Answers **a.** 8020 J **b.** 109 kJ **c.** 41 800 J **d.** 5.03 g

Guide to Calculations Using Heat of Vaporization	
Step 1	State the given and needed quantities.
Step 2	Write a plan to convert the given quantity to the needed quantity.
Step 3	Write the heat of vaporization factor and any metric factor needed.
Step 4	Set up the problem and calculate the needed quantity.

♦ Learning Exercise 10.5C

Calculate each of the following when a substance boils or condenses:

a. How many joules are needed to completely change 5.00 g of water to vapor at 100 °C?

b. How many kilojoules are released when 515 g of steam at 100 °C condenses to form liquid water at 100 °C?

c. How many grams of water can be converted to steam at 100 °C when 155 kJ of energy is absorbed?

d. How many grams of steam condense if 24 800 J is released at 100 °C?

Answers **a.** 11 300 J **b.** 1160 kJ **c.** 68.6 g **d.** 11.0 g

♦ Learning Exercise 10.5D

On each heating or cooling curve, indicate the portion that corresponds to a solid, liquid, or gas and the changes in state.

1. Draw a heating curve for water that begins at −20 °C and ends at 120 °C. Water has a melting point of 0 °C and a boiling point of 100 °C.

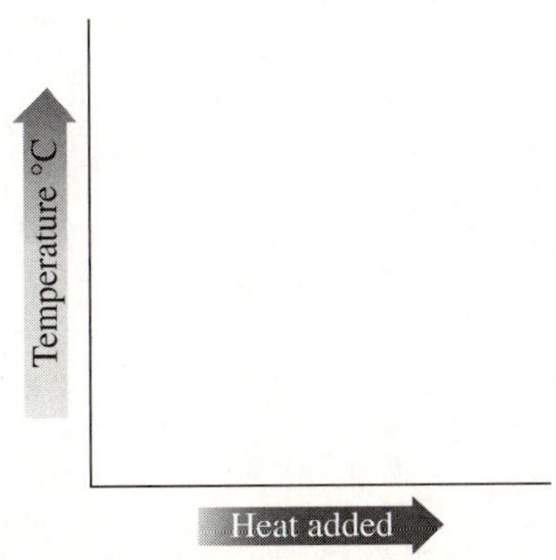

2. Draw a heating curve for bromine from −25 °C to 75 °C. Bromine has a melting point of −7 °C and a boiling point of 59 °C.

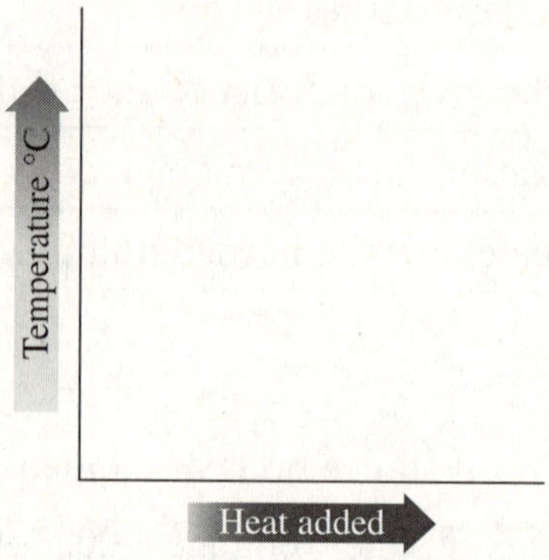

3. Draw a cooling curve for sodium from 1000 °C to 0 °C. Sodium has a freezing point of 98 °C and a boiling (condensation) point of 883 °C.

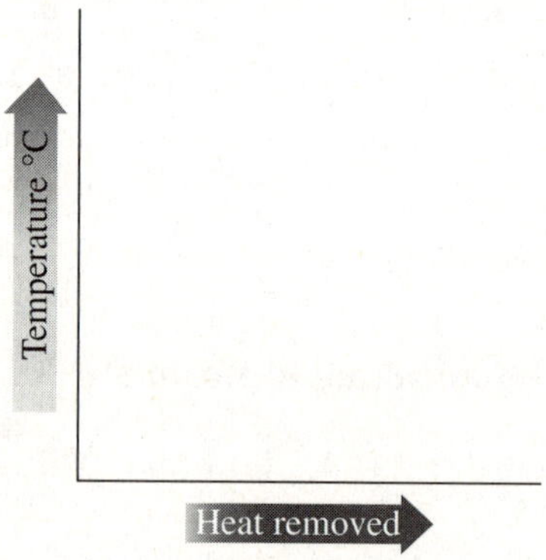

Answers

1.

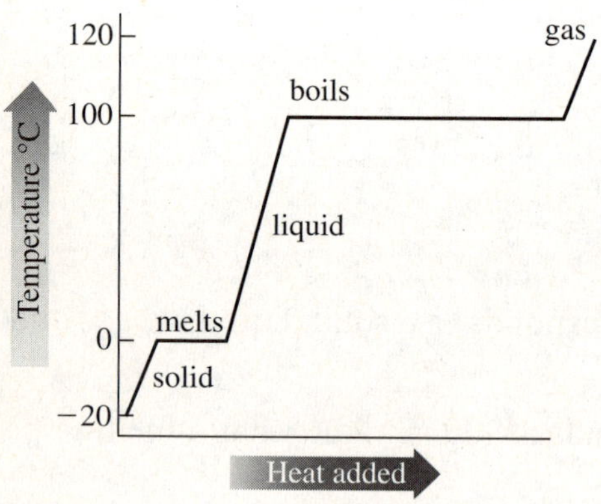

2.

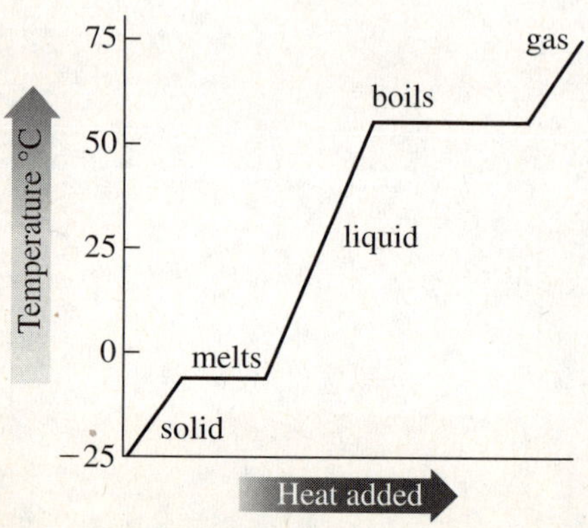

3.

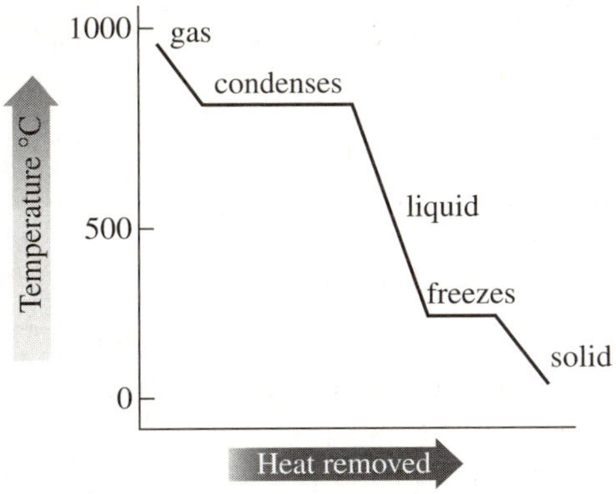

♦ Learning Check 10.5E

From the heat equation, we can calculate the energy change for a given mass, temperature change, and specific heat.

$$\text{Heat} = \text{mass (g)} \times \text{temperature change (°C)} \times \text{specific heat (J/g °C)}$$

a. How many kilojoules are released when 35.0 g of water at 65.0 °C is cooled to 0.0 °C and changed to solid? (*Hint:* Two steps are needed.)

b. How many kilojoules are needed to melt 15.0 g of ice at 0. °C, heat the water to 100. °C, and convert the water to gas at 100. °C? (*Hint:* Three steps are needed.)

c. Calculate the number of kilojoules released when 25.0 g of steam at 100. °C condenses, then cools, and finally freezes at 0. °C. (*Hint:* Three steps are needed.)

Answers **a.** 21.2 kJ **b.** 45.2 kJ **c.** 75.3 kJ

Checklist for Chapter 10

You are ready to take the Practice Test for Chapter 10. Be sure you have accomplished the following learning goals for this chapter. If not, review the section listed at the end of the goal. Then apply your new skills and understanding to the Practice Test.

After studying Chapter 10, I can successfully:

_____ Draw an electron-dot formula of a covalent compound or polyatomic ion with multiple bonds (10.1).

_____ Draw resonance structures for an electron-dot formula when two or more electron-dot formulas are possible (10.1).

_____ Predict the shape and bond angles for a molecule or polyatomic ion (10.2).

_____ Determine the electronegativity difference for two atoms in a compound (10.3).

_____ Classify a bond as nonpolar covalent, polar covalent, or ionic (10.3).

_____ Classify a molecule as polar or nonpolar (10.3).

_____ Identify the attractive forces between particles in a liquid or solid and predict relative melting and boiling points. (10.4).

_____ Calculate the heat change when a specific amount of a substance changes temperature (10.5).

_____ Calculate the heat change for a change of state for a specific amount of a substance (10.5).

_____ Calculate the total heat for a combination of change of temperature and change of state for a specific amount of a substance (10.5).

_____ Draw heating and cooling curves using the melting and boiling points of a substance (10.5).

Practice Test for Chapter 10

1. A polar covalent bond typically occurs between (10.3)
   A. two different nonmetals
   B. two of the same type of nonmetals
   C. two noble gases
   D. two different metals
   E. a metal and a nonmetal

For questions 2 through 7, indicate the type of bonding expected between the following elements: (10.3)

I. ionic **N.** nonpolar covalent **P.** polar covalent

2. ________ silicon and oxygen **3.** ________ barium and chlorine

4. ________ aluminum and fluorine **5.** ________ chlorine and chlorine

6. ________ sulfur and oxygen **7.** ________ nitrogen and oxygen

For questions 8 through 13, match each the following shapes with one of the descriptions: (10.2)

A. linear **B.** trigonal planar **C.** tetrahedral
D. trigonal pyramidal **E.** bent (120°) **F.** bent (109°)

8. has a central atom with three electron groups and three bonded atoms ________

9. has a central atom with two electron groups and two bonded atoms ________

10. has a central atom with four electron groups and three bonded atoms ________

11. has a central atom with three electron groups and two bonded atoms ________

12. has a central atom with four electron groups and four bonded atoms ________

13. has a central atom with four electron groups and two bonded atoms ________

For questions 14 through 18, determine the shape and bond angles for each of the following molecules: (10.2)

A. linear, 180° **B.** trigonal planar, 120° **C.** bent, 120°
D. tetrahedral, 109° **E.** trigonal pyramidal, 109° **F.** bent, 109°

14. PCl_3 **15.** CBr_4 **16.** H_2S **17.** BrO_2^- **18.** ClO_3^-

For questions 19 through 26, indicate the major type of interactive force (A-D) for each of the following: (10.4)

A. ionic bonds **B.** dipole—dipole attractions
C. hydrogen bonds **D.** dispersion forces

19. _____ KCl **20.** _____ NCl_3

21. _____ SF_2 **22.** _____ Cl_2

23. _____ HF **24.** _____ H_2O

25. _____ C_4H_{10} **26.** _____ Na_2O

For questions 27 through 31, match items A to E with one the statements: (10.5)

A. energy **B.** evaporation **C.** heat of fusion
D. heat of vaporization **E.** boiling

27. _________ the energy required to convert a gram of solid to liquid

28. _________ the heat needed to convert a gram of liquid to gas

29. _________ the conversion of liquid molecules to gas at the surface of a liquid

30. _________ the ability to do work

31. _________ the formation of a gas within the liquid as well as on the surface

32. Ice cools down a drink because (10.5)
A. the ice is warmer than the drink and heat flows out of the ice cubes
B. heat is absorbed from the drink to melt the ice cubes
C. the heat of fusion of the ice is higher than the heat of fusion for water
D. both A and B
E. none of the above

33. A can full of steam is tightly stoppered. As the can cools (10.5)
A. nothing will happen
B. the steam will blow up the can
C. the steam condenses and the can collapses
D. the steam condenses and the water formed will expand and explode the can
E. none of the above

34. The number of joules needed to convert 15.0 g of ice to liquid at 0 °C is (10.5)
A. 73 J **B.** 334 J **C.** 2260 J **D.** 4670 J **E.** 5010 J

35. The number of joules released when 2.0 g of water at 50. °C are cooled and frozen at 0. °C is (10.5)
A. 420 J **B.** 670 J **C.** 1100 J **D.** 2100 J **E.** 3400 J

36. What is the total number of kilojoules required to convert 25 g of ice at 0. °C to gas at 100. °C? (10.5)
A. 8.4 kJ **B.** 19 kJ **C.** 59 kJ **D.** 67 kJ **E.** 75 kJ

For questions 37 through 40, consider the heating curve below for p-toluidine. Answer the following questions when heat is added to p-toluidine at −20 °C, *where toluidine is below its melting point*: (10.5)

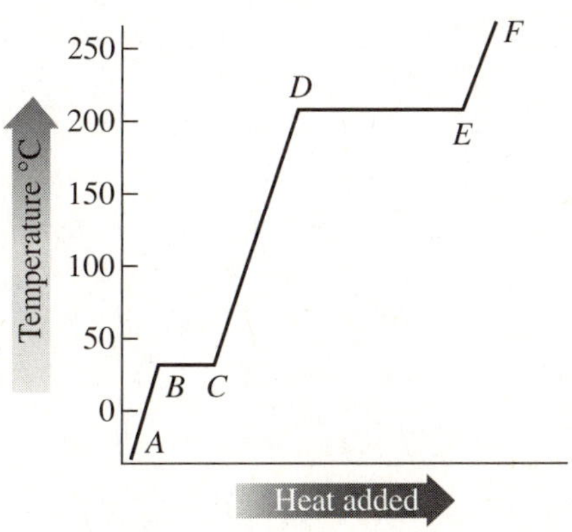

37. On the heating curve, segment ***BC*** indicates
A. solid **B.** melting **C.** liquid **D.** boiling **E.** gas

38. On the heating curve, segment ***CD*** shows toluidine as
A. solid **B.** melting **C.** liquid **D.** boiling **E.** gas

39. The boiling point of toluidine would be about
A. 20 °C **B.** 45 °C **C.** 100 °C **D.** 200 °C **E.** 250 °C

40. On the heating curve, segment ***EF*** shows toluidine as
A. solid **B.** melting **C.** liquid **D.** boiling **E.** gas

For questions 41 through 45, draw the electron-dot formula of CO_2: (10.1, 10.2, 10.3)

41. The correct number of valence electrons needed in the electron-dot formula of CO_2 is
A. 10 **B.** 12 **C.** 14 **D.** 16 **E.** 18

42. The correct number of single bonds is
A. 0 **B.** 1 **C.** 2 **D.** 3 **E.** 4

43. The correct number of double bonds is
A. 0 **B.** 1 **C.** 2 **D.** 3 **E.** 4

44. The molecule is
A. nonpolar covalent **B.** polar covalent **C.** ionic
D. trigonal planar **E.** tetrahedral

45. The shape of the molecule is
A. linear **B.** bent 120° **C.** bent 109°
D. trigonal planar **E.** tetrahedral

Answers to the Practice Test

1. A	**2.** P	**3.** I	**4.** I	**5.** N
6. P	**7.** P	**8.** B	**9.** A	**10.** D
11. E	**12.** C	**13.** F	**14.** E	**15.** D
16. F	**17.** F	**18.** E	**19.** A	**20.** B
21. B	**22.** D	**23.** C	**24.** C	**25.** D
26. A	**27.** C	**28.** D	**29.** B	**30.** A
31. E	**32.** B	**33.** C	**34.** E	**35.** C
36. E	**37.** B	**38.** C	**39.** D	**40.** E
41. D	**42.** A	**43.** C	**44.** A	**45.** A

11
Gases

After Whitney's asthma attack, she and her parents learned to identify and treat her asthma. Because her asthma is mild and intermittent, they follow a management plan to control the symptoms. Whitney participates in physical activity and exercise to maintain lung function and to prevent recurrent attacks. If her symptoms worsen, she uses a bronchodilator to help her breathing. If a child has a lung capacity of 2.0 L, a body temperature of 37 °C, and a lung pressure of 740 mmHg, how many grams of air are in the lungs if air has a molar mass of 29 g/mole?

11.1 Properties of Gases

Learning Goal: Describe the kinetic molecular theory of gases and the properties of gases.

- In a gas, particles are far apart and moving so fast that they are not attracted to each other.
- A gas is described by the physical properties of pressure (P), volume (V), temperature (T), and amount in moles (n).

♦ Learning Exercise 11.1

Answer true (T) or false (F) for each of the following:

a. ________ Gases are composed of small particles.

b. ________ Gas molecules are usually close together.

c. ________ Gas molecules move rapidly because they are strongly attracted.

d. ________ Distances between gas molecules are large.

e. ________ Gas molecules travel in straight lines until they collide.

f. ________ Pressure is the force of gas particles striking the walls of a container.

g. ________ Kinetic energy decreases with increasing temperature.

Answers **a.** T **b.** F **c.** F **d.** T
e. T **f.** T **g.** F

11.2 Gas Pressure

Learning Goal: Describe the units of measurement used for pressure, and change from one unit to another.

- A gas exerts pressure, which is the force that gas particles exert on the walls of a container.
- Units of gas pressure include torr, mmHg, Pascal (Pa), kilopascal (kPa), lb/in.^{2} (psi), and atm.
- Gas pressure can be converted from one unit to another using conversion factors.

♦ Learning Exercise 11.2

Complete the following:

a. 1.50 atm = ____________ mmHg

b. 150 mmHg = ____________ atm

c. 0.725 atm = ____________ kPa

d. 1520 mmHg = ____________ atm

e. 30.5 psi = ____________ mmHg

f. During the weather report on TV, the pressure is given as 29.4 in. (of mercury). What is this pressure in mmHg and in atm?

Answers **a.** 1140 mmHg **b.** 0.20 atm **c.** 73.5 kPa
d. 2.00 atm **e.** 1580 mmHg **f.** 747 mmHg; 0.983 atm

11.3 Pressure and Volume (Boyle's Law)

Learning Goal: Use the pressure—volume relationship (Boyle's law) to determine the final pressure or volume when the temperature and amount of gas are constant.

- According to Boyle's law, pressure increases if volume decreases and pressure decreases if volume increases.
- If two properties change in opposite directions, the properties have an *inverse relationship*.
- The volume (V) of a gas changes inversely with the pressure (P) of the gas when T and n are held constant: $P_1V_1 = P_2V_2$. The subscripts 1 and 2 represent initial and final conditions.

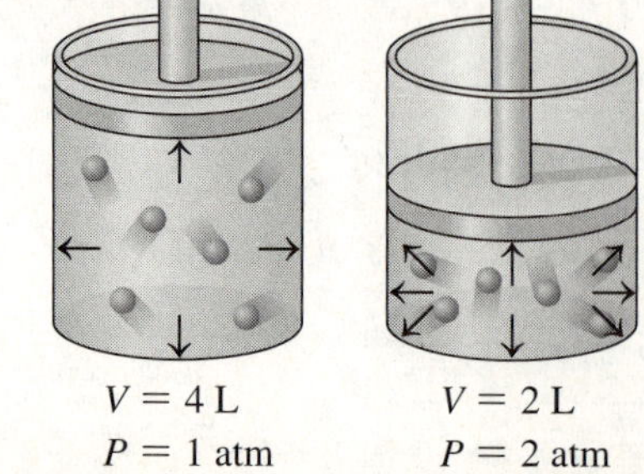

Boyle's law: As volume decreases, gas molecules become more crowded, causing the pressure to increase. Pressure and volume are inversely related. Gas molecules become more crowded, causing the pressure to increase. Pressure and volume are inversely related.

Guide to Using the Gas Laws	
Step 1	Organize the data in a table of initial and final conditions.
Step 2	Rearrange the gas law equation to solve for the unknown quantity.
Step 3	Substitute values into the gas law equation and claculate.

Key Terms for Sections 11.1–11.3

Match the key terms with the correct definition below.

a. kinetic molecular theory **b.** inverse relationship **c.** Boyle's law
d. pressure **e.** atmospheric pressure

1. ________ the pressure exerted by the atmosphere
2. ________ a force exerted by gas particles when they collide with the sides of a container
3. ________ relationship in which two properties change in opposite directions
4. ________ the volume of a gas varies inversely with the pressure of a gas (when T and n are constant)
5. ________ the model that explains the behavior of gaseous particles

Answers **1.** e **2.** d **3.** b **4.** c **5.** a

♦ Learning Exercise 11.3A

Complete each of the following with *increases* or *decreases*:

1. Gas pressure increases (T and n constant) when volume ________________.
2. Gas volume increases at constant T and n when pressure ________________.

Answers **1.** decreases **2.** decreases

♦ Learning Exercise 11.3B

CORE CHEMISTRY SKILL
Using the Gas Law

Calculate the unknown quantity in each of the following gas problems using Boyle's law:

a. A sample of 4.0 L of helium gas has a pressure of 800. mmHg. What is the new pressure, in millimeters of mercury, when the volume is reduced to 1.0 L (n and T constant)?

b. A gas occupies a volume of 360 mL at 750 mmHg. What volume, in milliliters, does it occupy at a pressure of (1) 1500 mmHg? (2) 375 mmHg (n and T constant)?

c. A gas sample at a pressure of 5.00 atm has a volume of 3.00 L. If the gas pressure is changed to 760. mmHg, what volume, in liters, will the gas occupy (n and T constant)?

d. A sample of 250. mL of nitrogen is initially at a pressure of 2.50 atm. If the pressure changes to 825 mmHg, what is the new volume, in mL (n and T constant)?

Answers **a.** 3200 mmHg **b.** (1) 180 mL (2) 720 mL **c.** 15.0 L **d.** 576 mL

11.4 Temperature and Volume (Charles's Law)

Learning Goal: Use the temperature—volume relationship (Charles's law) to determine the final temperature or volume when the pressure and amount of gas are constant.

- According to Charles's law, at constant pressure and quantity, the volume of a gas increases when the temperature of the gas increases, and the volume decreases when the temperature decreases.
- In a direct relationship, two properties increase or decrease together.
- The volume (V) of a gas is directly related to its Kelvin temperature (T) when there is no change in the pressure or moles of the gas:

$$\frac{V_1}{T_1} = \frac{V_2}{T_2}$$ The subscripts 1 and 2 represent initial and final conditions.

♦ Learning Exercise 11.4A

Complete each of the following with *increases* or *decreases*:

a. When the temperature of a gas increases at constant pressure and amount, its volume ________.

b. When the volume of a gas decreases at constant pressure and amount, its temperature ________.

Answers **a.** increases **b.** decreases

♦ Learning Exercise 11.4B

Solve each of the following gas law problems using Charles's law:

a. A balloon has a volume of 2.5 L at a temperature of 0 °C. What is the new volume, in liters, of the balloon when the temperature rises to 120 °C and the pressure and number of moles remain constant?

b. Consider a balloon filled with helium to a volume of 6600 L at a temperature of 223 °C. To what temperature, in degrees Celsius, must the gas be cooled to decrease the volume to 4800 L (P and n constant)?

c. A sample of 750 mL of neon is heated from 120 °C to 350 °C. If pressure and amount remain constant, what is the new volume, in milliliters?

d. What is the final temperature, in degrees Celsius, of 350. mL of oxygen gas at 22 °C if its volume increases to 0.800 L (P and n constant)?

Answers **a.** 3.6 L **b.** 88 °C **c.** 1200 mL **d.** 401 °C

11.5 Temperature and Pressure (Gay-Lussac's Law)

Learning Goal: Use the temperature—pressure relationship (Gay-Lussac's law) to determine the final temperature or pressure when the volume and amount of gas are constant.

- According to Gay-Lussac's law, at constant volume and amount, the gas pressure increases when the temperature of the gas increases, and the pressure decreases when the temperature decreases.
- The pressure (P) of a gas is directly related to its Kelvin temperature (T), when n and V are constant:

  $$\frac{P_1}{T_1} = \frac{P_2}{T_2}$$ The subscripts 1 and 2 represent initial and final conditions.

  This relationship means that an increase in temperature increases the pressure of a gas, or a decrease in the temperature decreases the pressure, as long as the volume and amount of gas stays constant.
- Vapor pressure is the pressure of the gas that forms when a liquid evaporates. At the boiling point of a liquid, the vapor pressure equals the atmospheric pressure.

♦ Learning Exercise 11.5A

Solve the following gas law problems using Gay-Lussac's law:

a. A sample of helium gas has a pressure of 860. mmHg at a temperature of 225 K. What is the new pressure, in millimeters of mercury, when the helium sample reaches a temperature of 675 K (V and n constant)?

b. A balloon contains a gas with a pressure of 580. mmHg and a temperature of 227 °C. What is the new pressure, in millimeters of mercury, of the gas when the temperature drops to 27 °C (*V* and *n* constant)?

c. A spray can contains a gas with a pressure of 3.0 atm at a temperature of 17 °C. What is the pressure, in atmospheres, inside the container if the temperature rises to 110 °C (*V* and *n* constant)?

d. A gas has a pressure of 1200. mmHg at 300. °C. What will the temperature, in degrees Celsius, be when the pressure falls to 1.10 atm (*V* and *n* constant)?

Answers **a.** 2580 mmHg **b.** 348 mmHg **c.** 4.0 atm **d.** 126 °C

♦ Learning Exercise 11.5B

Explain how a liquid can have a boiling point of 80 °C at sea level and a boiling point of 74 °C at an altitude of 1000 m.

Answer The vapor pressure required for boiling is lower at higher altitude because the atmospheric pressure is lower. As a result, the substance boils at a lower temperature.

11.6 The Combined Gas Law

Learning Goal: Use the combined gas law to find the final pressure, volume, or temperature of a gas when changes in two of these properties are given and the amount of gas is constant.

- At constant amount of a gas, the gas laws can be combined into a relationship of pressure (*P*), volume (*V*), and temperature (*T*):

$$\frac{P_1V_1}{T_1} = \frac{P_2V_2}{T_2}$$ Subscripts 1 and 2 represent initial and final conditions.

Key Terms for Section 11.4–11.6

Match the key terms with the correct definition below.

a. vapor pressure **b.** direct relationship **c.** Charles's law
d. combined gas law **e.** Gay-Lussac's law

1. ______ a gas law that states that the volume of a gas changes directly with a change in Kelvin temperature when pressure and amount (moles) of the gas do not change

2. ______ pressure exerted by the particles of vapor above a liquid

3. _____ relationship in which two properties change in the same direction

4. _____ a gas law stating that the pressure of a gas changes directly with a change in Kelvin temperature when the number of moles of a gas and its volume remain constant

5. _____ relationship that combines several gas laws relating pressure, volume, and temperature when the amount of gas remains constant

Answers **1.** c **2.** a **3.** b **4.** e **5.** d

♦ Learning Exercise 11.6

Solve each of the following using the combined gas law (n is constant):

a. A 4.0-L sample of nitrogen gas has a pressure of 1200 mmHg at 220 K. What is the pressure, in millimeters of mercury, of the sample when the volume increases to 20. L at 440 K?

b. A 25.0-mL bubble forms at the ocean depths where the pressure is 10.0 atm and the temperature is 5.0 °C. What is the volume, in milliliters, of that bubble at the ocean surface where the pressure is 760. mmHg and the temperature is 25 °C?

c. A 35.0-mL sample of argon gas has a pressure of 1.0 atm and a temperature of 15 °C. What is the final volume, in milliliters, if the pressure goes to 2.0 atm and the temperature to 45 °C?

d. A weather balloon with a volume of 315 L is launched at the Earth's surface where the temperature is 12 °C and the pressure is 0.930 atm. What is the volume, in liters, of the balloon in the upper atmosphere where the pressure is 116 mmHg and the temperature is −35 °C?

e. A 10.0-L sample of gas is emitted from a volcano with a pressure of 1.20 atm and a temperature of 150.0 °C. What is the volume, in liters, of the gas when its pressure is 0.900 atm and the temperature is −40.0 °C?

Answers **a.** 480 mmHg **b.** 268 mL **c.** 19 mL
d. 1.60×10^3 L **e.** 7.34 L

11.7 Volume and Moles (Avogadro's Law)

Learning Goal: Use Avogadro's law to determine the amount or volume of a gas when the pressure and temperature are constant.

- If the number of moles of gas increases, the volume increases; if the number of moles of gas decreases, the volume decreases.
- Avogadro's law states that equal volumes of gases at the same temperature and pressure contain the same number of moles. The volume (V) of a gas is directly related to the number of moles of the gas when the pressure and temperature of the gas do not change:

$$\frac{V_1}{n_1} = \frac{V_2}{n_2}$$ Subscripts 1 and 2 represent initial and final conditions.

- At STP conditions, standard temperature (0 °C) and pressure (1 atm), one mole of a gas occupies a volume of 22.4 L, the *molar volume*.

Guide to Using Molar Volume	
Step 1	State the given and needed quantities.
Step 2	Write a plan to calculate the needed quantity.
Step 3	Write the equalities and conversion factors including 22.4 L/mole at STP.
Step 4	Set up the problem with factors to cancel units.

Example: How many liters would 2.00 moles of N_2 occupy at STP?

Solution

Step 1: State the given and needed quantities.

Analyze the Problem

Given 2.00 moles of N_2 at STP **Need** liters of N_2 at STP

Step 2: Write a plan to calculate the needed quantity.

moles of N_2 $\xrightarrow{\text{Molar volume}}$ liters of N_2 (STP)

Step 3: Write the equalities and conversion factors including 22.4 L/mole at STP.

1 mole of N_2 = 22.4 L of N_2(STP)

$$\frac{22.4\text{ L }N_2\text{ (STP)}}{1\text{ mole }N_2} \quad\text{and}\quad \frac{1\text{ mole }N_2}{22.4\text{ L }N_2\text{ (STP)}}$$

Step 4: Set up the problem with factors to cancel units.

$$2.00\ \cancel{\text{moles }N_2} \times \frac{22.4\text{ L }N_2\text{ (STP)}}{1\ \cancel{\text{mole }N_2}} = 44.8\text{ L of }N_2\text{ (STP)}$$

♦ Learning Exercise 11.7

Use Avogadro's law to solve the following gas problems.

a. A gas containing 0.50 mol of helium has a volume of 4.0 L. What is the new volume, in liters, when 1.0 mol of nitrogen is added to the container (P and T constant)?

b. A balloon containing 1.0 mol oxygen has a volume of 15 L. What is the new volume, in liters, of the balloon when 2.0 mol of helium is added (P and T constant)?

c. What is the volume, in liters, occupied by 28.0 g of nitrogen (N_2) at STP (P and T constant)?

d. What is the mass, in grams, of 4.48 L of O_2 at STP?

Answers **a.** 12 L **b.** 45 L **c.** 22.4 L **d.** 6.40 g

11.8 The Ideal Gas Law

Learning Goal: Use the ideal gas law equation to solve for P, V, T, or n of a gas when given three of the four values in the ideal gas law equation. Calculate density, molar mass, or volume of a gas in a chemical reaction.

- The ideal gas law $PV = nRT$ gives the relationship among the four variables: pressure, volume, moles, and temperature. When any three variables are given, the fourth can be calculated.
- R is the universal gas constant whose value depends on the units used:

$$0.0821\ \text{L}\cdot\text{atm/mol}\cdot\text{K} \quad \text{or} \quad 62.4\ \text{L}\cdot\text{mmHg/mol}\cdot\text{K}$$

- The ideal gas law and the balanced chemical equation can be used to determine the mass or volume of a gas in a chemical reaction.
- The molar mass and density of a gas can be determined using the ideal gas law.

Guide to Using the Ideal Gas Law	
Step 1	State the given and needed quantities.
Step 2	Rearrange the ideal gas law to solve for the needed quantity.
Step 3	Substitute the gas data into the equation and calculate the needed quantity.

Study Note

Identify the three known variables for the ideal gas law, and arrange the equation to solve for the unknown variable.

Example: Solve the ideal gas law for P. $PV = nRT \quad P = \dfrac{nRT}{V}$

CORE CHEMISTRY SKILL

Using the Ideal Gas Law

♦ Learning Exercise 11.8A

Use the ideal gas law to solve for the unknown variable in each of the following:

a. What volume, in liters, is occupied by 0.250 mol of nitrogen gas (N_2) at 0 °C and 1.50 atm?

b. What is the temperature, in °C, of 0.500 mol of helium that occupies a volume of 15.0 L at a pressure of 1250 mmHg?

c. What is the pressure, in atm, of 1.50 mol of neon in a 5.00-L steel container at a temperature of 125 °C?

d. What is the pressure, in atm, of 8.0 g of oxygen (O_2) that has a volume of 245 mL at a temperature of 22 °C?

Answers **a.** 3.74 L **b.** 328 °C **c.** 9.80 atm **d.** 25 atm

Guide to Calculating the Molar Mass of a Gas	
Step 1	State the given and needed quantities.
Step 2	Rearrange the ideal gas law to solve for the number of moles.
Step 3	Obtain the molar mass by dividing the given number of grams by the number of moles.

♦ Learning Exercise 11.8B

Use gas laws to determine the molar mass of a gas. (Hint: Determine the moles (n) of gas and then calculate the mass/mole ratio to obtain molar mass.)

a. A gas has a mass of 0.650 g and a volume of 560 mL at STP. What is the molar mass of the gas?

b. A sample of gas with a mass of 0.412 g has a volume of 273 mL and a pressure of 746 mmHg at 25 °C. What is the molar mass of the gas?

Answers **a.** 26.0 g/mol **b.** 37.6 g/mol

11.9 Gas Laws and Chemical Reactions

Learning Goal: Determine the mass or volume of a gas that reacts or forms in a chemical reaction.

- The ideal gas law or molar volume at STP is used to determine the number of moles (and therefore, mass using the molar mass) or volume of a gas in a chemical reaction.
- Mole-mole factors are then used to calculate moles of other substances in a reaction.

Guide to Reactions Involving the Ideal Gas Law	
Step 1	State the given and needed quantities.
Step 2	Write a plan to convert the given quantity to the needed moles.
Step 3	Write the equalities and conversion factors for molar mass and mole–mole factors.
Step 4	Set up the problem to calculate moles of needed quantity.
Step 5	Convert the moles of needed quantity to mass or volume using the molar mass or the ideal gas law equation.

♦ Learning Exercise 11.9

CORE CHEMISTRY SKILL

Calculating Mass or Volume of a Gas in a Chemical Reaction

Use gas laws to determine the quantity of a reactant or product in each of the following chemical reactions:

a. How many liters of hydrogen gas at STP are produced when 12.5 g of magnesium react?

$$Mg(s) + 2HCl(aq) \longrightarrow MgCl_2(aq) + H_2(g)$$

b. How many grams of KNO_3 must decompose to produce 35.8 L of O_2 at 28 °C and 745 mmHg?

$$2KNO_3(s) \longrightarrow 2KNO_2(s) + O_2(g)$$

c. At a temperature of 325 °C and a pressure of 1.20 atm, how many liters of CO_2 can be produced when 50.0 g of propane undergoes combustion?

$$C_3H_8(g) + 5O_2(g) \longrightarrow 3CO_2(g) + 4H_2O(g)$$

Answers **a.** 11.5 L of H_2 **b.** 287 g of KNO_3 **c.** 139 L of CO_2

11.10 Partial Pressures (Dalton's Law)

Learning Goal: Use Dalton's law of partial pressures to calculate the total pressure of a mixture of gases.

- In a mixture of two or more gases, the total pressure is the sum of the partial pressures (subscripts 1, 2, 3 . . . represent the partial pressures of the individual gases).

$$P_{total} = P_1 + P_2 + P_3 + \cdots$$

- The partial pressure of a gas in a mixture is the pressure it would exert if it were the only gas in the container.

Key Terms for Sections 11.8–11.10

Match the key terms with the correct definition below.

a. molar volume **b.** STP **c.** Dalton's law
d. ideal gas law **e.** partial pressure **f.** Avogadro's law

1. ________ gas law stating that the total pressure exerted by a mixture of gases in a container is the sum of the partial pressures that each gas would exert alone

2. ________ pressure exerted by a single gas in a gas mixture

3. ________ volume of 22.4 L occupied by 1 mole of a gas at STP conditions of 0 °C (273 K) and 1 atm

4. ________ law that combines the four measured properties of a gas in the equation $PV = nRT$

5. ________ gas law stating that the volume of a gas is directly related to the number of moles of gas when pressure and temperature do not change

6. ________ standard conditions of 0 °C (273 K) and 1 atm, used for the comparison of gases

Answers **1.** c **2.** e **3.** a **4.** d **5.** f **6.** b

Guide to Solving for Partial Pressure	
Step 1	Write the equation for the sum of partial pressures.
Step 2	Rearrange the equation to solve for the unknown pressure.
Step 3	Substitute known pressures into the equation and calculate the unknown partial pressure.

CORE CHEMISTRY SKILL

Calculating Partial Pressure

♦ Learning Exercise 11.10A

Use Dalton's law to solve the following problems about gas mixtures:

a. What is the total pressure, in millimeters of mercury, of a sample of gases that contains oxygen at 0.500 atm, nitrogen (N_2) at 132 torr, and helium at 224 mmHg?

b. What is the total pressure, in atmospheres, of a gas sample that contains helium at 285 mmHg and oxygen (O_2) at 1.20 atm?

c. A gas sample containing nitrogen (N_2) and oxygen (O_2) has a pressure of 1500 mmHg. If the partial pressure of the nitrogen is 0.900 atm, what is the partial pressure, in millimeters of mercury, of the oxygen gas in the mixture?

Answers **a.** 736 mmHg **b.** 1.58 atm **c.** 816 mmHg

♦ Learning Exercise 11.10B

Fill in the blanks by writing I (*increases*) or D (*decreases*) for a gas in a closed container:

	Pressure	Volume	Moles	Temperature
a.	______	Increases	Constant	Constant
b.	Increases	Constant	______	Constant
c.	Constant	Decreases	______	Constant
d.	______	Constant	Constant	Increases
e.	Constant	______	Constant	Decreases
f.	______	Constant	Increases	Constant

Answers **a.** D **b.** I **c.** D
d. I **e.** D **f.** I

♦ Learning Exercise 11.10C

Complete the following table for typical blood gas values for partial pressures:

Gas	Alveoli	Oxygenated blood	Deoxygenated blood	Tissues
O_2	________	________	________	________
CO_2	________	________	________	________

Answers

O_2	100 mmHg	100 mmHg	40 mmHg or less	30 mmHg or less
CO_2	40 mmHg	40 mmHg	46 mmHg	50 mmHg or greater

Checklist for Chapter 11

You are ready to take the Practice Test for Chapter 11. Be sure you have accomplished the following learning goals for this chapter. If not, review the section listed at the end of the goal. Then apply your new skills and understanding to the Practice Test.

After studying Chapter 11, I can successfully:

_____ Describe the kinetic molecular theory of gases (11.1).

_____ Change the units of pressure from one to another (11.2).

_____ Use the pressure–volume relationship (Boyle's law) to determine the new pressure or volume of a fixed amount of gas at constant temperature (11.3).

_____ Use the temperature–volume relationship (Charles's law) to determine the new temperature or volume of a fixed amount of gas at a constant pressure (11.4).

_____ Use the temperature–pressure relationship (Gay-Lussac's law) to determine the new temperature or pressure of a certain amount of gas at a constant volume (11.5).

_____ Use the combined gas law to find the new pressure, volume, or temperature of a certain amount of gas when changes in two of these properties are given (11.6).

_____ Describe the relationship between the amount of a gas and its volume (Avogadro's law) at constant P and T, and use this relationship in calculations (11.7).

_____ Use the ideal gas law to solve for pressure, volume, temperature, amount of a gas, molar mass, or density (11.8).

_____ Use the gas laws to calculate the quantity of a reactant or product in a chemical reaction (11.9).

_____ Calculate the total pressure of a gas mixture from the partial pressures (11.10).

Practice Test for Chapter 11

For questions 1 through 5, answer using T (true) or F (false): (11.1)

1. _____ A gas does not have its own volume or shape.

2. _____ The molecules of a gas are moving extremely fast.

3. _____ The collisions of gas molecules with the walls of their container create pressure.

4. _____ Gas molecules are close together and move in straight lines.

5. ____ Gas molecules have no attractions between them.

6. When a gas is heated in a closed metal container, the (11.2)
   A. pressure increases
   B. pressure decreases
   C. volume increases
   D. volume decreases
   E. number of molecules increases

7. The pressure of a gas will increase when (11.3)
   A. the volume increases
   B. the temperature decreases
   C. more molecules of gas are added
   D. molecules of gas are removed
   E. none of these

8. If the temperature of a gas is increased, (11.4, 11.5)
   A. the pressure will decrease
   B. the volume will increase
   C. the volume will decrease
   D. the number of molecules will increase
   E. none of these

9. The relationship that the volume of a gas is inversely related to its pressure at constant temperature is known as (11.3)
   A. Boyle's law   B. Charles's law   C. Gay-Lussac's law
   D. Dalton's law   E. Avogadro's law

10. What is the pressure, in atmospheres, of a gas with a pressure of 1200 mmHg? (11.2)
    A. 0.63 atm   B. 0.79 atm   C. 1.2 atm
    D. 1.6 atm   E. 2.0 atm

11. A 6.00-L sample of oxygen has a pressure of 660. mmHg. When the volume is reduced to 2.00 L at constant temperature and moles, it will have a new pressure of (11.3)
    A. 1980 mmHg   B. 1320 mmHg   C. 330. mmHg
    D. 220. mmHg   E. 110. mmHg

12. A sample of nitrogen gas at 110 K has a pressure of 1.0 atm. When the temperature is increased to 360 K at constant moles and volume, the new pressure will be (11.5)
    A. 0.50 atm   B. 1.0 atm   C. 1.5 atm
    D. 3.3 atm   E. 4.0 atm

13. If two gases have the same volume, temperature, and pressure, they also have the same (11.7, 11.8)
    A. density   B. number of molecules   C. molar mass
    D. speed   E. size molecules

14. A gas sample with a volume of 4.0 L has a pressure of 750 mmHg and a temperature of 77 °C. What is its new volume at 277 °C and 250 mmHg (*n* constant)? (11.6)
    A. 7.6 L   B. 19 L   C. 2.1 L
    D. 0.00056 L   E. 3.3 L

15. If the temperature and amount of a gas does not change, but its volume doubles, its pressure will (11.3)
    A. double
    B. triple
    C. decrease to one-half the original pressure
    D. decrease to one-fourth the original pressure
    E. not change

16. A sample of oxygen with a pressure of 220 mmHg contains 2.0 mol of gas and has a volume of 4.0 L. What will the new pressure be when the volume expands to 5.0 L and 3.0 mol of helium gas is added while temperature is constant? (11.6)
A. 110 mmHg **B.** 220 mmHg **C.** 440 mmHg
D. 550 mmHg **E.** 700. mmHg

17. A sample of 2.00 mol of gas initially at STP is converted to a volume of 5.0 L and a temperature of 27 °C. What is its new pressure in atm? (11.6, 11.7)
A. 0.12 atm **B.** 5.5 atm **C.** 7.5 atm
D. 9.8 atm **E.** 10. atm

18. The conditions for standard temperature and pressure (STP) are (11.7)
A. 0 K, 1 atm **B.** 0 °C, 10 atm **C.** 25 °C, 1 atm
D. 273 K, 1 atm **E.** 273 K, 0.5 atm

19. The volume occupied by 1.50 mol of CH_4 at STP is (11.7)
A. 44.10 L **B.** 33.6 L **C.** 22.4 L
D. 11.2 L **E.** 5.60 L

20. How many grams of oxygen gas (O_2) are present in 44.1 L of oxygen at STP? (11.7)
A. 10.0 g **B.** 16.0 g **C.** 32.0 g
D. 410.0 g **E.** 63.0 g

21. What is the volume, in liters, of 0.50 mol of nitrogen gas (N_2) at 25 °C and 2.0 atm? (11.8)
A. 0.51 L **B.** 1.0 L **C.** 4.2 L
D. 6.1 L **E.** 24 L

22. A gas mixture contains helium with a partial pressure of 0.100 atm, oxygen with a partial pressure of 445 mmHg, and nitrogen with a partial pressure of 235 mmHg. What is the total pressure, in atm, for the gas mixture? (11.10)
A. 0.995 atm **B.** 1.39 atm **C.** 1.69 atm
D. 2.00 atm **E.** 10.0 atm

23. A mixture of oxygen and nitrogen has a total pressure of 1040 mmHg. If the oxygen has a partial pressure of 510 mmHg, what is the partial pressure of the nitrogen? (11.10)
A. 240 mmHg **B.** 530 mmHg **C.** 770 mmHg
D. 1040 mmHg **E.** 1350 mmHg

24. 3.00 mol of He in a steel container has a pressure of 12.0 atm. What is the new pressure after 4.00 mol of He is added (T is constant)? (11.10)
A. 5.14 atm **B.** 16.0 atm **C.** 28.0 atm
D. 32.0 atm **E.** 45.0 atm

25. 2.50 g of $KClO_3$ decomposes by the equation: $2KClO_3(s) \longrightarrow 2KCl(s) + 3O_2(g)$. What volume of O_2 is produced at 25 °C and a pressure of 750. mmHg? (11.9)
A. 0.00998 L **B.** 0.759 L **C.** 0.0636 L
D. 0.506 L **E.** 1.32 L

26. A gas sample with a mass of 0.440 g has a volume of 0.224 L at STP. The gas has a molar mass of (11.7, 11.8)
A. 19.6 g/mol **B.** 22.4 g/mol **C.** 44.0 g/mol
D. 44.8 g/mol **E.** 50.9 g/mol

27. The exchange of gases between the alveoli, blood, and tissues of the body is a result of (11.10)
A. pressure gradients **B.** different molecular masses **C.** shapes of molecules
D. altitude **E.** all of these

28. Oxygen moves into the tissues from the blood because its partial pressure (11.10)
A. in arterial blood is higher than in the tissues
B. in venous blood is higher than in the tissues
C. in arterial blood is lower than in the tissues
D. in venous blood is lower than in the tissues
E. is equal in the blood and in the tissues

Answers to the Practice Test

1. T	**2.** T	**3.** T	**4.** F	**5.** T
6. A	**7.** C	**8.** B	**9.** A	**10.** D
11. A	**12.** D	**13.** B	**14.** B	**15.** C
16. C	**17.** D	**18.** D	**19.** B	**20.** E
21. D	**22.** A	**23.** B	**24.** C	**25.** B
26. C	**27.** A	**28.** A		

12
Solutions

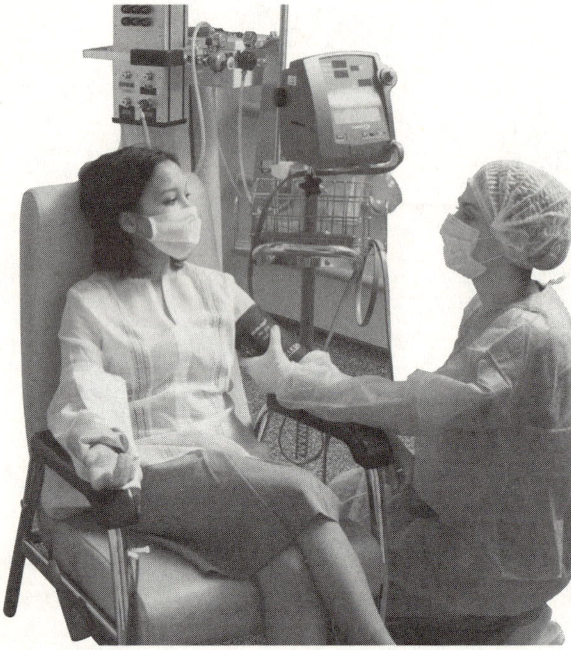

Michelle has continued to have hemodialysis treatment three times a week for about 4 h each visit. The treatment removes harmful waste products and extra salt and fluid. Her blood pressure is maintained because the electrolytes potassium and sodium are adjusted each time. Michelle measures the amount of fluid she takes in each day. Her diet is low in protein and includes fruits and vegetables that are low in potassium. She and her husband are now learning about performing hemodialysis at home. Soon she will have her own hemodialysis machine. If 65 mEq of potassium ion is removed using a dialysate with a low potassium concentration, how many moles of potassium ion are removed?

12.1 Solutions

Learning Goal: Identify the solute and solvent in a solution; describe the formation of a solution.

- A solution is a homogeneous mixture that forms when a solute dissolves in a solvent. The solvent is the substance that is present in a greater amount.
- Solvents and solutes can be solids, liquids, or gases. The solution is the same physical state as the solvent.
- A polar solute is soluble in a polar solvent; a nonpolar solute is soluble in a nonpolar solvent.
- Water molecules form hydrogen bonds because the partial positive charge of the hydrogen in one water molecule is attracted to the partial negative charge of oxygen in another water molecule.
- An ionic solute dissolves in water, a polar solvent, because the polar water molecules attract and pull the positive and negative ions into solution. In solution, water molecules surround the ions in a process called hydration.

♦ Learning Exercise 12.1A

Water is polar, and hexane is nonpolar. In which solvent is each of the following solutes soluble?

a. bromine, Br_2, nonpolar ________ **b.** HCl, polar ________

c. cholesterol, nonpolar ________ **d.** vitamin D, nonpolar ________

e. vitamin C, polar ________

Answers **a.** hexane **b.** water **c.** hexane **d.** hexane **e.** water

♦ Learning Exercise 12.1B

Indicate the solute and solvent in each of the following:

	Solute	Solvent
a. 10 g of KCl dissolved in 100 g of water	________	________
b. soda water: $CO_2(g)$ dissolved in water	________	________
c. an alloy composed of 80% Zn and 20% Cu	________	________
d. a mixture of O_2 (200 mmHg) and He (500 mmHg)	________	________
e. a solution of 40 mL of CCl_4 and 2 mL of Br_2	________	________

Answers **a.** KCl; water **b.** CO_2; water **c.** Cu; Zn
d. O_2; He **e.** Br_2; CCl_4

12.2 Electrolytes and Nonelectrolytes

Learning Goal: Identify solutes as electrolytes or nonelectrolytes.

- Electrolytes conduct an electrical current because they produce ions in aqueous solutions.
- Strong electrolytes are completely dissociated into ions, whereas weak electrolytes are partially dissociated into ions.
- Nonelectrolytes do not form ions in solution but dissolve as molecules.

♦ Learning Exercise 12.2A

Write an equation for the formation of an aqueous solution of each of the following strong electrolytes:

a. LiCl

b. $Mg(NO_3)_2$

c. Na_3PO_4

d. K_2SO_4

e. $MgCl_2$

Answers

a. $LiCl(s) \xrightarrow{H_2O} Li^+(aq) + Cl^-(aq)$
b. $Mg(NO_3)_2(s) \xrightarrow{H_2O} Mg^{2+}(aq) + 2NO_3^-(aq)$
c. $Na_3PO_4(s) \xrightarrow{H_2O} 3Na^+(aq) + PO_4^{3-}(aq)$
d. $K_2SO_4(s) \xrightarrow{H_2O} 2K^+(aq) + SO_4^{2-}(aq)$
e. $MgCl_2(s) \xrightarrow{H_2O} Mg^{2+}(aq) + 2Cl^-(aq)$

♦ Learning Exercise 12.2B

Indicate whether an aqueous solution of each of the following contains ions only, molecules only, or mostly molecules with some ions. Write an equation for the formation of the solution.

a. glucose, $C_6H_{12}O_6$, a nonelectrolyte

b. NaOH, a strong electrolyte

c. KCl, a strong electrolyte

d. HF, a weak electrolyte

Answers

a. $C_6H_{12}O_6(s) \xrightarrow{H_2O} C_6H_{12}O_6(aq)$	molecules only
b. $NaOH(s) \xrightarrow{H_2O} Na^+(aq) + OH^-(aq)$	ions only
c. $KCl(s) \xrightarrow{H_2O} K^+(aq) + Cl^-(aq)$	ions only
d. $HF(aq) + H_2O(l) \rightleftharpoons H_3O^+(aq) + F^-(aq)$	mostly molecules and a few ions

12.3 Solubility

Learning Goal: Define solubility; distinguish between an unsaturated and a saturated solution. Identify a salt as soluble or insoluble.

- The amount of solute that dissolves depends on the nature of the solute and solvent.
- Solubility describes the maximum amount of a solute that dissolves in exactly 100 g solvent at a given temperature.
- A saturated solution contains the maximum amount of dissolved solute at a certain temperature whereas an unsaturated solution contains less than this amount.
- An increase in temperature increases the solubility of most solids, but decreases the solubility of gases in water.
- Henry's law states that the solubility of gases in liquids is directly related to the partial pressure of that gas above the liquid.
- The solubility rules describe the kinds of ionic combinations that are soluble and insoluble in water. If a salt contains Li^+, Na^+, K^+, NO_3^-, or NH_4^+ ion, it is soluble in water. Most halides and sulfates are soluble.

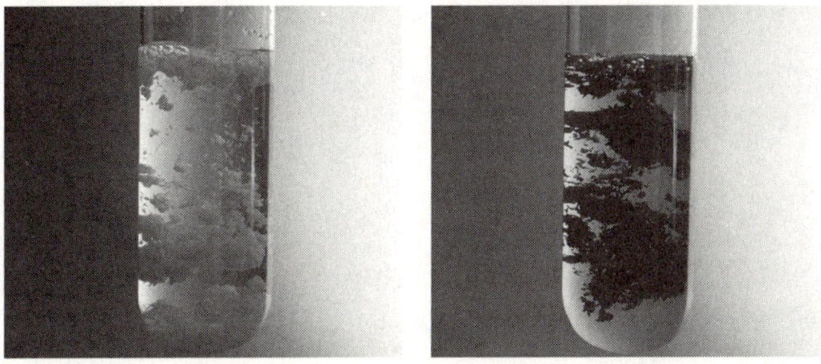

CdS FeS

If a salt contains a combination of a cation and an anion such as Cd^{2+} and S^{2-} or Fe^{2+} and S^{2-}, that are not soluble, that salt is insoluble.

♦ Learning Exercise 12.3A

Identify each of the following as a saturated solution (S) or an unsaturated solution (U):

1. A sugar cube dissolves when added to a cup of coffee. ________
2. A KCl crystal added to a KCl solution does not change in size. ________
3. A layer of sugar forms in the bottom of a glass of iced tea. ________
4. The rate of crystal formation is equal to the rate of dissolving. ________
5. Upon heating, all the sugar in a solution dissolves. ________

Answers **1.** U **2.** S **3.** S **4.** S **5.** U

♦ Learning Exercise 12.3B

Use the solubility data for $NaNO_3$ listed below to answer each of the following problems:

Temperature (°C)	Solubility (g of $NaNO_3$/100 g of H_2O)
40	100. g
60	120. g
80	150. g
100	200. g

a. How many grams of $NaNO_3$ will dissolve in 100. g water at 40 °C?

b. How many grams of $NaNO_3$ will dissolve in 300. g of water at 60 °C?

c. A solution is prepared using 200. g of water and 350. g of $NaNO_3$ at 80 °C. Will any solute remain undissolved? If so, how much?

d. Will 150. g of $NaNO_3$ all dissolve when added to 100. g of water at 100 °C?

Answers **a.** 100. g **b.** 360. g
c. 300. g of $NaNO_3$ will dissolve leaving 50. g of $NaNO_3$ that will not dissolve.
d. Yes, all 150. g of $NaNO_3$ will dissolve.

♦ Learning Exercise 12.3C

CORE CHEMISTRY SKILL
Using Solubility Rules

Predict whether the following salts are soluble (S) or insoluble (I) in water:

1. ________ NaCl **2.** ________ $AgNO_3$ **3.** ________ $PbCl_2$
4. ________ Ag_2S **5.** ________ $BaSO_4$ **6.** ________ Na_2CO_3
7. ________ K_2S **8.** ________ $MgCl_2$ **9.** ________ BaS

Answers **1.** S **2.** S **3.** I **4.** I **5.** I
6. S **7.** S **8.** S **9.** I

Guide to Writing an Equation for the Formation of an Insoluble Salt	
Step 1	Write the ions of the reactants.
Step 2	Write the combinations of ions and determine if any are insoluble.
Step 3	Write the ionic equation including any solid.
Step 4	Write the net ionic equation.

Example: Write the ionic and net ionic equations for the reaction that occurs when solutions of $AgNO_3(aq)$ and $Na_2SO_4(aq)$ are mixed.

Solution:

Step 1: Write the ions of the reactants.

$$AgNO_3(aq) = Ag^+(aq) + NO_3^-(aq) \qquad Na_2SO_4(aq) = 2Na^+ + SO_4^{2-}(aq)$$

Step 2: Write the combinations of ions and determine if any are insoluble.

$$Na^+(aq) + NO_3^-(aq) \longrightarrow NaNO_3(aq) \quad \text{a soluble salt}$$
$$2Ag^+(aq) + SO_4^{2-}(aq) \longrightarrow Ag_2SO_4(s) \quad \text{an insoluble salt}$$

Step 3: Write the ionic equation including any solid. By breaking up all the ionic formulas into ions, we can write

$$Na^+(aq) + NO_3^-(aq) + 2Ag^+(aq) + SO_4^{2-}(aq) \longrightarrow Na^+(aq) + NO_3^-(aq) + Ag_2SO_4(s)$$

Step 4: Write the net ionic equation. Remove the spectator ions, which appear on both sides of the ionic equation, to write the net ionic equation.

$$Na^+(aq) + NO_3^-(aq) + \mathbf{2Ag^+(aq) + SO_4^{2-}(aq)} \longrightarrow Na^+(aq) + NO_3^-(aq) + \mathbf{Ag_2SO_4(s)}$$
$$2Ag^+(aq) + SO_4^{2-}(aq) \longrightarrow Ag_2SO_4(s) \quad \text{net ionic equation}$$

♦ Learning Exercise 12.3D

Predict whether an insoluble salt forms in the following mixtures of soluble salts. If so, write the formula of the solid.

1. $NaCl(aq) + Pb(NO_3)_2(aq) \longrightarrow$
2. $BaCl_2(aq) + Na_2SO_4(aq) \longrightarrow$
3. $K_3PO_4(aq) + NaNO_3(aq) \longrightarrow$
4. $Na_2S(aq) + AgNO_3(aq) \longrightarrow$

Answers **1.** Yes; $PbCl_2$ **2.** Yes; $BaSO_4$ **3.** None **4.** Yes; Ag_2S

♦ Learning Exercise 12.3E

For the mixtures that form insoluble salts in Learning Exercise 12.3D, write the ionic equation and the net ionic equation.

1.

2.

3.

4.

Answers

1. $2Na^+(aq) + 2Cl^-(aq) + Pb^{2+}(aq) + 2NO_3^-(aq) \longrightarrow 2Na^+(aq) + PbCl_2(s) + 2NO_3^-(aq)$ (ionic)

   $2Cl^-(aq) + Pb^{2+}(aq) \longrightarrow PbCl_2(s)$ (net ionic equation)
2. $Ba^{2+}(aq) + 2Cl^-(aq) + 2Na^+(aq) + SO_4^{2-}(aq) \longrightarrow 2Na^+(aq) + BaSO_4(s) + 2Cl^-(aq)$ (ionic)

   $Ba^{2+}(aq) + SO_4^{2-}(aq) \longrightarrow BaSO_4(s)$ (net ionic equation)
3. No insoluble salt forms.
4. $2Na^+(aq) + S^{2-}(aq) + 2Ag^+(aq) + 2NO_3^-(aq) \longrightarrow 2Na^+(aq) + Ag_2S(s) + 2NO_3^-(aq)$ (ionic)

   $S^{2-}(aq) + 2Ag^+(aq) \longrightarrow Ag_2S(s)$ (net ionic equation)

12.4 Concentration of Solutions

Learning Goal: Calculate the concentration of a solute in a solution; use concentration units to calculate the amount of solute or solution.

- The concentration of a solution is the relationship between the amount of solute, in grams or milliliters, and the amount of solution, in grams, milliliters, or liters.
- A mass percent (m/m) expresses the ratio of the mass of solute to the mass of solution multiplied by 100. Mass units must be the same such as grams of solute/grams of solution × 100%.

$$\text{Mass percent (m/m)} = \frac{\text{mass of solute}}{\text{mass of solution}} \times 100\%$$

- A volume percent (v/v) expresses the ratio of the volume of solute to volume of solution multiplied by 100. Volumes units must be the same such as milliliters of solute/milliliters of solution × 100%.

$$\text{Volume percent (v/v)} = \frac{\text{volume of solute}}{\text{volume of solution}} \times 100\%$$

- A mass/volume percent (m/v) expresses the ratio of the mass of solute to the volume of solution multiplied by 100.

$$\text{Mass/volume percent (m/v)} = \frac{\text{grams of solute}}{\text{milliliters of solution}} \times 100\%$$

- Molarity is a concentration term that indicates the number of moles of solute dissolved in 1 L (1000 mL) of solution.

$$\text{Molarity (M)} = \frac{\text{moles of solute}}{\text{L solution}}$$

- Concentrations can be used as conversion factors to relate the amount of solute and the volume or mass of the solution.

$$\frac{\text{amount solute}}{\text{amount solution}} \quad \text{and} \quad \frac{\text{amount solution}}{\text{amount solute}}$$

Key Terms for Sections 12.3–12.4

Match the following terms with the correct definition below:

a. mass percent (m/m) **b.** concentration **c.** solubility
d. unsaturated **e.** molarity

1. _____ the ratio of the number of moles of solute in 1 L of solution

2. _____ the ratio of the amount of solute dissolved in a certain amount of solution

3. _____ the maximum amount of solute that can dissolve in a solution at a certain temperature

4. _____ a solution with less than the maximum amount of solute that can dissolve

5. _____ concentration of a solution in terms of mass of solute in 100 g of solution

Answers **1.** e **2.** b **3.** c **4.** d **5.** a

Guide to Calculating Solution Concentration	
Step 1	State the given and needed quantities.
Step 2	Write the concentration expression.
Step 3	Substitute solute and solution quantities into the expression and calculate.

Example: What is the mass percent (m/m) when 2.4 g of $NaHCO_3$ dissolves in water to make 120 g of solution?

Step 1: State the given and needed quantities.

Analyze the Problem

Given 2.4 g of $NaHCO_3$ solute; 120 g of $NaHCO_3$ solution

Need mass percent (m/m) concentration

Step 2: Write the concentration expression.

$$\text{mass percent (m/m)} = \frac{\text{mass of solute}}{\text{mass of solution}} \times 100\% \text{ mass percent}$$

Step 3: Substitute solute and solution quantities into the expression and calculate.

$$\text{mass percent (m/m)} = \frac{2.4 \text{ g } NaHCO_3}{120 \text{ g } NaHCO_3 \text{ solution}} \times 100\% = 2.0\% \text{ (m/m)}$$

CORE CHEMISTRY SKILL
Calculating Concentration

♦ Learning Exercise 12.4A

Calculate the percent concentration for each of the following solutions:

a. mass percent (m/m) for 18.0 g of NaCl in 90.0 g of solution

b. mass/volume percent (m/v) for 4.0 g of KOH in 50.0 mL of solution

c. mass/volume percent (m/v) for 5.0 g of KCl in 2.0 L of solution

d. volume percent (v/v) for 18 mL of ethanol in 350 mL of mouthwash solution

Answers **a.** 20.0% (m/m) **b.** 8.0% (m/v) **c.** 0.25% (m/v) **d.** 5.1% (v/v)

♦ Learning Exercise 12.4B

Calculate the molarity of the following solutions:

a. 1.45 mol of HCl in 0.250 L of HCl solution

b. 10.0 mol of glucose ($C_6H_{12}O_6$) in 2.50 L of glucose solution

c. 80.0 g of NaOH in 1.60 L of NaOH solution (*Hint:* Find moles of NaOH.)

d. 38.8 g of NaBr in 175 mL of NaBr solution

Answers **a.** 5.80 M HCl solution **b.** 4.00 M glucose solution
c. 1.25 M NaOH solution **d.** 2.15 M NaBr solution

Guide to Using Concentration to Calculate Mass or Volume	
Step 1	State the given and needed quantities.
Step 2	Write a plan to calculate the mass or volume.
Step 3	Write equalities and conversion factors.
Step 4	Set up the problem to calculate the mass or volume.

Example: How many grams of KI are needed to prepare 225 g of a 4.0% (m/m) KI solution?

Solution:

Step 1: State the given and needed quantities.

Analyze the Problem

Given 225 g of a 4.0% (m/m) KI solution **Need** grams of KI

Step 2: Write a plan to calculate mass or volume.

grams of KI solution ⟶ grams of KI

Step 3: Write equalities and conversion factors.

100 g of KI solution = 4.0 g of KI

$$\frac{4.0 \text{ g KI}}{100 \text{ g KI solution}} \quad \text{and} \quad \frac{100 \text{ g KI solution}}{4.0 \text{ g KI}}$$

Step 4: Set up problem to calculate mass or volume.

$$225 \cancel{\text{g solution}} \times \frac{4.0 \text{ g KI}}{100 \cancel{\text{g solution}}} = 9.0 \text{ g of KI}$$

♦ Learning Exercise 12.4C

CORE CHEMISTRY SKILL

Using Concentration as a Conversion Factor

Calculate the number of grams of solute needed to prepare each of the following solutions:

a. How many grams of glucose are needed to prepare 480 g of a 5.0% (m/m) solution?

b. How many grams of lidocaine hydrochloride are needed to prepare 50.0 g of a 2.0% (m/m) solution?

c. How many grams of KCl are needed to prepare 1250 g of a 4.00% (m/m) KCl solution?

d. How many grams of NaCl are needed to prepare 75.6 g of a 1.50% (m/m) solution?

Answers **a.** 24 g **b.** 1.0 g **c.** 50.0 g of KCl **d.** 1.13 g of NaCl

♦ Learning Exercise 12.4D

Use percent concentration factors to calculate the mass of each solution that contains the amount of solute stated in each problem:

a. 8.75 g of NaCl from a 3.00% (m/m) NaCl solution

b. 24 g of glucose from a 5.0% (m/m) glucose solution

c. 10.5 g of KCl from a 2.5% (m/m) KCl solution

d. 12.0 g of NaOH from a 5.00% (m/m) NaOH solution

Answers **a.** 292 g **b.** 480 g **c.** 420 g **d.** 240. g

Study Note

Molarity is used as a conversion factor to convert between the amount of solute and the volume of solution. How many grams of NaOH are in 0.250 L of a 5.00 M NaOH solution?

Solution:

Step 1 State the given and needed quantities.

Given	0.250 L of a 5.00 M NaOH solution
Need	grams of NaOH

Step 2 Write a plan to calculate mass or volume.

liters of solution ⟶ moles of solution ⟶ grams of solute

Step 3 Write equalities and conversion factors. The concentration 5.00 M can be expressed as conversion factors.

$$1 \text{ L of NaOH solution} = 5.00 \text{ mol of NaOH}$$

$$\frac{5.00 \text{ mol NaOH}}{1 \text{ L NaOH solution}} \quad \text{and} \quad \frac{1 \text{ L NaOH solution}}{5.00 \text{ mol NaOH}}$$

Step 4 Set up problem to calculate mass or volume.

$$0.250 \text{ L } \cancel{\text{solution}} \times \frac{5.00 \text{ } \cancel{\text{mol NaOH}}}{1 \text{ L } \cancel{\text{solution}}} \times \frac{40.00 \text{ g NaOH}}{1 \text{ } \cancel{\text{mol NaOH}}} = 50.0 \text{ g of NaOH}$$

♦ Learning Exercise 12.4E

Calculate the quantity of solute in each of the following solutions:

a. How many moles of HCl are in 1.50 L of a 6.00 M HCl solution?

b. How many moles of KOH are in 0.750 L of a 10.0 M KOH solution?

c. How many grams of NaOH are needed to prepare 0.500 L of a 4.40 M NaOH solution? (*Hint:* Find moles of NaOH.)

d. How many grams of NaCl are in 285 mL of a 1.75 M NaCl solution?

Answers **a.** 9.00 mol of HCl **b.** 7.50 mol of KOH **c.** 88.0 g of NaOH **d.** 29.2 g of NaCl

♦ Learning Exercise 12.4F

Calculate the volume, in milliliters, needed of each solution to obtain the given quantity of solute:

a. 1.50 mol of $Mg(OH)_2$ from a 2.00 M $Mg(OH)_2$ solution

b. 0.150 mol of glucose from a 2.20 M glucose solution

c. 18.5 g of KI from a 3.00 M KI solution

d. 18.0 g of NaOH from a 6.00 M NaOH solution

Answers **a.** 750 mL **b.** 68.2 mL **c.** 37.1 mL **d.** 75.0 mL

12.5 Dilution and Chemical Reactions in Solution

Learning Goal: Calculate the new concentration or volume of a diluted solution. Given the volume and concentration of a solution, calculate the amount of another reactant or product in a reaction.

- Dilution is the process of mixing a solution with solvent to obtain a solution with a lower concentration.
- For dilutions, use the expression $C_1V_1 = C_2V_2$ or $M_1V_1 = M_2V_2$ and solve for the unknown value.
- The volume and molarity of a solution are used to determine the moles or liters of a substance required or produced in a chemical reaction.
- The balanced equation can be used to convert moles of one substance to moles of another substance.
- Molar mass is used to convert moles to grams or grams to moles.
- The number of moles of a solute and the molarity of the solution are used to determine the volume of a solution in a chemical reaction.

Guide to Calculating Dilution Quantities	
Step 1	Prepare a table of the concentrations and volumes of the solutions.
Step 2	Rearrange the dilution expression to solve for the unknown quantity.
Step 3	Substitute the known quantities into the dilution expression and calculate.

Example: What is the final molarity after 150. mL of a 2.00 M NaCl solution is diluted to a volume of 400. mL?

Solution:

Step 1: Prepare a table of the concentrations and volumes of the solutions.

Analyze the Problem

Concentrated Solution	**Diluted Solution**	**Know**	**Predict**
$M_1 = 2.00$ M	$M_2 = ?$		M decreases
$V_1 = 150.$ mL	$V_2 = 400.$ mL	V increases	

Step 2: Rearrange the dilution expression to solve for the unknown quantity.

$$M_1V_1 = M_2V_2 \qquad M_2 = M_1 \times \frac{V_1}{V_2}$$

Step 3: Substitute the known quantities into the dilution expression and calculate.

$$M_2 = 2.00\text{ M} \times \frac{150.\ \cancel{\text{mL}}}{400.\ \cancel{\text{mL}}} = 0.750\text{ M}$$

♦ Learning Exercise 12.5A

Solve each of the following dilution problems (assume the volumes are additive):

a. What is the final molarity after 100. mL of a 5.0 M KCl solution is diluted with water to give a final volume of 200. mL?

b. What is the final molarity of the diluted solution if 5.0 mL of a 1.5 M KCl solution is diluted to 25 mL?

c. What is the final molarity after 250 mL of an 8.0 M NaOH solution is diluted with 750 mL of water?

d. 160. mL of water is added to 40. mL of a 1.0 M NaCl solution. What is the final molarity? (Assume that the volumes add.)

e. What volume of 6.0 M HCl is needed to prepare 300. mL of a 1.0 M HCl solution? How much water must be added? (Assume that the volumes add.)

Answers **a.** 2.5 M **b.** 0.30 M **c.** 2.7 M **d.** 0.20 M
e. $V_1 = 50.$ mL; add 250. mL of water

Guide to Calculations Involving Solutions in Chemical Reactions	
Step 1	State the given and needed quantities.
Step 2	Write a plan to calculate needed quantity or concentration.
Step 3	Write equalities and conversion factors including mole–mole and concentration factors.
Step 4	Set up problem to calculate the needed quantity or concentration.

CORE CHEMISTRY SKILL

Calculating the Quantity of a Reactant or Proaduct for a Chemical Reaction in a Solution

♦ Learning Exercise 12.5B

For the reaction, $2AgNO_3(aq) + H_2SO_4(aq) \longrightarrow Ag_2SO_4(s) + 2H_2O(l)$

a. How many milliliters of a 1.5 M $AgNO_3$ solution will react with 40.0 mL of a 1.0 M H_2SO_4 solution?

b. How many grams of Ag_2SO_4 will be produced?

Answers

a. $40.0\ \cancel{\text{mL H}_2\text{SO}_4} \times \frac{1\ \cancel{\text{L H}_2\text{SO}_4}}{1000\ \cancel{\text{mL H}_2\text{SO}_4}} \times \frac{1.0\ \cancel{\text{mol H}_2\text{SO}_4}}{1\ \cancel{\text{L H}_2\text{SO}_4}} \times \frac{2\ \cancel{\text{mol AgNO}_3}}{1\ \cancel{\text{mol H}_2\text{SO}_4}}$

$\times \frac{1000\ \text{mL AgNO}_3}{1.5\ \cancel{\text{mol AgNO}_3}} = 53\ \text{mL of AgNO}_3\ \text{solution}$

b. $40.0\ \cancel{\text{mL H}_2\text{SO}_4} \times \frac{1\ \cancel{\text{L H}_2\text{SO}_4}}{1000\ \cancel{\text{mL H}_2\text{SO}_4}} \times \frac{1.0\ \cancel{\text{mol H}_2\text{SO}_4}}{1\ \cancel{\text{L H}_2\text{SO}_4}} \times \frac{1\ \cancel{\text{mol Ag}_2\text{SO}_4}}{1\ \cancel{\text{mol H}_2\text{SO}_4}}$

$\times \frac{311.9\ \text{g Ag}_2\text{SO}_4}{1\ \cancel{\text{mol Ag}_2\text{SO}_4}} = 12\ \text{g of Ag}_2\text{SO}_4$

♦ Learning Exercise 12.5C

For the following reaction, calculate the number of milliliters of a 1.80 M KOH solution that react with 18.5 mL of a 2.20 M HCl solution.

$$HNO_3(aq) + KOH(aq) \longrightarrow KNO_3(aq) + H_2O(l)$$

Answer

$18.5\ \cancel{\text{mL HCl}} \times \frac{1\ \cancel{\text{L HCl}}}{1000\ \cancel{\text{mL HCl}}} \times \frac{2.20\ \cancel{\text{mol HCl}}}{1\ \cancel{\text{L HCl}}} \times \frac{1\ \cancel{\text{mol KOH}}}{1\ \cancel{\text{mol HCl}}} \times \frac{1000\ \text{mL KOH}}{1.80\ \cancel{\text{mol KOH}}} = 22.6\ \text{mL of KOH solution}$

12.6 Properties of Solutions

Learning Goal: Identify a mixture as a solution, a colloid, or a suspension. Describe how the number of particles in a solution affects the freezing point, boiling point, and osmotic pressure.

- Colloids contain particles that do not settle out and pass through filters but not through semipermeable membranes.
- Suspensions are composed of large particles that settle out of solution.
- The freezing point of a solution is lower than that of the solvent and boiling point is higher and depend only on the number of particles of solute in the solution.
- The freezing point and boiling point of a solution depends on the molal (m) concentration of the solution.

$$\text{Molality } (m) = \frac{\text{moles of solute}}{\text{kilogram of solvent}}$$

- The freezing point and freezing point depression (ΔT_f) of a solution can be calculated based on the molality of particles in the solution and the freezing point depression constant (K_f) for the solvent.

$$\Delta T_f = mK_f$$

- The boiling point and boiling point elevation (ΔT_b) of a solution can be calculated based on the molality of particles in the solution and the boiling point elevation constant (K_b) for the solvent

$$\Delta T_b = mK_b$$

- In the process of osmosis, water (solvent) moves through a semipermeable membrane from the solution that has a lower solute concentration to a solution where the solute concentration is higher.
- Osmotic pressure is the pressure that prevents the flow of water into a more concentrated solution.
- Isotonic solutions have osmotic pressures equal to that of body fluids. A hypotonic solution has a lower osmotic pressure than body fluids; a hypertonic solution has a higher osmotic pressure.
- A red blood cell maintains its volume in an isotonic solution, but it swells (hemolysis) in a hypotonic solution and shrinks (crenation) in a hypertonic solution.
- In dialysis, water and small solute particles can pass through a dialyzing membrane, while larger particles such as blood cells cannot.

♦ Learning Exercise 12.6A

Identify each of the following as a solution, colloid, or suspension:

1. ________ the solute consists of single atoms, ions, or small molecules

2. ________ settles out with gravity

3. ________ retained by filters

4. ________ cannot diffuse through a cellular membrane

5. ________ contains large particles that are visible

Answers **1.** solution **2.** suspension **3.** suspension
4. colloid **5.** suspension

♦ Learning Exercise 12.6B

Fill in the blanks:

In osmosis, the direction of solvent flow is from the (1) [higher/lower] solute concentration to the (2) [higher/lower] solute concentration. A semipermeable membrane separates 5% (m/v) and 10% (m/v) sucrose solutions. The (3) ________ % (m/v) solution has the greater osmotic pressure. Water will move from the (4) ________ % (m/v) solution into the (5) ________ % (m/v) solution. The compartment that contains the (6) ________ % (m/v) solution increases in volume.

Answers **1.** lower **2.** higher **3.** 10 **4.** 5 **5.** 10 **6.** 10

♦ Learning Exercise 12.6C

A semipermeable membrane separates a 2% starch solution from a 10% starch solution. Complete each of the following with 2% or 10%.

a. Water will flow from the ________ starch solution to the ________ starch solution.

b. The volume of the ________ starch solution will increase and the volume of the ________ starch solution will decrease.

Answers **a.** 2%, 10% **b.** 10%, 2%

Guide to Using Molality	
Step 1	State the given and needed quantities.
Step 2	Write the expression for the change in freezing or boiling point.
Step 3	Substitute molality into the expression and calculate.

Example: Calculate the freezing point lowering and freezing point of a solution containing 125 grams of KCl in 500. g of water.

Solution:

Step 1: State the given and needed quantities

Given 125 g of KCl; 500. g of water (solvent) = 0.500 kg of water

Need freezing point lowering ΔT_f and freezing point of KCl solution

Step 2: Write the expression for the change in freezing or boiling point.

$$\text{grams of KCl} \longrightarrow \text{moles of KCl} \times \frac{2 \text{ mole particles}}{1 \text{ mole KCl}} = \text{moles of particles}$$

$$m = \frac{\text{moles of particles}}{\text{kilogram of water}} \quad \text{and} \quad m \times \text{freezing point factor} = \Delta T_f$$

$$\text{Final temperature} = 0.0\ °\text{C} - \Delta T_f$$

$$1 \text{ mole of KCl} = 74.55 \text{ g of KCl} \qquad 1\ m = 1.86\ °\text{C}$$

$$\frac{1 \text{ mole KCl}}{74.55 \text{ g KCl}} \text{ and } \frac{74.55 \text{ g KCl}}{1 \text{ mole KCl}} \qquad \frac{1\ m}{1.86\ °\text{C}} \text{ and } \frac{1.86\ °\text{C}}{1\ m}$$

Step 3: Substitute molality into the expression and calculate.

$$\text{Moles of particles} = 125 \cancel{\text{g KCl}} \times \frac{1 \cancel{\text{mole KCl}}}{74.55 \cancel{\text{g KCl}}} \times \frac{2 \text{ mole particles}}{1 \cancel{\text{mole KCl}}} = 3.35 \text{ moles of particles (ions)}$$

$$\text{Molality } (m) = \frac{3.35 \text{ mole particles}}{0.500 \text{ kg water}} = 6.70\ m$$

$$\Delta T_f = 6.70 \cancel{m} \times \frac{1.86\ °\text{C}}{1 \cancel{m}} = 12.5\ °\text{C}$$

$$\text{Freezing point} = 0.0\ °\text{C} - 12.5\ °\text{C} = -12.5\ °\text{C}$$

♦ Learning Exercise 12.6D

CORE CHEMISTRY SKILL

Calculating Temperature Change

One mole of particles lowers the freezing point of 1000 g of water by 1.86 °C and raises the boiling point by 0.52 °C. For the following solutes each in 1000 g of water, indicate:

(1) the number of moles of particles

(2) the freezing point change and freezing point of the solution

(3) the boiling point change and boiling point of the solution

Solute	Number of Moles of Particles	Freezing Point Change and Freezing Point	Boiling Point Change and Boiling Point
a. 1.00 mole of fructose (nonelectrolyte)			
b. 1.50 mole of KCl (strong electrolyte)			
c. 0.650 mole of $Ca(NO_3)_2$ (strong electrolyte)			

Answers

Solute	Number of Moles of Particles	Freezing Point Change and Freezing Point	Boiling Point Change and Boiling Point
a. 1.00 mole of fructose (nonelectrolyte)	1.00 mole	Decreases 1.86 °C; fp = −1.86 °C	Increases 0.52 °C; bp = 100.52 °C
b. 1.50 mole of KCl (strong electrolyte)	3.00 mole	Decreases 5.58 °C; fp = −5.58 °C	Increases 1.6 °C; bp = 101.6 °C
c. 0.650 mole of $Ca(NO_3)_2$ (strong electrolyte)	1.95 mole	Decreases 3.63 °C; fp = −3.63 °C	Increases 1.0 °C; bp = 101.0 °C

♦ Learning Exercise 12.6E

Fill in the blanks:

A (1) ________ % (m/v) NaCl solution and a (2) ________ % (m/v) glucose solution are isotonic to the body fluids. A red blood cell placed in these solutions does not change in volume because these solutions are (3) ________ tonic. When a red blood cell is placed in water, it undergoes (4) ________ because water is (5) ________ tonic. A 20% (m/v) glucose solution will cause a red blood cell to undergo (6) ________ because the 20% (m/v) glucose solution is (7) ________ tonic.

Answers **1.** 0.9 **2.** 5 **3.** iso **4.** hemolysis
5. hypo **6.** crenation **7.** hyper

♦ Learning Exercise 12.6F

Indicate whether the following are

a. hypotonic **b.** hypertonic **c.** isotonic

1. ____ 5% (m/v) glucose **2.** ____ 3% (m/v) NaCl **3.** ____ 2% (m/v) glucose

4. ____ water **5.** ____ 0.9% (m/v) NaCl **6.** ____ 10% (m/v) glucose

Answers **1.** c **2.** b **3.** a **4.** a **5.** c **6.** b

♦ Learning Exercise 12.6G

Indicate whether the following will cause a red blood cell to undergo

a. crenation **b.** hemolysis **c.** no change (stays the same)

1. ____ 10% (m/v) NaCl **2.** ____ 1% (m/v) glucose **3.** ____ 5% (m/v) glucose

4. ____ 0.5% (m/v) NaCl **5.** ____ 10% (m/v) glucose **6.** ____ water

Answers **1.** a **2.** b **3.** c **4.** b **5.** a **6.** b

♦ Learning Exercise 12.6H

A dialysis bag contains starch, glucose, NaCl, protein, and urea.

a. When the dialysis bag is placed in water, what components would you expect to dialyze through the bag? Why?

b. Which components will stay inside the dialysis bag? Why?

Answers

a. Glucose, NaCl, and urea are solution particles. Solution particles will pass through semipermeable membranes.

b. Starch and protein form colloidal particles; because of their size colloidal particles are retained by semipermeable membranes.

Checklist for Chapter 12

You are ready to take the Practice Test for Chapter 12. Be sure you have accomplished the following learning goals for this chapter. If not, review the section listed at the end of the goal. Then apply your new skills and understanding to the Practice Test.

After studying Chapter 12, I can successfully:

____ Identify the solute and solvent in a solution and describe the process of dissolving an ionic solute in water (12.1).

____ Identify the components in solutions of electrolytes and nonelectrolytes (12.2).

____ Identify a saturated and an unsaturated solution (12.3).

____ Identify a salt as soluble or insoluble (12.3).

____ Write a chemical equation (or ionic or net ionic equation) to show the formation of an insoluble salt (12.3).

____ Describe the effects of temperature and nature of a solute on its solubility in a solvent (12.3).

____ Determine the solubility of an ionic compound (salt) in water (12.3).

____ Calculate the percent concentration, m/m,v/v, and m/v, of a solute in a solution, and use percent concentration to calculate the amount of solute or solution (12.4).

_____ Calculate the molarity of a solution (12.4).

_____ Use molarity as a conversion factor to calculate the moles (or grams) of a solute or the volume of the solution (12.4).

_____ Calculate the new concentration or new volume when a solution is diluted (12.5).

_____ Use the volume and molarity of a solution to calculate the grams or moles of a substance produced in a chemical reaction (12.5).

_____ Calculate the molality of a solution (12.6).

_____ Use the concentration of particles in a solution to calculate the freezing point or boiling point of the solution (12.6).

_____ Identify a solution as isotonic, hypotonic, or hypertonic (12.6).

_____ Identify a mixture as a solution, a colloid, or a suspension (12.6).

_____ Explain the processes of osmosis and dialysis (12.6).

Practice Test for Chapter 12

For questions 1 through 4, indicate if each of the following is more soluble in (W) water, a polar solvent, or (B) benzene, a nonpolar solvent: (12.1)

1. $I_2(s)$, nonpolar

2. $NaBr(s)$, polar

3. $KI(s)$, polar

4. C_6H_{12}, nonpolar

5. When dissolved in water, $Ca(NO_3)_2(s)$ dissociates into (12.2)
A. $Ca^{2+}(aq) + (NO_3)_2^{2-}(aq)$
B. $Ca^{+}(aq) + NO_3^{-}(aq)$
C. $Ca^{2+}(aq) + 2NO_3^{-}(aq)$
D. $Ca^{2+}(aq) + 2N^{5+}(aq) + 2O_3^{6-}(aq)$
E. $CaNO_3^{+}(aq) + NO_3^{-}(aq)$

6. Ethanol, $CH_3—CH_2—OH$, is a nonelectrolyte. When placed in water it (12.2)
A. dissociates completely.
B. dissociates partially.
C. does not dissociate.
D. makes the solution acidic.
E. makes the solution basic.

7. The solubility of NH_4Cl is 46 g in 100. g of water at 40 °C. How much NH_4Cl can dissolve in 500. g of water at 40 °C? (12.3)
A. 9.2 g **B.** 46 g **C.** 100 g **D.** 184 g **E.** 230 g

For questions 8 through 11, indicate if each of the following are soluble (S) or insoluble (I) in water: (12.3)

8. NaCl **9.** AgCl **10.** $BaSO_4$ **11.** FeO

12. A solution containing 1.20 g of sucrose in 50.0 g of solution has a percent (m/m) concentration of (12.4)
A. 0.600% **B.** 1.20% **C.** 2.40% **D.** 30.0% **E.** 41.6%

13. The amount of lactose in 250 g of a 3.0% (m/m) lactose solution for infant formula is (12.4)
A. 0.15 g **B.** 1.2 g **C.** 6.0 g **D.** 7.5 g **E.** 30 g

14. The mass of solution needed to obtain 0.40 g of glucose from a 5.0% (m/m) glucose solution is (12.4)
A. 1.0 g B. 2.0 g C. 4.0 g D. 5.0 g E. 8.0 g

15. The amount of NaCl needed to prepare 50.0 g of a 4.00% (m/m) NaCl solution is (12.4)
A. 20.0 g B. 15.0 g C. 10.0 g D. 4.00 g E. 2.00 g

16. A solution containing 6.0 g of NaCl in 1500 g of solution has a mass percent concentration of (12.4)
A. 0.40% (m/m) B. 0.25% (m/m) C. 4.0% (m/m) D. 0.90% (m/m) E. 2.5% (m/m)

17. The number of moles of KOH needed to prepare 2400 mL of a 2.0 M KOH solution is (12.4)
A. 1.2 mol B. 2.4 mol C. 4.8 mol D. 12 mol E. 48 mol

18. The number of grams of NaOH needed to prepare 7.5 mL of a 5.0 M NaOH is (12.4)
A. 1.5 g B. 3.8 g C. 6.7 g D. 15 g E. 38 g

For questions 19 through 21, consider a 20.0-g sample of a solution that contains 2.0 g of NaOH: (12.4)

19. The percent (m/m) concentration of the solution is
A. 1.0% B. 4.0% C. 5.0% D. 10.% E. 20.%

20. The number of moles of NaOH in the sample is
A. 0.050 mol B. 0.40 mol C. 1.0 mol D. 2.5 mol E. 4.0 mol

21. If the solution has a volume of 0.025 L, what is the molarity of the sample?
A. 0.10 M B. 0.5 M C. 1.0 M D. 1.5 M E. 2.0 M

22. Which of the following is soluble in water? (12.3)
A. AgCl B. $BaCO_3$ C. K_2SO_4 D. PbS E. MgO

23. The insoluble salt that forms when a solution of NaCl mixes with a $Pb(NO_3)_2$ solution is (12.3)
A. Na_2Pb B. $ClNO_3$ C. $NaNO_3$ D. $PbCl_2$ E. no insoluble salt forms

24. Which of the following salts is insoluble in water? (12.3)
A. $CuCl_2$ B. $Pb(NO_3)_2$ C. K_2CO_3 D. $(NH_4)SO_4$ E. $CaCO_3$

25. A 20.-mL sample of 5.0 M HCl is diluted with water to give 100. mL of solution. The final concentration of the HCl solution is (12.5)
A. 10 M B. 5.0 M C. 2.0 M D. 1.0 M E. 0.50 M

26. Water is added to 200. mL of a 4.00 M KNO_3 solution to give 400. mL of solution. The final concentration of the diluted solution is (12.5)
A. 1.00 M B. 2.00 M C. 4.00 M D. 0.500 M E. 0.100 M

27. 5.0 mL of a 2.0 M KOH solution is diluted with water to give 50.0 mL of solution. The final concentration of the KOH solution is (12.5)
A. 1.5 M B. 1.0 M C. 20 M D. 0.20 M E. 0.50 M

28. What mass of Ag_2SO_4 is formed when 25.0 mL of a 0.111 M $AgNO_3$ solution reacts with excess H_2SO_4? (12.5)

$$2AgNO_3(aq) + H_2SO_4(aq) \longrightarrow Ag_2SO_4(s) + 2H_2O(l)$$

A. 433 g B. 1.74 g C. 0.866 g D. 0.433 g E. 2.78 g

For questions 29 through 33, indicate whether each statement describes a (12.6)
A. solution **B.** colloid **C.** suspension

29. ____ contains single atoms, ions, or small molecules of solute

30. ____ settles out upon standing

31. ____ can be separated by filtering

32. ____ can be separated by semipermeable membranes

33. ____ passes through semipermeable membranes

34. Two solutions that have identical osmotic pressures are (12.6)
A. hypotonic **B.** hypertonic **C.** isotonic **D.** isotopic **E.** hyperactive

35. In osmosis, the net flow of water is (12.6)
A. between solutions of equal concentrations
B. from higher solute concentration to lower solute concentration
C. from lower solute concentration to higher solute concentration
D. from a colloid to a solution of equal concentration
E. from lower solvent concentration to higher solvent concentration

36. A red blood cell undergoes hemolysis when placed in a solution that is (12.6)
A. isotonic **B.** hypotonic **C.** hypertonic **D.** colloidal **E.** semitonic

37. A solution that has the same osmotic pressure as body fluids is (12.6)
A. 0.1% (m/v) NaCl **B.** 0.9 % (m/v) NaCl **C.** 5% (m/v) NaCl
D. 10% (m/v) glucose **E.** 15% (m/v) glucose

38. If a solution contains 0.50 mole of $CaCl_2$ in 1000 g of water, what is the freezing point of the solution? (12.6)
A. 0 °C **B.** 2.8 °C **C.** −2.8 °C **D.** 0.93 °C **E.** −0.93 °C

Answers to the Practice Test

1. B	**2.** W	**3.** W	**4.** B	**5.** C
6. C	**7.** E	**8.** S	**9.** I	**10.** I
11. I	**12.** C	**13.** D	**14.** E	**15.** E
16. A	**17.** C	**18.** A	**19.** D	**20.** A
21. E	**22.** C	**23.** D	**24.** E	**25.** D
26. B	**27.** D	**28.** D	**29.** A	**30.** C
31. C	**32.** B	**33.** A	**34.** C	**35.** C
36. B	**37.** B	**38.** C		

13
Reaction Rates and Chemical Equilibrium

Peter, a chemical oceanographer, has worked for six months obtaining samples to test the levels of CO_2 dissolved in the seawater. The oceans absorb more than a quarter of the carbon dioxide that humans put into the air. From the data he collects, Peter hopes to determine the impact of dissolved CO_2 on conditions in the ocean. Equilibrium is established between the dissolved CO_2 and H_2CO_3, carbonic acid, which it forms with water.

$$CO_2(aq) + H_2O(l) \rightleftarrows H_2CO_3(aq)$$

Carbonic acid, a weak acid, dissociates to yield hydrogen ions and carbonate.

$$H_2CO_3(aq) \rightleftarrows 2H^+(aq) + CO_3^{2-}(aq)$$

The carbonate interacts with cations in seawater. According to solubility rules, carbonate forms insoluble salts by reacting with certain cations in seawater. For example, Ca^{2+} and carbonate ions form calcium carbonate, which is the main constituent of clam shells, pearls, and coral reefs. Write the equilibrium constant expression for the formation of carbonic acid from carbon dioxide and water.

13.1 Rates of Reaction

Learning Goal: Describe how temperature, concentration, and catalysts affect the rate of a reaction.

- The rate of a reaction is the speed at which reactants are consumed or products are formed.
- Collision theory is used to explain reaction rates and the factors that affect them.
- The rate of a reaction depends on temperature, concentration, and the presence of a catalyst.
- At higher temperatures, reaction rates increase because reactants move faster, collide more often, and produce more collisions with the required energy of activation. The opposite effects are observed at lower temperatures.
- Increasing the concentrations of reactants often increases the rate of a reaction because collisions occur more often.
- The reaction rate is increased by the addition of a catalyst since a catalyst provides an alternate reaction pathway with a lower energy of activation.

♦ Learning Exercise 13.1A

Indicate the effect of each of the following on the rate of a chemical reaction:

increase (I) decrease (D) no effect (N)

1. _____ adding a catalyst
2. _____ running the reaction at a lower temperature
3. _____ doubling the concentrations of the reactants
4. _____ removing a catalyst
5. _____ running the experiment in a different laboratory with the same environmental conditions
6. _____ increasing the temperature
7. _____ using a container with a different shape
8. _____ using lower concentrations of reactants

Answers **1.** I **2.** D **3.** I **4.** D
5. N **6.** I **7.** N **8.** D

♦ Learning Exercise 13.1B

For the reaction, $NO_2(g) + CO(g) \longrightarrow NO(g) + CO_2(g)$, indicate the effect of each of the following conditions as:

increase (I) decrease (D) no effect (N)

1. _____ adding CO
2. _____ running the experiment under the same conditions two days later
3. _____ removing NO_2
4. _____ adding a catalyst
5. _____ adding NO_2

Answers **1.** I **2.** N **3.** D **4.** I **5.** I

13.2 Chemical Equilibrium

Learning Goal: Use the concept of reversible reactions to explain chemical equilibrium.

- A reversible reaction proceeds in both the forward and reverse directions.
- Chemical equilibrium is achieved when the rate of the forward reaction becomes equal to the rate of the reverse reaction.
- In a system at equilibrium, there is no change in the concentrations of reactants and products.
- At equilibrium, the concentrations of reactants can be greater than, less than, or in some cases, equal to the concentrations of products.

♦ Learning Exercise 13.2

Indicate if each of the following indicates a system at equilibrium (E) or not at equilibrium (NE):

a. _____ The rate of the forward reaction is faster than the rate of the reverse reaction.

b. _____ There is no change in the concentrations of reactants and products.

c. _____ The rate of the forward reaction is equal to the rate of the reverse reaction.

d. _____ The concentrations of reactants are decreasing.

e. _____ The concentrations of products are increasing.

Answers **a.** NE **b.** E **c.** E **d.** NE **e.** NE

13.3 Equilibrium Constants

Learning Goal: Calculate the equilibrium constant for a reversible reaction given the concentrations of reactants and products at equilibrium.

- The equilibrium constant expression for a system at equilibrium is the ratio of the concentrations of the products to the concentrations of the reactants with the concentration of each substance raised to a power equal to its coefficient in the equation.
- For the general equation, $a\text{A} + b\text{B} \rightleftarrows c\text{C} + d\text{D}$, the equilibrium constant is written:

$$K_c = \frac{[\text{C}]^c[\text{D}]^d}{[\text{A}]^a[\text{B}]^b}$$

- Because the concentrations of liquids and solids remain constant, their concentrations are not included in the equilibrium constant expression for heterogeneous systems.

Guide to Writing the Equilibrium Constant Expression	
Step 1	Write the balanced equilibrium equation.
Step 2	Write the concentrations of the products as the numerator and the reactants as the denominator.
Step 3	Write any coefficient in the equation as an exponent.

CORE CHEMISTRY SKILL

Writing the Equilibrium Constant Expression

♦ Learning Exercise 13.3A

Write the equilibrium constant expression (K_c) for each of the following reactions:

a. $2SO_3(g) \rightleftarrows 2SO_2(g) + O_2(g)$

b. $2NO(g) + Br_2(g) \rightleftarrows 2NOBr(g)$

c. $N_2(g) + 3H_2(g) \rightleftarrows 2NH_3(g)$

d. $2NO_2(g) \rightleftarrows N_2O_4(g)$

Answers

a. $K_c = \dfrac{[SO_2]^2[O_2]}{[SO_3]^2}$ **b.** $K_c = \dfrac{[NOBr]^2}{[NO]^2[Br_2]}$

c. $K_c = \dfrac{[NH_3]^2}{[N_2][H_2]^3}$ **d.** $K_c = \dfrac{[N_2O_4]}{[NO_2]^2}$

♦ Learning Exercise 13.3B

Write the equilibrium constant expression (K_c) for each of the following heterogeneous systems at equilibrium:

a. $H_2(g) + S(l) \rightleftharpoons H_2S(g)$

b. $H_2O(g) + C(s) \rightleftharpoons H_2(g) + CO(g)$

c. $2PbS(s) + 3O_2(g) \rightleftharpoons 2PbO(s) + 2SO_2(g)$

d. $SiH_4(g) + 2O_2(g) \rightleftharpoons SiO_2(s) + 2H_2O(g)$

Answers

a. $K_c = \dfrac{[H_2S]}{[H_2]}$ **b.** $K_c = \dfrac{[H_2][CO]}{[H_2O]}$

c. $K_c = \dfrac{[SO_2]^2}{[O_2]^3}$ **d.** $K_c = \dfrac{[H_2O]^2}{[SiH_4][O_2]^2}$

Guide to Calculating the K_c Value	
Step 1	State the given and needed quantities.
Step 2	Write the K_c expression for the equilibrium.
Step 3	Substitute equilibrium (molar) concentrations and calculate K_c.

♦ Learning Exercise 13.3C

CORE CHEMISTRY SKILL

Calculating an Equilibrium Constant

Calculate the K_c value for each of the following equilibrium concentrations:

a. $H_2(g) + I_2(g) \rightleftharpoons 2HI(g)$

$[H_2] = 0.28$ M $[I_2] = 0.28$ M $[HI] = 2.0$ M

b. $2NO_2(g) \rightleftharpoons N_2(g) + 2O_2(g)$
$[NO_2] = 0.60\ M$ $[N_2] = 0.010\ M$ $[O_2] = 0.020\ M$

c. $N_2(g) + 3H_2(g) \rightleftharpoons 2NH_3(g)$
$[N_2] = 0.50\ M$ $[H_2] = 0.20\ M$ $[NH_3] = 0.80\ M$

Answers

a. $$K_c = \frac{[HI]^2}{[H_2][I_2]} = \frac{[2.0]^2}{[0.28][0.28]} = 51$$

b. $$K_c = \frac{[N_2][O_2]^2}{[NO_2]^2} = \frac{[0.010][0.020]^2}{[0.60]^2} = 1.1 \times 10^{-5}$$

c. $$K_c = \frac{[NH_3]^2}{[N_2][H_2]^3} = \frac{[0.80]^2}{[0.50][0.20]^3} = 1.6 \times 10^2$$

13.4 Using Equilibrium Constants

Learning Goal: Use an equilibrium constant to predict the extent of reaction and to calculate equilibrium concentrations.

- A large K_c results when a reaction at equilibrium has more product(s) than reactant(s); a small K_c indicates that a reaction at equilibrium has more reactant(s) than product(s).
- The concentration of one component in an equilibrium mixture is calculated from the K_c and the concentrations of all other components.

♦ Learning Exercise 13.4A

Consider the reaction $2NOBr(g) \rightleftharpoons 2NO(g) + Br_2(g)$.

a. Write the equilibrium constant expression for the reaction at equilibrium.

b. If the equilibrium constant is 2×10^3, does the equilibrium mixture contain mostly reactants, mostly products, or about equal amounts of both reactants and products? Explain.

Answers

a. $$K_c = \frac{[NO]^2[Br_2]}{[NOBr]^2}$$

b. A large $K_c (>1)$ means the equilibrium mixture contains mostly products.

♦ Learning Exercise 13.4B

Consider the reaction $2HI(g) \rightleftarrows H_2(g) + I_2(g)$.

a. Write the expression for the equilibrium constant for the reaction.

b. If the equilibrium constant is 1.6×10^{-2}, does the equilibrium mixture contain mostly reactants, mostly products, or both reactants and products? Explain.

Answers

a. $K_c = \dfrac{[H_2][I_2]}{[HI]^2}$

b. A small K_c (<1) means that the equilibrium mixture contains mostly reactants.

Guide to Using the Equilibrium Constant	
Step 1	State the given and needed quantities.
Step 2	Write the K_c expression for the equilibrium and solve for the needed concentration.
Step 3	Substitute the equilibrium (molar) concentrations and calculate the needed concentration.

♦ Learning Exercise 13.4C

CORE CHEMISTRY SKILL
Calculating Equilibrium Concentrations

Calculate the concentration of the indicated component for each of the following at equilibrium:

a. $PCl_5(g) \rightleftarrows PCl_3(g) + Cl_2(g)$ $\quad K_c = 1.2 \times 10^{-2}$

$[PCl_5] = 2.50$ M $\quad [PCl_3] = 0.50$ M $\quad [Cl_2] = ?$

b. $CO(g) + H_2O(g) \rightleftarrows CO_2(g) + H_2(g)$ $\quad K_c = 1.6$

$[CO] = 1.0$ M $\quad [H_2O] = 0.80$ M $\quad [CO_2] = ?$ $\quad [H_2] = 1.2$ M

Answers

a. $K_c = \dfrac{[PCl_3][Cl_2]}{[PCl_5]} \qquad [Cl_2] = \dfrac{K_c[PCl_5]}{[PCl_3]} = \dfrac{1.2 \times 10^{-2}[2.50]}{[0.50]} = 6.0 \times 10^{-2}\text{ M}$

b. $K_c = \dfrac{[CO_2][H_2]}{[CO][H_2O]} \qquad [CO_2] = \dfrac{K_c[CO][H_2O]}{[H_2]} = \dfrac{1.6[1.0][0.80]}{[1.2]} = 1.1\text{ M}$

13.5 Changing Equilibrium Conditions: Le Châtelier's Principle

Learning Goal: Use Le Châtelier's principle to describe the changes made in equilibrium concentrations when reaction conditions change.

- Le Châtelier's principle states when a stress (change in conditions) is placed on a reaction at equilibrium, the equilibrium will shift in a direction (forward or reverse) that relieves the stress.
- A change in the concentration of a reactant or product or the temperature of any reaction will cause the reaction to shift to relieve the stress.
- A change in the volume of a reaction involving gases will cause a shift if the reaction involves different number of moles of gaseous reactants and products.
- The addition of a catalyst does not affect the position of an equilibrium reaction.

♦ Learning Exercise 13.5A

CORE CHEMISTRY SKILL
Using Le Châtelier's Principle

Identify the effect of each of the following on the reaction at equilibrium:

$$N_2(g) + O_2(g) + 180\text{ kJ} \rightleftarrows 2NO(g)$$

a. shift toward products **b.** shift toward reactants **c.** no change

1. ____ adding $O_2(g)$

2. ____ removing $N_2(g)$

3. ____ removing $NO(g)$

4. ____ adding heat

5. ____ reducing the container volume

6. ____ increasing the container volume

7. ____ adding a platinum catalyst

Answers **1.** a **2.** b **3.** a **4.** a **5.** c **6.** c **7.** c

♦ Learning Exercise 13.5B

Identify the effect of the each of the following on the equilibrium of the reaction:

$$2NOBr(g) \rightleftarrows 2NO(g) + Br_2(g) + 340\text{ kJ}$$

a. shift toward products **b.** shift toward reactants **c.** no change

1. ____ adding $NO(g)$

2. ____ removing $Br_2(g)$

3. ____ removing $NOBr(g)$

4. ____ adding a catalyst

5. ____ lowering the temperature

6. ____ increasing the volume of the container

Answers **1.** b **2.** a **3.** b **4.** c **5.** a **6.** a

13.6 Equilibrium in Saturated Solutions

Learning Goal: Write the solubility product expression for a slightly soluble salt and calculate the K_{sp}; use the K_{sp} to determine the solubility.

- The solubility of a slightly soluble salt in an aqueous solution is represented by an equilibrium expression called the solubility product constant (K_{sp}).
- As in other heterogeneous equilibria, the concentration of the solid is constant and not included in the K_{sp} expression. For example, the K_{sp} of AgBr is: $K_{sp} = [Ag^+][Br^-]$

- The K_{sp} can be used to calculate the solubility of the salt or the solubility can be used to calculate the K_{sp}.
- The solubility of a salt is affected by the addition of an ion (common) already present in the equilibrium.

Key Terms for Sections 13.1–13.6

Match each of the following key terms with the correct definition:

a. activation energy
b. equilibrium
c. Le Châtelier's principle
d. catalyst
e. equilibrium constant expression
f. collision theory
g. heterogeneous equilibrium
h. solubility product constant

1. ____ a substance that lowers the activation energy and increases the rate of reaction

2. ____ equilibrium components are present in at least two different states

3. ____ states that a stress placed on a reaction at equilibrium causes the equilibrium to shift in a direction that relieves the stress

4. ____ describes the equilibrium of a slightly soluble salt in an aqueous solution

5. ____ the ratio of the concentrations of products to reactants with each component raised to an exponent equal to its coefficient

6. ____ the energy that is required in a collision to break the bonds in the reactants

7. ____ a reaction requires that reactants collide to form products

8. ____ the condition in which the rate of the forward reaction is equal to the rate of the reverse reaction

Answers **1.** d **2.** g **3.** c **4.** h **5.** e **6.** a **7.** f **8.** b

♦ Learning Exercise 13.6A

CORE CHEMISTRY SKILL
Writing the Solubility Product Expression

Write the equilibrium equation and the K_{sp} expression for each of the following slightly soluble salts:

a. BaF_2

b. NiS

c. Ag_2CO_3

d. Ag_3PO_4

Answers

a. $BaF_2(s) \rightleftarrows Ba^{2+}(aq) + 2F^-(aq)$ $\quad K_{sp} = [Ba^{2+}][F^-]^2$
b. $NiS(s) \rightleftarrows Ni^{2+}(aq) + S^{2-}(aq)$ $\quad K_{sp} = [Ni^{2+}][S^{2-}]$
c. $Ag_2CO_3(s) \rightleftarrows 2Ag^+(aq) + CO_3^{2-}(aq)$ $\quad K_{sp} = [Ag^+]^2[CO_3^{2-}]$
d. $Ag_3PO_4(s) \rightleftarrows 3Ag^+(aq) + PO_4^{3-}(aq)$ $\quad K_{sp} = [Ag^+]^3[PO_4^{3-}]$

Guide to Calculating K_{sp}	
Step 1	State the given and needed quantities.
Step 2	Write the equilibrium equation for the dissociation of the slightly soluble salt.
Step 3	Write the solubility product expression (K_{sp}).
Step 4	Substitute the molar concentration of each ion into the K_{sp} expression and calculate.

♦ Learning Exercise 13.6B

CORE CHEMISTRY SKILL

Calculating the Solubility Product Constant

Calculate the K_{sp} value for each of the following slightly soluble salts given the equilibrium concentration of the ions:

a. CdS with $[Cd^{2+}] = [S^{2-}] = 3 \times 10^{-14}$ M

b. $AgIO_3$ with $[Ag^+] = [IO_3^-] = 1.7 \times 10^{-4}$ M

c. SrF_2 with $[Sr^{2+}] = 8.5 \times 10^{-4}$ and $[F^-] = 1.7 \times 10^{-3}$

d. Ag_2SO_3 with $[Ag^+] = 3.2 \times 10^{-5}$ and $[SO_3^{2-}] = 1.6 \times 10^{-5}$

Answers

a. $K_{sp} = 9 \times 10^{-28}$ $\quad$ b. $K_{sp} = 2.9 \times 10^{-8}$
c. $K_{sp} = 2.5 \times 10^{-9}$ $\quad$ d. $K_{sp} = 1.6 \times 10^{-14}$

Guide to Calculating Molar Solubility from K_{sp}	
Step 1	State the given and needed quantities.
Step 2	Write the equilibrium equation for the dissociation of the slightly soluble salt.
Step 3	Write the solubility product expression, K_{sp}, using S.
Step 4	Calculate the molar solubility S.

♦ Learning Exercise 13.6C

CORE CHEMISTRY SKILL
Calculating the Molar Solubility

Calculate the molar solubility (S) of each of the following slightly soluble salts:

a. $CaSO_4$ $\quad K_{sp} = 9 \times 10^{-6}$

b. $ZnCO_3$ $\quad K_{sp} = 1.4 \times 10^{-11}$

c. FeS $\quad K_{sp} = 8 \times 10^{-19}$

d. CuI $\quad K_{sp} = 1.3 \times 10^{-12}$

Answers
a. $CaSO_4$ $\quad S = 3 \times 10^{-3}$ M $\qquad$ **b.** $ZnCO_3$ $\quad S = 3.7 \times 10^{-6}$ M
c. FeS $\quad S = 9 \times 10^{-10}$ M $\qquad$ **d.** CuI $\quad S = 1.1 \times 10^{-6}$ M

♦ Learning Exercise 13.6D

What is the $[Ca^{2+}]$ in each of the following solutions, given that for $CaSO_4$ the $K_{sp} = 9 \times 10^{-6}$:

a. A solution with $[SO_4^{2-}] = 1.0 \times 10^{-1}$ M

b. A solution with $[SO_4^{2-}] = 2.7 \times 10^{-2}$ M

Answers **a.** $[Ca^{2+}] = 9 \times 10^{-5}$ M $\qquad$ **b.** $[Ca^{2+}] = 3 \times 10^{-4}$ M

Checklist for Chapter 13

You are ready to take the Practice Test for Chapter 13. Be sure you have accomplished the following learning goals for this chapter. If not, review the section listed at the end of the goal. Then apply your new skills and understanding to the Practice Test.

After studying Chapter 13, I can successfully:

_____ Describe the factors that increase or decrease the rate of a reaction (13.1).

_____ Write the forward and reverse reactions of a reversible reaction (13.2).

_____ Explain how equilibrium occurs when the rate of a forward reaction is equal to the rate of a reverse reaction (13.2).

_____ Write the equilibrium constant expression for a reaction system at equilibrium (13.3).

_____ Calculate the equilibrium constant from the equilibrium concentrations (13.3).

_____ Use the equilibrium constant to determine whether an equilibrium favors the reactants or products (13.4).

_____ Use the equilibrium constant expression to determine the equilibrium concentration of a component in the reaction (13.4).

_____ Use Le Châtelier's principle to describe the shift in a system at equilibrium when stress is applied to the system (13.5).

_____ Calculate the solubility product for a saturated salt solution (13.6).

_____ Use the solubility product expression to calculate the solubility of a slightly soluble salt (13.6).

_____ Use the solubility product expression to calculate the solubility of a slightly soluble salt in the presence of a common ion (13.6).

Practice Test for Chapter 13

1. The number of effective molecular collisions increases when (13.1)
A. more reactants are added.
B. products are removed.
C. the energy of collision is below the energy of activation.
D. the reaction temperature is lowered.
E. the reacting molecules have an incorrect orientation upon impact.

2. The energy of activation is lowered when (13.1)
A. more reactants are added.
B. products are removed.
C. a catalyst is used.
D. the reaction temperature is lowered.
E. the reaction temperature is raised.

3. Food deteriorates more slowly in a refrigerator because (13.1)
A. more reactants are added.
B. products are removed.
C. the energy of activation is higher.
D. fewer collisions have the energy of activation.
E. collisions have the wrong orientation upon impact.

4. A reaction reaches equilibrium when (13.2)
A. the rate of the forward reaction is faster than the rate of the reverse reaction.
B. the rate of the reverse reaction is faster than the rate of the forward reaction.
C. the concentrations of reactants and products are changing.
D. fewer collisions have the energy of activation.
E. the rate of the forward reaction is equal to the rate of the reverse reaction.

5. The equilibrium constant expression for the following reaction is equal to (13.3)
$2NOCl(g) \rightleftharpoons 2NO(g) + Cl_2(g)$

A. $\frac{[NO][Cl_2]}{[NOCl]}$ **B.** $\frac{[NOCl_2]^2}{[NO]^2[Cl_2]}$ **C.** $\frac{[NOCl_2]}{[NO][Cl_2]}$

D. $\frac{[NO^2][Cl_2]}{[NOCl]}$ **E.** $\frac{[NO]^2[Cl_2]}{[NOCl]^2}$

6. The equilibrium constant expression for the following reaction is equal to (13.3)
$MgO(s) \rightleftharpoons CO_2(g) + MgCO_3(s)$

A. $[CO_2]$ **B.** $\frac{[CO_2][MgCO_3]}{[MgO]}$ **C.** $\frac{[MgO]}{[CO_2][MgCO_3]}$

D. $\frac{1}{[CO_2]}$ **E.** $\frac{[CO_2]}{[MgO]}$

7. The equilibrium constant expression for the following reaction is equal to (13.3)
$2PbS(s) + 3O_2(g) \rightleftharpoons 2PbO(s) + 2SO_2(g)$

A. $\frac{[PbO][SO_2]}{[PbS][O_2]}$ **B.** $\frac{[PbO]^2[SO_2]^2}{[PbS]^2[O_2]^3}$ **C.** $\frac{[SO_2]^2}{[O_2]^3}$

D. $\frac{[SO_2]}{[O_2]}$ **E.** $\frac{[O_2]^3}{[SO_2]^2}$

8. The equilibrium equation that has the following equilibrium constant expression is (13.3)
$\frac{[H_2S]^2}{[H_2]^2[S_2]}$

A. $H_2S(g) \rightleftharpoons H_2(g) + S_2(g)$
B. $2H_2S(g) \rightleftharpoons H_2(g) + S_2(g)$
C. $2H_2(g) + S_2(g) \rightleftharpoons 2H_2S(g)$
D. $2H_2S(g) \rightleftharpoons 2H_2(g)$
E. $2H_2(g) \rightleftharpoons 2H_2S(g)$

9. Use the given equilibrium concentrations to calculate the value of the equilibrium constant for the following equilibrium is (13.3)
$COBr_2(g) \rightleftharpoons CO(g) + Br_2(g)$
$[COBr_2] = 0.93$ M $[CO] = [Br_2] = 0.42$ M
A. 0.19 **B.** 0.39 **C.** 0.42
D. 2.2 **E.** 5.3

10. Use the given equilibrium concentrations to calculate the value of the equilibrium constant for the following equilibrium is (13.3)
$2NO(g) + O_2(g) \rightleftharpoons 2NO_2(g)$
$[NO] = 2.7$ M $[O_2] = 1.0$ $[NO_2] = 3.0$ M
A. 0.81 **B.** 1.1 **C.** 1.2
D. 8.1 **E.** 9.0

11. Use the given equilibrium concentrations to calculate the $[PCl_5]$ for the decomposition of PCl_5 that has a $K_c = 0.050$. (13.4)
$PCl_5(g) \rightleftharpoons PCl_3(g) + Cl_2(g)$
$[PCl_3] = [Cl_2] = 0.20$ M
A. 0.01 M **B.** 0.050 M **C.** 0.20 M
D. 0.40 M **E.** 0.80 M

12. The reaction that has a much greater concentration of products at equilibrium has a K_c value of (13.4)
A. 1.6×10^{-15} **B.** 2×10^{-11} **C.** 1.2×10^{-5}
D. 3×10^{-3} **E.** 1.4×10^{5}

13. The reaction that has a much greater concentration of reactants at equilibrium has a K_c value of (13.4)
A. 1.1×10^{-11} **B.** 2×10^{-2} **C.** 1.2×10^{2}
D. 2×10^{4} **E.** 1.3×10^{12}

14. The reaction that has about the same concentration of reactant and products at equilibrium has a K_c value of (13.4)
A. 1.4×10^{-12} **B.** 2×10^{-8} **C.** 1.2
D. 3×10^{2} **E.** 1.3×10^{7}

For questions 15 through 19, indicate how each of the following affects the equilibrium of the reaction shown:
$PCl_5(g) + \text{heat} \rightleftharpoons PCl_3(g) + Cl_2(g)$ (13.5)

A. shifts toward products **B.** shifts toward reactants **C.** no change

15. add more Cl_2 **16.** cool the reaction **17.** remove some PCl_3

18. add more PCl_5 **19.** remove some PCl_5

For questions 20 through 24, indicate whether the equilibrium will (13.5)

A. shift toward products **B.** shift toward reactants **C.** no change

$$2NO(g) + O_2(g) \rightleftharpoons 2NO_2(g) + \text{heat}$$

20. add more NO **21.** increase temperature **22.** add a catalyst

23. add some O_2 **24.** remove NO_2

25. The solubility product expression for the insoluble salt $Ca_3(PO_4)_2$ is equal to (13.6)
A. $[Ca^{2+}][PO_4^{3-}]$ **B.** $[Ca^{2+}]_3[PO_4^{3-}]_2$ **C.** $[Ca^{2+}]^3[PO_4^{3-}]^2$
D. $[Ca^{2+}]^2[PO_4^{3-}]^3$ **E.** $\dfrac{[Ca^{2+}][PO_4^{3-}]}{[Ca_3(PO_4^{3-})_2]}$

26. The K_{sp} of CuI when a saturated solution has $[Cu^+] = 1 \times 10^{-6}$ M and $[I^-] = 1 \times 10^{-6}$ M is equal to (13.6)
A. 1×10^{-12} **B.** 1×10^{-6} **C.** 1
D. 2×10^{-6} **E.** 2×10^{-12}

27. What is the molar solubility of $SrCO_3$ if it has a K_{sp} of 5.6×10^{-10}? (13.6)
A. 2.8×10^{-10} M **B.** 2.8×10^{-5} M **C.** 5.6×10^{-10} M
D. 2.3×10^{-20} M **E.** 2.4×10^{-5} M

28. What is the molar solubility of $SrCO_3$ in a solution with $[CO_3^{2-}] = 0.10$ M? $SrCO_3$ has a K_{sp} of 5.6×10^{-10}. (13.6)
A. 0.10 M **B.** 2.4×10^{-5} M **C.** 5.6×10^{-9} M
D. 5.6×10^{-10} M **E.** 5.6×10^{-11} M

Answers to the Practice Test

1. A	2. C	3. D	4. E	5. E
6. A	7. C	8. C	9. A	10. C
11. E	12. E	13. A	14. C	15. B
16. B	17. A	18. A	19. B	20. A
21. B	22. C	23. A	24. A	25. C
26. A	27. E	28. C		

14
Acids and Bases

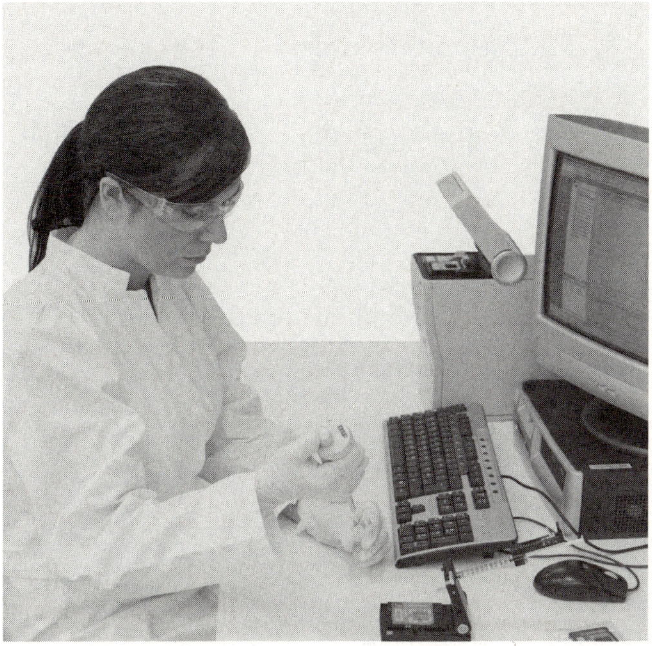

A 55-year-old woman has come into the emergency room complaining of being tired and short of breath. She has a history of smoking one or more packs of cigarettes a day for 25 years. She is somewhat overweight and coughs frequently. A chest examination indicates rales (dry crackles) and wheezes. At rest, her respiratory rate is more than 20 breaths per minute. Based on the blood samples analyzed by Brianna, a laboratory clinician, the oxygen concentration in the patient's arterial blood is 60 mmHg, which is below the normal oxygen level of 80–100 mmHg. The blood pH is 7.55, above the normal range of 7.35–7.45. When the test results are returned to the emergency room doctor, he diagnoses the patient as having chronic bronchitis and asthma, forms a of chronic obstructive pulmonary disease, COPD. COPD, also known as airflow obstruction, is a condition that makes it difficult to move air in and out of the lungs. To relieve the symptoms, the patient's doctor prescribes Spiriva and an inhaler with formoterol and budesonide (Symbicort). What is the pH of a sample of blood that has $[H_3O^+] = 3.5 \times 10^{-8}$ M?

14.1 Acids and Bases

Learning Goal: Describe and name Arrhenius acids and bases and organic acids.

- In water, an Arrhenius acid produces hydrogen ions (H^+) and an Arrhenius base produces hydroxide ions (OH^-).
- The names of inorganic acids are based on the anion present.
- Arrhenius bases are named as metal or ammonium hydroxides.
- In the IUPAC system, a carboxylic acid is named by replacing the *e* ending of the alkane name with *oic acid*. Simple acids are named by the common naming system using the prefixes form- (1C), acet- (2C), propion- (4C), butyr- (4C), followed by *ic acid*: HCOOH methanoic acid (formic acid), $CH_3{-}COOH$ ethanoic acid (acetic acid), $CH_3{-}CH_2{-}COOH$ propanoic acid (propionic acid), $CH_3{-}CH_2{-}CH_2{-}COOH$ butanoic acid (butyric acid).

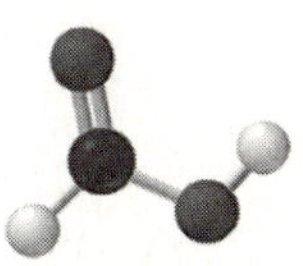

Methanoic acid (formic acid)

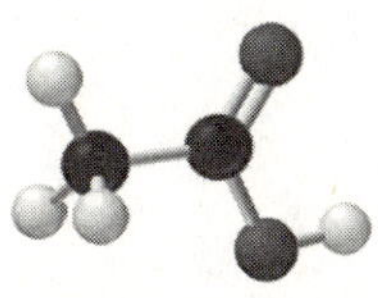

Ethanoic acid (acetic acid)

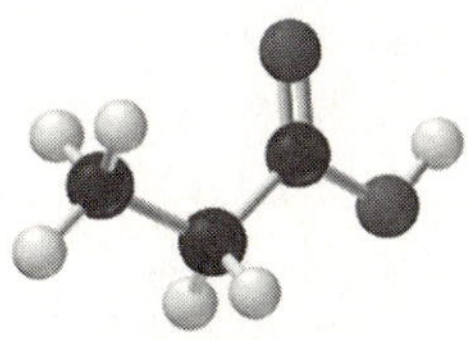

Propanoic acid (propionic acid)

♦ Learning Exercise 14.1A

Indicate if each of the following characteristics describe an (A) acid or (B) base:

1. ____ turns blue litmus red
2. ____ tastes sour
3. ____ contains more OH^- than H_3O^+
4. ____ neutralizes bases
5. ____ tastes bitter
6. ____ turns red litmus blue
7. ____ contains more H_3O^+ than OH^-
8. ____ neutralizes acids

Answers **1.** A **2.** A **3.** B **4.** A **5.** B **6.** B **7.** A **8.** B

♦ Learning Exercise 14.1B

Fill in the blanks with the formula or name of an acid or base.

1. HCl ____________
2. ________ sodium hydroxide
3. ________ sulfurous acid
4. ________ nitric acid
5. $Ca(OH)_2$ ____________
6. H_2CO_3 ____________
7. $Al(OH)_3$ ____________
8. ________ potassium hydroxide
9. $HClO_4$ ____________
10. H_3PO_3 ____________

Answers
1. hydrochloric acid
2. NaOH
3. H_2SO_3
4. HNO_3
5. calcium hydroxide
6. carbonic acid
7. aluminum hydroxide
8. KOH
9. perchloric acid
10. phosphorous acid

♦ Learning Exercise 14.1C

Give the IUPAC and common names or condensed structural formula for each of the following:

1. ethanoic acid ____________
2. $CH_3—CH_2—\overset{\overset{\large O}{\|}}{C}—OH$ ____________
3. butanoic acid ____________
4. formic acid ____________
5. $CH_3—CH_2—CH_2—CH_2—COOH$ ____________

Answers

1. $CH_3-\overset{\overset{\displaystyle O}{\|}}{C}-OH$
2. propanoic acid; propionic acid
3. $CH_3-CH_2-CH_2-\overset{\overset{\displaystyle O}{\|}}{C}-OH$
4. $H-\overset{\overset{\displaystyle O}{\|}}{C}-OH$
5. pentanoic acid

14.2 Brønsted–Lowry Acids and Bases

Learning Goal: Identify conjugate acid–base pairs for Brønsted–Lowry acids and bases.

- According to the Brønsted–Lowry theory, acids are proton (H^+) donors, and bases are proton acceptors.
- In water, protons bond to polar water molecules to form hydronium ions (H_3O^+).
- Conjugate acid–base pairs are molecules or ions linked by the loss and gain of one H^+.
- Every proton transfer reaction involves two acid–base conjugate pairs.

♦ Learning Exercise 14.2A

Complete the following conjugate acid–base pairs:

	Conjugate Acid	Conjugate Base
1.	H_2O	______
2.	HSO_4^-	______
3.	______	F^-
4.	______	CO_3^{2-}
5.	HNO_3	______
6.	NH_4^+	______
7.	______	HS^-
8.	______	$H_2PO_4^-$

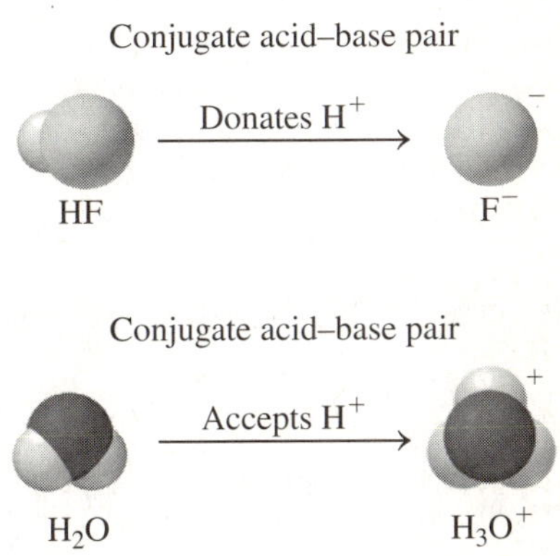

Answers

1. OH^-	2. SO_4^{2-}	3. HF	4. HCO_3^-
5. NO_3^-	6. NH_3	7. H_2S	8. H_3PO_4

Guide to Writing Conjugate Acid–Base Pairs	
Step 1	Identify the reactant that loses H^+ as the conjugate acid.
Step 2	Identify the reactant that gains H^+ as the conjugate base.
Step 3	Write the conjugate acid–base pairs for each.

♦ Learning Exercise 14.2B

CORE CHEMISTRY SKILL

Identifying Conjugate Acid–Base Pairs

Identify the conjugate acid–base pairs in each of the following reactions:

1. $HF(aq) + H_2O(l) \rightleftarrows H_3O^+(aq) + F^-(aq)$

2. $NH_4^+(aq) + SO_4^{2-}(aq) \rightleftarrows NH_3(aq) + HSO_4^-(aq)$

3. $NH_3(aq) + H_2O(l) \rightleftarrows NH_4^+(aq) + OH^-(aq)$

4. $HNO_3(aq) + OH^-(aq) \rightleftarrows H_2O(l) + NO_3^-(aq)$

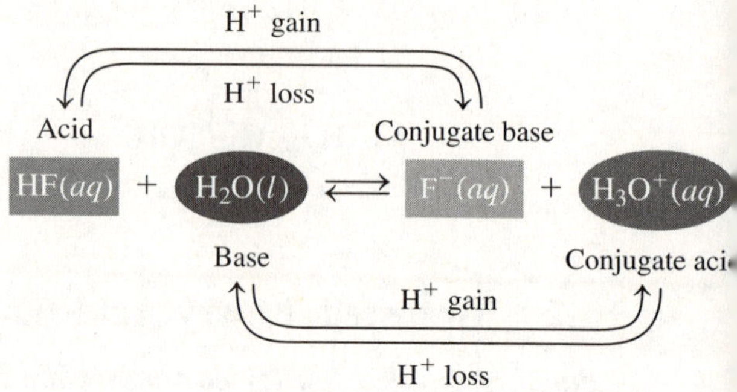

Answers

1. HF/F^- and H_3O^+/H_2O
2. NH_4^+/NH_3 and HSO_4^-/SO_4^{2-}
3. NH_4^+/NH_3 and H_2O/OH^-
4. HNO_3/NO_3^- and H_2O/OH^-

♦ Learning Exercise 14.2C

Write an equation with conjugate acid–base pairs starting with each of the following reactants:

1. HBr (acid) and CO_3^{2-} (base)

2. HSO_4^- (acid) and OH^- (base)

3. NH_4^+ (acid) and H_2O (base)

4. HCl (acid) and SO_4^{2-} (base)

Answers

1. $HBr(aq) + CO_3^{2-}(aq) \rightleftarrows HCO_3^-(aq) + Br^-(aq)$
2. $HSO_4^-(aq) + OH^-(aq) \rightleftarrows H_2O(l) + SO_4^{2-}(aq)$
3. $NH_4^+(aq) + H_2O(l) \rightleftarrows H_3O^+(aq) + NH_3(aq)$
4. $HCl(aq) + SO_4^{2-}(aq) \rightleftarrows Cl^-(aq) + HSO_4^-(aq)$

14.3 Strengths of Acids and Bases

Learning Goal: Write equations for the dissociation of strong and weak acids; identify the direction of reaction.

- In aqueous solution, a strong acid donates nearly all its protons to water, whereas a weak acid donates only a small percentage of protons to water.
- Most hydroxides of Groups 1A (1) and 2A (2) are strong bases, which dissociate almost completely in water.
- In an aqueous ammonia solution, a weak base, NH_3, accepts only a small percentage of protons to form the conjugate acid, NH_4^+.
- As the strength of an acid decreases, the strength of the conjugate base increases. By comparing relative strengths, the direction of an acid-base reaction can be predicted.

Study Note

Only six common acids are strong acids; other acids are weak acids.

HCl	HNO_3
HBr	H_2SO_4 (first H)
HI	$HClO_4$

Example: Is H_2S a strong or a weak acid?

Solution: H_2S is a weak acid because it is not one of the six strong acids.

♦ Learning Exercise 14.3A

Identify each of the following as a strong or weak acid or base:

1. HNO_3 ________ **2.** H_2CO_3 ________

3. $H_2PO_4^-$ ________ **4.** NH_3 ________

5. LiOH ________ **6.** H_3BO_3 ________

7. $Ca(OH)_2$ ________ **8.** H_2SO_4 ________

Answers **1.** strong acid **2.** weak acid **3.** weak acid **4.** weak base **5.** strong base **6.** weak acid **7.** strong base **8.** strong acid (first H^+)

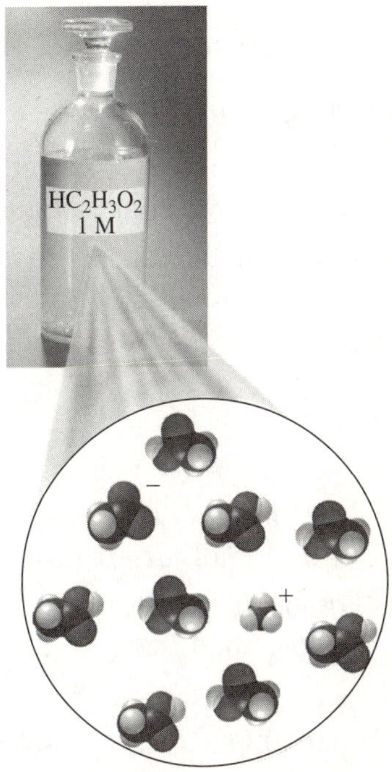

Acetic acid ($HC_2H_3O_2$) is a weak acid because it dissociates slightly in water, producing only a small percentage of H_3O^+.

♦ Learning Exercise 14.3B

Using Table 14.4 (in textbook), identify the stronger acid in each of the following pairs of acids:

1. HCl or H_2CO_3 ________ **2.** HNO_2 or HCN ________

3. H_2S or HBr ________ **4.** H_2SO_4 or HSO_4^- ________

5. HF or H_3PO_4 ________

Answers **1.** HCl **2.** HNO_2 **3.** HBr **4.** H_2SO_4 **5.** H_3PO_4

Study Note

In an acid–base reaction, the relative strengths of the two acids or two bases indicate whether the equilibrium contain mostly reactants or products.

Example: Does the equilibrium mixture for the following reaction contain mostly reactants or products?

$$NO_3^-(aq) + H_2O(l) \rightleftarrows HNO_3(aq) + OH^-(aq)$$

Solution: Because the reactants contain the weaker base and acid, NO_3^- and H_2O, the equilibrium mixture contains mostly reactants.

$$NO_3^-(aq) + H_2O(l) \rightleftarrows HNO_3(aq) + OH^-(aq)$$

♦ Learning Exercise 14.3C

Indicate whether each of the following reactions contains mostly reactants or products:

1. $HNO_3(aq) + H_2O(l) \rightleftarrows H_3O^+(aq) + NO_3^-(aq)$
2. $I^-(aq) + H_3O^+(aq) \rightleftarrows H_2O(l) + HI(aq)$
3. $NH_3(aq) + H_2O(l) \rightleftarrows NH_4^+(aq) + OH^-(aq)$
4. $HCl(aq) + CO_3^{2-}(aq) \rightleftarrows Cl^-(aq) + HCO_3^-(aq)$

Answers **1.** products **2.** reactants **3.** reactants **4.** products

14.4 Dissociation Constants for Acids and Bases

Learning Goal: Write the expression for the dissociation constant of a weak acid or weak base.

- An acid or base dissociation constant is the equilibrium constant when the concentration of water is considered constant and is omitted from the equilibrium constant expression.
- For an acid dissociation constant, as with other equilibrium expressions, the molar concentrations of the products are divided by the molar concentrations of the reactants.

$$\mathbf{HA}(aq) + \mathbf{H_2O}(l) \rightleftarrows \mathbf{H_3O^+}(aq) + \mathbf{A^-}(aq)$$

$$K_a = K_c[H_2O] = \frac{[\mathbf{H_3O^+}][\mathbf{A^-}]}{[\mathbf{HA}]}$$

- An acid or a base with a large dissociation constant is more fully dissociated than an acid or base with a small dissociation constant.
- Dissociation constants greater than one ($K > 1$) contain mostly products, whereas constants smaller than one ($K < 1$) contain mostly reactants.

♦ Learning Exercise 14.4A

Write the acid dissociation constant expression for the ionization of each of the following weak acids:

1. HCN
2. HNO_2

3. H_2CO_3 (first ionization only)

4. H_2S (first ionization only)

Answers

1. $K_a = \frac{[H_3O^+][CN^-]}{[HCN]}$

2. $K_a = \frac{[H_3O^+][NO_2^-]}{[HNO_2]}$

3. $K_a = \frac{[H_3O^+][HCO_3^-]}{[H_2CO_3]}$

4. $K_a = \frac{[H_3O^+][HS^-]}{[H_2S]}$

♦ Learning Exercise 14.4B

For each of the following pairs of acid dissociation constants, indicate the dissociation constant of the weaker acid:

1. 5.2×10^{-5} or 3.8×10^{-3} ____________

2. 3.0×10^{-8} or 1.6×10^{-10} ____________

3. 4.5×10^{-2} or 7.2×10^{-6} ____________

Answers **1.** 5.2×10^{-5} **2.** 1.6×10^{-10} **3.** 7.2×10^{-6}

♦ Learning Exercise 14.4C

Indicate whether the equilibrium mixture contains mostly reactants or products for each of the following dissociation constants:

1. 5.2×10^{-5} ____________

2. 3.0×10^{8} ____________

3. 4.5×10^{-5} ____________

4. 7.2×10^{15} ____________

Answers **1.** reactants **2.** products **3.** reactants **4.** products

14.5 Ionization of Water

Learning Goal: Use the ion product of water to calculate the $[H_3O^+]$ and $[OH^-]$ in an aqueous solution.

- In pure water, a few water molecules transfer protons (H^+) to other water molecules, producing small, but equal, amounts of each ion such that $[H_3O^+]$ and $[OH^-]$ each $= 1.0 \times 10^{-7}$ M.

$$H_2O(l) + H_2O(l) \rightleftharpoons H_3O^+(aq) + OH^-(aq)$$

- K_w, the ion product constant of water, $[H_3O^+][OH^-] = [1.0 \times 10^{-7}][1.0 \times 10^{-7}] = 1.0 \times 10^{-14}$, applies to all aqueous solutions.
- In acidic solutions, the $[H_3O^+]$ is greater than the $[OH^-]$. In basic solutions, the $[OH^-]$ is greater than the $[H_3O^+]$.

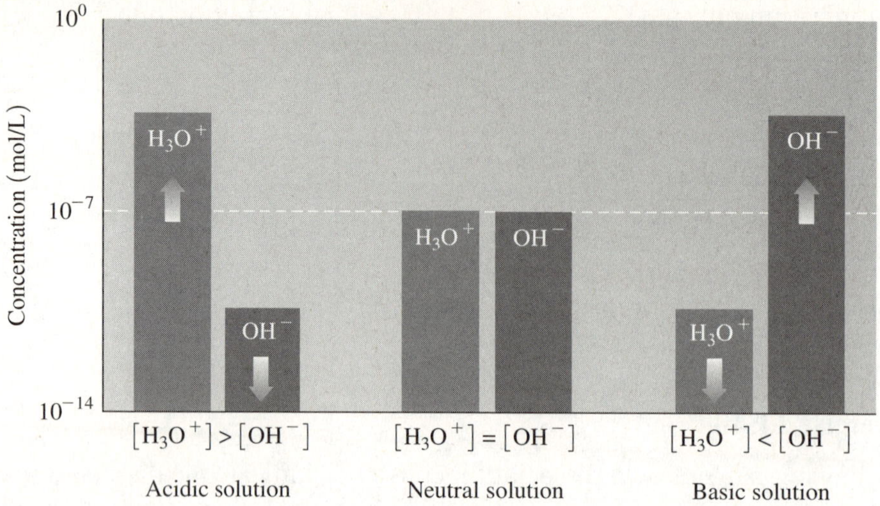

Guide to Calculating $[H_3O^+]$ and $[OH^-]$ in Aqueous Solutions	
Step 1	State the given and needed quantities.
Step 2	Write the K_w for water and solve for the unknown $[H_3O^+]$ or $[OH^-]$.
Step 3	Substitute the known $[H_3O^+]$ or $[OH^-]$ into the equation and calculate.

Example: What is the $[H_3O^+]$ in a solution that has $[OH^-] = 2.0 \times 10^{-9}$ M?

Step 1: State the given and needed quantities.

Solution: **Analyze the Problem**

Given $[OH^-] = 2.0 \times 10^{-9}$ **M** **Need** $[H_3O^+]$ $K_w = [H_3O^+][OH^-] = 1.0 \times 10^{-14}$

Step 2: Write the K_w for water and solve for the unknown $[H_3O^+]$ or $[OH^-]$.

$$[H_3O^+][OH^-] = 1.0 \times 10^{-14}$$

$$[H_3O^+] = \frac{1.0 \times 10^{-14}}{[OH^-]}$$

Step 3: Substitute the known $[H_3O^+]$ or $[OH^-]$ into the equation and calculate.

$$[H_3O^+] = \frac{1.0 \times 10^{-14}}{2.0 \times 10^{-9}} = 5.0 \times 10^{-6}\text{ M}$$

CORE CHEMISTRY SKILL

Calculating $[H_3O^+]$ and $[OH^-]$ in Solutions

♦ Learning Exercise 14.5A

Use the K_w to calculate the $[H_3O^+]$ when the $[OH^-]$ has each of the following values:

a. $[OH^-] = 1.0 \times 10^{-10}$ M $[H_3O^+] =$ ______________

b. $[OH^-] = 2.0 \times 10^{-5}$ M $[H_3O^+] =$ ______________

c. $[OH^-] = 4.5 \times 10^{-7}$ M $[H_3O^+] =$ ______________

d. $[OH^-] = 8.0 \times 10^{-4}$ M $\quad [H_3O^+] =$ ______________

e. $[OH^-] = 5.5 \times 10^{-8}$ M $\quad [H_3O^+] =$ ______________

Answers **a.** 1.0×10^{-4} M **b.** 5.0×10^{-10} M **c.** 2.2×10^{-8} M
d. 1.3×10^{-11} M **e.** 1.8×10^{-7} M

♦ Learning Exercise 14.5B

Use the K_w to determine the $[OH^-]$ when the $[H_3O^+]$ has each of the following values:

a. $[H_3O^+] = 1.0 \times 10^{-3}$ M $\quad [OH^-] =$ ______________

b. $[H_3O^+] = 3.0 \times 10^{-10}$ M $\quad [OH^-] =$ ______________

c. $[H_3O^+] = 4.0 \times 10^{-6}$ M $\quad [OH^-] =$ ______________

d. $[H_3O^+] = 2.8 \times 10^{-13}$ M $\quad [OH^-] =$ ______________

e. $[H_3O^+] = 8.6 \times 10^{-7}$ M $\quad [OH^-] =$ ______________

Answers **a.** 1.0×10^{-11} M **b.** 3.3×10^{-5} M **c.** 2.5×10^{-9} M
d. 3.6×10^{-2} M **e.** 1.2×10^{-8} M

14.6 The pH Scale

Learning Goal: Calculate pH from $[H_3O^+]$; given the pH, calculate the $[H_3O^+]$ and $[OH^-]$ of a solution.

- The pH scale is a range of numbers generally running from 0 to 14 related to the $[H_3O^+]$ of the solution.
- A neutral solution has a pH of 7.0. In acidic solutions, the pH is below 7.0, and in basic solutions, the pH is above 7.0.
- Mathematically, pH is the negative logarithm of the hydronium ion concentration: $pH = -\log[H_3O^+]$.
- The pOH is similar to pH; $pOH = -\log[OH^-]$. The pH plus the pOH are equal to 14.00.

♦ Learning Exercise 14.6A

State whether the following pH values are acidic, basic, or neutral:

1. ____ plasma, pH = 7.4 **2.** ____ soft drink, pH = 2.8

3. ____ maple syrup, pH = 6.8 **4.** ____ beans, pH = 5.0

5. ____ tomatoes, pH = 4.2 **6.** ____ lemon juice, pH = 2.2

7. ____ saliva, pH = 7.0 **8.** ____ egg white, pH = 7.8

9. ____ lye, pH = 12.4 **10.** ____ strawberries, pH = 3.0

Answers **1.** basic **2.** acidic **3.** acidic **4.** acidic **5.** acidic
6. acidic **7.** neutral **8.** basic **9.** basic **10.** acidic

Guide to Calculating pH of an Aqueous Solution	
Step 1	State the given and needed quantities.
Step 2	Enter the $[H_3O^+]$ into the pH equation.
Step 3	Press the *log* key and change the sign. Adjust the number of SFs on the *right* of the decimal point.

KEY MATH SKILL
Calculating $[H_3O^+]$ *from pH*

♦ Learning Exercise 14.6B

Calculate the pH of each of the following solutions:

a. $[H_3O^+] = 1.0 \times 10^{-8}$ M ______ **b.** $[OH^-] = 1 \times 10^{-12}$ M ______

c. $[H_3O^+] = 1 \times 10^{-3}$ M ______ **d.** $[OH^-] = 1 \times 10^{-10}$ M ______

e. $[H_3O^+] = 3.4 \times 10^{-8}$ M ______ **f.** $[OH^-] = 7.8 \times 10^{-2}$ M ______

Answers **a.** 8.00 **b.** 2.0 **c.** 3.0 **d.** 4.0 **e.** 7.47 **f.** 12.89

♦ Learning Exercise 14.6C

Calculate the pH of each of the following solutions:

a. $[H_3O^+] = 5.0 \times 10^{-3}$ M **b.** $[OH^-] = 4.0 \times 10^{-6}$ M

c. $[H_3O^+] = 7.5 \times 10^{-8}$ M **d.** $[OH^-] = 2.5 \times 10^{-10}$ M

Answers **a.** 2.30 **b.** 8.60 **c.** 7.12 **d.** 4.40

Guide to Calculating $[H_3O^+]$ from pH	
Step 1	State the given and needed quantities.
Step 2	Enter the pH value into the inverse log equation and change the sign.

♦ Learning Exercise 14.6D

Complete the following table:

	$[H_3O^+]$	$[OH^-]$	pH
a.	______	1×10^{-12} M	______
b.	______	______	8.0
c.	5.0×10^{-11} M	______	______
d.	______	______	7.80
e.	______	______	4.25
f.	2.0×10^{-10} M	______	______

Answers

	$[H_3O^+]$	$[OH^-]$	pH
a.	1×10^{-2} M	1×10^{-12} M	2.0
b.	1×10^{-8} M	1×10^{-6} M	8.0
c.	5.0×10^{-11} M	2.0×10^{-4} M	10.30
d.	1.6×10^{-8} M	6.3×10^{-7} M	7.80
e.	5.6×10^{-5} M	1.8×10^{-10} M	4.25
f.	2.0×10^{-10} M	5.0×10^{-5} M	9.70

♦ Learning Exercise 14.6E

Calculate the pOH of each of the following solutions:

a. $[OH^-] = 1 \times 10^{-4}$ M ________ **b.** $[H_3O^+] = 1.0 \times 10^{-2}$ M ________

c. pH = 8.15 ________ **d.** $[OH^-] = 4.0 \times 10^{-10}$ M ________

e. $[H_3O^+] = 3.4 \times 10^{-8}$ M ________ **f.** pH = 2.22 ________

Answers **a.** 4.0 **b.** 12.00 **c.** 5.85 **d.** 9.40 **e.** 6.53 **f.** 11.78

14.7 Reactions of Acids and Bases

Learning Goal: Write balanced equations for reactions of acids with metals, carbonates, and bases.

- Acids react with many metals to yield hydrogen gas, H_2, and the salt of the metal.
- Acids react with carbonates and bicarbonates to yield CO_2, H_2O, and the salt of the metal.
- Acids neutralize bases in a reaction that produces water and a salt.
- The net ionic equation for any strong acid–strong base neutralization is $H^+(aq) + OH^-(aq) \longrightarrow H_2O(l)$.
- In a balanced neutralization equation, an equal number of moles of H^+ and OH^- must react.
- The concentration of an acid solution can be determined by titration.

♦ Learning Exercise 14.7A

CORE CHEMISTRY SKILL

Writing Equations for Reactions of Acids and Bases

Complete and balance each of the following reactions of acids:

1. ____ $Zn(s)$ + ____ $HCl(aq) \longrightarrow$ ____ $ZnCl_2(aq)$ + ____

2. ____ $HCl(aq)$ + ____ $Li_2CO_3(aq) \longrightarrow$ ____ + ____ + ____

3. ____ $HCl(aq)$ + ____ $NaHCO_3(aq) \longrightarrow$ ____ $CO_2(g)$ + ____ $H_2O(l)$ + ____

4. ____ $Al(s)$ + ____ $H_2SO_4(aq) \longrightarrow$ ____ $Al_2(SO_4)_3(aq)$ + ____

Answers

1. $Zn(s) + 2HCl(aq) \longrightarrow ZnCl_2(aq) + H_2(g)$

2. $2HCl(aq) + Li_2CO_3(aq) \longrightarrow CO_2(g) + H_2O(l) + 2LiCl(aq)$

3. $HCl(aq) + NaHCO_3(aq) \longrightarrow CO_2(g) + H_2O(l) + NaCl(aq)$

4. $2Al(s) + 3H_2SO_4(aq) \longrightarrow Al_2(SO_4)_3(aq) + 3H_2(g)$

Guide to Balancing an Equation for Neutralization	
Step 1	Write the reactants and products.
Step 2	Balance the H^+ in the acid with the OH^- in the base.
Step 3	Balance the H_2O with the H^+ and the OH^-.
Step 4	Write the salt from the remaining ions.

♦ Learning Exercise 14.7B

Balance each of the following neutralization reactions:

1. ____ $NaOH(aq)$ + ____ $H_2SO_4(aq) \longrightarrow$ ____ $Na_2SO_4(aq)$ + ____ $H_2O(l)$
2. ____ $Mg(OH)_2(aq)$ + ____ $HCl(aq) \longrightarrow$ ____ $MgCl_2(aq)$ + ____ $H_2O(l)$
3. ____ $Al(OH)_3(aq)$ + ____ $HNO_3(aq) \longrightarrow$ ____ $Al(NO_3)_3(aq)$ + ____ $H_2O(l)$
4. ____ $Ca(OH)_2(aq)$ + ____ $H_3PO_4(aq) \longrightarrow$ ____ $Ca_3(PO_4)_2(s)$ + ____ $H_2O(l)$

Answers

1. $2NaOH(aq) + H_2SO_4(aq) \longrightarrow Na_2SO_4(aq) + 2H_2O(l)$
2. $Mg(OH)_2(aq) + 2HCl(aq) \longrightarrow MgCl_2(aq) + 2H_2O(l)$
3. $Al(OH)_3(aq) + 3HNO_3(aq) \longrightarrow Al(NO_3)_3(aq) + 3H_2O(l)$
4. $3Ca(OH)_2(aq) + 2H_3PO_4(aq) \longrightarrow Ca_3(PO_4)_2(s) + 6H_2O(l)$

♦ Learning Exercise 14.7C

Complete each of the following neutralization reactions and then balance:

a. ____ $KOH(aq)$ + ____ $H_3PO_4(aq) \longrightarrow$ ________ + ____ $H_2O(l)$

b. ____ $NaOH(aq)$ + ________ $\longrightarrow$ ____ $Na_2SO_4(aq)$ + ________

c. ________ + ________ $\longrightarrow$ ____ $AlCl_3(aq)$ + ________

d. ________ + ________ $\longrightarrow$ ____ $Fe_2(SO_4)_3(aq)$ + ________

Answers

a. $3KOH(aq) + H_3PO_4(aq) \longrightarrow K_3PO_4(aq) + 3H_2O(l)$

b. $2NaOH(aq) + H_2SO_4(aq) \longrightarrow Na_2SO_4(aq) + 2H_2O(l)$

c. $Al(OH)_3(s) + 3HCl(aq) \longrightarrow AlCl_3(aq) + 3H_2O(l)$

d. $2Fe(OH)_3(s) + 3H_2SO_4(aq) \longrightarrow Fe_2(SO_4)_3(aq) + 6H_2O(l)$

14.8 Acid–Base Titration

Learning Goal: Calculate the molarity or volume of an acid or base from titration information.

- In a laboratory procedure called a titration, an acid sample is neutralized with a known amount of a base.
- From the volume and molarity of the base, the concentration of the acid is calculated.
- A titration is used to determine the volume or concentration of an acid or a base from the laboratory data.

Guide to Calculations for an Acid–Base Titration	
Step 1	State the given and needed quantities and concentrations.
Step 2	Write a plan to calculate molarity or volume.
Step 3	State equalities and conversion factors, including concentrations.
Step 4	Set up the problem to calculate the needed quantity.

♦ Learning Exercise 14.8

CORE CHEMISTRY SKILL

Calculating Molarity or Volume of an Acid or Base in a Titration

Solve the following problems using the titration data given:

a. A 5.00-mL sample of HCl is placed in a flask. In the titration, 15.0 mL of a 0.200 M NaOH solution is required for neutralization. What is the molarity of the HCl in the sample?

$$HCl(aq) + NaOH(aq) \longrightarrow NaCl(aq) + H_2O(l)$$

b. How many milliliters of a 0.200 M NaOH solution are required to neutralize completely 8.50 mL of a 0.500 M H_2SO_4 solution?

$$H_2SO_4(aq) + 2NaOH(aq) \longrightarrow Na_2SO_4(aq) + 2H_2O(l)$$

c. A 10.0-mL sample of H_3PO_4 solution is placed in a flask. If titration requires 42.0 mL of a 0.100 M NaOH solution to reach the endpoint, what is the molarity of the H_3PO_4 solution?

$$H_3PO_4(aq) + 3NaOH(aq) \longrightarrow Na_3PO_4(aq) + 3H_2O(l)$$

d. A 24.6-mL sample of HCl solution reacts with 33.0 mL of a 0.222 M NaOH solution. What is the molarity of the HCl solution (see reaction in part **a**)?

e. A 15.7-mL sample of H_2SO_4 reacts with 27.7 mL of a 0.187 M NaOH solution. What is the molarity of the H_2SO_4 solution (see reaction in part **b**)?

Answers **a.** 0.600 M HCl solution **b.** 42.5 mL **c.** 0.140 M H_3PO_4 solution
d. 0.298 M HCl solution **e.** 0.165 M H_2SO_4 solution

14.9 Buffers

Learning Goal: Describe the role of buffers in maintaining the pH of a solution; calculate the pH of a buffer.

- A buffer solution resists a change in pH when small amounts of acid or base are added.
- A buffer contains either (1) a weak acid and its salt or (2) a weak base and its salt.
- The weak acid reacts with added OH^-, and the anion of the salt reacts with added H_3O^+, which maintains the pH of the solution.

- The weak base reacts with added H_3O^+, and the cation of the salt reacts with added OH^-, which maintains the pH of the solution.
- The pH of a buffer can be calculated by rearranging and solving the K_a for $[H_3O^+]$, and then converting to pH.

Key Terms for Sections 4.1–4.9

Match each of the following key terms with the correct definition:

a. acid **b.** base **c.** pH **d.** neutralization
e. buffer **f.** K_a **g.** endpoint **h.** dissociation

1. _____ a substance that forms hydroxide ions (OH^-) in water and/or accepts protons (H^+)

2. _____ a reaction between an acid and a base to form a salt and water

3. _____ a substance that forms hydrogen ions (H^+) in water

4. _____ a mixture of a weak acid (or base) and its salt that maintains the pH of a solution

5. _____ a measure of the acidity (H_3O^+) of a solution

6. _____ separation of a substance into ions in water

7. _____ product of the molar concentrations of the ions from a weak acid divided by the molar concentration of that weak acid

8. _____ the point in a titration where the indicator changes color

Answers **1.** b **2.** d **3.** a **4.** e **5.** c **6.** h **7.** f **8.** g

♦ Learning Exercise 14.9A

State whether each of the following represents a buffer system and explain why:

a. HCl + NaCl

b. K_2SO_4

c. H_2CO_3

d. $H_2CO_3 + NaHCO_3$

Answers
a. No. A strong acid and its salt do not make a buffer.
b. No. A salt alone cannot act as a buffer.
c. No. A weak acid alone cannot act as a buffer.
d. Yes. A weak acid and its salt act as a buffer system.

Guide to Calculating pH of a Buffer	
Step 1	State the given and needed quantities.
Step 2	Write the K_a expression and rearrange for $[H_3O^+]$.
Step 3	Substitute [HA] and $[A^-]$ into the K_a expression.
Step 4	Use $[H_3O^+]$ to calculate pH.

Study Note

Rearranging the K_a to solve for $[H_3O^+]$ gives the following expression:

$$[H_3O^+] = K_a \times \frac{[HA]}{[A^-]}$$

Using the $[H_3O^+]$, write the equation to calculate the pH of the buffer.

$$pH = -\log[H_3O^+]$$

CORE CHEMISTRY SKILL

Calculating the pH of a Buffer

♦ Learning Exercise 14.9B

The K_a for acetic acid, CH_3COOH, is 1.8×10^{-5}.

a. What is the pH of a buffer that contains 1.0 M CH_3COOH and 1.0 M $NaCH_3COO$?

b. What is the pH of a buffer made from 0.20 M CH_3COOH and 0.10 M $NaCH_3COO$?

c. What is the pH of a buffer made from 0.10 M CH_3COOH and 1.0 M $NaCH_3COO$?

Answers

a. $[H_3O^+] = 1.8 \times 10^{-5} \times \frac{[1.0\ \cancel{M}]}{[1.0\ \cancel{M}]} = 1.8 \times 10^{-5}$ pH = 4.74

b. $[H_3O^+] = 1.8 \times 10^{-5} \times \frac{[0.20\ \cancel{M}]}{[0.10\ \cancel{M}]} = 3.6 \times 10^{-5}$ pH = 4.44

c. $[H_3O^+] = 1.8 \times 10^{-5} \times \frac{[0.10\ \cancel{M}]}{[1.0\ \cancel{M}]} = 1.8\ 10^{-6}$ pH = 5.74

Checklist for Chapter 14

You are ready to take the Practice Test for Chapter 14. Be sure you have accomplished the following learning goals for this chapter. If not, review the section listed at the end of the goal. Then apply your new skills and understanding to the Practice Test.

After studying Chapter 14, I can successfully:

_____ Describe the properties of Arrhenius acids and bases and write their names (14.1).

_____ Write the IUPAC and common names and draw structural formulas of carboxylic acids (14.1).

_____ Describe the Brønsted–Lowry concept of acids and bases (14.2).

_____ Write conjugate acid–base pairs for an acid–base reaction (14.2).

_____ Write equations for the ionization of strong and weak acids and bases (14.3).

_____ Write a dissociation constant for a weak acid or base (14.4).

_____ Use the ion product constant of water to calculate $[H_3O^+]$ and $[OH^-]$ (14.5).

____ Calculate pH or pOH from the $[H_3O^+]$ or $[OH^-]$ of a solution (14.6).

____ Write a balanced equation for the reactions of acids with metals, carbonates, and bases (14.7).

____ Calculate the molarity or volume of an acid or base from titration information (14.8).

____ Describe the role of buffers in maintaining the pH of a solution, and calculate the pH of a buffer solution (14.9).

Practice Test for Chapter 14

1. An acid is a compound that when placed in water yields this characteristic ion. (14.1)
A. H_3O^+ **B.** OH^- **C.** Na^+ **D.** Cl^- **E.** CO_3^{2-}

2. $MgCl_2$ would be classified as a(n) (14.1)
A. acid **B.** base **C.** salt **D.** buffer **E.** nonelectrolyte

3. $Mg(OH)_2$ would be classified as a (14.1)
A. weak acid **B.** strong base **C.** salt **D.** buffer **E.** nonelectrolyte

4. In the K_w expression for pure H_2O, the $[H_3O^+]$ has the value (14.5)
A. 1.0×10^{-7} M **B.** 1.0×10^{-1} M **C.** 1.0×10^{-14} M
D. 1.0×10^{-6} M **E.** 1.0×10^{12} M

5. Of the following pH values, which is the most acidic? (14.6)
A. 8.0 **B.** 5.5 **C.** 1.5 **D.** 3.2 **E.** 9.0

6. Of the following pH values, which is the most basic? (14.6)
A. 10.0 **B.** 4.0 **C.** 2.2 **D.** 11.5 **E.** 9.0

For questions 7 through 9, consider a solution with $[H_3O^+] = 1 \times 10^{-11}$ M. (14.6)

7. The pH of the solution is
A. 1.0 **B.** 2.0 **C.** 3.0 **D.** 11.0 **E.** 14.0

8. The hydroxide ion concentration is
A. 1×10^{-1} M **B.** 1×10^{-3} M **C.** 1×10^{-4} M **D.** 1×10^{-7} M **E.** 1×10^{-11} M

9. The solution is
A. acidic **B.** basic **C.** neutral **D.** a buffer **E.** neutralized

For questions 10 through 13, consider a solution with $[OH^-] = 1 \times 10^{-5}$ M. (14.6)

10. The hydrogen ion concentration of the solution is
A. 1×10^{-5} M **B.** 1×10^{-7} M **C.** 1×10^{-9} M **D.** 1×10^{-10} M **E.** 1×10^{-14} M

11. The pH of the solution is
A. 2.0 **B.** 5.0 **C.** 9.0 **D.** 11 **E.** 14

12. The pOH of the solution is
A. 2.0 **B.** 5.0 **C.** 9.0 **D.** 11.0 **E.** 14

13. The solution is
A. acidic **B.** basic **C.** neutral **D.** a buffer **E.** neutralized

14. Acetic acid is a weak acid because (14.3)
A. it forms a dilute acid solution **B.** it is isotonic
C. it is slightly ionized in water **D.** it is a nonpolar molecule
E. it can form a buffer

15. A weak base when added to water (14.3)
A. makes the solution slightly basic
B. does not affect the pH
C. dissociates completely
D. does not dissociate
E. makes the solution slightly acidic

16. Which is an equation for neutralization of an acid and a base? (14.7)
A. $CaCO_3(s) \longrightarrow CaO(s) + CO_2(g)$
B. $Na_2SO_4(s) \longrightarrow 2Na^+(aq) + SO_4^{2-}(aq)$
C. $H_2SO_4(aq) + 2NaOH(aq) \longrightarrow Na_2SO_4(aq) + 2H_2O(l)$
D. $Na_2O(s) + SO_3(g) \longrightarrow Na_2SO_4(aq)$
E. $H_2CO_3(aq) \longrightarrow CO_2(g) + H_2O(l)$

17. What is the name given to components in the body that keep blood pH within its normal 7.35 to 7.45 range? (14.9)
A. nutrients **B.** buffers **C.** metabolites **D.** regufluids **E.** neutralizers

18. A buffer system (14.9)
A. maintains a pH of 7.0
B. contains a weak base
C. contains a salt
D. contains a strong acid and its salt
E. maintains the pH of a solution

19. Which of the following would act as a buffer system? (14.9)
A. HCl
B. Na_2CO_3
C. $NaOH + NaNO_3$
D. NH_4OH
E. $NaHCO_3 + H_2CO_3$

20. What is the molarity of a 10.0-mL sample of HCl that is neutralized by 15.0 mL of a 2.0 M NaOH solution? (14.8)
A. 0.50 M HCl **B.** 1.0 M HCl **C.** 1.5 M HCl **D.** 2.0 M HCl **E.** 3.0 M HCl

21. In a titration, 6.0 mol NaOH will completely neutralize ________ mol of H_2SO_4. (14.8)
A. 1.0 **B.** 2.0 **C.** 3.0 **D.** 6.0 **E.** 11

22. Which of the following pairs is a conjugate acid–base pair? (14.2)
A. HCl/HNO_3 **B.** HNO_2/NO_2^- **C.** $NaOH/KOH$
D. HSO_4^-/HCO_3^- **E.** Cl^-/F^-

23. The conjugate base of HSO_4^- is (14.2)
A. SO_4^{2-} **B.** H_2SO_4 **C.** HS^- **D.** H_2S **E.** SO_3^{2-}

24. In which of the following reactions does H_2O act as an acid? (14.2)
A. $H_3PO_4(aq) + H_2O(l) \longrightarrow H_3O^+(aq) + H_2PO_4^-(aq)$
B. $H_2SO_4(aq) + H_2O(l) \longrightarrow H_3O^+(aq) + HSO_4^-(aq)$
C. $H_2O(l) + HS^-(aq) \longrightarrow H_3O^+(aq) + S^{2-}(aq)$
D. $NaOH(aq) + HCl(aq) \longrightarrow NaCl(aq) + H_2O(l)$
E. $NH_3(g) + H_2O(l) \longrightarrow NH_4^+(aq) + OH^-(aq)$

25. Which of the following acids has the smallest K_a value? (14.4)
A. HNO_3 **B.** H_2SO_4 **C.** HCl **D.** H_2CO_3 **E.** HBr

26. Using the following K_a values, identify the strongest acid in the group. (14.3)
A. 7.5×10^{-3} **B.** 1.8×10^{-5} **C.** 4.5×10^{8} **D.** 4.9×10^{-10} **E.** 3.2×10^{4}

27. What is the $[H_3O^+]$ of a solution that has a pH = 8.7? (14.6)
A. 7×10^{-8} **B.** 2×10^{-9} **C.** 2×10^{9} **D.** 8×10^{-7} **E.** 2×10^{-8}

28. A buffer is made with 1.0 M HF and 1.0 M NaF. If HF has a K_a of 7.2×10^{-4} what is the pH of the buffer? (14.9)
A. 3.14 **B.** 4.00 **C.** 4.14 **D.** 4.72 **E.** 7.20

29. Which of the following would turn litmus blue? (14.1)
A. HCl **B.** NH_4Cl **C.** Na_2SO_4 **D.** KOH **E.** $NaNO_3$

30. What is the name of $HClO_3$? (14.1)
A. hydrochloric acid **B.** chloric acid **C.** chlorous acid
D. chloric trioxide acid **E.** perchloric acid

31. What is the name of NH_4OH? (14.1)
A. ammonium oxide
B. nitrogen tetrahydride hydroxide
C. ammonium hydroxide
D. perammonium hydroxide
E. amine oxide hydride

32. What is the name of $CH_3—CH_2—COOH$? (14.1)
A. methyl peroxidic acid
B. propanoic acid
C. acetic acid
D. methyl acetic acid
E. butanoic acid

Answers to the Practice Test

1. A	**2.** C	**3.** B	**4.** A	**5.** C
6. D	**7.** D	**8.** B	**9.** B	**10.** C
11. C	**12.** B	**13.** B	**14.** C	**15.** A
16. C	**17.** B	**18.** E	**19.** E	**20.** E
21. C	**22.** B	**23.** A	**24.** E	**25.** D
26. C	**27.** B	**28.** A	**29.** D	**30.** B
31. C	**32.** B			

15
Oxidation and Reduction

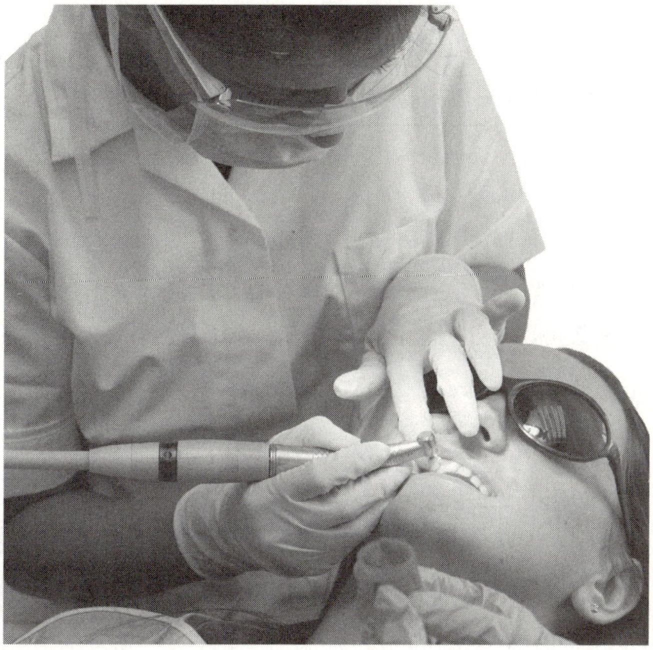

Kimberly has been delighted with the teeth whitening process and how her teeth look. The hydrogen peroxide gel caused a chemical reaction in the enamel coating of her teeth, oxidizing the brown and yellow stains on the teeth. Kimberly now avoids food or drinks, such as dark tea and coffee, that can stain her teeth. She now uses a teeth-whitening toothpaste once a week and will return to her dental hygienist in a year for a touch-up treatment. If hydrogen peroxide forms oxygen gas and water when it reacts, what is the balanced equation?

15.1 Oxidation and Reduction

Learning Goal: Identify what is oxidized and what is reduced in an oxidation–reduction reaction.

- In an oxidation–reduction reaction, electrons are transferred from one reactant to another.
- In an oxidation, electrons are lost. In a reduction, there is a gain of electrons.

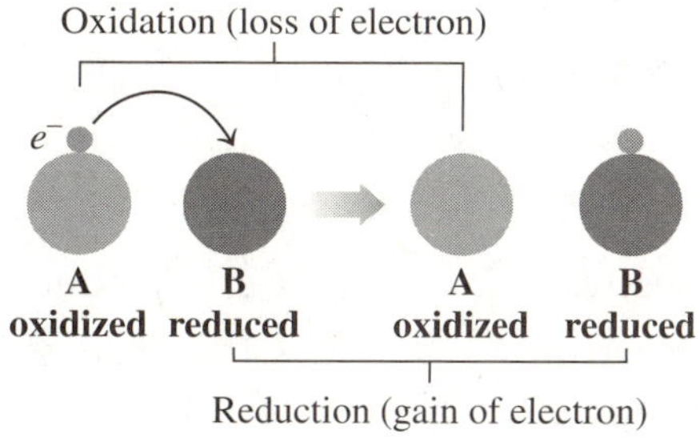

- In any oxidation–reduction reaction, the number of electrons lost must be equal to the number of electrons gained.
- A reducing agent (oxidized) provides the electrons, and an oxidizing agent (reduced) accepts electrons.

Study Note

Oxidation **i**s the **l**oss of electrons (OIL). **R**eduction **i**s the **g**ain of electrons (RIG).

A balanced oxidation–reduction equation contains an oxidation of one reactant and a reduction of the other reactant. $Cu(s) + 2Ag^+(aq) \longrightarrow Cu^{2+}(aq) + 2Ag(s)$

Oxidation: $Cu(s) \longrightarrow Cu^{2+}(aq) + 2\,e^-$

Reduction: $2Ag^+(aq) + 2\,e^- \longrightarrow 2Ag(s)$

In a balanced oxidation–reduction equation, the loss of electrons must be equal to the gain of electrons.

♦ Learning Exercise 15.1A

Identify each of the following as oxidation (O) or reduction (R):

1. ________ loss of two electrons

2. ________ $Zn(s) \longrightarrow Zn^{2+}(aq) + 2\,e^-$

3. ________ $Cu^{2+}(aq) + e^- \longrightarrow Cu^+(aq)$

4. ________ gain of electrons

Answers **1.** O **2.** O **3.** R **4.** R

CORE CHEMISTRY SKILL

Identifying Oxidized and Reduced Substances

♦ Learning Exercise 15.1B

For each of the following, indicate whether the underlined element is *oxidized* or *reduced* and indicate if there is a loss or gain of electrons:

a. $4\underline{Al}(s) + 3O_2(g) \longrightarrow 2Al_2O_3(s)$ Al is ____________

b. $\underline{Fe}^{3+}(aq) + e^- \longrightarrow Fe^{2+}(aq)$ Fe^{3+} is ____________

c. $\underline{Cu}O(s) + H_2(g) \longrightarrow Cu(s) + H_2O(l)$ Cu^{2+} is ____________

d. $2\underline{Cl}^-(aq) \longrightarrow Cl_2(g) + 2\,e^-$ Cl^- is ____________

e. $2H\underline{Br}(aq) + Cl_2(g) \longrightarrow 2HCl(aq) + Br_2(g)$ Br^- is ____________

f. $2\underline{Na}(s) + Cl_2(g) \longrightarrow 2NaCl(aq)$ Na is ____________

g. $\underline{Cu}Cl_2(aq) + Zn(s) \longrightarrow ZnCl_2(aq) + Cu(s)$ Cu^{2+} is ____________

Answers

a. Al^0 is oxidized to Al^{3+}; loss of electrons.

b. Fe^{3+} is reduced to Fe^{2+}; gain of electrons.

c. Cu^{2+} is reduced to Cu^0; gain of electrons.

d. Cl^- is oxidized to Cl^0; loss of electrons.

e. Br^- is oxidized to Br^0; loss of electrons.

f. Na^0 is oxidized to Na^+; loss of electrons.

g. Cu^{2+} is reduced to Cu^0; gain of electrons.

♦ Learning Exercise 15.1C

For each of the following oxidation–reduction reactions, identify the reactant that is oxidized and the reactant that is reduced:

a. $2Na(s) + Cl_2(g) \longrightarrow 2NaCl(aq)$

b. $Zn(s) + 2H^+(aq) \longrightarrow Zn^{2+}(aq) + H_2(g)$

c. $H_2S(aq) + I_2(s) \longrightarrow 2HI(aq) + S(s)$

d. $2Mg(s) + O_2(g) \longrightarrow 2MgO(s)$

e. $B_2O_3(s) + 3Mg(s) \longrightarrow 3MgO(s) + 2B(s)$

Answers

a. Na loses electrons and is oxidized; Cl_2 gains electrons and is reduced.
b. Zn loses electrons and is oxidized; H^+ gains an electrons and is reduced.
c. S^{2-} in H_2S loses electrons and is oxidized; I_2 gains electrons and is reduced.
d. Mg loses electrons and is oxidized; O_2 gains electrons and is reduced.
e. Mg loses electrons and is oxidized; B in B_2O_3 gains electrons and is reduced.

15.2 Oxidation Numbers

Learning Goal: Assign and use oxidation numbers to identify elements that are oxidized or reduced.

- An oxidation number is assigned to atoms to determine which substance is oxidized and which is reduced. These numbers do not necessarily represent actual charges.
- Oxidation numbers are assigned based on a set of rules.
- An increase in oxidation number indicates an oxidation; a decrease in oxidation number indicates a reduction.

Table 15.1 Rules for Assigning Oxidation Numbers

Rule
1. The sum of the oxidation numbers in a molecule is equal to zero (0), or for a polyatomic ion the sum of the oxidation numbers is equal to its charge.
2. The oxidation number of an element (monatomic or diatomic) is zero (0).
3. The oxidation number of a monatomic ion is equal to its charge.
4. In compounds, the oxidation number of Group 1A (1) elements is +1 and of Group 2A (2) elements is +2.
5. In compounds, the oxidation number of fluorine is always −1. Other elements in Group 7A (17) also are −1 except when combined with oxygen or fluorine.
6. In compounds, the oxidation number of oxygen is usually −2 except in OF_2.
7. In compounds with nonmetals, the oxidation number of hydrogen is +1; in compounds with metals, the oxidation number of hydrogen is −1.

♦ Learning Exercise 15.2A

CORE CHEMISTRY SKILL
Assigning Oxidation Numbers

Assign oxidation numbers to the elements in each of the following:

a. PbO **b.** Fe **c.** NO_2 **d.** $CuCl_2$

e. H_2CO_3 **f.** HNO_3 **g.** K_3PO_3 **h.** $Cr_2O_7^{2-}$

Answers

a. PbO
+2−2

b. Fe
0

c. NO_2
+4−2
$N + 2(-2) = 0$
$N = +4$

d. $CuCl_2$
+2−1
$Cu + 2(-1) = 0$
$Cu = +2$

e. H_2CO_3
+1+4−2
$2(+1) + C + 3(-2) = 0$
$C = -2 + 6 = +4$

f. HNO_3
+1 +5 −2
$+1 + N + 3(-2) = 0$
$N = -1 + 6 = +5$

g. K_3PO_3
+1+3−2
$3(+1) + P + 3(-2)$
$P = -3 + 6 = +3$

h. $Cr_2O_7^{2-}$
+6 −2
$2Cr + 7(-2) = -2$
$2Cr = -2 + 14 = +12$
$Cr = +6$

Study Note

The oxidizing agent gains electrons from the oxidation. The reducing agent provides electrons for the reduction.

♦ Learning Exercise 15.2B

Complete each of the following statements about oxidation–reduction reactions by using the term *loses* or *gains*:

a. In reduction, an element ________ electrons.

b. The oxidizing agent ________ electrons.

c. The element that ________ electrons is oxidized.

d. The reducing agent is the element that ________ electrons.

e. The oxidation number increases when an element ________ electrons.

f. The oxidizing agent is the substance that ________ electrons.

g. The oxidation number decreases when an element ________ electrons.

h. In oxidation, an element ________ electrons.

Answers	**a.** gains	**b.** gains	**c.** loses	**d.** loses
	e. loses	**f.** gains	**g.** gains	**h.** loses

♦ Learning Exercise 15.2C

CORE CHEMISTRY SKILL
Using Oxidation Numbers

For each of the following, assign oxidation numbers, identify the elements in the reactants that are oxidized and reduced, and identify the reactant that is the oxidizing agent and the reactant that is the reducing agent:

a. $PbO(s) + CO(g) \longrightarrow Pb(s) + CO_2(g)$

substance oxidized ________ substance reduced ________

reducing agent ________ oxidizing agent ________

b. $Fe_2O_3(s) + 3C(s) \longrightarrow 2Fe(s) + 3CO(g)$

CORE CHEMISTRY SKILL
Identifying Oxidizing and Reducing Agents

substance oxidized ________ substance reduced ________

reducing agent ________ oxidizing agent ________

c. $Cu(s) + 4HNO_3(aq) \longrightarrow Cu(NO_3)_2(aq) + 2NO_2(g) + 2H_2O(l)$

substance oxidized ________ substance reduced ________

reducing agent ________ oxidizing agent ________

Answers

a. $PbO(s) + CO(g) \longrightarrow Pb(s) + CO_2(g)$

+2−2   +2−2   0   +4−2

element oxidized	C(+2 ⟶ +4)	element reduced	Pb(+2 ⟶ 0)
reducing agent	CO	oxidizing agent	PbO

b. $Fe_2O_3(s) + 3C(s) \longrightarrow 2Fe(s) + 3CO(g)$

+3 −2   0   0   +2 −2

element oxidized	C(0 ⟶ +2)	element reduced	Fe(+3 ⟶ 0)
reducing agent	C	oxidizing agent	Fe_2O_3

c. $Cu(s) + 4HNO_3(aq) \longrightarrow Cu(NO_3)_2(aq) + 2NO_2(g) + 2H_2O(l)$

0   +1+5−2   +2 +5−2   +4−2   +1−2

element oxidized	Cu(0 ⟶ +2)	element reduced	N(+5 ⟶ +4)
reducing agent	Cu	oxidizing agent	HNO_3

15.3 Balancing Oxidation–Reduction Equations Using Half-Reactions

Learning Goal: Balance oxidation–reduction equations using the half-reaction method.

- The half-reaction method of balancing oxidation–reduction equations is typically used to balance oxidation–reduction equations that are written in the ionic form.
- Using the half-reaction method for balancing redox equations, the elements are balanced, O is balanced by adding H_2O, and H is balanced by adding H^+. Then charge is balanced by adding electrons, and each half-reaction is multiplied by an integer to equalize electron loss and gain, and then the half-reactions are added together.

Guide to Balancing Redox Equations Using Half-Reactions	
Step 1	Write two half-reactions for the equation.
Step 2	For each half-reaction, balance the elements other than H and O. If necessary, balance O by adding H_2O, and H by adding H^+.
Step 3	Balance each half-reaction for charge by adding electrons.
Step 4	Multiply each half-reaction by factors that equalize the loss and gain of electrons.
Step 5	Add half-reactions and cancel electrons and any identical ions or molecules.

Example: Consider the following unbalanced oxidation–reduction equation in an acidic solution:

$$I^-(aq) + SO_4^{2-}(aq) \longrightarrow H_2S(g) + I_2(s)$$

Step 1: Write two half-reactions for the equation.

$$2I^-(aq) \longrightarrow I_2(s) \qquad SO_4^{2-}(aq) \longrightarrow H_2S(g)$$

Step 2: For each half-reaction, balance the elements other than H and O. If necessary, balance O by adding H_2O, and H by adding H^+.

$$SO_4^{2-}(aq) \longrightarrow H_2S(g) + \mathbf{4H_2O}(l) \qquad SO_4^{2-}(aq) + \mathbf{10H^+}(aq) \longrightarrow H_2S(g) + 4H_2O(l)$$

Step 3: Balance each half-reaction for charge by adding electrons.

$$2I^-(aq) \longrightarrow I_2(s) + \mathbf{2\,e^-} \qquad SO_4^{2-}(aq) + \mathbf{10H^+}(aq) + \mathbf{8\,e^-} \longrightarrow H_2S(g) + 4H_2O(l)$$

Step 4: Multiply each half-reaction by factors that equalize the loss and gain of electrons.

$$[2I^-(aq) \longrightarrow I_2(s) + 2\,e^-] \times 4 \; (= 8\,e^- \text{ loss})$$

$$8I^-(aq) \longrightarrow 4I_2(g) + \mathbf{8\,e^-} \qquad SO_4^{2-}(aq) + 10H^+(aq) + \mathbf{8\,e^-} \longrightarrow H_2S(g) + 4H_2O(l)$$

Step 5: Add half-reactions and cancel electrons and any identical ions or molecules. Check balance of atoms and charge.

$$8I^-(aq) + SO_4^{2-}(aq) + 10H^+(aq) + \cancel{8e^-} \longrightarrow 4I_2(s) + H_2S(g) + 4H_2O(l) + \cancel{8e^-}$$

$$8I^-(aq) + SO_4^{2-}(aq) + 10H^+(aq) \longrightarrow 4I_2(s) + H_2S(g) + 4H_2O(l) \text{ balanced}$$

♦ Learning Exercise 15.3A

CORE CHEMISTRY SKILL

Using Half-Reactions to Balance Redox Equations

Use half-reactions to balance the following oxidation–reduction equations in acidic solution:

a. $NO_3^-(aq) + S(s) \longrightarrow NO_2(g) + H_2SO_4(aq)$

b. $MnO_4^-(aq) + Cl^-(aq) \longrightarrow Mn^{2+}(aq) + Cl_2(g)$

c. $Cr_2O_7^{2-}(aq) + Fe^{2+}(aq) \longrightarrow Cr^{3+}(aq) + Fe^{3+}(aq)$

d. $NO_2(g) + ClO^-(aq) \longrightarrow NO_3^-(aq) + Cl^-(aq)$

e. $BrO_3^-(aq) + MnO_2(s) \longrightarrow Br^-(aq) + MnO_4^-(aq)$

Answers

a. $6NO_3^-(aq) + S(s) + 6H^+(aq) \longrightarrow 6NO_2(g) + H_2SO_4(aq) + 2H_2O(l)$
b. $2MnO_4^-(aq) + 10Cl^-(aq) + 16H^+(aq) \longrightarrow 2Mn^{2+}(aq) + 5Cl_2(g) + 8H_2O(l)$
c. $Cr_2O_7^{2-}(aq) + 6Fe^{2+}(aq) + 14H^+(aq) \longrightarrow 2Cr^{3+}(aq) + 6Fe^{3+}(aq) + 7H_2O(l)$
d. $2NO_2(g) + ClO^-(aq) + H_2O(l) \longrightarrow 2NO_3^-(aq) + Cl^-(aq) + 2H^+(aq)$
e. $BrO_3^-(aq) + 2MnO_2(s) + H_2O(l) \longrightarrow Br^-(aq) + 2MnO_4^-(aq) + 2H^+(aq)$

♦ Learning Exercise 15.3B

1. Oxalic acid, $H_2C_2O_4$, present in plants such as spinach, reacts with permanganate, MnO_4^-,

$$H_2C_2O_4(aq) + MnO_4^-(aq) \longrightarrow Mn^{2+}(aq) + CO_2(g)$$

   a. Balance the equation in acidic solution.

   b. If 25.0 mL of a 0.0200 M MnO_4^- solution (from $KMnO_4$) is required to react with 5.00 mL of a $H_2C_2O_4$ solution, what is the molarity of the $C_2O_4^{2-}$ solution?

2. Iodine reacts with a thiosulfate, $S_2O_3^{2-}$, solution: $I_2(s) + S_2O_3^{2-}(aq) \longrightarrow I^-(aq) + S_4O_6^{2-}(aq)$

   a. Balance the equation in acidic solution.

   b. If 37.6 mL of 0.250 M $S_2O_3^{2-}$ solution is required to completely react with I_2, how many grams of I_2 are present in the sample?

Answers

1. a. $5H_2C_2O_4(aq) + 6H^+(aq) + 2MnO_4^-(aq) \longrightarrow 2Mn^{2+}(aq) + 10CO_2(g) + 8H_2O(l)$

b. $0.0250\ \cancel{L} \times \dfrac{0.0200\ \cancel{\text{mole } MnO_4^-}}{1\ \cancel{L}} \times \dfrac{5\ \text{mole } H_2C_2O_4}{2\ \cancel{\text{mole } MnO_4^-}} \times \dfrac{1}{0.00500\ L} = 0.250\ M$

2. a. $I_2(s) + 2S_2O_3^{2-}(aq) \longrightarrow 2I^-(aq) + S_4O_6^{2-}(aq)$

b. $0.0376\ \cancel{L} \times \dfrac{0.0250\ \cancel{\text{mole } S_2O_3^{2-}}}{1\ \cancel{L}} \times \dfrac{1\ \cancel{\text{mole } I_2}}{2\ \cancel{\text{mole } S_2O_3^{2-}}} \times \dfrac{253.8\ g\ I_2}{1\ \cancel{\text{mol } I_2}} = 1.19\ \text{g of } I_2$

15.4 Electrical Energy from Oxidation–Reduction Reactions

Learning Goal: Use the activity series to determine if an oxidation–reduction reaction is spontaneous. Write the half-reactions that occur in a voltaic cell and the cell notation.

- The activity series arranges elements in order of their ability to oxidize spontaneously when combined with the ions of any metal lower on the list.
- In an electrochemical cell, the half-reactions of an oxidation–reduction reaction are physically separated, so the electrons flow through an external circuit.
- Oxidation takes place at the anode; reduction takes place at the cathode.
- Electrons flow from the anode to the cathode.
- In a voltaic cell, the flow of electrons through an external circuit generates electrical energy.
- Batteries are examples of electrochemical cells that provide electrical energy.

Study Note

We can diagram the cell using a shorthand notation as follows:

$$Zn(s)\,|\,Zn^{2+}(aq)\,||\,Cu^{2+}(aq)\,|\,Cu(s)$$

The components of the oxidation half-cell (anode) are written on the left side in this shorthand notation, and the components of the reduction half-cell (cathode) are written on the right. A single vertical line separates the solid Zn anode from the ionic Zn^{2+} solution and the Cu^{2+} solution from the Cu cathode. Double vertical lines || indicate the salt bridge, which separates the two half-cells.

♦ Learning Exercise 15.4A

Use the activity series to identify the more active metal in each of the following pairs:

a. $Fe(s)$ or $Pb(s)$ **b.** $Ag(s)$ or $Na(s)$ **c.** $Zn(s)$ or $Ca(s)$

d. $Al(s)$ or $Pb(s)$ **e.** $Fe(s)$ or $Ni(s)$

Answers **a.** $Fe(s)$ **b.** $Na(s)$ **c.** $Ca(s)$ **d.** $Al(s)$ **e.** $Fe(s)$

♦ Learning Exercise 15.4B

CORE CHEMISTRY SKILL
Identifying Spontaneous Reactions

Use the activity series, predict whether each of the following reactions occurs spontaneously or not, and explain:

a. $Cu(s) + Sn^{2+}(aq) \longrightarrow Cu^{2+}(aq) + Sn(s)$

b. $Pb^{2+}(aq) + Mg(s) \longrightarrow Mg^{2+}(aq) + Pb(s)$

c. $2Ag(s) + Ni^{2+}(aq) \longrightarrow Ni(s) + 2Ag^{+}(aq)$

d. $Zn(s) + 2H^{+}(aq) \longrightarrow Zn^{2+}(aq) + H_2(g)$

e. $2Cr(s) + 3Ca^{2+}(aq) \longrightarrow 2Cr^{3+}(aq) + 3Ca(s)$

Answers

a. not spontaneous: $Cu(s)$ is less active than $Sn(s)$
b. spontaneous: $Mg(s)$ is more active than $Pb(s)$
c. not spontaneous: $Ag(s)$ is less active than $Ni(s)$
d. spontaneous: $Zn(s)$ is more active than $H_2(g)$
e. not spontaneous: $Cr(s)$ is less active than $Ca(s)$

♦ Learning Exercise 15.4C

For each of the following voltaic cells, diagram the cell, write the cell notation, write the half-reactions at the anode and cathode and the balanced equation, and indicate the direction of electron flow:

a. Mg anode in a solution containing Mg^{2+}, and a Ag cathode in a solution containing Ag^+

b. Cr cathode in a solution containing Cr^{3+}, and an Al anode in a solution containing Al^{3+}

c. Fe anode in a solution containing Fe^{2+}, and a Ni cathode in a solution containing Ni^{2+}

Answers

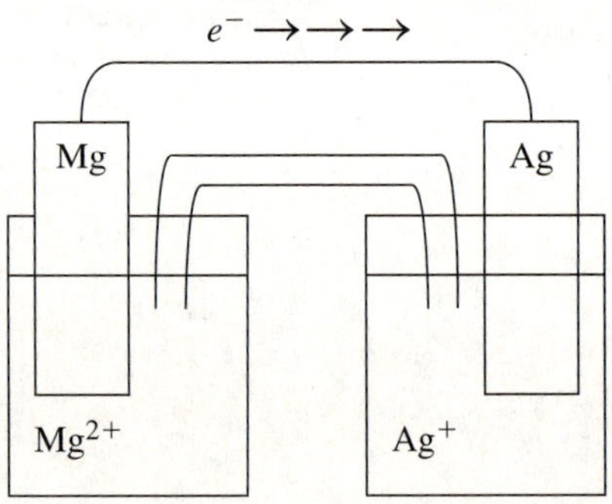

a.

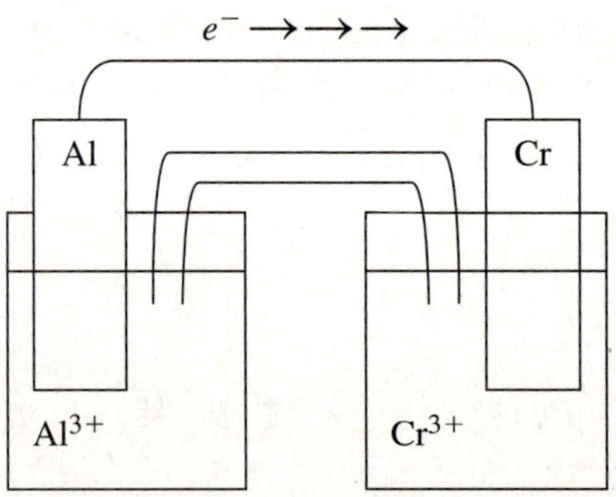

b.

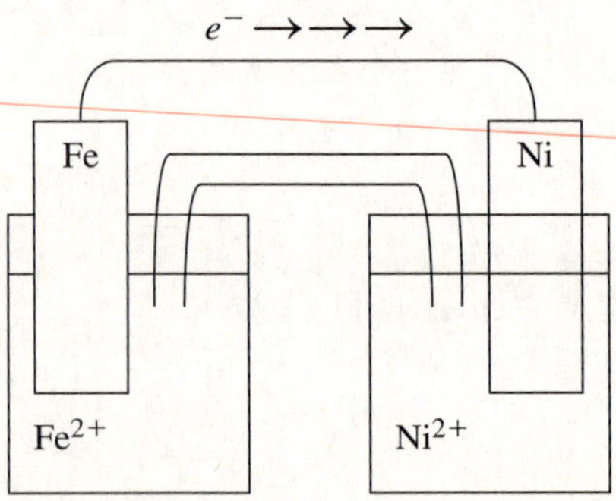

c.

a. $Mg(s) \mid Mg^{2+}(aq) \parallel Ag^{+}(aq) \mid Ag(s)$
Anode: $Mg(s) \longrightarrow Mg^{2+}(aq) + 2\,e^{-}$
Cathode: $[Ag^{+}(aq) + e^{-} \longrightarrow Ag(s)] \times 2$
$Mg(s) + 2Ag^{+}(aq) \longrightarrow Mg^{2+}(aq) + 2Ag(s)$

b. $Al(s) \mid Al^{3+}(aq) \parallel Cr^{3+}(aq) \mid Cr(s)$
Anode: $Al(s) \longrightarrow Al^{3+}(aq) + 3\,e^{-}$
Cathode: $Cr^{3+}(aq) + 3\,e^{-} \longrightarrow Cr(s)$
$Cr^{3+}(aq) + Al(s) \longrightarrow Cr(s) + Al^{3+}(aq)$

c. $Fe(s) \mid Fe^{2+}(aq) \parallel Ni^{2+}(aq) \mid Ni(s)$
Anode: $Fe(s) \longrightarrow Fe^{2+}(aq) + 2\,e^{-}$
Cathode: $Ni^{2+}(aq) + 2\,e^{-} \longrightarrow Ni(s)$
$Fe(s) + Ni^{2+}(aq) \longrightarrow Fe^{2+}(aq) + Ni(s)$

♦ Learning Exercise 15.4D

Write the half-reactions for the anode and cathode and the overall cell reaction from the following cell notations:

a. $Sr(s) \mid Sr^{2+}(aq) \parallel Sn^{4+}(aq), Sn^{2+}(aq) \mid Pt(s)$

b. $Cr(s) \mid Cr^{3+}(aq) \parallel Ni^{2+}(aq) \mid Ni(s)$

c. $Ag(s) \mid Ag^{+}(aq) \parallel Au^{3+}(aq) \mid Au(s)$

Answers

a. Anode: $Sr(s) \longrightarrow Sr^{2+}(aq) + 2\,e^{-}$
Cathode: $Sn^{4+}(aq) + 2\,e^{-} \longrightarrow Sn^{2+}(aq)$
$Sr(s) + Sn^{4+}(aq) \longrightarrow Sr^{2+}(aq) + Sn^{2+}(aq)$

b. Anode: $[Cr(s) \longrightarrow Cr^{3+}(aq) + 3\,e^{-}] \times 2$
Cathode: $[Ni^{2+}(aq) + 2\,e^{-} \longrightarrow Ni(s)] \times 3$
$2Cr(s) + 3Ni^{2+}(aq) \longrightarrow 2Cr^{2+}(aq) + 3Ni(s)$

c. Anode: $[Ag(s) \longrightarrow Ag^{+}(aq) + e^{-}] \times 3$
Cathode: $[Au^{3+}(aq) + 3\,e^{-} \longrightarrow Au(s)] \times 1$
$3Ag(s) + Au^{3+}(aq) \longrightarrow 3Ag^{+}(aq) + Au(s)$

15.5 Oxidation–Reduction Reactions That Require Electrical Energy

Learning Goal: Describe the half-cell reactions and the overall reactions that occur in electrolysis.

- An oxidation–reduction reaction is not spontaneous when a less active metal is combined with the ions of a more active metal.
- In an electrolytic cell, a process called electrolysis uses electrical energy is used to drive a nonspontaneous reaction.
- Electrolytic cells are used in the electrolysis of molten salts and electroplating.

♦ Learning Exercise 15.5

An electrolytic cell is used to electroplate nickel on the surface an iron object. If the electrolytic cell uses a nickel bar as the anode, and the iron object as the cathode, indicate the following:

a. What happens at the anode when electrical energy flows through the cell?

b. What is the equation for the half-reaction that occurs at the anode?

c. What happens at the cathode when electrical energy flows through the cell?

d. What is the equation for the half-reaction that occurs at the cathode?

Answers

a. At the anode, the nickel metal in the bar is oxidized to nickel(II) ions to make a Ni^{2+} solution.
b. $Ni(s) \longrightarrow Ni^{2+}(aq) + 2\,e^-$
c. At the cathode, the nickel(II) ions in the solution are reduced to nickel metal that plates on the surface of the iron object.
d. $Ni^{2+}(aq) + 2\,e^- \longrightarrow Ni(s)$

15.6 Oxidation of Alcohols: Extended Topic

Learning Goal: Classify alcohols as primary, secondary, or tertiary; write equations for the oxidation of alcohols.

- For organic and biochemical compounds, oxidation typically involves the addition of oxygen or the loss of hydrogen. Reduction can be the loss of oxygen or the gain of hydrogen.
- In organic chemistry, oxidation occurs when there is an increase in the number of carbon to oxygen bonds and reduction when there is a decrease in the number of carbon to oxygen bonds.
- Alcohols are classified as primary, secondary, or tertiary. A primary alcohol has one carbon group attached to the carbon bonded to the hydroxyl group (—OH). Methanol, CH_3OH, is considered a primary alcohol. A secondary alcohol has two carbon groups attached to the carbon atom bonded to the hydroxyl group (—OH). A tertiary alcohol has three carbon groups attached to the carbon atom bonded to the hydroxyl group (—OH).
- Primary (1°) alcohols oxidize to aldehydes, and secondary (2°) alcohols oxidize to ketones, but tertiary (3°) alcohols do not oxidize.

Key Terms for Sections 15.1–15.6

Match each of the following key terms with the correct definition:

a. cathode **b.** oxidation **c.** reduction **d.** oxidizing agent
e. reducing agent **f.** voltaic cell **g.** secondary alcohol

1. _____ the reactant that is reduced
2. _____ the gain of electrons by a substance
3. _____ a cell that uses an oxidation–reduction reaction to produce electrical energy
4. _____ the loss of electrons by a substance
5. _____ the reactant that is oxidized
6. _____ the electrode in an electrochemical cell where reduction takes place
7. _____ a compound in which the carbon atom with —OH group is bonded to two other carbon atoms

Answers **1.** d **2.** c **3.** f **4.** b **5.** e **6.** a **7.** g

Study Note

Example: Identify the following as primary, secondary, or tertiary alcohols.

Solution: Determine the number of carbon groups attached to the carbon atom that is attached to the hydroxyl (—OH) group.

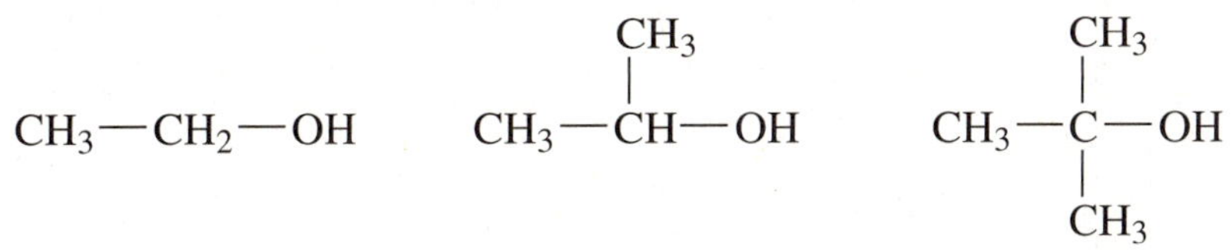

Primary (1°) Secondary (2°) Tertiary (3°)

♦ Learning Exercise 15.6A

Match the following terms with the statements shown:

a. primary alcohol **b.** secondary alcohol **c.** tertiary alcohol
d. aldehyde **e.** ketone

1. _____ an organic compound with one carbon group bonded to the carbon with the —OH group
2. _____ an alcohol that oxidizes to a ketone
3. _____ the oxidation product of a secondary alcohol
4. _____ an organic compound with three carbon groups bonded to the carbon with the —OH group
5. _____ the oxidation product of a primary alcohol

Answers **1.** a **2.** b **3.** e **4.** c **5.** d

♦ Learning Exercise 15.6B

Classify each of the following alcohols as primary (1°), secondary (2°), or tertiary (3°):

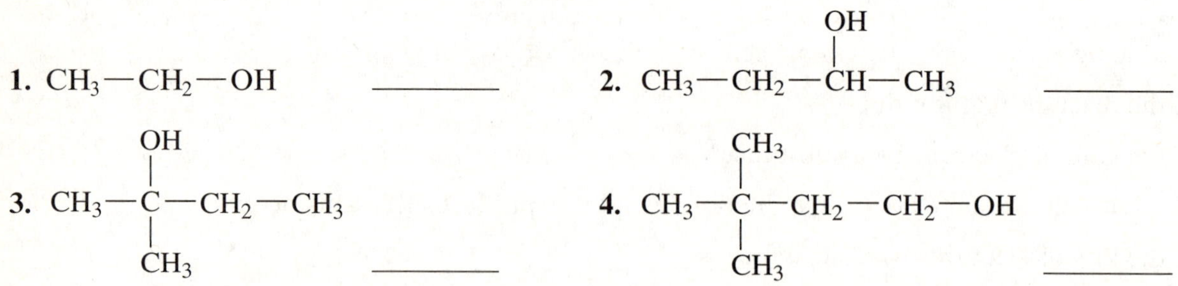

Answers **1.** primary (1°) **2.** secondary (2°) **3.** tertiary (3°) **4.** primary (1°)

♦ Learning Exercise 15.6C

Draw the condensed structural formulas of the products expected in the oxidation reaction of each of the following reactants:

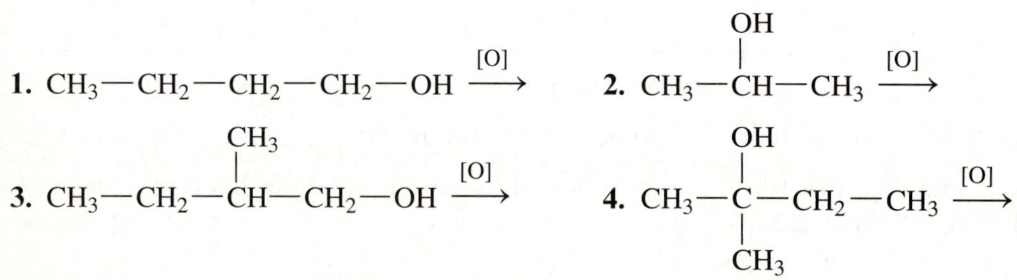

Answers

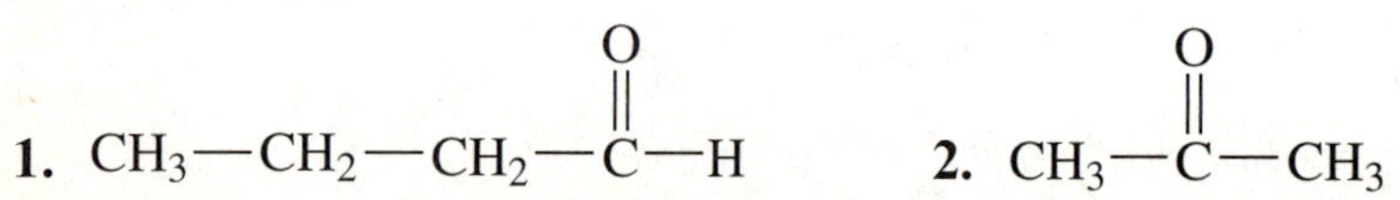

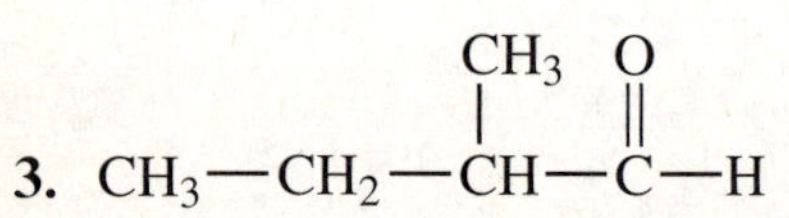

4. No reaction occurs with a tertiary (3°) alcohol.

Checklist for Chapter 15

You are ready to take the Practice Test for Chapter 15. Be sure you have accomplished the following learning goals for this chapter. If not, review the section listed at the end of the goal. Then apply your new skills and understanding to the Practice Test.

After studying Chapter 15, I can successfully:

_____ Identify what is oxidized and reduced in an oxidation–reduction reaction (15.1).

_____ Assign oxidation numbers to the atoms in an equation (15.2).

_____ Use oxidation numbers to determine the oxidized element, reduced element, oxidizing agent, and reducing agent (15.2).

_____ Balance an oxidation–reduction equation using half-reactions (15.3).

_____ Use the activity series for metals to determine whether an oxidation–reduction reaction is spontaneous (15.4).

_____ Write the half-reactions for the anode and the cathode of a voltaic cell (15.4).

_____ Write the shorthand cell notation for the half-reactions for the anode and the cathode of a voltaic cell (15.4).

_____ Draw the cell diagram for a voltaic cell (15.4).

_____ Write the half-cell reactions for an electrolytic cell (15.5).

_____ Describe the process taking place in electroplating (15.5).

_____ Classify an alcohol as primary, secondary, or tertiary (15.6).

_____ Draw the condensed structural formulas of the products formed by the oxidation of an alcohol (15.6).

Practice Test for Chapter 15

Identify whether the element in each of the following is oxidized (O) or reduced (R): (15.1)

1. _________ gains electrons

2. _________ an oxidizing agent

3. _________ loses electrons

4. _________ a reducing agent

5. _________ oxidation number increases

For questions 6 through 10, indicate the oxidation number of each of the elements: (15.2)

A. +1 **B.** +2 **C.** +3 **D.** +4 **E.** +5

6. _________ N in HNO_2

7. _________ P in P_2O_5

8. _________ C in HCO_3^-

9. _________ Mn in MnO_2

10. _________ Ca in $CaSO_4$

For questions 11 through 15, indicate the oxidation number of the indicated element as (15.2)

A. 0 **B.** −1 **C.** −2 **D.** −3 **E.** −4

11. _________ P in Na_3P

12. _________ N in N_2H_4

13. _________ C in CH_4

14. _________ Br in Br_2

15. _________ F in OF_2

For questions 16 through 20, use the following unbalanced equation: (15.2)

$NO(g) + Br_2(g) \longrightarrow NOBr(g)$

16. The oxidation numbers of N in the reactants and products are
A. +1, +2 **B.** +2, +3 **C.** +3, +2 **D.** −2, −3 **E.** +2, −3

17. The oxidation numbers of Br in the reactants and products are
A. 0, 0 **B.** −1, 0 **C.** 0, +1 **D.** −1, −1 **E.** 0, −1

18. The oxidation numbers of O in the reactants and products are
A. 0, 0 **B.** −2, 0 **C.** +3, +2 **D.** −2, −2 **E.** 0, −2

19. The change in the oxidation number of N is
A. a decrease of 1 **B.** a decrease of 2 **C.** an increase of 1
D. an increase of 2 **E.** no change

20. The oxidizing agent is
A. NO **B.** Br_2 **C.** NOBr

For questions 21 through 23, consider the following half-reaction in acidic solution: (15.3)

$ClO_2^- \longrightarrow Cl^-$

21. What is the number of H_2O in the balanced half-reaction?
A. 1 **B.** 2 **C.** 3 **D.** 4 **E.** 5

22. What is the number of H^+ in the balanced half-reaction?
A. 1 **B.** 2 **C.** 3 **D.** 4 **E.** 5

23. How many electrons are in the balanced half-reaction?
A. 1 **B.** 2 **C.** 3 **D.** 4 **E.** 5

For questions 24 through 26, consider the following half-reaction in acidic solution: (15.3)

$Cr_2O_7^{2-} \longrightarrow Cr^{3+}$

24. What is the number of H_2O in the balanced half-reaction?
A. 2 **B.** 4 **C.** 5 **D.** 7 **E.** 14

25. What is the number of H^+ in the balanced half-reaction?
A. 2 **B.** 6 **C.** 7 **D.** 12 **E.** 14

26. How many electrons are in the balanced half-reaction?
A. 2 **B.** 4 **C.** 6 **D.** 10 **E.** 14

27. Which equation has the following cell notation? (15.4)
$Al(s)\,|\,Al^{3+}(aq)\,||\,Fe^{3+}(aq)\,|\,Fe(s)$?
A. $Al(s) + Fe^{3+}(aq) \longrightarrow Al^{3+}(aq) + Fe(s)$
B. $Al^{3+}(aq) + Fe(s) \longrightarrow Al(s) + Fe^{3+}(aq)$
C. $Al^{3+}(aq) + Fe(s) \longrightarrow Al(s) + Fe^{3+}(aq)$
D. $Al^{3+}(aq) + Fe^{3+}(aq) \longrightarrow Fe(s) + Al(s)$
E. $Al(s) + Fe(s) \longrightarrow Fe^{3+}(aq) + Al^{3+}(aq)$

For questions 28 through 30, consider the following metals and ions in order from most to least active:

$Mg(s) \longrightarrow Mg^{2+}(aq) + 2\,e^-$
$Cr(s) \longrightarrow Cr^{3+}(aq) + 3\,e^-$
$Fe(s) \longrightarrow Fe^{2+}(aq) + 2\,e^-$
$Pb(s) \longrightarrow Pb^{2+}(aq) + 2\,e^-$
$Ag(s) \longrightarrow Ag^+(aq) + e^-$

28. Which of the following reactions is spontaneous? (15.4)

A. $3Pb(s) + 2Cr^{3+}(aq) \longrightarrow 3Pb^{2+}(aq) + 2Cr(s)$
B. $Mg(s) + 2Ag^{+}(aq) \longrightarrow Mg^{2+}(aq) + 2Ag(s)$
C. $Fe^{2+}(aq) + Cu(s) \longrightarrow Fe(s) + Cu^{2+}(aq)$
D. $2Ag(s) + Pb^{2+}(aq) \longrightarrow Pb(s) + 2Ag^{+}(aq)$
E. $3Ag(s) + Cr^{3+}(aq) \longrightarrow 3Ag^{+}(aq) + Cr(s)$

29. Which of the following reactions requires an external source of energy? (15.5)

A. $3Mg(s) + 2Cr^{3+}(aq) \longrightarrow 2Cr(s) + 3Mg^{2+}(aq)$
B. $Pb(s) + 2Ag^{+}(aq) \longrightarrow Pb^{2+}(aq) + 2Ag(s)$
C. $Mg(s) + Pb^{2+}(aq) \longrightarrow Mg^{2+}(aq) + Pb(s)$
D. $Mg(s) + 2Ag^{+}(aq) \longrightarrow Mg^{2+}(aq) + 2Ag(s)$
E. $3Pb(s) + 2Cr^{3+}(aq) \longrightarrow 3Pb^{2+}(aq) + 2Cr(s)$

30. In the electroplating of a bowl, an aluminum anode is placed in an Al^{3+} solution, $Al(NO_3)_3$. The bowl to be aluminum-plated is the cathode. The anode and cathode are wired to a battery. The reaction at the anode is (15.5)

A. $Al^{3+}(aq) + 3e^{-} \longrightarrow Al(s)$
B. $Al^{3+}(aq) \longrightarrow Al(s) + 3e^{-}$
C. $Al(s) + 3e^{-} \longrightarrow Al^{3+}(aq)$
D. $Al(s) \longrightarrow Al^{3+}(aq) + 3e^{-}$
E. $Al(s) + Al^{3+}(aq) \longrightarrow 2Al(s)$

31. The oxidation of organic compounds involves (15.6)

A. addition of H
B. loss of O
C. decrease in the number of C—O bonds
D. increase in the number of C—O bonds
E. gain of electrons

32. When a secondary alcohol is oxidized, the product is (15.6)

A. a ketone
B. an aldehyde
C. a carboxylic acid
D. a primary alcohol
E. an ester

Answers to the Practice Test

1. R	**2.** R	**3.** O	**4.** O	**5.** O
6. C	**7.** E	**8.** D	**9.** D	**10.** B
11. D	**12.** C	**13.** E	**14.** A	**15.** B
16. B	**17.** E	**18.** D	**19.** C	**20.** B
21. B	**22.** D	**23.** D	**24.** D	**25.** E
26. C	**27.** A	**28.** B	**29.** E	**30.** D
31. D	**32.** A			

16
Nuclear Chemistry

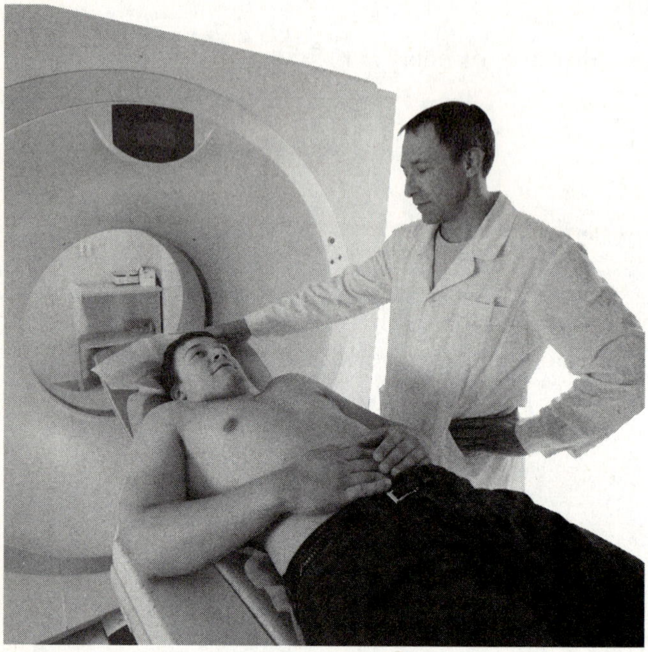

After having CT scans of his liver and other laboratory tests, Luke was diagnosed with Wilson disease, which is caused by a genetic mutation that prevents his liver from eliminating extra copper. High levels of copper can cause damage to the brain, kidneys, and eyes. Luke is now being treated for Wilson disease by taking the drug trientine hydrochloride (Syprine), which removes excess copper from the liver. Luke has also added zinc in the form of zinc acetate to block absorption of copper in the digestive tract. To reduce copper intake in foods, Luke no longer eats shellfish, mushrooms, chocolate, or liver, all of which have high levels of copper. It is anticipated that due to early detection and treatment, Luke will have good health. What is the balanced equation for the radioactive decay of Tc-99m when it emits gamma radiation?

16.1 Natural Radioactivity

Learning Goal: Describe alpha, beta, positron, and gamma radiation.

- Radioactive isotopes have unstable nuclei that break down (decay), spontaneously emitting alpha (α), beta (β), positron (β^+), or gamma (γ) radiation.
- An alpha particle is the same as a helium nucleus; it contains 2 protons and 2 neutrons.
- A beta particle is a high-energy electron and a positron is a high-energy positive particle. A gamma ray is very high-energy radiation.
- Because radiation can damage cells in the body, proper protection must be used: shielding, time limitation, and distance.

Study Note

It is important to learn the symbols for the radiation particles in order to describe the different types of radiation:

^1_1H or p	^1_0n or n	$^0_{-1}e$ or β	^4_2He or α	$^0_{+1}e$ or β^+	$^0_0\gamma$ or γ
proton	neutron	beta particle	alpha particle	positron	gamma ray

♦ Learning Check 16.1A

Match the description in column B with the terms in column A:

A		B
1. ____	$^{18}_{8}O$	a. symbol for a beta particle
2. ____	γ	b. symbol for an alpha particle
3. ____	β^+	c. an atom that emits radiation
4. ____	radioisotope	d. symbol for a positron
5. ____	$^{4}_{2}He$	e. symbol for an isotope of oxygen
6. ____	β	f. symbol for gamma radiation

Answers **1.** e **2.** f **3.** d **4.** c **5.** b **6.** a

♦ Learning Check 16.1B

Discuss some things you can do to minimize the amount of radiation received if you work with a radioactive substance. Describe how each method helps to limit the amount of radiation you would receive.

Answer

Three ways to minimize exposure to radiation are: (1) use shielding, (2) keep time short in the radioactive area, and (3) keep as much distance as possible from the radioactive materials. Shielding such as clothing and gloves stop alpha and beta particles from reaching your skin, whereas lead or concrete will absorb gamma rays. Limiting the time spent near radioactive samples reduces exposure time. Increasing the distance from a radioactive source reduces the intensity of radiation. Wearing a film badge will monitor the amount of radiation you receive.

♦ Learning Check 16.1C

What type(s) of radiation (alpha, beta, and/or gamma) would each of the following shielding materials protect you from?

a. heavy clothing ____________ **b.** skin ____________

c. paper ____________ **d.** concrete ____________

e. lead wall ____________

Answers **a.** alpha, beta **b.** alpha **c.** alpha
d. alpha, beta, gamma **e.** alpha, beta, gamma

16.2 Nuclear Equations

Learning Goal: Write a balanced nuclear equation showing mass numbers and atomic numbers for radioactive decay.

- A balanced nuclear equation represents the changes in the nuclei of radioisotopes.
- The new isotopes and the type of radiation emitted can be determined from the symbols that show the mass numbers and atomic numbers of the isotopes in the nuclear reaction.

$$\text{Radioisotope} \longrightarrow \text{new nucleus} + \text{radiation}$$

Total of the mass numbers is equal on both sides of the equation.

$${}^{11}_{6}\text{C} \longrightarrow {}^{7}_{4}\text{Be} + {}^{4}_{2}\text{He}$$

Total of the atomic numbers is equal on both sides of the equation.

- A new radioactive isotope is produced when a nonradioactive isotope is bombarded by a small particle such as a proton, an alpha particle, a beta particle, or a neutron.

${}^{10}_{5}\text{B}$	+	${}^{4}_{2}\text{He}$	$\longrightarrow$	${}^{13}_{7}\text{N}$	+	${}^{1}_{0}n$
Stable nucleus		*Bombarding particle* (α)		*New nucleus*		*Neutron emitted*

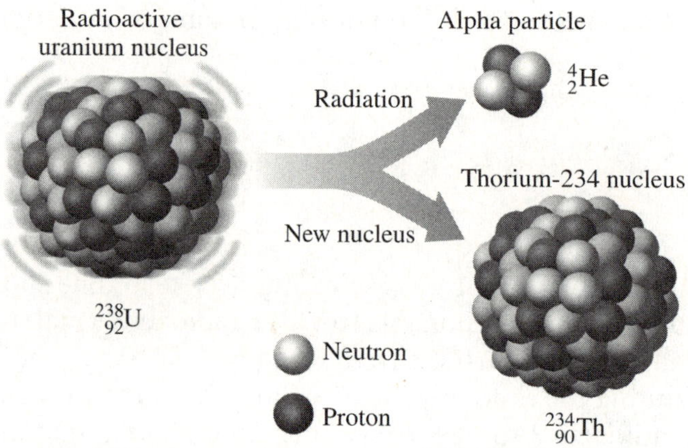

146 neutrons 92 protons		144 neutrons 90 protons		2 neutrons 2 protons
${}^{238}_{92}\text{U}$	$\longrightarrow$	${}^{234}_{90}\text{Th}$	+	${}^{4}_{2}\text{He}$
Radioactive nucleus		New nucleus		Alpha particle

In alpha decay, the mass number of the new nucleus decreases by 4, and its atomic number decreases by 2.

Guide to Completing a Nuclear Equation	
Step 1	Write the incomplete nuclear equation.
Step 2	Determine the missing mass number.
Step 3	Determine the missing atomic number.
Step 4	Determine the symbol of the new nucleus.
Step 5	Complete the nuclear equation.

Study Note

When balancing nuclear equations for radioactive decay, be sure that

1. The mass number of the reactant is equal to the sum of the mass number of the products.
2. The atomic number of the reactant is equal to the sum of the atomic number of the products.

♦ Learning Check 16.2A

CORE CHEMISTRY SKILL
Writing Nuclear Equations

Write a nuclear symbol that completes each of the following nuclear equations:

a. ${}^{66}_{29}\text{Cu} \longrightarrow {}^{66}_{30}\text{Zn} + ?$ ____________

b. ${}^{127}_{53}\text{I} \longrightarrow ? + {}^{1}_{0}n$ ____________

c. ${}^{238}_{92}\text{U} \longrightarrow {}^{4}_{2}\text{He} + ?$ ____________

d. ${}^{24}_{11}\text{Na} \longrightarrow ? + {}^{0}_{-1}e$ ____________

e. $? \longrightarrow {}^{30}_{14}\text{Si} + {}^{0}_{-1}e$ ____________

Answers **a.** ${}^{0}_{-1}e$ **b.** ${}^{126}_{53}\text{I}$ **c.** ${}^{234}_{90}\text{Th}$ **d.** ${}^{24}_{12}\text{Mg}$ **e.** ${}^{30}_{13}\text{Al}$

Study Note

In balancing nuclear transmutation equations (bombardment by small particles),

1. The sum of the mass numbers of the reactants must equal the sum of the mass numbers of the products.
2. The sum of the atomic numbers of the reactants must equal the sum of the atomic numbers of the products.

♦ Learning Check 16.2B

Complete each of the following equations for bombardment reactions:

a. $? + {}^{40}_{20}\text{Ca} \longrightarrow {}^{40}_{19}\text{K} + {}^{1}_{1}\text{H}$ ____________

b. ${}^{1}_{0}n + {}^{27}_{13}\text{Al} \longrightarrow {}^{24}_{11}\text{Na} + ?$ ____________

c. ${}^{1}_{0}n + {}^{10}_{5}\text{B} \longrightarrow ? + {}^{4}_{2}\text{He}$ ____________

d. $? + {}^{23}_{11}\text{Na} \longrightarrow {}^{23}_{12}\text{Mg} + {}^{1}_{0}n$ ____________

e. ${}^{1}_{1}\text{H} + {}^{197}_{79}\text{Au} \longrightarrow ? + {}^{1}_{0}n$ ____________

Answers **a.** ${}^{1}_{0}n$ **b.** ${}^{4}_{2}\text{He}$ **c.** ${}^{7}_{3}\text{Li}$ **d.** ${}^{1}_{1}\text{H}$ **e.** ${}^{197}_{80}\text{Hg}$

16.3 Radiation Measurement

Learning Goal: Describe the detection and measurement of radiation.

- A Geiger counter is used to detect radiation. When radiation passes through the gas in the counter tube, some atoms of gas are ionized, producing an electrical current.
- The activity of a radioactive sample measures the number of nuclear transformations/s. The curie (Ci) is equal to 3.7×10^{10} disintegrations/s. The SI unit is the becquerel (Bq), which is equal to 1 disintegration/s.
- The radiation dose absorbed by a gram of body tissue is measured in the unit rad (radiation absorbed dose) or the SI unit, gray (Gy), which is equal to 100 rad.
- The biological damage to the body caused by different types of radiation is measured in the unit rem (radiation equivalent) or the SI unit, sievert (Sv), which is equal to 100 rem.

♦ Learning Check 16.3A

Match each type of measurement unit with the radiation process measured.

a. curie **b.** becquerel **c.** rad **d.** gray **e.** rem

1. _____ an activity of one disintegration/s

2. _____ the amount of radiation absorbed by 1 g of material

3. _____ an activity of 3.7×10^{10} disintegrations/s

4. _____ a unit measuring the biological damage caused by different kinds of radiation

5. _____ a unit that measures the absorbed dose of radiation, which is equal to 100 rad

Answers **1.** b **2.** c **3.** a **4.** e **5.** d

♦ Learning Check 16.3B

a. A sample of Ir-192 has an activity of 35 μCi. What is its activity in becquerels?

b. If an absorbed dose is 15 mrem, what is the absorbed dose in sieverts?

Answers **a.** 1.3×10^6 Bq **b.** 1.5×10^{-4} Sv

16.4 Half-Life of a Radioisotope

Learning Goal: Given the half-life of a radioisotope, calculate the amount of radioisotope remaining after one or more half-lives.

- The half-life of a radioactive sample is the time required for one-half of the sample to decay (emit radiation).
- Most radioisotopes used in medicine, such as Tc-99m and I-131, have short half-lives. By comparison, many naturally occurring radioisotopes, such as C-14, Ra-226, and U-238, have long half-lives. For example, potassium-42 has a half-life of 12 h, whereas potassium-40 takes 1.3×10^9 y for one-half of the radioactive sample to decay.

Guide to Using Half-Lives	
Step 1	State the given and needed quantities.
Step 2	Write a plan to calculate the unknown quantity.
Step 3	Write the half-life equality and conversion factors.
Step 4	Set up the problem to calculate the needed quantity.

Example: Chromium-51 has a half-life of 28 days. How much of a 16-μg sample of chromium-51 remains after 84 days?

Solution:

Step 1: State the given and needed quantities.

Given 16-μg sample of chromium-51; 84 days; half-life = 28 days
Need μg of Cr-51 remaining

Step 2: Write a plan to calculate the unknown quantity.

Plan $84 \text{ days} \xrightarrow{\textit{half-life}} \text{Number of half-lives}$

$16\text{-}\mu\text{g of Cr-51} \xrightarrow{\textit{Number of half-lives}} \mu\text{g of Cr-51 remaining}$

Step 3: Write the half-life equality and conversion factors.

$$1 \text{ half-life Cr-51} = 28 \text{ days}$$

$$\frac{28 \text{ days}}{1 \text{ half-life}} \quad \text{and} \quad \frac{1 \text{ half-life}}{28 \text{ days}}$$

Step 4: Set up problem to calculate the needed quantity.

$$\text{Number of half-lives} = 84 \cancel{\text{days}} \times \frac{1 \text{ half-life}}{28 \cancel{\text{days}}} = 3 \text{ half-lives}$$

Now, we can calculate the quantity of Cr-51 that remains active.

$$16 \ \mu\text{g} \xrightarrow{1 \textit{ half-life}} 8.0 \ \mu\text{g} \xrightarrow{1 \textit{ half-life}} 4.0 \ \mu\text{g} \xrightarrow{1 \textit{ half-life}} 2.0 \ \mu\text{g}$$

♦ **Learning Check 16.4**

CORE CHEMISTRY SKILL

Using Half-Lives

a. Suppose you have an 80. mg sample of iodine-125. If iodine-125 has a half-life of 60. days, how many mg are radioactive

1. after one half-life?

2. after two half-lives?

3. after 240 days?

b. $^{99m}_{43}\text{Tc}$ has a half-life of 6 h. If a technician picked up a 16-mg sample at 8 A.M., how much of the radioactive sample remained at 8 P.M. that same day?

c. Phosphorus-32 has a half-life of 14 days. How much of a 240-μg sample will be radioactive after 56 days?

d. Iodine-131 has a half-life of 8.0 days. How many days will it take for 80. mg of I-131 to decay to 5 mg?

e. Suppose a group of archaeologists digs up some pieces of a wooden boat at an ancient site. When a sample of the wood is analyzed for C-14, scientists determine that 12.5%, or 1/8, of the original amount of C-14 remains. If the half-life of carbon-14 is 5730 years, how long ago was the boat made?

Answers **a.** **1.** 40. mg **2.** 20. mg **3.** 5.0 mg **b.** 4.0 mg
c. 15 μg **d.** 32 days **e.** 17 200 y

16.5 Medical Applications Using Radioactivity

Learning Goal: Describe the use of radioisotopes in medicine.

- Nuclear medicine uses radioactive isotopes that go to specific sites in the body.
- For diagnostic work, radioisotopes are used that emit gamma rays and produce nonradioactive products.
- By detecting the radiation emitted by medical radioisotopes, evaluations can be made about the location and extent of an injury, disease, tumor, blood flow, or level of function of a particular organ.

♦ Learning Check 16.5

Write the nuclear symbol for each of the following radioactive isotopes:

a. _____ iodine-131, used to study thyroid gland activity

b. _____ phosphorus-32, used to locate brain tumors

c. _____ sodium-24, used to determine blood flow and to locate a blood clot or embolism

d. _____ nitrogen-13, used in positron emission tomography

Answers **a.** $^{131}_{53}I$ **b.** $^{32}_{15}P$ **c.** $^{24}_{11}Na$ **d.** $^{13}_{7}N$

16.6 Nuclear Fission and Fusion

Learning Goal: Describe the processes of nuclear fission and fusion.

- In fission, a large nucleus breaks apart into smaller nuclei, releasing one or more types of radiation and a great amount of energy.
- A chain reaction is a fission reaction that will continue once initiated.
- In fusion, small nuclei combine to form a larger nucleus, which releases great amounts of energy.
- Nuclear fission is currently used to produce electrical power while nuclear fusion applications are still in the experimental stage.

Key Terms for Sections 16.1–16.6

Match each of the following key terms with the correct definition:

a. radiation **b.** half-life **c.** curie **d.** nuclear fission
e. alpha particle **f.** positron **g.** rem

1. ____ a particle identical to a helium nucleus produced in a radioactive nucleus
2. ____ the time required for one-half of a radioactive sample to undergo radioactive decay
3. ____ a unit of radiation measurement equal to 3.7×10^{10} disintegrations per second
4. ____ a process in which large nuclei split into smaller nuclei with the release of energy
5. ____ energy or particles released by radioactive atoms
6. ____ a measure of biological damage caused by radiation
7. ____ produced when a proton is transformed into a neutron

Answers **1.** e **2.** b **3.** c **4.** d **5.** a **6.** g **7.** f

♦ Learning Check 16.6A

Discuss the nuclear processes of fission and fusion for the production of energy.

Answer

Nuclear fission is a splitting of the atom into two or more nuclei accompanied by the release of large amounts of energy and radiation. In the process of *nuclear fusion*, two or more nuclei combine to form a heavier nucleus and release a large amount of energy. However, fusion requires a considerable amount of energy to initiate the process. Because the extremely high temperatures are difficult to maintain, fusion technology is still in the experimental stage.

♦ Learning Check 16.6B

Balance each of the following nuclear equations by writing the nuclear symbol for the missing particle and state if each is a fission or fusion reaction.

a. ${}^{235}_{92}\text{U} + {}^{1}_{0}n \longrightarrow {}^{143}_{54}\text{Xe} + ? + 3{}^{1}_{0}n$ ______________

b. ${}^{2}_{1}\text{H} + {}^{1}_{1}\text{H} \longrightarrow ?$ ______________

c. $^{3}_{1}H + ? \longrightarrow ^{4}_{2}He + ^{1}_{0}n$ ________________

d. $^{235}_{92}U + ^{1}_{0}n \longrightarrow ^{91}_{36}Kr + ? + 3^{1}_{0}n$ ________________

Answers **a.** $^{90}_{38}Sr$, fission **b.** $^{3}_{1}H$, fusion **c.** $^{2}_{1}H$, fusion **d.** $^{142}_{56}Ba$, fission

Checklist for Chapter 16

You are ready to take the Practice Test for Chapter 16. Be sure you have accomplished the following learning goals for this chapter. If not, review the section listed at the end of the goal. Then apply your new skills and understanding to the Practice Test.

After studying Chapter 16, I can successfully:

_____ Describe alpha, beta, and gamma radiation (16.1).

_____ Write a nuclear equation showing mass numbers and atomic numbers for radioactive decay (16.2).

_____ Write a nuclear equation for the formation of a radioactive isotope (16.2).

_____ Describe the detection and measurement of radiation (16.3).

_____ Calculate the amount of radioisotope remaining after one or more half-lives (16.4).

_____ Calculate the amount of time for a specific amount of a radioisotope to decay given its half-life (16.4).

_____ Describe the uses of radioisotopes in medicine (16.5).

_____ Describe the processes of nuclear fission and fusion (16.6).

Practice Test for Chapter 16

1. The correctly written symbol for an isotope of sulfur is (16.1)
A. $^{30}_{16}Su$ **B.** $^{14}_{30}Si$ **C.** $^{30}_{16}S$ **D.** $^{30}_{16}Si$ **E.** $^{16}_{30}S$

2. Alpha particles are composed of (16.1)
A. protons **B.** neutrons **C.** electrons
D. protons and electrons **E.** protons and neutrons

3. Gamma radiation is a type of radiation that (16.1)
A. originates in the electron shells
B. is most dangerous
C. is least dangerous
D. is the heaviest
E. travels the shortest distance

4. The charge on an alpha particle is (16.1)
A. −1 **B.** +1 **C.** −2 **D.** +2 **E.** +4

5. Beta particles formed in a radioactive nucleus are (16.1)
A. protons **B.** neutrons **C.** electrons
D. protons and electrons **E.** protons and neutrons

For questions 6 through 10, select from the following: (16.1)

A. ${}_{-1}^{0}X$ **B.** ${}_{2}^{4}X$ **C.** ${}_{1}^{1}X$ **D.** ${}_{0}^{1}X$ **E.** ${}_{0}^{0}X$

6. an alpha particle

7. a beta particle

8. a gamma ray

9. a proton

10. a neutron

11. Shielding from gamma rays is provided by (16.1)
A. skin **B.** paper **C.** clothing **D.** lead **E.** air

12. The skin will provide shielding from (16.1)
A. alpha particles **B.** beta particles **C.** gamma rays
D. ultraviolet rays **E.** X rays

13. The radioisotope iodine-131 is used as a radioactive tracer for studying thyroid gland activity. The symbol for iodine-131 is (16.5)
A. I **B.** ${}_{131}I$ **C.** ${}_{53}^{131}I$ **D.** ${}_{131}^{53}I$ **E.** ${}_{53}^{78}I$

14. When an atom emits an alpha particle, its mass number will (16.2)
A. increase by 1 **B.** increases by 2 **C.** increase by 4
D. decrease by 4 **E.** not change

15. When a nucleus emits a positron, the atomic number of the new nucleus (16.2)
A. increases by 1 **B.** increases by 2 **C.** decreases by 1
D. decreases by 2 **E.** does not change

16. When a nucleus emits a gamma ray, the atomic number of the new nucleus (16.2)
A. increases by 1 **B.** increases by 2 **C.** decreases by 1
D. decreases by 2 **E.** does not change

For questions 17 through 20, select the particle that completes each of the following equations: (16.2)

A. neutron **B.** alpha particle **C.** beta particle **D.** gamma ray

17. ${}_{50}^{126}Sn \longrightarrow {}_{51}^{126}Sb + ?$

18. ${}_{30}^{69}Zn \longrightarrow {}_{31}^{69}Ga + ?$

19. ${}_{43}^{99m}Tc \longrightarrow {}_{43}^{99}Tc + ?$

20. ${}_{62}^{149}Sm \longrightarrow {}_{60}^{145}Nd + ?$

21. What symbol completes the following reaction? (16.2)
${}_{0}^{1}n + {}_{7}^{14}N \longrightarrow ? + {}_{1}^{1}H$
A. ${}_{8}^{15}O$ **B.** ${}_{6}^{15}C$ **C.** ${}_{8}^{14}O$ **D.** ${}_{6}^{14}C$ **E.** ${}_{7}^{15}N$

22. To complete this nuclear equation, you need to write (16.2)
$? + {}_{26}^{54}Fe \longrightarrow {}_{28}^{57}Ni + {}_{0}^{1}n$
A. an alpha particle **B.** a beta particle **C.** gamma
D. neutron **E.** proton

23. The name of the unit used to measure the number of disintegrations/s is (16.3)
A. curie **B.** rad **C.** rem **D.** gray **E.** sievert

24. The rem and the sievert are units used to measure (16.3)
A. activity of a radioactive sample
B. biological damage of different types of radiation
C. radiation absorbed
D. background radiation
E. half-life of a radioactive sample

25. Radiation can cause (16.3)
A. nausea **B.** a lower white cell count **C.** fatigue
D. hair loss **E.** all of these

26. Radioisotopes used in medical diagnosis (16.5)
A. have short half-lives **B.** emit only gamma rays **C.** locate in specific organs
D. produce nonradioactive nuclei **E.** all of these

27. The time required for a radioisotope to decay is measured by its (16.4)
A. half-life **B.** protons **C.** activity **D.** fusion **E.** radioisotope

28. Oxygen-15 used in imaging has a half-life of 2 min. How many half-lives have occurred in the 10 minutes it takes to prepare the sample? (16.4)
A. 2 **B.** 3 **C.** 4 **D.** 5 **E.** 6

29. Iodine-131 has a half-life of 8.0 days. How long will it take for a 160-mg sample to decay to 10 mg? (16.4)
A. 8.0 days **B.** 16 days **C.** 32 days **D.** 40. days **E.** 48 days

30. Phosphorus-32 has a half-life of 14.3 days. After 28.6 days, how many milligrams of a 120-mg sample will still be radioactive? (16.4)
A. 240 mg **B.** 120 mg **C.** 60. mg **D.** 30. mg **E.** 15 mg

31. The "splitting" of a large nucleus to form smaller particles accompanied by a release of energy is called (16.6)
A. radioisotope **B.** nuclear fission **C.** nuclear fusion **D.** rem **E.** half-life

32. The process of combining small nuclei to form larger nuclei is (16.6)
A. radioisotope **B.** fission **C.** fusion **D.** rem **E.** half-life

33. A fusion reaction (16.6)
A. occurs in the Sun
B. forms larger nuclei from smaller nuclei
C. requires extremely high temperatures
D. releases a large amount of energy
E. all of the these

Answers to the Practice Test

1. C	**2.** E	**3.** B	**4.** D	**5.** C
6. B	**7.** A	**8.** E	**9.** C	**10.** D
11. D	**12.** A	**13.** C	**14.** D	**15.** C
16. E	**17.** C	**18.** C	**19.** D	**20.** B
21. D	**22.** A	**23.** A	**24.** B	**25.** E
26. E	**27.** A	**28.** D	**29.** C	**30.** D
31. B	**32.** C	**33.** E		

17
Organic Chemistry

Last week, Jack was part of a firefighting team called in to put out a fire that started in a campsite in Rocky Mountain National Park. Jack used a driptorch to set a backfire that consumed fuel ahead of the main fire. The driptorch consisted of a fuel canister that had a spout with a loop to prevent fire from entering the canister, a breather valve that allowed air to replace fuel as it is used, and a wick that transferred flaming fuel to the ground. Typically, the fuel is a mixture of hydrocarbons such as kerosene, gasoline, and diesel. Sometimes heavier oils are used to increase adhesion of the liquid fuel to the vegetation as well as to increase burn time and heat. In forest and prairie management, the driptorch is the most common tool used to ignite prescribed burns, which are used to remove excess fuel buildup or to re-create natural cycles of fire in an ecosystem. Write a balanced equation for the combustion of a compound in kerosene, $C_{11}H_{24}$, used in a driptorch for setting backfires.

17.1 Alkanes and Naming Substituents

Learning Goal: Write the IUPAC names and draw the condensed structural formulas for alkanes with substituents.

- The IUPAC (International Union of Pure and Applied Chemistry) system for naming organic compounds is used.
- The longest continuous chain is named as the parent alkane, and the substituents attached to the chain are numbered and listed alphabetically in front of the name of the parent chain.

$$\begin{array}{l} \quad\quad\;\; CH_3 \\ \quad\quad\quad | \\ CH_3-CH-CH_2-CH_3 \end{array} \quad \text{2-methylbutane}$$

Guide to Naming Alkanes	
Step 1	Write the alkane name of the longest chain of carbon atoms.
Step 2	Number the carbon atoms from the end nearer a substituent.
Step 3	Give the location and name of each substituent (alphabetical order) as a prefix to the name of the main chain.

♦ Learning Exercise 17.1A

Write the IUPAC name for each of the following compounds:

1. 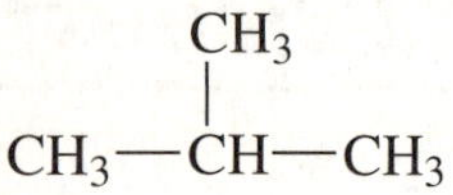

2. 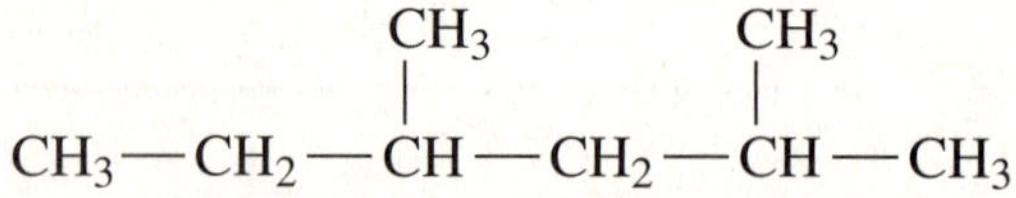

3. 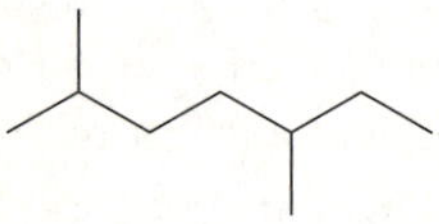

4. 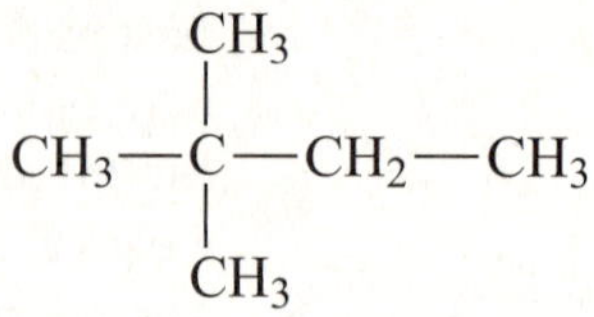

5. $Cl-CH_2-CH_2-CH_2-Cl$

Answers	**1.** methylpropane	**2.** 2,4-dimethylhexane
	3. 2,5-dimethylheptane	**4.** 2,2-dimethylbutane
	5. 1,3-dichloropropane	

Guide to Drawing Alkane Formulas	
Step 1	Draw the main chain of carbon atoms.
Step 2	Number the chain and place the substituents on the carbons indicated by the numbers.
Step 3	Add the correct number of hydrogen atoms to give four bonds to each C atom.

♦ Learning Exercise 17.1B

Draw the condensed structural formulas for 1, 2, and 3 and the skeletal formulas for 4, 5, and 6:

1. pentane

2. 2-methylpentane

3. 4-ethyl-2-methylhexane

4. 2,2,4-trimethylhexane

5. 1,2-dichlorobutane

6. 2-bromo-4-methylheptane

Answers

1. $CH_3—CH_2—CH_2—CH_2—CH_3$

2. $CH_3—CH(CH_3)—CH_2—CH_3—CH_3$

3. $CH_3—CH(CH_3)—CH_2—CH(CH_2—CH_3)—CH_2—CH_3$

4. 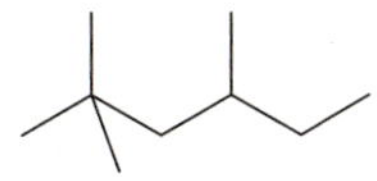

5. 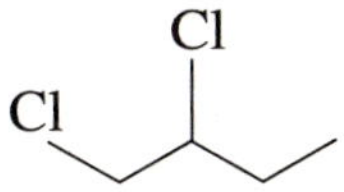

6. 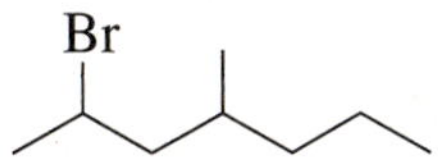

17.2 Alkenes, Alkynes, and Polymers

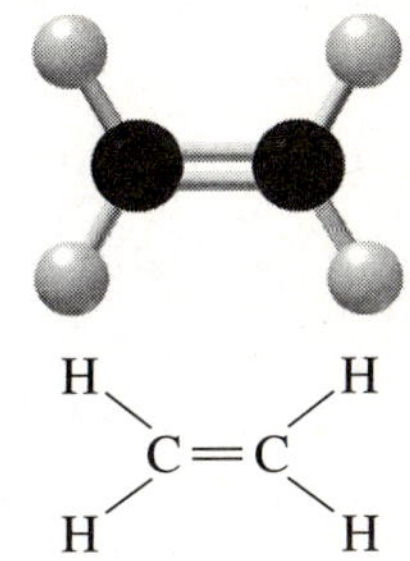

Ethene

Learning Goal: Write the IUPAC names and formulas for alkenes and alkynes; draw the condensed structural formulas for the monomers that form a polymer.

- The alkenes are a family of unsaturated hydrocarbon that contain a double bond. The IUPAC names of alkenes are derived by changing the *ane* ending of the parent alkane to *ene*.
- In alkenes, the longest carbon chain containing the double bond is numbered from the end nearer the double bond.

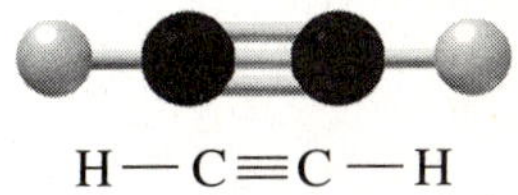

Ethyne

$CH_3—CH{=}CH_2$	$CH_2{=}CH—CH_2—CH_3$	$CH_3—CH{=}C(CH_3)—CH_3$
propene (propylene)	1-butene	2-methyl-2-butene

- The alkynes are a family of unsaturated hydrocarbons that contain a triple bond. They use naming rules similar to the alkenes, but the parent chain ends with *yne*.

$HC{\equiv}CH$	$CH_3—C{\equiv}CH$
Ethyne	Propyne

- *Polymers* are large molecules that consist of repeated units called *monomers*.
- Many synthetic polymers are made from small alkene monomers.

Guide to Naming Alkenes and Alkynes	
Step 1	Name the longest carbon chain that contains the double or triple bond.
Step 2	Number the carbon chain from the end nearer the double or triple bond.
Step 3	Give the location and name of each substituent (alphabetical order) as a prefix to the alkene or alkyne name.

♦ Learning Exercise 17.2A

Write the IUPAC name for each of the following alkenes:

1. $CH_3-CH=CH_2$

2. 

3. 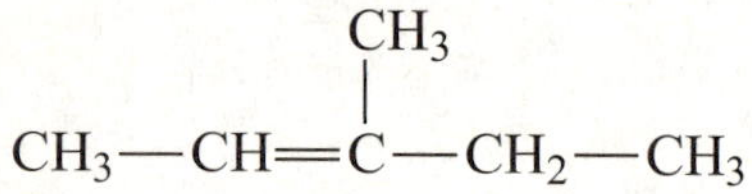

4. 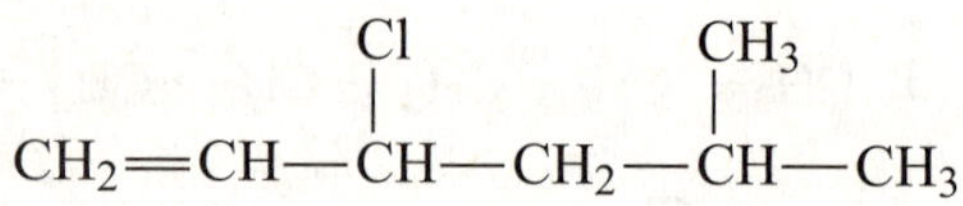

Answers

1. propene
2. 1-butene
3. 3-methyl-2-pentene
4. 3-chloro-5-methyl-l-hexene

♦ Learning Exercise 17.2B

Write the IUPAC and common name (if any) of each of the following alkynes:

1. $HC\equiv CH$

2. $CH_3-C\equiv CH$

3. $CH_3-CH_2-C\equiv CH$

4. 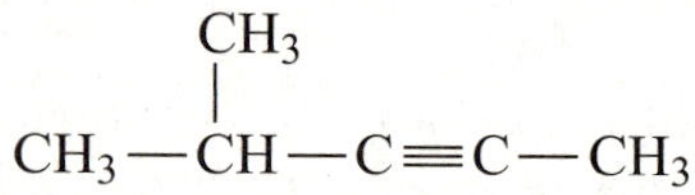

Answers

1. ethyne (acetylene)
2. propyne
3. 1-butyne
4. 4-methyl-2-pentyne

♦ Learning Exercise 17.2C

Draw the condensed structural formulas for 1, 2, and 3 and the skeletal formulas for 4, 5, and 6.

1. ethyl chloride

2. bromomethane

3. 3-bromo-1-chloropentane

4. 2,3-dibromohexane

5. 2,2,3-trichlorobutane

6. 2,4-dibromo-2,4-dichloropentane

Answers

1. CH_3-CH_2-Cl

2. CH_3-Br

3. 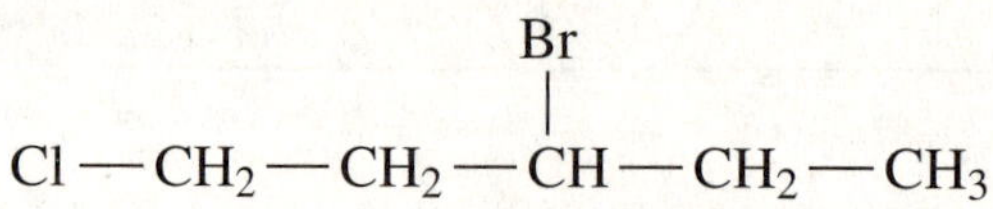

4. 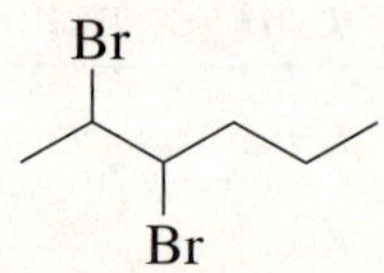

5. 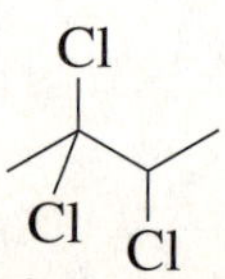

6. 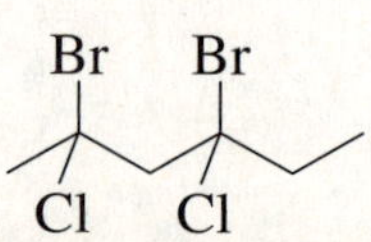

♦ Learning Check 17.2D

Write the formula of the alkene monomer that would be used for each of the following polymers:

1. 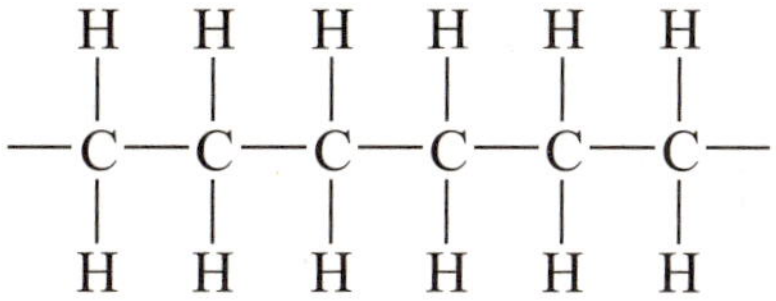

2. 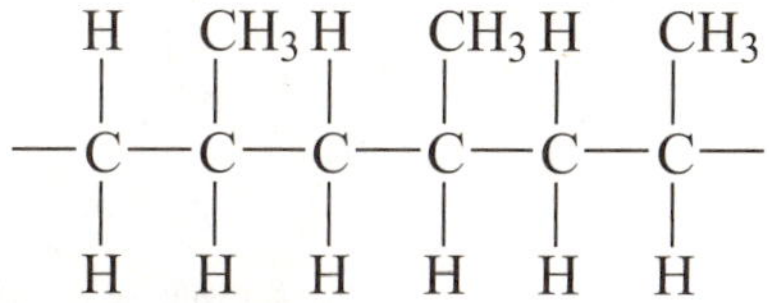

3. 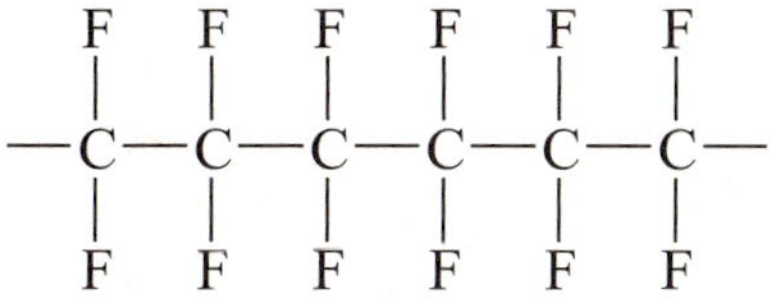

Tables, benches, trash receptacles, and pipes used for irrigation systems are made from recycled plastics.

Answers **1.** $H_2C{=}CH_2$ **2.** 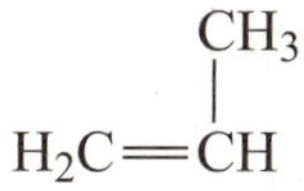 **3.** $F_2C{=}CF_2$

♦ Learning Check 17.2E

Write three sections of the polymer that would result when 1,1-difluoroethene is the monomer unit.

Answer

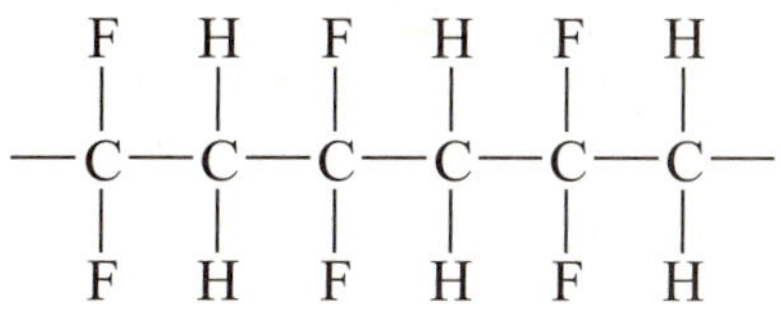

17.3 Aromatic Compounds

Learning Goal: Describe the bonding in benzene; name aromatic compounds and draw their condensed structural formulas.

- Most aromatic compounds contain benzene, a cyclic structure containing six CH units. The structure of benzene can be represented as a hexagon with alternating single and double bonds or with a circle in the center.

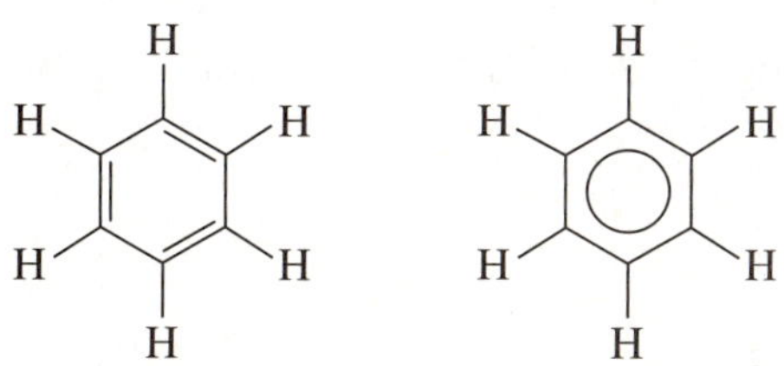

Structural representations for benzene

- The names of many aromatic compounds use the parent name benzene, although many common names were retained as IUPAC names, such as toluene.

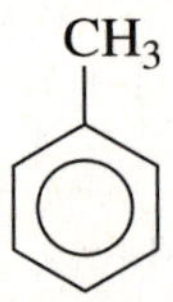

Toluene
(methylbenzene)

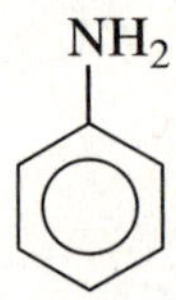

Aniline
(aminobenzene)

Phenol
(hydroxybenzene)

- When benzene has only one substituent, the ring is not numbered. For two or more substituents, the ring is numbered to give the lower numbers to the substituents.

♦ Learning Exercise 17.3

Write the IUPAC (or common name) for each of the following:

1. 

2. 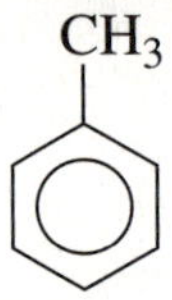

3. 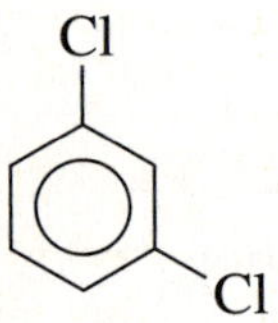

4. 

Answers

1. benzene	**2.** methylbenzene; toluene
3. 1,3-dichlorobenzene	**4.** 4-bromotoluene

17.4 Alcohols, Phenols, and Ethers

Learning Goal: Write the IUPAC and common names for alcohols, phenols, and ethers; draw the condensed structural formulas when given their names.

- An alcohol contains the hydroxyl —OH group attached to a carbon chain.
- In the IUPAC system, alcohols are named by replacing the *ane* of the alkane name with *ol*. The location of the —OH group is given by numbering the carbon chain. Simple alcohols are generally named by their common names, with the alkyl name preceding the term *alcohol*. For example, CH_3—OH is methyl alcohol, and CH_3—CH_2—OH is ethyl alcohol.
- When a hydroxyl group is attached to a benzene ring, the compound is a phenol.
- An ether contains an oxygen atom attached by single bonds to two carbon atoms, —O—.

- Simple ethers are named using the names of the alkyl groups attached to the oxygen atom.

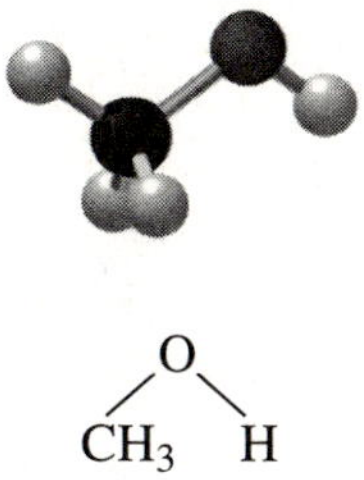

Methanol

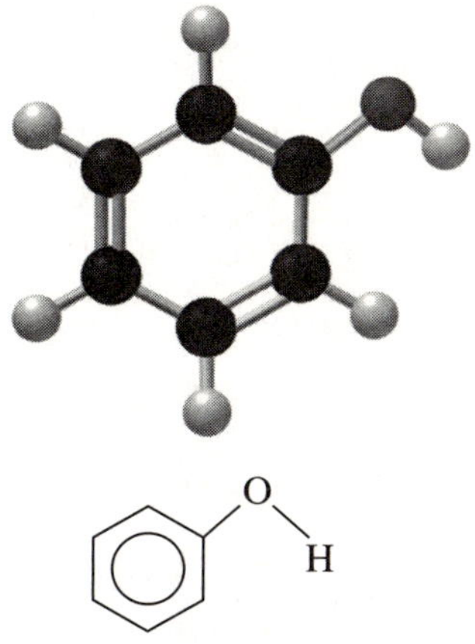

Phenol

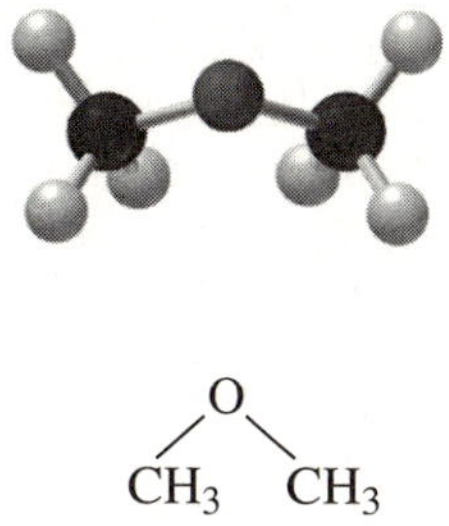

Dimethyl ether

Guide to Naming Alcohols	
Step 1	Name the longest carbon chain attached to the —OH group by replacing the *e* in the corresponding alkane name with *ol*. Name an aromatic alcohol as a *phenol*.
Step 2	Number the chain starting at the end nearer to the —OH group.
Step 3	Give the location and name of each substituent relative to the —OH group.

♦ Learning Exercise 17.4A

Give the correct IUPAC and common name (if any) for each of the following compounds:

1. $CH_3—CH_2—OH$

2. $CH_3—CH_2—CH_2—OH$

3. $CH_3—\underset{}{CH}(OH)—CH_2—CH_2—CH_3$

4. 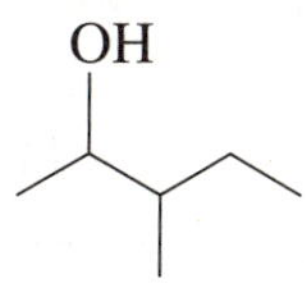

5. 

6. $CH_3—CH_2—CH_2—O—CH_3$

Answers

1. ethanol (ethyl alcohol)
2. 1-propanol (propyl alcohol)
3. 2-pentanol
4. 3-methyl-2-pentanol
5. phenol
6. methyl propyl ether (common)

♦ Learning Exercise 17.4B

Draw the condensed structural formula for each of the following compounds.

1. 2-butanol

2. 2-chloro-1-propanol

3. 2,4-dimethyl-3-pentanol

4. 3-methylphenol

5. ethyl methyl ether

Answers

1. 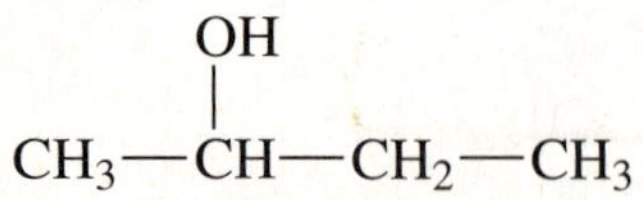

2. 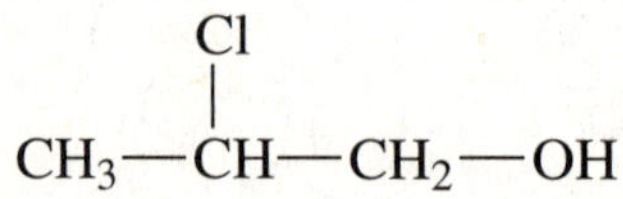

3. 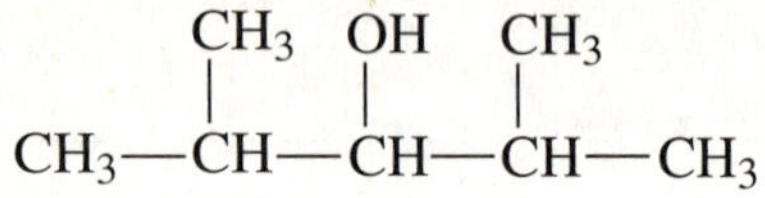

4. 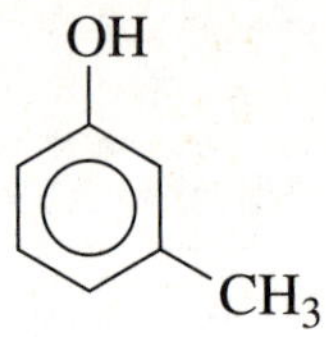

5. $CH_3—CH_2—O—CH_3$

17.5 Aldehydes and Ketones

Learning Goal: Write the IUPAC and common names for aldehydes and ketones; draw the condensed structural formulas when given their names.

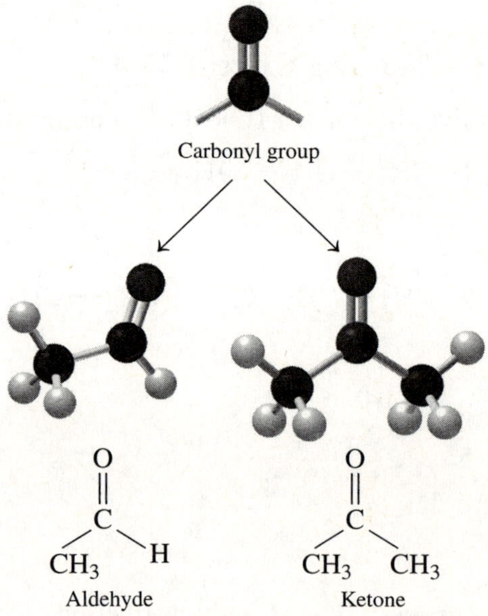

- In an aldehyde, the carbonyl group (C=O) appears at the end of a carbon chain attached to at least one hydrogen atom.
- In a ketone, the carbonyl group (C=O) occurs between carbon groups and has no hydrogens attached to it.
- In the IUPAC system, aldehydes and ketones are named by replacing the *e* in the longest chain containing the carbonyl group with *al* for aldehydes and *one* for ketones. The location of the carbonyl group in a ketone is given if there are more than four carbon atoms in the chain.
- In the IUPAC system, the location of the aldehyde carbon atom is carbon 1 but is not specified in the name.

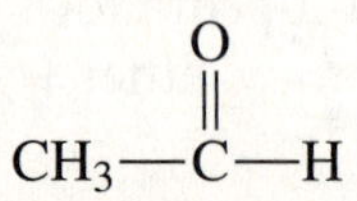

Ethanal
(acetaldehyde)

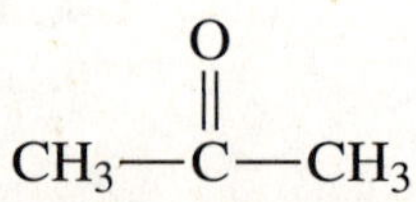

Propanone
(dimethyl ketone, acetone)

Guide to Naming Aldehydes	
Step 1	Name the longest carbon chain by replacing the *e* in the alkane name with *al*.
Step 2	Name and number substituents by counting the carbonyl group as carbon 1.

Guide to Naming Ketones	
Step 1	Name the longest carbon chain by replacing the *e* in the alkane name with *one*.
Step 2	Number the carbon chain from the end nearer the carbonyl group, and indicate its location.
Step 3	Name and number any substituents on other carbons in the chain.

♦ Learning Exercise 17.5A

Write the correct IUPAC (or common name) for the following aldehydes and ketones:

1. $CH_3-\overset{\overset{O}{\|}}{C}-H$

2. $CH_3-CH_2-CH_2-CH_2-\overset{\overset{O}{\|}}{C}-H$

3. $CH_3-\overset{\overset{O}{\|}}{C}-CH_3$

4. 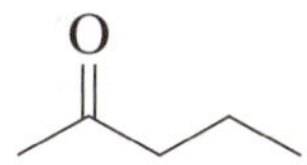

5. $CH_3-CH_2-\overset{\overset{CH_3}{|}}{CH}-CH_2-CH_2-\overset{\overset{O}{\|}}{C}-H$

6. $CH_3-CH_2-\overset{\overset{O}{\|}}{C}-CH_2-CH_3$

Answers
1. ethanal; acetaldehyde
2. pentanal
3. propanone; dimethyl ketone, acetone
4. 2-pentanone; methyl propyl ketone
5. 4-methylhexanal
6. 3-pentanone; diethyl ketone

♦ Learning Exercise 17.5B

Draw the condensed structural formulas for **1** to **4** and the skeletal formulas for **5** and **6**.

1. ethanal

2. 2-methylbutanal

3. 2-chloropropanal

4. ethyl methyl ketone

5. 3-hexanone

6. benzaldehyde

Answers

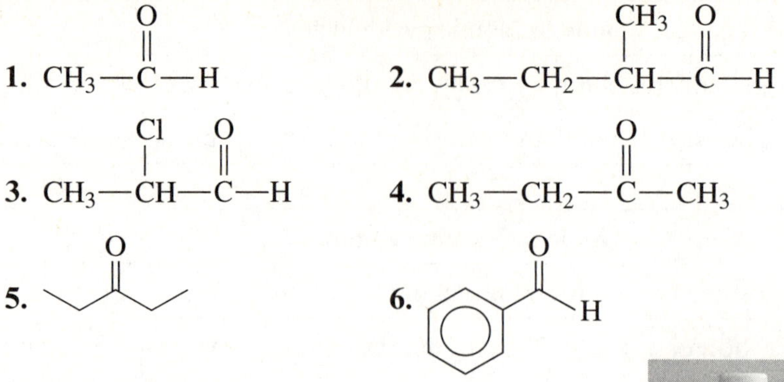

The sour taste of vinegar is due to ethanoic acid (acetic acid)

17.6 Carboxylic Acids and Esters

Learning Goal: Write the IUPAC and common names for carboxylic acids and esters; draw the condensed structural formulas when given their names.

- In the IUPAC system, a carboxylic acid is named by replacing the *e* ending of the alkane name with *oic acid.* The carboxylic acid carbon atom is carbon 1 but it is not specified in the name. Simple acids usually are named by the common naming system using the prefixes *form-* (1C), *acet-* (2C), *propion-* (4C), *butyr-* (4C), followed by *ic acid.*

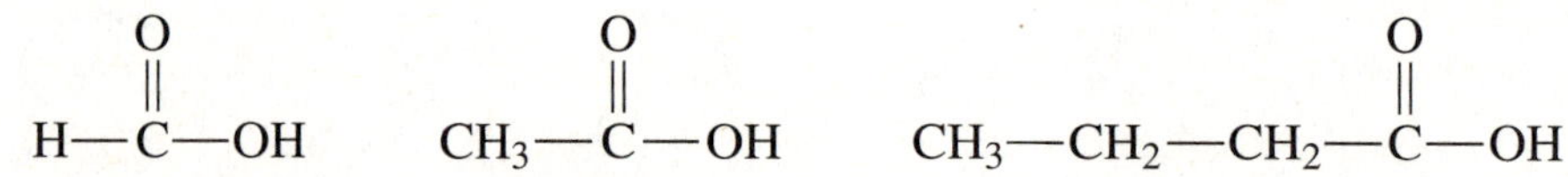

methanoic acid (formic acid) | ethanoic acid (acetic acid) | butanoic acid (butyric acid)

- In the presence of a strong acid, carboxylic acids react with alcohols to produce esters and water (esterification reaction).

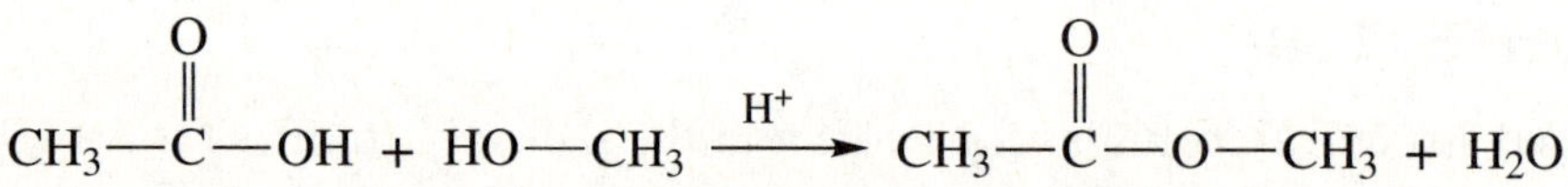

Guide to Naming Carboxylic Acids	
Step 1	Identify the carbon chain and replace the *e* in the alkane name with *oic acid.*
Step 2	Give the location and name of each substituent by counting the carboxyl carbon as 1.

methyl ethanoate (methyl acetate) | butyl ethanoate (butyl acetate)

Guide to Naming Esters	
Step 1	Write the name of the carbon chain from the alcohol as an *alkyl* group.
Step 2	Change the *ic acid* of the acid name to *ate.*

♦ Learning Exercise 17.6A

Give the IUPAC and common names, if any, for each of the following carboxylic acids:

1. $CH_3-\overset{O}{\overset{\|}{C}}-OH$

2. $CH_3-CH_2-\overset{O}{\overset{\|}{C}}-OH$

3. 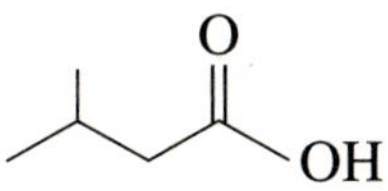

4. 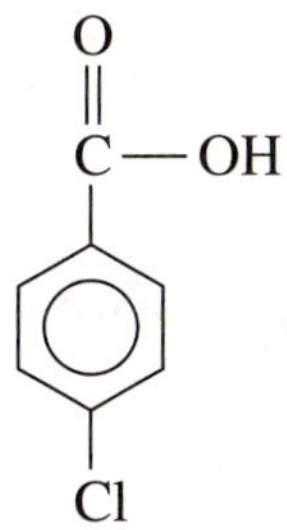

Answers

1. ethanoic acid (acetic acid)
2. propanoic acid (propionic acid)
3. 3-methylbutanoic acid
4. 4-chlorobenzoic acid

♦ Learning Exercise 17.6B

Draw the condensed structural formula for 1 through 3 and the skeletal formula for 4 through 6.

1. ethanoic acid

2. 3-chloropropanoic acid

3. formic acid

4. 2-bromobutanoic acid

5. benzoic acid

6. 3-methylpentanoic acid

Answers

1. $CH_3-\overset{O}{\overset{\|}{C}}-OH$

2. $Cl-CH_2-CH_2-\overset{O}{\overset{\|}{C}}-OH$

3. $H-\overset{O}{\overset{\|}{C}}-OH$

4. 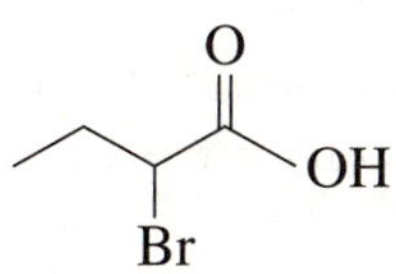

5. 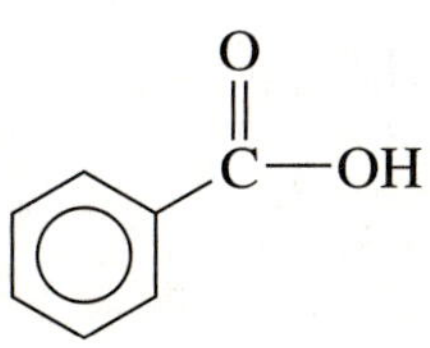

6. 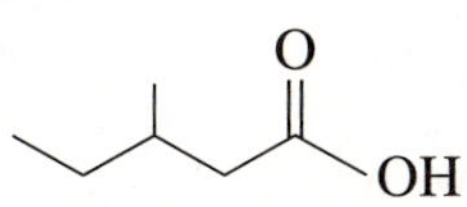

♦ Learning Exercise 17.6C

Draw the condensed structural formula for the products of each of the following esterification reactions:

1. $CH_3-\overset{\overset{\displaystyle O}{\|}}{C}-OH + CH_3-OH \xrightarrow{H^+}$

2. $H-\overset{\overset{\displaystyle O}{\|}}{C}-OH + CH_3-CH_2-OH \xrightarrow{H^+}$

3. $CH_3-CH_2-\overset{\overset{\displaystyle O}{\|}}{C}-OH + CH_3-CH_2-OH \xrightarrow{H^+}$

Answers

1. $CH_3-\overset{\overset{\displaystyle O}{\|}}{C}-O-CH_3 + H_2O$
2. $H-\overset{\overset{\displaystyle O}{\|}}{C}-O-CH_2-CH_3 + H_2O$
3. $CH_3-CH_2-\overset{\overset{\displaystyle O}{\|}}{C}-O-CH_2-CH_3 + H_2O$

17.7 Amines and Amides

Learning Goal: Write the common names for amines and the IUPAC and common names for amides; draw the condensed structural formulas when given their names.

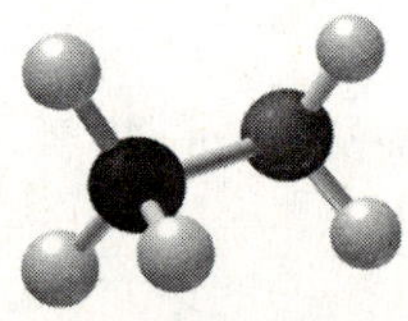

Methylamine

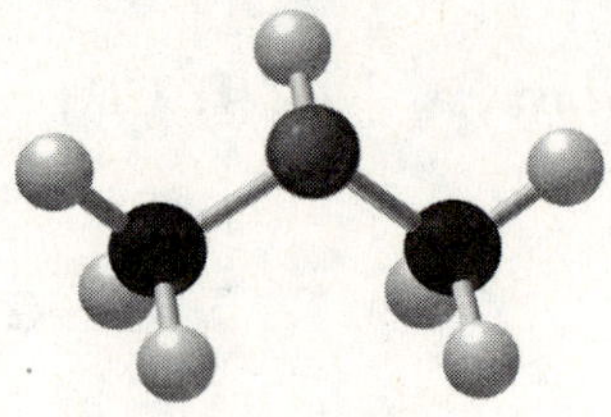

Dimethylamine

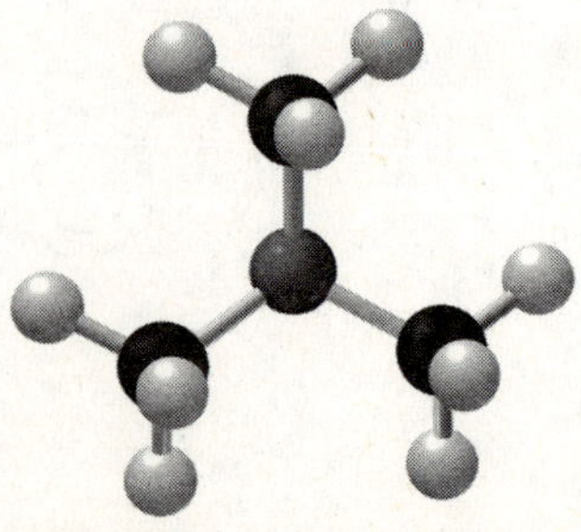

Trimethylamine

- Amines are derivatives of ammonia (NH_3), in which alkyl or aromatic groups replace one or more hydrogen atoms.
- Simple amines are named with common names in which the names of the alkyl groups are listed alphabetically preceding the suffix *amine*.

CH_3-NH_2	$CH_3-NH-CH_3$	$CH_3-\underset{}{\overset{\overset{\displaystyle CH_3}{|}}{N}}-CH_3$
Methylamine	Dimethylamine	Trimethylamine

- Aromatic amines are named as derivatives of aniline (aminobenzene).

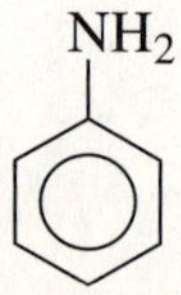

Aniline

- Amides are derivatives of carboxylic acids in which an amine group replaces the —OH group in the acid.
- Amides are named by replacing the *ic acid* or *oic acid* ending by *amide*.

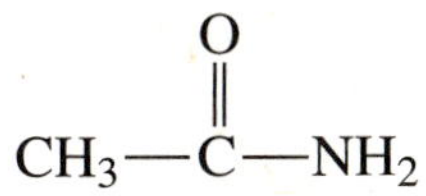

ethanamide
(acetamide)

- An amide is produced in an amidation reaction in which a carboxylic acid reacts with ammonia or an amine.

Key Terms for Sections 17.1–17.7

Match the following terms with the descriptions shown:

a. branched alkane **b.** monomer **c.** substituent **d.** isomers
e. esterification **f.** amidation **g.** amine

1. _____ organic compounds in which identical molecular formulas have different arrangements of atoms

2. _____ the reaction of a carboxylic acid and an amine

3. _____ an organic compound that contains an amino group

4. _____ the formation of an ester from a carboxylic acid and an alcohol with the elimination of a molecule of water in the presence of an acid catalyst

5. _____ an alkane containing a hydrocarbon substituent bonded to the main chain

6. _____ the small organic molecule that is repeated many times in a polymer

7. _____ groups of atoms such as an alkyl group or a halogen bonded to the main chain or ring of carbon atoms

Answers **1.** d **2.** f **3.** g **4.** e **5.** a **6.** b **7.** c

♦ Learning Exercise 17.7A

Give the common name for each of the following amines:

1. 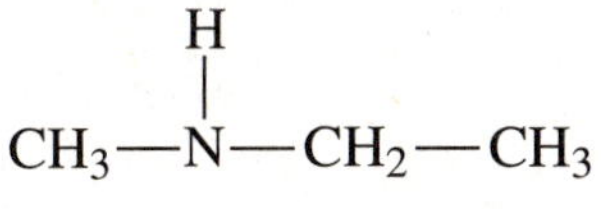

2. 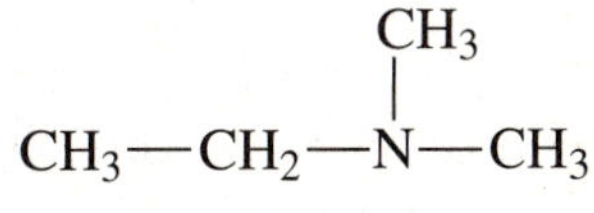

3. 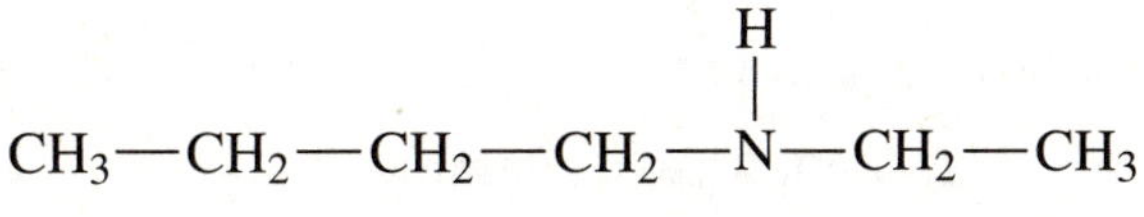

4. 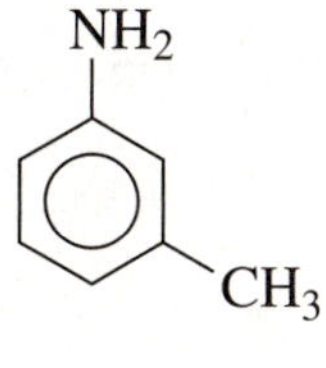

Answers **1.** ethylmethylamine **2.** ethyldimethylamine
3. butylethylamine **4.** 3-methylaniline

♦ Learning Exercise 17.7B

Draw the condensed structural formula for each of the following amines:

1. propylamine

2. ethylmethylamine

3. dimethylamine

4. aniline

Answers

1. $CH_3—CH_2—CH_2—NH_2$

2. $CH_3—NH—CH_2—CH_3$

3. $CH_3—NH—CH_3$

4. 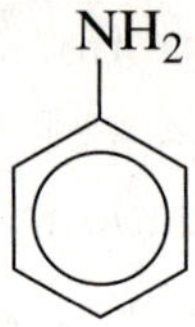

♦ Learning Exercise 17.7C

Write the IUPAC name and common name, if any, for each of the following amides:

1. $CH_3—CH_2—\overset{\overset{\Large O}{\|}}{C}—NH_2$

2. 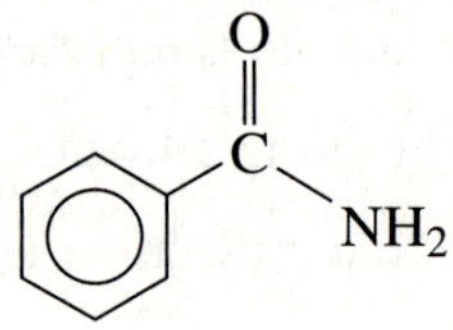

3. $CH_3—CH_2—CH_2—CH_2—\overset{\overset{\Large O}{\|}}{C}—NH_2$

4. $CH_3—\overset{\overset{\Large O}{\|}}{C}—NH_2$

Answers

1. propanamide (propionamide)

2. benzamide

3. pentanamide

4. ethanamide (acetamide)

♦ Learning Exercise 17.7D

Draw the condensed structural formula for each of the following amides:

1. propanamide

2. 2-methylbutanamide

3. 3-chloropentanamide

4. benzamide

Answers

1. $CH_3-CH_2-\overset{\overset{\displaystyle O}{\|}}{C}-NH_2$
2. $CH_3-CH_2-\overset{\overset{\displaystyle CH_3}{|}}{CH}-\overset{\overset{\displaystyle O}{\|}}{C}-NH_2$
3. $CH_3-CH_2-\overset{\overset{\displaystyle Cl}{|}}{CH}-CH_2-\overset{\overset{\displaystyle O}{\|}}{C}-NH_2$
4. 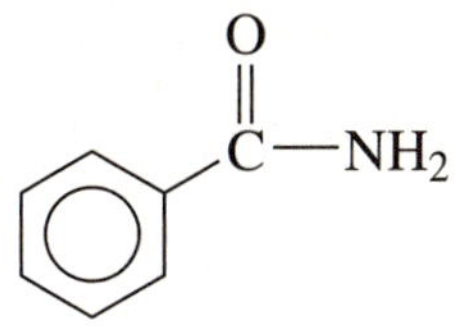

♦ **Learning Exercise 17.7E**

Draw the condensed structural formula of the amide formed in each of the following reactions:

1. $CH_3-CH_2-\overset{\overset{\displaystyle O}{\|}}{C}-OH + NH_3 \xrightarrow{\text{heat}}$

2. $CH_3-\overset{\overset{\displaystyle O}{\|}}{C}-OH + NH_3 \xrightarrow{\text{heat}}$

Answers 1. $CH_3-CH_2-\overset{\overset{\displaystyle O}{\|}}{C}-NH_2$ 2. $CH_3-\overset{\overset{\displaystyle O}{\|}}{C}-NH_2$

Checklist for Chapter 17

You are ready to take the Practice Test for Chapter 17. Be sure you have accomplished the following learning goals for this chapter. If not, review the section listed at the end of the goal. Then apply your new skills and understanding to the Practice Test.

After studying Chapter 17, I can successfully:

_____ Name and draw condensed structural formulas of alkanes with substituents (17.1).

_____ Identify the structural features of alkenes and alkynes (17.2).

_____ Name alkenes and alkynes using IUPAC rules and draw their condensed structural formulas (17.2).

_____ Describe the process of forming polymers from alkene monomers (17.2).

_____ Write the names and draw the condensed structural formulas for compounds that contain a benzene ring (17.3).

_____ Write the IUPAC or common name of an alcohol or phenol and draw the condensed structural formula from the name (17.4).

_____ Write the common name of a simple ether and draw the condensed structural formula from the name (17.4).

_____ Write the IUPAC and common names and draw the condensed structural formulas of aldehydes and ketones (17.5).

_____ Write the IUPAC and common names and draw condensed structural formulas of carboxylic acids (17.6).

_____ Write the IUPAC or common names and draw the condensed structural formulas of esters (17.6).

_____ Draw the condensed structural formula of the product of the esterification reaction between an alcohol and a carboxylic acid (17.6).

_____ Write the IUPAC or common names and draw the condensed structural formulas of esters (17.6).

_____ Write the common names of amines and amides and draw their condensed structural formulas (17.7).

_____ Draw the condensed structural formula of the product of the amidation reaction between a carboxylic acid and ammonia or an amine (17.7).

Practice Test for Chapter 17

For questions 1 and 2, match the name of the hydrocarbon with each structure: (17.1)

A. hexane **B.** 3-methylbutane **C.** 2-methylbutane
D. butane **E.** 2,4-dimethylhexane

1. $CH_3—CH_2—CH(CH_3)—CH_3$ (with CH_3 branch on the CH)

2. $CH_3—CH(CH_3)—CH_2—CH(CH_3)—CH_2—CH_3$ (with CH_3 branches on each CH)

In questions 3 through 5, match the name of the alkene or alkyne with the structural formula: (17.2)

A. 1-butene **B.** 2-butene **C.** 3-butene
D. 2-butyne **E.** 1-butyne

3. _________ $CH_3—CH_2—C \equiv CH$

4. _________ $CH_3—CH_2—CH=CH_2$

5. _________ $CH_3—CH=CH—CH_3$

In questions 6 through 9, match the name of each aromatic compound with its structural formula: (17.3)

A. 
B. 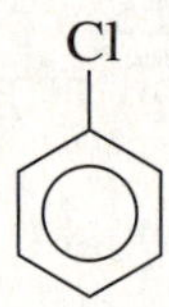
C. 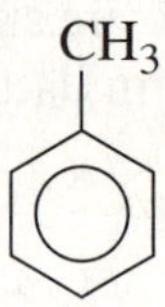
D. 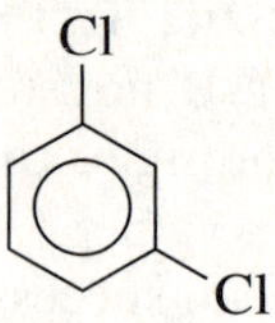

6. _________ chlorobenzene

7. _________ benzene

8. _________ toluene

9. _________ 1,3-chlorobenzene

For questions 10 through 13, match the names with one of the following condensed structural formulas: (17.4)

A. 1-propanol **B.** 3-propanol **C.** 2-propanol
D. ethyl methyl ether **E.** diethyl ether

10. $CH_3-\overset{\overset{OH}{|}}{CH}-CH_3$

11. $CH_3-CH_2-CH_2-OH$

12. $CH_3-O-CH_2-CH_3$

13. $CH_3-CH_2-O-CH_2-CH_3$

For questions 14 through 17, match each of the condensed structural formulas with its name: (17.6)

A. $CH_3-\overset{\overset{OH}{|}}{CH}-CH_3$ **B.** $CH_3-\overset{\overset{O}{\|}}{C}-O-CH_2-CH_3$

C. $CH_3-\overset{\overset{O}{\|}}{C}-OH$ **D.** $CH_3-CH_2-CH_2-\overset{\overset{O}{\|}}{C}-OH$

E. $CH_3-CH_2-CH_2-\overset{\overset{O}{\|}}{C}-O-CH_3$

14. ________ butyric acid

15. ________ methyl butanoate

16. ________ ethyl acetate

17. ________ acetic acid

18. Identify the carboxylic acid and alcohol needed to produce (17.6)

$CH_3-CH_2-CH_2-\overset{\overset{O}{\|}}{C}-O-CH_2-CH_3$

A. propanoic acid and ethanol **B.** acetic acid and 1-pentanol
C. acetic acid and 1-butanol **D.** butanoic acid and ethanol
E. hexanoic acid and methanol

For questions 19 through 22, match the amines and amides with the following names: (17.7)

A. ethyldimethylamine **B.** ethanamide **C.** propanamide
D. diethylamine **E.** butanamide

19. ________ $CH_3-CH_2-\overset{\overset{CH_3}{|}}{N}-CH_3$

20. ________ $CH_3-CH_2-CH_2-\overset{\overset{O}{\|}}{C}-NH_2$

21. ________ $CH_3-CH_2-NH-CH_2-CH_3$

22. ________ $CH_3-CH_2-\overset{\overset{O}{\|}}{C}-NH_2$

23. Indicate the carboxylic acid and ammonia or amine needed to produce the following: (17.7)

$$CH_3—CH_2—CH_2—C(=O)—NH_2$$

A. propanoic acid and ammonia
B. propanoic acid and methylamine
C. acetic acid and ammonia
D. butanoic acid and methylamine
E. butanoic acid and ammonia

24. Indicate the monomer needed to produce polyethylene. (17.2)
A. propylene
B. chloroethene
C. tetrafluoroethene
D. ethylene
E. styrene

25. Indicate the IUPAC name of the following: (17.5)

$$CH_3—CH_2—CH_2—CH_2—C(=O)—H$$

A. methylbutanal
B. pentanal
C. pentaldehyde
D. hexanal
E. pentanone

26. The ester produced from the reaction of 1-butanol and propanoic acid is (17.6)
A. butyl propanoate
B. butyl propanone
C. propyl butyrate
D. propyl butanone
E. heptanoate

Answers to the Practice Test

1. C	**2.** E	**3.** E	**4.** A	**5.** B
6. B	**7.** A	**8.** C	**9.** D	**10.** C
11. A	**12.** D	**13.** E	**14.** D	**15.** E
16. B	**17.** C	**18.** D	**19.** A	**20.** E
21. D	**22.** C	**23.** E	**24.** D	**25.** B
26. A				

18
Biochemistry

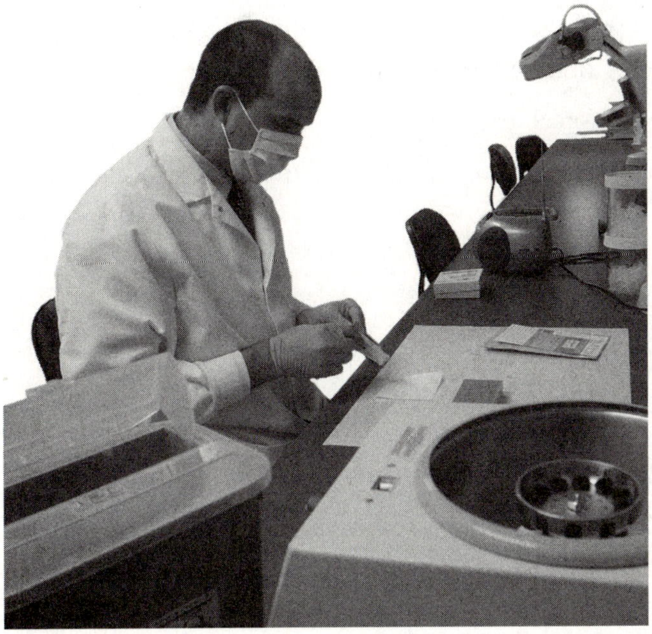

At a recent crime scene, investigators collected samples believed to be blood. The samples were taken to a forensic laboratory where Ron, a forensic scientist, used a luminol spray. Ron observed that the sample emitted a blue glow when sprayed with luminol, indicating that it contained blood. If human blood, the blood type, A and B antigens, and Rh factor can be determined to help identify the victim. Ron also isolated DNA (deoxyribonucleic acid) from the blood cells to determine the DNA pattern in the blood and compared it with DNA patterns of the victim. What type of bond holds amino acids in a protein together?

18.1 Carbohydrates

Learning Goal: Classify a carbohydrate as an aldose or ketose; draw the open-chain and Haworth structures for glucose, galactose, and fructose.

- Carbohydrates are classified as monosaccharides (simple sugars), disaccharides (two monosaccharide units), and polysaccharides (many monosaccharide units).
- Monosaccharides are aldehydes (aldoses) or ketones (ketoses) with hydroxyl groups on all other carbon atoms.
- Monosaccharides can be classified based on the number of carbons atoms, such as triose, tetrose, pentose, or hexose.
- Important monosaccharides are the aldohexoses glucose and galactose and the ketohexose fructose.
- The predominant form of monosaccharides is the cyclic form of five or six atoms. The Haworth structure forms by a reaction between an —OH on carbon 5 in hexoses with the carbonyl group of the same molecule.
- The formation of a new hydroxyl group on carbon 1 (or 2 in fructose) gives α and β forms (anomers) of the Haworth structure of a monosaccharide.

Key Terms for Sections 18.1–18.2

Match the following key terms with the descriptions shown:

a. carbohydrate **b.** glucose **c.** disaccharide **d.** cellulose **e.** polysaccharide

1. ____ a simple or complex sugar composed of a carbon chain with an aldehyde or ketone group and several hydroxyl groups

2. ____ a carbohydrate that contains many monosaccharides linked by glycosidic bonds

3. ____ an unbranched polysaccharide that cannot be digested by humans

4. ____ an aldohexose that is the most prevalent monosaccharide in the diet

5. ____ a carbohydrate that contains two monosaccharides linked by a glycosidic bond

Answers **1.** a **2.** e **3.** d **4.** b **5.** c

♦ Learning Exercise 18.1A

Indicate the number of monosaccharide units (one, two, or many) in each of the following carbohydrates:

1. sucrose, a disaccharide ________ **2.** cellulose, a polysaccharide ________

3. glucose, a monosaccharide ________ **4.** amylose, a polysaccharide ________

5. maltose, a disaccharide ________

Answers **1.** two **2.** many **3.** one **4.** many **5.** two

♦ Learning Exercise 18.1B

Identify the following monosaccharides as aldo- or ketotrioses, tetroses, pentoses, or hexoses:

1.
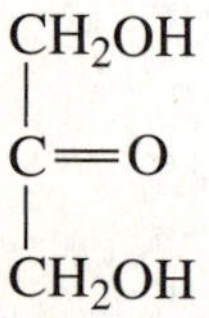

2.
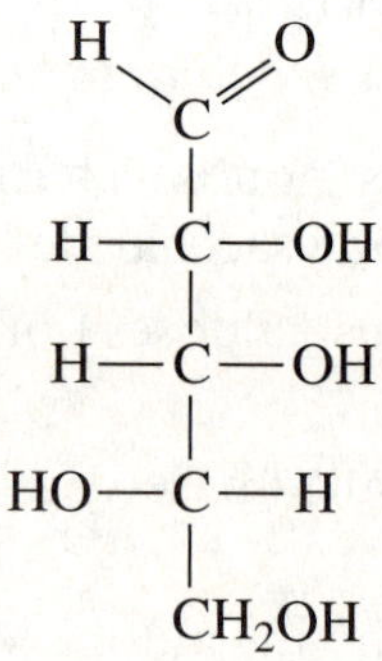

3.
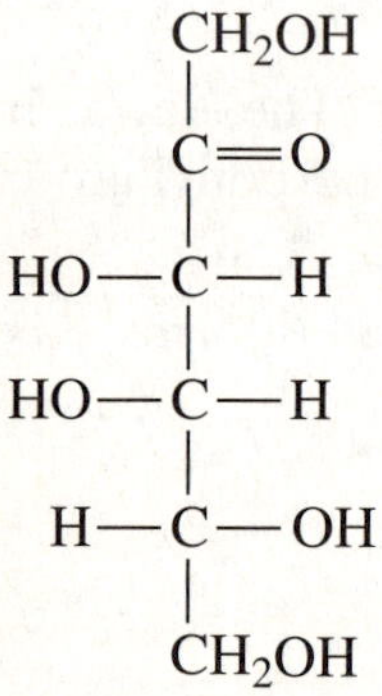

4.
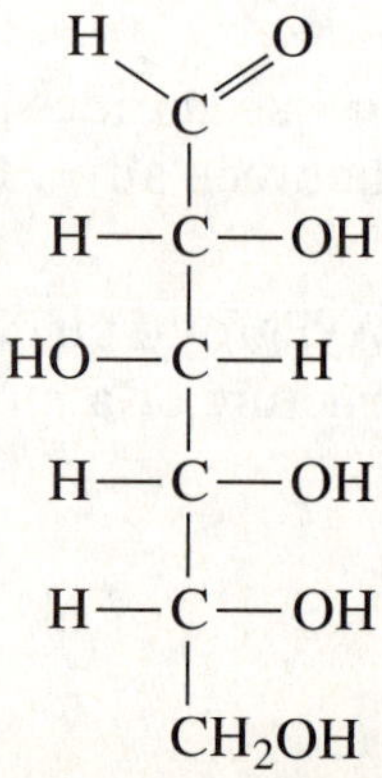

5. 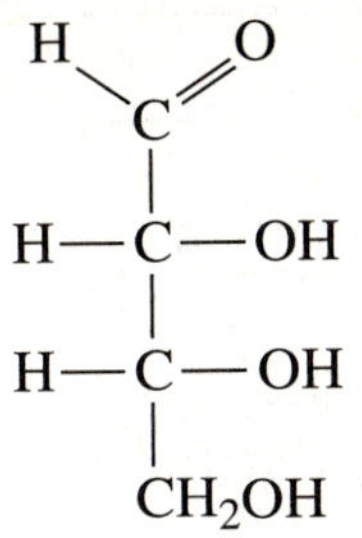

1. ______________________ 2. ______________________ 3. ______________________

4. ______________________ 5. ______________________

Answers **1.** ketotriose **2.** aldopentose **3.** ketohexose
4. aldohexose **5.** aldotetrose

♦ Learning Exercise 18.1C

Draw the condensed structural formula for each of the following monosaccharides:

glucose galactose fructose

Answers

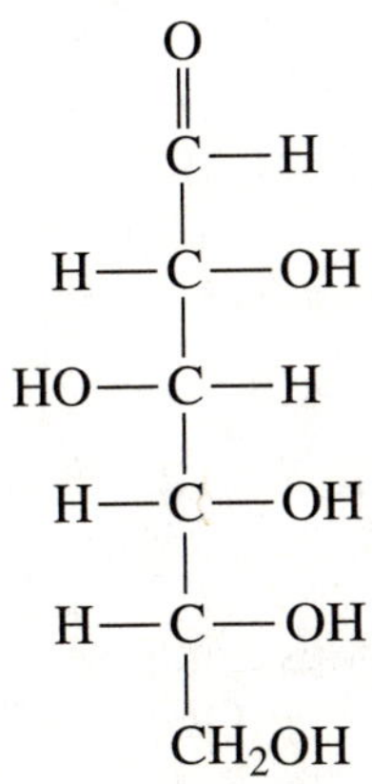

D-Glucose

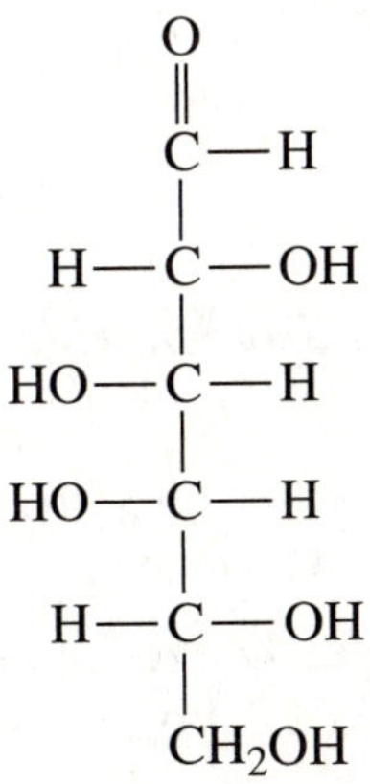

D-Galactose

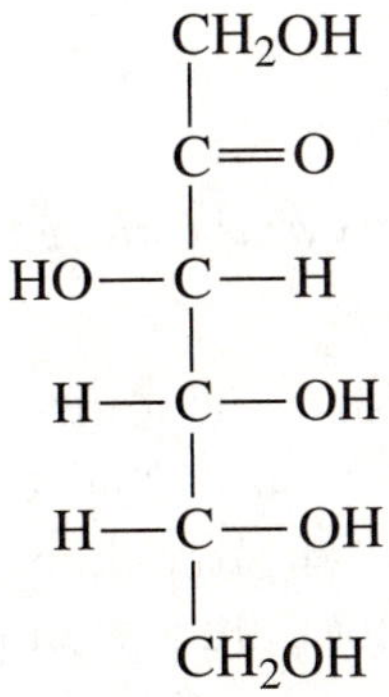

D-Fructose

Guide to Drawing Haworth Structures	
Step 1	Turn the open-chain structure clockwise by 90°.
Step 2	Fold the horizontal carbon chain into a hexagon and bond the O on carbon 5 to the carbonyl group to form the Haworth structure.
Step 3	Complete the Haworth structure by drawing the —OH group on carbon 1 below the ring to give the α form or above the ring to give the β form.

♦ **Learning Exercise 18.1D**

Draw the Haworth structures (α forms) for each of the following:

1. D-glucose **2.** D-galactose **3.** D-fructose

Answers

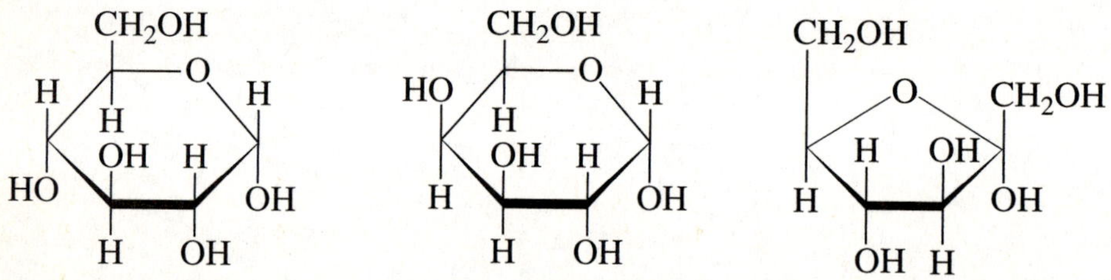

α-D-Glucose α-D-Galactose α-D-Fructose

18.2 Disaccharides and Polysaccharides

Learning Goal: Describe the monosaccharide units and linkages in disaccharides and polysaccharides.

- Disaccharides are glycosides, two monosaccharide units joined together by a glycosidic bond: monosaccharide (1) + monosaccharide (2) ⟶ disaccharide + H_2O
- In maltose, two glucose units are linked by an α-1,4-glycosidic bond. The α-1,4 indicates that the —OH of the alpha form at carbon 1 in the first glucose molecule is bonded to the —OH on carbon 4 of the other glucose molecule.

  Glucose + glucose ⟶ maltose + H_2O
  Glucose + galactose ⟶ lactose + H_2O
  Glucose + fructose ⟶ sucrose + H_2O

- Polysaccharides are polymers of monosaccharide units.

- Starches consist of amylose, an unbranched chain of glucose, and amylopectin which is a branched polymer of glucose. Glycogen, the storage form of glucose in animals, is similar to amylopectin, but has more branching.
- Cellulose is also a polymer of glucose, but in cellulose the glycosidic bonds are β bonds rather than α bonds as in the starches. Humans can digest starches to obtain energy, but not cellulose. However, cellulose is important as a source of fiber in our diets.

♦ Learning Exercise 18.2A

For the following disaccharides, state (a) the monosaccharide units, (b) the type of glycosidic bond, and (c) the name of the disaccharide:

1. 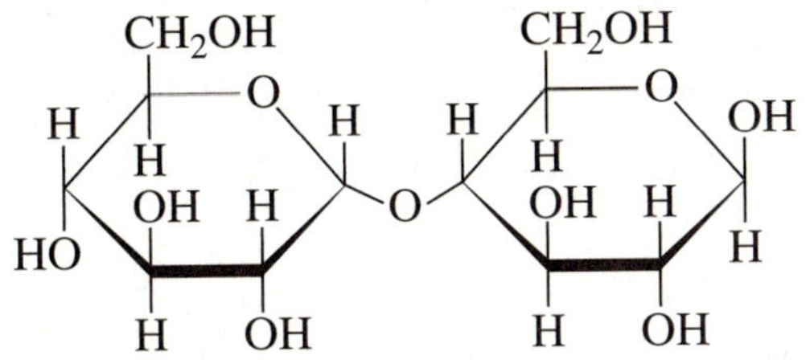

2. 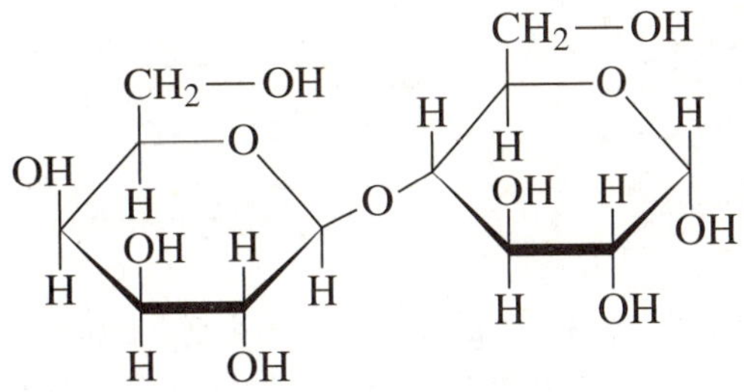

3. 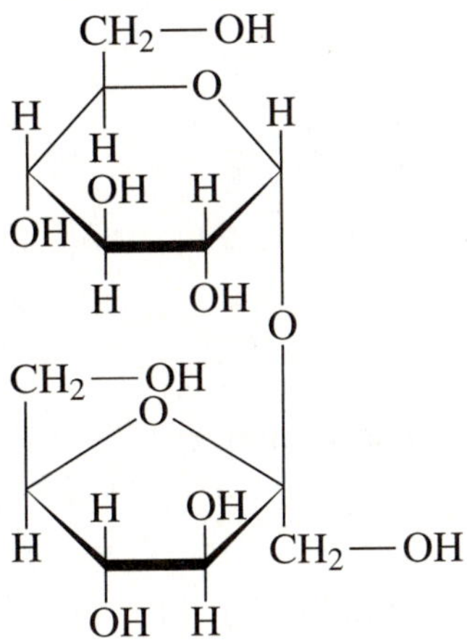

4. 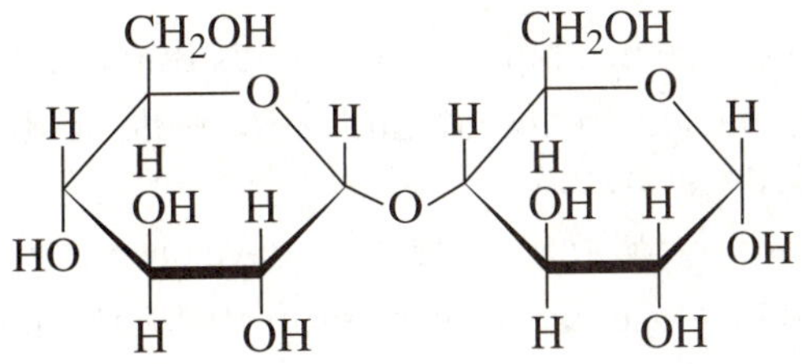

	a. Monosaccharide(s)	b. Type of Glycosidic Bond	c. Name of Disaccharide
1.			
2.			
3.			
4.			

Answers

	(a)	(b)	(c)
1.	two D-glucose units	α-1,4-glycosidic bond	β-maltose
2.	D-galactose + D-glucose	β-1,4-glycosidic bond	α-lactose
3.	D-fructose + D-glucose	α, β-1,2 glycosidic bond	sucrose
4.	two D-glucose units	α-1,4-glycosidic bond	α-maltose

♦ Learning Exercise 18.2B

List the monosaccharides and describe the glycosidic bonds in each of the following carbohydrates:

	Monosaccharides	Type(s) of Glycosidic Bonds
1. amylose	______________	______________________
2. amylopectin	______________	______________________
3. glycogen	______________	______________________
4. cellulose	______________	______________________

Answers

1. D-glucose; α-1,4-glycosidic bonds
2. D-glucose; α-1,4- and α-1,6-glycosidic bonds
3. D-glucose; α-1,4- and α-1,6-glycosidic bonds
4. D-glucose; β-1,4-glycosidic bonds

18.3 Lipids

Learning Goal: Describe some properties of lipids.

- Lipids are nonpolar compounds that are not soluble in water.
- Fatty acids are long-chain carboxylic acids that may be saturated, monounsaturated, or polyunsaturated.
- Unsaturated fatty acids can have cis and trans isomers.
- The hydrogenation of unsaturated fatty acids converts carbon to carbon double bonds to single bonds.
- In saponification, a fat heated with a strong base produces glycerol and the salts of the fatty acids or soaps.
- The triacylglycerols in fats and oils are esters of glycerol and three fatty acids.
- Fats from animal sources contain more saturated fatty acids and have higher melting points than fats found in most vegetable oils.
- Steroids are lipids containing the steroid nucleus, which is a fused structure of four rings. Cholesterol is one of the most important and abundant steroids.

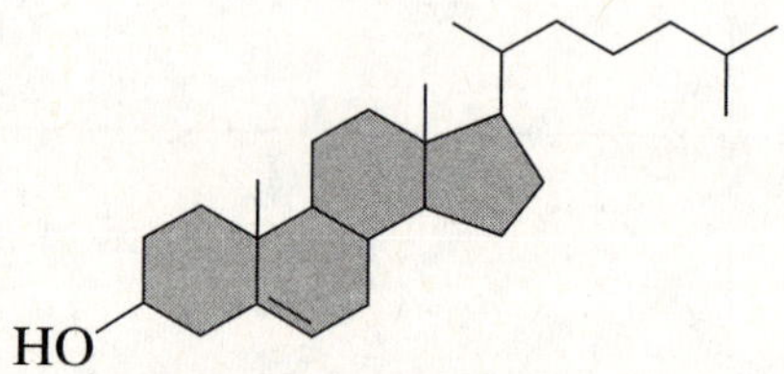

Cholesterol

Key Terms for Section 18.3

Match each of the following terms with the correct statement:

a. lipid **b.** fatty acid **c.** triacylglycerol **d.** saponification **e.** steroid

1. _____ a type of compound that is not soluble in water, but is in nonpolar solvents

2. _____ the reaction of a triacylglycerol with a strong base producing salts, called soaps, and glycerol

3. _____ a lipid consisting of glycerol bonded to three fatty acids

4. _____ a lipid composed of a fused four-ring system

5. _____ long-chain carboxylic acid found in triacylglycerols

Answers **1.** a **2.** d **3.** c **4.** e **5.** b

♦ Learning Exercise 18.3A

Draw the condensed structural formulas of linoleic acid, stearic acid, and oleic acid.

1. linoleic acid

2. stearic acid

3. oleic acid

4. Which of the three fatty acids (1, 2, and/or 3)

a. is the most saturated? _____

b. is the most unsaturated? _____

c. has the lowest melting point? _____

d. has the highest melting point? _____

e. is found in vegetables? _____

f. is from animal sources? _____

Answers

1. linoleic acid

$$CH_3-(CH_2)_4-CH{=}CH-CH_2-CH{=}CH-(CH_2)_7-\overset{\overset{\displaystyle O}{\|}}{C}-OH$$

2. stearic acid

$$CH_3-(CH_2)_{16}-\overset{\overset{\displaystyle O}{\|}}{C}-OH$$

3. oleic acid

$$CH_3—(CH_2)_7—CH{=}CH—(CH_2)_7—\overset{\overset{\displaystyle O}{\|}}{C}—OH$$

4. **a.** 2 **b.** 1 **c.** 1 **d.** 2 **e.** 1 and 3 **f.** 2

♦ Learning Exercise 18.3B

Consider the following fatty acid called oleic acid:

$$CH_3—(CH_2)_7—CH{=}CH—(CH_2)_7—\overset{\overset{\displaystyle O}{\|}}{C}—OH$$

1. Why is oleic acid considered an acid?

2. Is oleic acid a saturated or unsaturated compound? Why?

3. Is oleic acid likely to be a solid or a liquid at room temperature?

4. Why is oleic acid not soluble in water?

Answers

1. contains a carboxylic acid group

2. unsaturated; has a carbon to carbon double bond

3. liquid

4. It has a long hydrocarbon chain.

♦ Learning Exercise 18.3C

Draw the condensed structural formula of the triacylglycerol formed from the following:

1. glycerol and three palmitic acids, $CH_3—(CH_2)_{14}—\overset{\overset{\displaystyle O}{\|}}{C}—OH$

2. glycerol and three myristic acids, $CH_3-(CH_2)_{12}-\overset{\overset{\displaystyle O}{\|}}{C}-OH$

Answers

1.

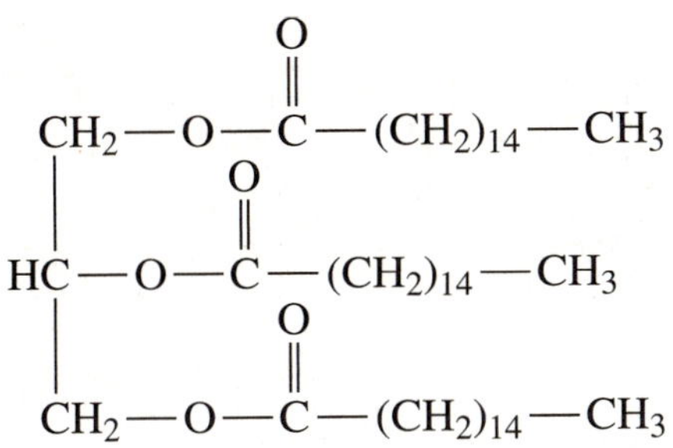

2.

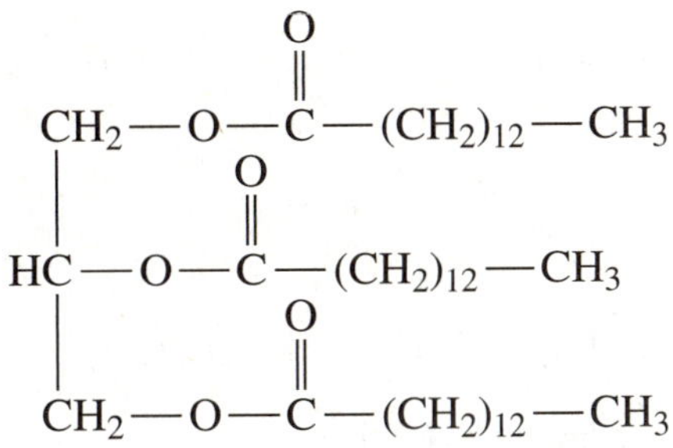

♦ Learning Exercise 18.3D

Draw the condensed structural formula for each of the following triacylglycerols:

1. glyceryl tristearate (tristearin)

2. glyceryl trioleate (triolein)

Answers

1.

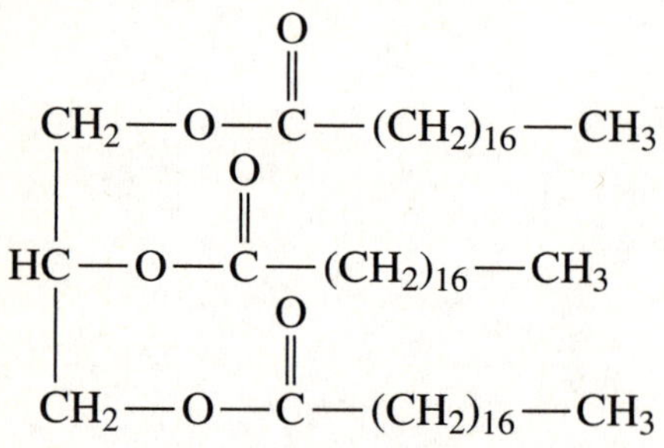

Glyceryl tristearate (tristearin)

2.

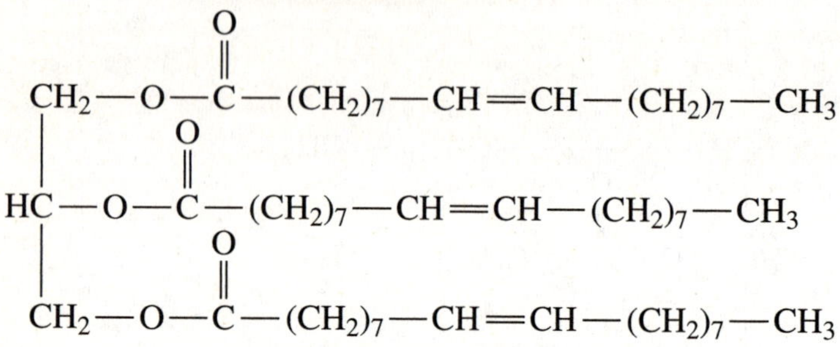

Glyceryl trioleate (triolein)

♦ Learning Exercise 18.3E

Write the equations for the following reactions of glyceryl trioleate (triolein):

1. hydrogenation with a nickel catalyst

2. saponification with NaOH

Answers

1.

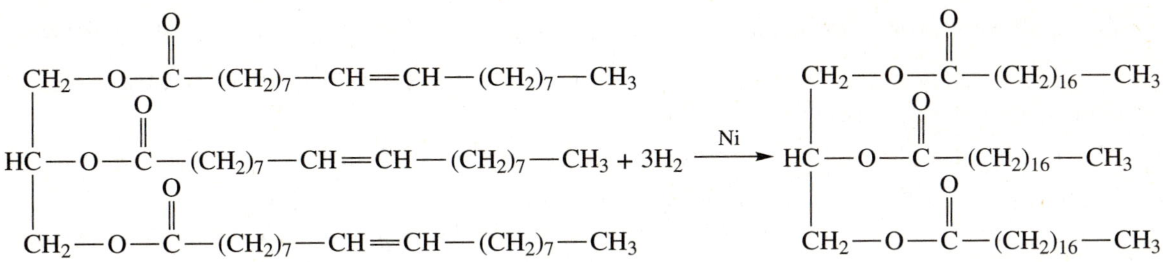

2.

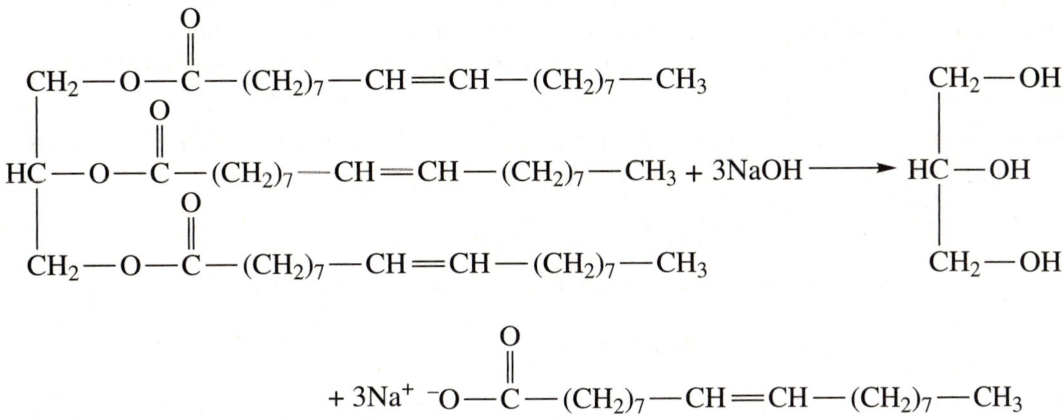

♦ Learning Exercise 18.3F

1. Draw the structure of the steroid nucleus.
2. Draw the structure of cholesterol.

Answers

1.

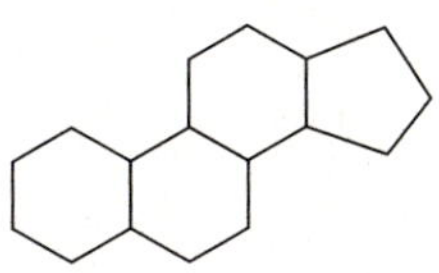

2.

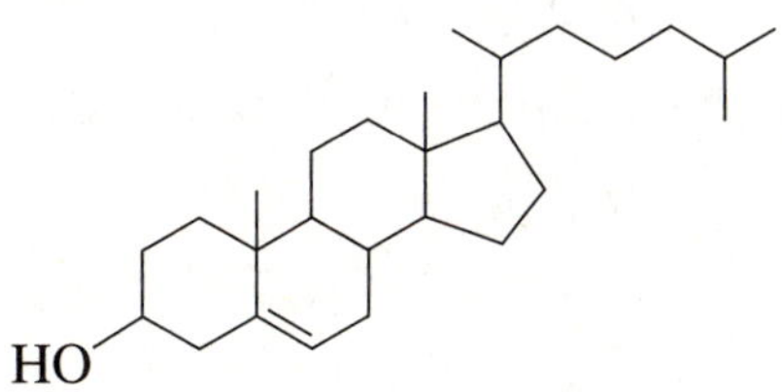

Cholesterol

18.4 Proteins

Learning Goal: Describe protein functions and draw structures for amino acids and dipeptides.

- Some proteins are enzymes or hormones, whereas others are important in structure, transport, protection, storage, and muscle contraction.
- The molecular building blocks of proteins are a group of 20 amino acids.
- In an amino acid, a central (alpha) carbon is attached to an amino ($-NH_3^+$) group, a carboxyl ($-COO^-$) group, and a side chain (R group), which is a characteristic group for each amino acid.
- The particular side chain makes each amino acid polar, nonpolar, acidic, or basic. Nonpolar amino acids contain hydrocarbon side chains, whereas polar amino acids contain electronegative atoms such as oxygen ($-OH$) or sulfur ($-SH$). Acidic side chains contain a carboxyl group and basic side chains contain an amino group.
- A peptide bond is an amide bond between the carboxyl group of one amino acid and the amino group of a second amino acid.

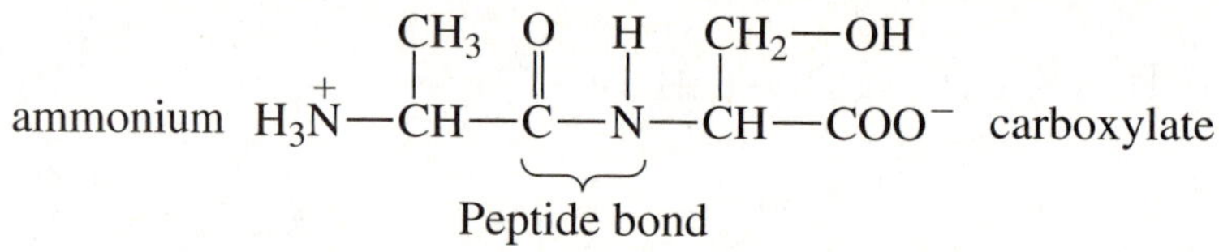

- Short chains of amino acids are called peptides. Long chains of amino acids are called proteins.
- In the name of a peptide, each amino acid beginning from the N terminal end (written on the left) has the *ine* (or *ic acid*) replaced by *yl*. The last amino acid at the C terminal end (written on the right) of the peptide uses its full name.
- Structures of peptides can be shown using the three letter and single letter designations for the component amino acids written from the N terminal to the C terminal (left to right).

Key Terms for Section 18.4

Match one of the following functions of a protein with the examples:

a. structural **b.** contractile **c.** storage **d.** transport
e. hormonal **f.** enzyme **g.** protection

1. ____ hemoglobin carries oxygen in blood

2. ____ amylase hydrolyzes starch

3. ____ egg albumin, a protein in egg white

4. ____ human growth factor

5. ____ collagen in connective tissue

6. ____ immunoglobulin

7. ____ keratin, a major protein of hair

8. ____ lipoproteins carry lipids in blood

Answers	**1.** d	**2.** f	**3.** c	**4.** e
	5. a	**6.** g	**7.** a	**8.** d

♦ Learning Exercise 18.4A

Using the appropriate side chain, complete the structural formula of each of the following amino acids. Indicate whether the amino acid would be polar or nonpolar.

Glycine

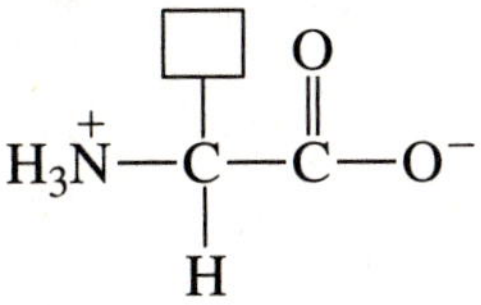

Alanine

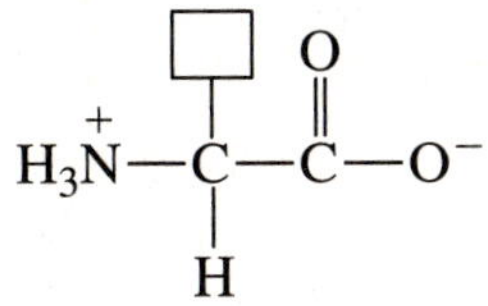

Serine

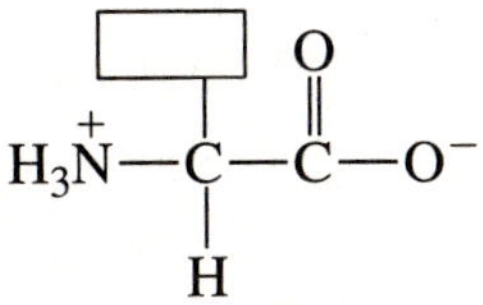

Aspartic acid

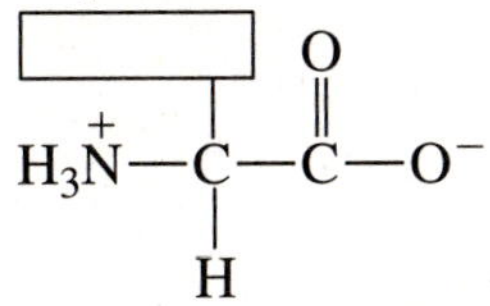

Answers

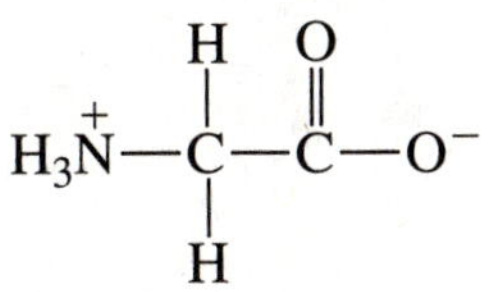

nonpolar

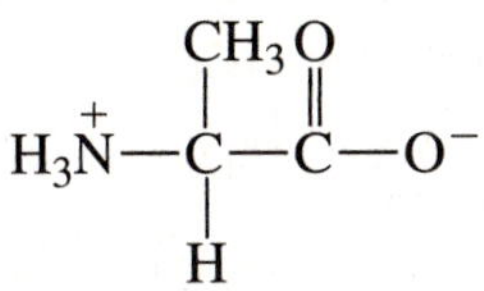

nonpolar

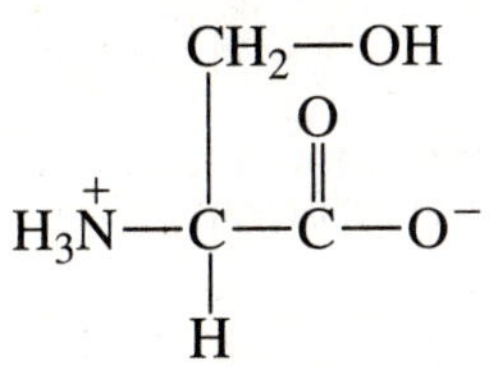

polar

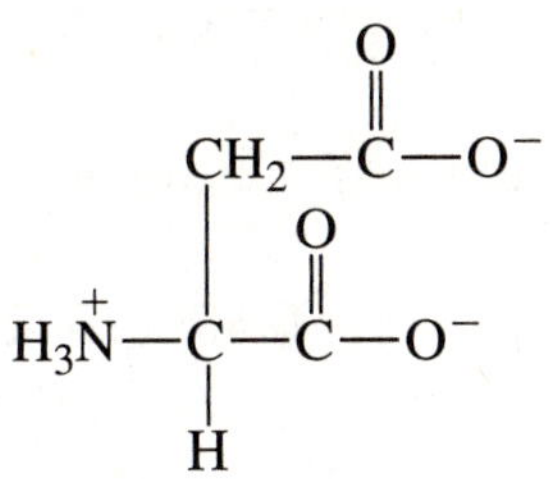

polar

♦ Learning Exercise 18.4B

Draw the structural formulas of the following di- and tripeptides:

1. Ser–Gly

2. CV

3. Gly–Ser–Cys

Answers

1. 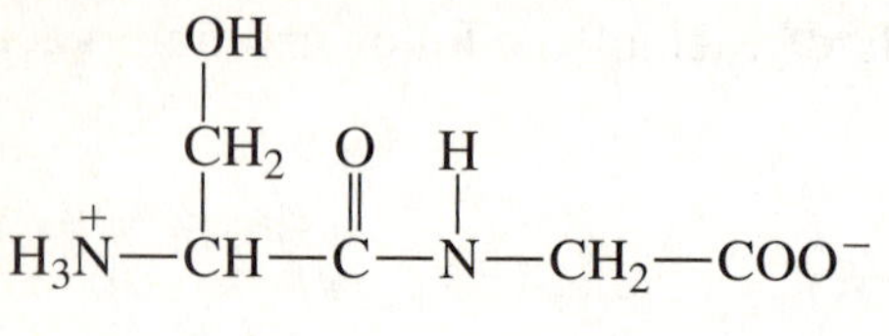

2. 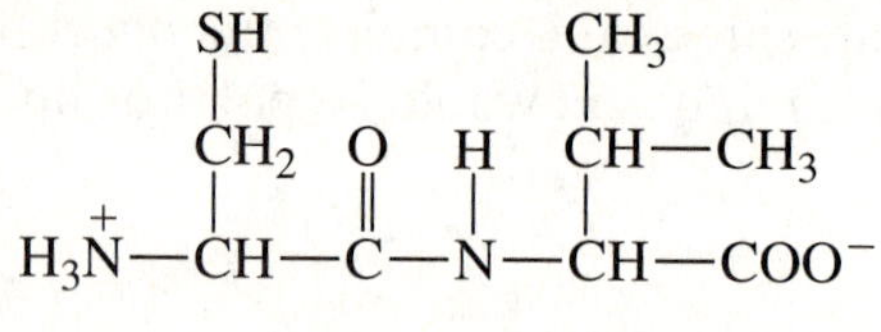

3. 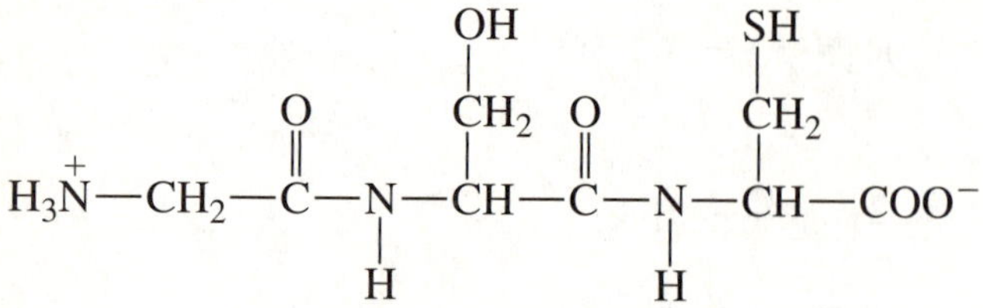

18.5 Protein Structure

Learning Goal: Identify the levels of structure of a protein.

- The primary structure of a protein is the sequence of amino acids.
- In the secondary structure, hydrogen bonds between different sections of the polypeptide or between different polypeptides produce a characteristic shape such as an α helix, β-pleated sheet, or a triple helix.
- In the tertiary structure, the polypeptide chain folds upon itself to form a three-dimensional shape with hydrophobic side groups on the inside and hydrophilic side groups on the outside surface. The tertiary structure is stabilized by interactions between side groups.
- The interactions can be hydrophobic, hydrophilic, salt bridges, hydrogen bonds, or disulfide bonds.
- In a quaternary structure, two or more subunits combine for biological activity.

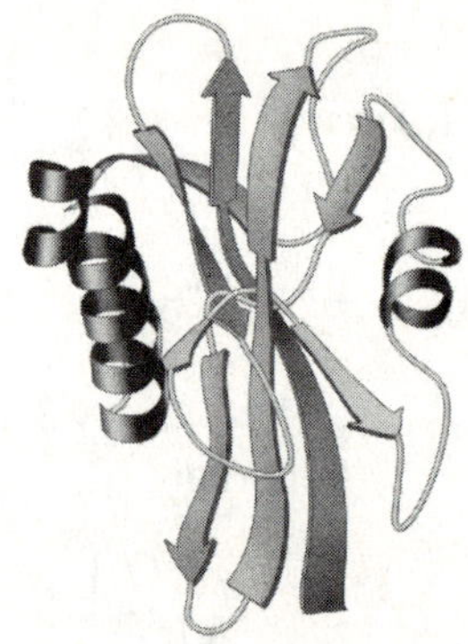

Protein with alpha helices and beta-pleated sheets

♦ Learning Exercise 18.5A

Identify the following as descriptions of primary or secondary protein structure:

1. ________ Hydrogen bonding forms an alpha (α) helix.
2. ________ Hydrogen bonding occurs between C=O and N—H within a peptide chain.
3. ________ The sequence of amino acids is linked by peptide bonds.
4. ________ Hydrogen bonds between protein chains form a pleated-sheet structure.

Answers **1.** secondary **2.** secondary **3.** primary **4.** secondary

♦ Learning Exercise 18.5B

Identify the following descriptions of protein structure as tertiary or quaternary:

1. ________ a disulfide bond joining distant parts of a peptide
2. ________ the combination of four protein subunits

3. ________ hydrophilic side groups attracted to water
4. ________ a salt bridge forms between two oppositely charged side chains

Answers **1.** tertiary **2.** quaternary **3.** tertiary **4.** tertiary

18.6 Proteins as Enzymes

Learning Goal: Describe enzymes and their role in enzyme-catalyzed reactions.

- Enzymes are proteins that accelerate the rate of biological reactions by providing a reaction pathway with lower activation energy.
- Within the structure of the enzyme, a small pocket called the active site has a specific shape that fits a specific substrate.
- In the models of enzyme action, an enzyme and substrate form an enzyme–substrate complex, so the reaction of the substrate can be catalyzed at the active site.
- In the induced-fit model, the active site is flexible so that both the configurations of the active site and substrate are adjusted for the best fit.

Key Terms for Section 18.6

Match the following terms with the following descriptions:

a. active site **b.** substrate **c.** enzyme–substrate complex **d.** induced-fit

1. ____ the combination of an enzyme with a substrate
2. ____ has a tertiary structure that fits the structure of the active site
3. ____ a model of enzyme action in which the shape of the active site adjusts to fit the shape of a substrate
4. ____ the portion of an enzyme that binds to the substrate and catalyzes the reaction

Answers **1.** c **2.** b **3.** d **4.** a

♦ Learning Exercise 18.6

Indicate whether each of the following characteristics of an enzyme is true (T) or false (F).
An enzyme

1. ________ is a biological catalyst.
2. ________ is smaller than a substrate.
3. ________ does not change the equilibrium of a reaction.
4. ________ must be obtained from the diet.
5. ________ increases the rate of a cellular reaction.
6. ________ is needed for every reaction that takes place in the cell.
7. ________ provides an alternate pathway for the biological reaction with a lower activation energy.

Answers **1.** T **2.** F **3.** T **4.** F **5.** T **6.** T **7.** T

18.7 Nucleic Acids

Learning Goal: Describe the structure of the nucleic acids in DNA and RNA.

- Five nitrogenous bases, two five-carbon sugars, and phosphate groups are the basic building blocks form which nucleic acids are assembled.
- Deoxyribonucleic acid (DNA) and ribonucleic acid (RNA) are polymers of nucleotides. Each nucleotide consists of a base, a sugar, and a phosphate group.

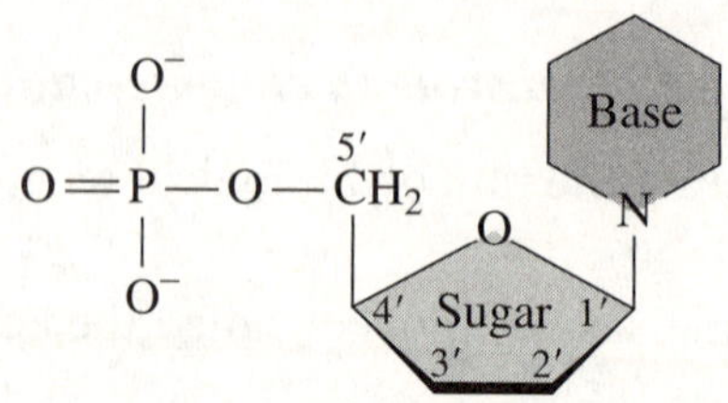

- In DNA, the bases are adenine, thymine, guanine, or cytosine. In RNA, uracil replaces thymine. In DNA, the sugar is deoxyribose; in RNA, the sugar is ribose.
- The two strands of DNA are held together in a double helix by hydrogen bonds between complementary base pairs: A with T, and G with C.
- During DNA replication, new DNA strands form along each original DNA strand.
- Complementary base pairing ensures the correct pairing of bases to give identical copies of the original DNA.

Key Terms for Section 18.7

Match each of the following items with the statements below:

a. DNA **b.** RNA **c.** double helix
d. adenine **e.** hydrogen bonds **f.** complementary base pair
g. daughter DNA

1. ____ the shape of DNA with a sugar-phosphate backbone and base pairs linked in the center

2. ____ the genetic material comprised of nucleotides containing adenine, cytosine, guanine, or thymine

3. ____ the nucleic acid that is a single strand of nucleotides containing adenine, cytosine, guanine, or uracil

4. ____ the attractions between base pairs that connect the two DNA strands

5. ____ the bases guanine and cytosine in the double helix

6. ____ the new DNA strand that forms during DNA replication

7. ____ the base that bonds to thymine

Answers **1.** c **2.** a **3.** b **4.** e **5.** f **6.** g **7.** d

♦ Learning Exercise 18.7A

1. Write the names and abbreviations for the bases in each of the following:

DNA ____

RNA ____

2. Write the name of the sugar in each of the following nucleotides:

DNA ____________________

RNA ____________________

Answers

1. DNA: adenine (A), thymine (T), guanine (G), cytosine (C)
RNA: adenine (A), uracil (U), guanine (G), cytosine (C)

2. DNA: deoxyribose
RNA: ribose

♦ Learning Exercise 18.7B

Name and write the abbreviation for each of the following bases:

a.

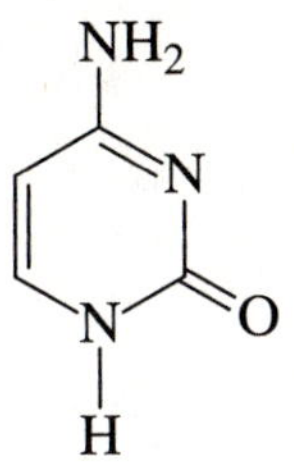

b.

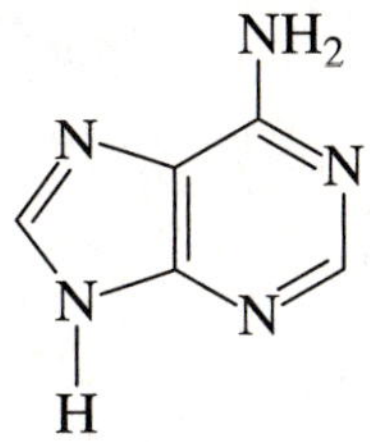

c.

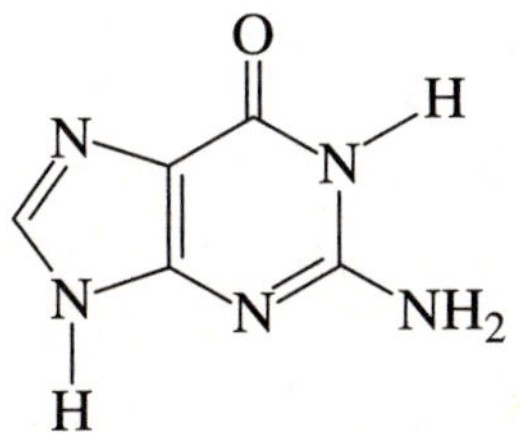

d.

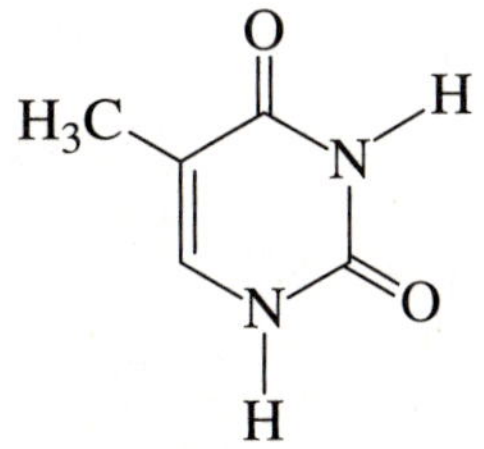

Answers

1. cytosine (C)
2. adenine (A)
3. guanine (G)
4. thymine (T)

♦ Learning Exercise 18.7C

Draw the structural formula for deoxycytidine-5′-monophosphate (dCMP).

Answer

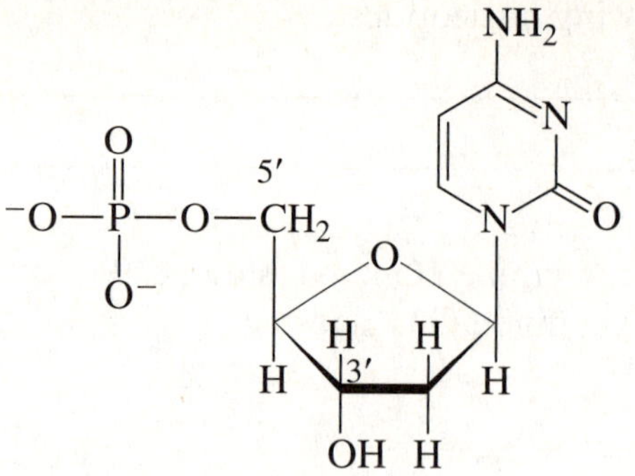

♦ **Learning Exercise 18.7D**

Identify each of the nucleotides in the following:

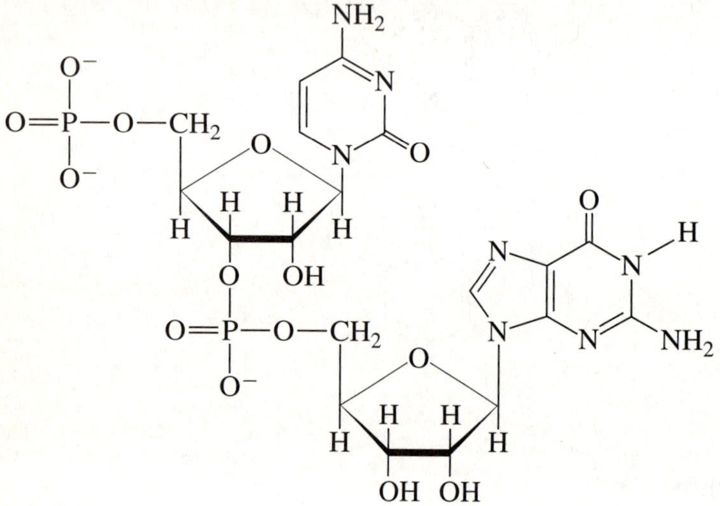

Answer Cytidine-5′-monophosphate and guanosine-5′-monophosphate

♦ **Learning Exercise 18.7E**

Complete each DNA section by writing the complementary strand:

1. —T A C G A A C C G A G G—

2. —A A A T T T C C C G G G—

3. —G C G C T C A A A T G C—

Answers

1. —A U G C U U G G C U C C—
2. —T T T A A A G G G C C C—
3. —C G C G A G T T T A C G—

♦ Learning Exercise 18.7F

How does the replication of DNA produce identical copies of the DNA?

Answer In the replication process, the bases on each strand of the separated parent DNA are paired with their complementary bases. Because each complementary base is specific for a base in DNA, the new DNA daughter strands are complements of the original DNA strands and when they pair to form the double helix produce an identical copy of the original DNA.

18.8 Protein Synthesis

Learning Goal: Describe the synthesis of protein from mRNA.

- The three types of RNA differ by function in the cell: ribosomal RNA makes up most of the structure of the ribosomes, messenger RNA carries genetic information from the DNA to the ribosomes, and transfer RNA places the correct amino acids in the protein.
- Transcription is the process by which RNA polymerase produces mRNA from one strand of DNA.
- The bases in the mRNA are complementary to the DNA, except U is paired with A in DNA.
- In mRNA, the genetic code consists of a sequence of three bases (triplet or codon) which code for a specific amino acid and the sequence of codons specify the order of the amino acids in the protein.
- The codon AUG signals the start of transcription, and codons UAG, UGA, and UAA signal the stop.
- Proteins are synthesized in a translation process in which tRNA molecules bring the appropriate amino acids to the ribosome to which mRNA is bound. Each amino acid is bonded by a peptide bond to a growing peptide chain.
- When the polypeptide is released, it takes on its secondary and tertiary structures to become a functional protein in the cell.

♦ Learning Exercise 18.8A

Indicate the type of RNA (mRNA, tRNA, or rRNA) described by each of the following:

1. ________ most abundant type in a cell

2. ________ has the shortest chain of nucleotides

3. ________ carries information from DNA to the ribosomes for protein synthesis

4. ________ major component of ribosomes

5. ________ carries specific amino acids to the ribosome for protein synthesis

Answers **1.** rRNA **2.** tRNA **3.** mRNA **4.** rRNA **5.** tRNA

♦ Learning Exercise 18.8B

Write the corresponding section of an mRNA produced from each of the following sections of DNA:

1. —C A T T C G G T A—

2. —G T A C C T A A C G T C C G—

Answers
1. —G U A A G C C A U—
2. —C A U G G A U U G C A G G C—

♦ Learning Exercise 18.8C

Write the mRNA that would form for the following sections of DNA:

1. DNA strand: —C C C T C A G G G C G C—
   mRNA: —____ ____ ____ ____—
2. DNA strand: —A T A G C C T T T G G C A A C—
   mRNA: —____ ____ ____ ____ ____—

Answers
1. mRNA: —G G G A G U C C C G C G—
2. mRNA: —U A U C G G A A A C C G U U G—

♦ Learning Exercise 18.8D

Give the abbreviation for each of the amino acids coded for by the following mRNA codons:

1. UUU ________	2. GCG ________
3. AGC ________	4. CCA ________
5. GGA ________	6. ACA ________
7. AUG ________	8. CUC ________
9. CAU ________	10. GUU ________

Answers

1. Phe	2. Ala	3. Ser	4. Pro	5. Gly
6. Thr	7. Start/Met	8. Leu	9. His	10. Val

♦ Learning Exercise 18.8E

Write the mRNA that would form for the following section of DNA. For each codon in the mRNA, write the amino acid that would be placed in the protein by a tRNA.

1. DNA strand: —C C C T C A G G G C G C—
   mRNA: —____ ____ ____ ____—
   Amino acids: —____ ____ ____ ____—

2. DNA: —ATA GCC TTT GGC AAC—

mRNA: —____ ____ ____ ____ ____—

Amino acids: —____ ____ ____ ____ ____—

Answers

1. mRNA: —GGG AGU CCC GCG—
   Amino acids: —Gly—Ser—Pro—Ala—
2. mRNA: —UAU CGG AAA CCG UUG—
   Amino acids: —Tyr—Arg—Lys—Pro—Leu—

Checklist for Chapter 18

You are ready to take the Practice Test for Chapter 18. Be sure you have accomplished the following learning goals for this chapter. If not, review the section listed at the end of the goal. Then apply your new skills and understanding to the Practice Test.

After studying Chapter 18, I can successfully:

____ Classify carbohydrates as monosaccharides, disaccharides, and polysaccharides (18.1).

____ Classify a monosaccharide (aldose or ketose), and indicate the number of carbon atoms (18.1).

____ Draw the open-chain structures for glucose, galactose, and fructose (18.1).

____ Draw the cyclic structures of monosaccharides (18.1).

____ Describe the monosaccharide units and linkages in disaccharides (18.2).

____ Describe the structural features of amylose, amylopectin, glycogen, and cellulose (18.2).

____ Identify a fatty acid as saturated or unsaturated (18.3).

____ Draw the triacylglycerol produced by the reaction of glycerol and fatty acids (18.3).

____ Draw the product from the hydrogenation or saponification of a triacylglycerol (18.3).

____ Describe the structure of a steroid and cholesterol (18.3).

____ Classify proteins by their functions in the cells (18.4).

____ Draw the structure for an amino acid (18.4).

____ Classify amino acids by the characteristics of the side chain (R group) (18.4).

____ Describe a peptide bond and draw the structure for a peptide using amino acid abbreviations (18.4).

____ Distinguish between the primary and secondary structures of a protein (18.5).

____ Distinguish between the tertiary and quaternary structures of a protein (18.5).

____ Describe the lock-and-key and induced-fit models of enzyme action (18.6).

____ Identify the components of nucleic acids RNA and DNA (18.7).

____ Describe the nucleotides contained in DNA and RNA (18.7).

____ Describe the primary structure of nucleic acids (18.7).

____ Describe the structures of RNA and DNA (18.7).

____ Explain the process of DNA replication and write a complementary strand for a DNA template (18.7).

_____ Describe the structures and characteristics of the three types of RNA (18.8).

_____ Describe the synthesis of mRNA (transcription) and write the mRNA section for a DNA template (18.8).

_____ Describe the role of translation in protein synthesis and write the amino acid sequence for a mRNA section (18.8).

Practice Test for Chapter 18

1. The name *carbohydrate* came from the fact that (18.1)
   A. carbohydrates are hydrates of water.
   B. carbohydrates contain carbon, hydrogen and oxygen in a 1:2:1 ratio.
   C. carbohydrates contain a great quantity of water.
   D. all plants produce carbohydrates.
   E. carbon and hydrogen atoms are abundant in carbohydrates.

2. What functional group(s) is (are) found in the open chains of monosaccharides? (18.1)
   A. hydroxyl groups
   B. aldehyde groups
   C. ketone groups
   D. hydroxyl and carbonyl groups
   E. carbonyl group

3. What is the classification of the following sugar? (18.1)

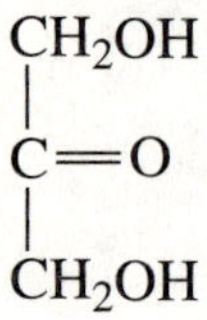

   A. aldotriose　　**B.** ketotriose　　**C.** aldotetrose
   D. ketotetrose　　**E.** ketopentose

4. The structure shown below is the Haworth structure of a (18.1)

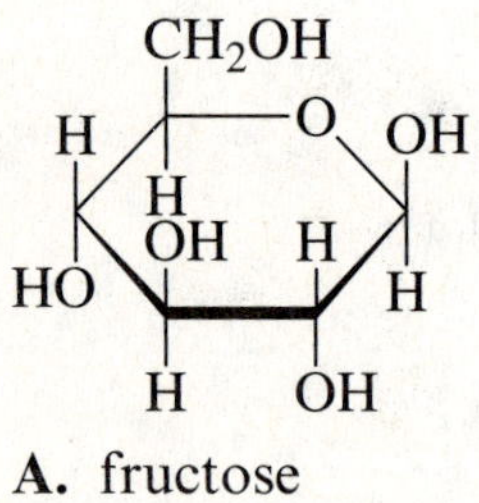

   A. fructose　　**B.** glucose　　**C.** ribose
   D. glyceraldehydes　　**E.** galactose

For questions 5 through 10, identify the carbohydrate that each of the following statements describe: (18.2)

A. amylose　　**B.** cellulose　　**C.** glycogen　　**D.** lactose　　**E.** sucrose

5. ________ composed of many glucose units linked by α-1,4-glycosidic bonds
6. ________ contains glucose and galactose
7. ________ composed of glucose units joined by both α-1,4- and α-1,6-glycosidic bonds
8. ________ composed of glucose units joined by β-1,4-glycosidic bonds

9. ________ produced as a storage form of energy in plants

10. ________ used for structural purposes by plants

11. A triacylglycerol is a (18.3)
A. carbohydrate **B.** lipid **C.** protein
D. oxyacid **E.** soap

12. A fatty acid that is unsaturated is usually (18.3)
A. from animal sources and liquid at room temperature
B. from animal sources and solid at room temperature
C. from vegetable sources and liquid at room temperature
D. from vegetable sources and solid at room temperature
E. from both vegetable and animal sources and solid at room temperature

For questions 13 through 15, consider the following compound: (18.3)

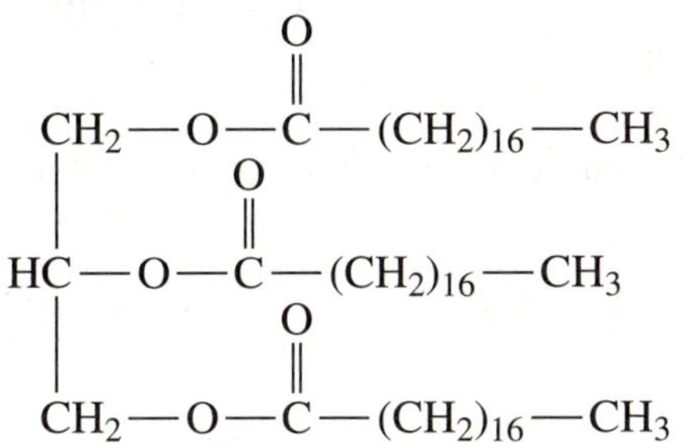

13. The molecule shown above was formed by
A. esterification **B.** combustion **C.** saponification
D. emulsification **E.** oxidation

14. If this molecule reacts with a strong base such as NaOH, the products are
A. glycerol and fatty acids
B. glycerol and water
C. glycerol and soap
D. an ester and salts of fatty acids
E. an ester and fatty acids

15. This compound would be
A. saturated, and a solid at room temperature.
B. saturated, and a liquid at room temperature.
C. unsaturated, and a solid at room temperature.
D. unsaturated, and a liquid at room temperature.
E. supersaturated, and a liquid at room temperature.

16. Which amino acid is nonpolar? (18.4)
A. serine **B.** aspartic acid **C.** valine **D.** cysteine **E.** glutamine

17. Which amino acid will form disulfide cross-links in a tertiary structure? (18.4)
A. serine **B.** aspartic acid **C.** valine **D.** cysteine **E.** glutamine

18. Which amino acid has a basic side chain? (18.4)
A. serine **B.** aspartic acid **C.** valine **D.** cysteine **E.** lysine

19. The sequence Tyr–Ala–Gly (18.5)
A. is a tripeptide
B. has two peptide bonds
C. has tyrosine with a free $—NH_3^+$ end
D. has glycine with the free $—COO^-$ end
E. all of these

20. What type of interaction is expected between lysine and aspartic acid? (18.5)
A. salt bridge (ionic bond) **B.** hydrogen bond **C.** disulfide bond
D. hydrophobic interaction **E.** hydrophilic attraction

21. What type of bond is used to form the α helix structure of a protein? (18.5)
A. peptide bond **B.** hydrogen bond **C.** ionic bond
D. disulfide bond **E.** hydrophobic interaction

22. What type of bonding places portions of the protein chain in the center of a tertiary structure? (18.5)
A. peptide bonds **B.** ionic bonds **C.** disulfide bonds
D. hydrophobic interaction **E.** hydrophilic attraction

For questions 23 through 27, identify the protein structural levels that each of the following statements describe: (18.5)

A. primary **B.** secondary **C.** tertiary **D.** quaternary **E.** pentenary

23. ________ peptide bonds

24. ________ a pleated sheet

25. ________ two or more protein subunits

26. ________ an α helix

27. ________ disulfide bonds

28. Enzymes are (18.6)
A. biological catalysts. **B.** polysaccharides. **C.** insoluble in water.
D. lipids. **E.** named with an *ose* ending.

29. The first step in enzyme action is (18.6)
E = enzyme; S = substrate; P = product
A. $S \longrightarrow P$ **B.** $EPS \longrightarrow E + P$ **C.** $E + S \longrightarrow ES$
D. $ES \longrightarrow E + P$ **E.** $EP \longrightarrow ES$

30. The final step in enzyme action is (18.6)
E = enzyme; S = substrate; P = product
A. $S \longrightarrow P$ **B.** $EP \longrightarrow E + P$ **C.** $E + S \longrightarrow ES$
D. $ES \longrightarrow E + P$ **E.** $EP \longrightarrow ES$

31. A nucleotide contains (18.7)
A. a base **B.** a base and a sugar
C. a phosphate and a sugar **D.** a base and a deoxyribose
E. a base, a sugar, and a phosphate

32. The double helix in DNA is held together by (18.7)
A. hydrogen bonds **B.** ester linkages **C.** peptide bonds
D. salt bridges **E.** disulfide bonds

33. The process of producing DNA in the nucleus is called (18.7)
A. complementation **B.** replication **C.** translation
D. transcription **E.** mutation

34. Which occurs in RNA but **not** in DNA? (18.7)

A. thymine **B.** cytosine **C.** adenine
D. phosphate **E.** uracil

For questions 35 through 42, select answers from the following nucleic acids: (18.8)

A. DNA **B.** mRNA **C.** tRNA **D.** rRNA

35. ________ A double helix consisting of two chains of nucleotides held together by hydrogen bonds between bases.

36. ________ A nucleic acid that uses deoxyribose as the sugar.

37. ________ A nucleic acid produced in the nucleus, which migrates to the ribosomes to direct the formation of a protein.

38. ________ A nucleic acid that brings the proper amino acid to the ribosome to build the peptide chain.

39. ________ A nucleic acid that contains adenine, cytosine, guanine, and thymine.

40. ________ Along with protein, it is a major component of the ribosomes.

41. ________ It contains a triplet called an anticodon.

42. ________ This nucleic acid is replicated during cellular division.

For questions 43 to 46, select answers from the following: (18.8)

A. —A G C C T A—
⋮ ⋮ ⋮ ⋮ ⋮ ⋮
—T C G G A T—

B. — AUU GCU C—

C. —A G T U G U—
⋮ ⋮ ⋮ ⋮ ⋮ ⋮
—T C A A C A—

D. UUU

E. —ATGTAT—

43. ________ a section of mRNA

44. ________ an impossible section of DNA

45. ________ a codon for phenylalanine

46. ________ a section from a DNA molecule

For questions 47 through 50, indicate the correct order of protein synthesis: (18.8)

A. DNA forms mRNA.
B. Protein is formed and breaks away.
C. tRNA picks up specific amino acids.
D. mRNA goes to the ribosomes.

47. ________ first step

48. ________ second step

49. ________ third step

50. ________ fourth step

Answers to the Practice Test

1. B	**2.** D	**3.** B	**4.** B	**5.** A
6. D	**7.** C	**8.** B	**9.** A	**10.** B
11. B	**12.** C	**13.** A	**14.** C	**15.** A
16. C	**17.** D	**18.** E	**19.** E	**20.** A
21. B	**22.** D	**23.** A	**24.** B	**25.** D
26. B	**27.** C	**28.** A	**29.** C	**30.** D
31. E	**32.** A	**33.** B	**34.** E	**35.** A
36. A	**37.** B	**38.** C	**39.** A	**40.** D
41. C	**42.** A	**43.** B	**44.** C	**45.** D
46. A	**47.** A	**48.** D	**49.** C	**50.** B

1

Chemistry in Our Lives

Learning Goals

- Define the term *chemistry* and identify substances as chemicals.
- Describe the activities that are part of the scientific method.
- Develop a study plan for learning chemistry.
- Review math concepts used in chemistry: place values, positive and negative numbers, percentages, solving equations, and interpreting line graphs.

Chapter Outline

Key Math Skills

- Identifying Place Values (1.4A)
- Using Positive and Negative Numbers in Calculations (1.4B)
- Calculating a Percentage (1.4C)
- Solving Equations (1.4D)
- Interpreting a Line Graph (1.4E)

Answers and Solutions to Text Problems

1.1 **a.** Chemistry is the study of the composition, structure, properties, and reactions of matter.
b. A chemical is a substance that has the same composition and properties wherever it is found.

1.3 Many chemicals are listed on a vitamin bottle such as vitamin A, vitamin B_3, vitamin B_{12}, vitamin C, and folic acid.

1.5 Typical items found in a medicine cabinet and some of the chemicals they contain are as follows:
Antacid tablets: calcium carbonate, cellulose, starch, stearic acid, silicon dioxide
Mouthwash: water, alcohol, thymol, glycerol, sodium benzoate, benzoic acid
Cough suppressant: menthol, beta-carotene, sucrose, glucose

1.7 An advantage of a pesticide is that it gets rid of insects that bite or damage crops. A disadvantage is that a pesticide can destroy beneficial insects or be retained in a crop that is eventually eaten by animals or humans.

1.9 **a.** A hypothesis proposes a possible explanation for a natural phenomenon. A hypothesis must be stated in a way that it can be tested by experiments.
b. An experiment is a procedure that tests the validity of a hypothesis.
c. A theory is a hypothesis that has been validated many times by many scientists.
d. An observation is a description or measurement of a natural phenomenon.

1.11 **a.** An observation (O) is a description or measurement of a natural phenomenon.
b. A hypothesis (H) proposes a possible explanation for a natural phenomenon.
c. An experiment (E) is a procedure that tests the validity of a hypothesis.
d. An observation (O) is a description or measurement of a natural phenomenon.
e. An observation (O) is a description or measurement of a natural phenomenon.
f. A theory (T) is developed from a hypothesis and is a possible explanation for a natural phenomenon.

1.13 There are several things you can do that will help you successfully learn chemistry, including forming a study group, going to lecture, working *Sample Problems* and *Study Checks*, working *Questions and Problems* and checking answers, reading the assignment ahead of class, going to the instructor's office hours, and keeping a problem notebook.

1.15 Ways you can enhance your learning of chemistry:
a. Form a study group.
c. Visit the professor during office hours.
e. Become an active learner.

1.17 **a.** The bolded 8 is in the thousandths place.
b. The bolded 6 is in the ones place.
c. The bolded 6 is in the hundreds place.

1.19 **a.** $15 - (-8) = 15 + 8 = 23$
b. $-8 + (-22) = -30$
c. $4 \times (-2) + 6 = -8 + 6 = -2$

1.21 **a.** $\dfrac{21 \text{ correct answers}}{25 \text{ total answers}} \times 100\% = 84\% \text{ correct}$
b. total grams of alloy = 56 g silver + 22 g copper = 78 g of alloy
$\dfrac{56 \text{ g silver}}{78 \text{ g alloy}} \times 100\% = 72\% \text{ silver}$
c. total number of coins = 11 nickels + 5 quarters + 7 dimes = 23 coins
$\dfrac{7 \text{ dimes}}{23 \text{ coins}} \times 100\% = 30\% \text{ dimes}$

1.23 **a.**
$$\begin{aligned} 4a + 4 &= 40 \\ 4a + \cancel{4} - \cancel{4} &= 40 - 4 \\ 4a &= 36 \\ \frac{\cancel{4}a}{\cancel{4}} &= \frac{36}{4} \\ a &= 9 \end{aligned}$$
b.
$$\begin{aligned} \frac{a}{6} &= 7 \\ \cancel{6}\left(\frac{a}{\cancel{6}}\right) &= 6(7) \\ a &= 42 \end{aligned}$$

1.25 **a.** The graph shows the relationship between the temperature of a cup of tea and elapsed time.
b. The vertical axis measures temperature, in °C.
c. The values on the vertical axis range from 20 °C to 80 °C.
d. As time increases, the temperature decreases.

1.27 No. All of these ingredients are chemicals.

1.29 Yes. Sherlock's investigation includes making observations (gathering data), formulating a hypothesis, testing the hypothesis, and modifying it until the hypothesis is validated.

1.31 **a.** Determination of a melting point with a thermometer is an observation (O).
b. Describing a reason for the extinction of dinosaurs is a hypothesis (H).
c. Measuring the completion time of a race is an observation (O).

1.33 **a.** When two negative numbers are added, the answer has a negative sign.
b. When a positive and negative number are multiplied, the answer has a negative sign.

1.35 A hypothesis is an important part of the scientific process because it is the initial generation of a proposed explanation for a natural phenomenon and can be tested by experiments.

1.37 If experimental results do not support your hypothesis, you should **b.** write another hypothesis, and **c.** do more experiments.

1.39 A successful study plan would include:
b. Working the *Sample Problems* as you go through a chapter.
c. Going to your professor's office hours.

1.41 **a.** $4 \times (-8) = -32$
b. $-12 - 48 = -60$
c. $\dfrac{-168}{-4} = 42$

1.43 total number of gumdrops $=$ 16 orange $+$ 8 yellow $+$ 16 black $=$ 40 gumdrops

a. $\dfrac{8 \text{ yellow gumdrops}}{40 \text{ total gumdrops}} \times 100\% = 20\%$ yellow gumdrops

b. $\dfrac{16 \text{ black gumdrops}}{40 \text{ total gumdrops}} \times 100\% = 40\%$ black gumdrops

1.45 **a.** An observation (O) is a description or measurement of a natural phenomenon.
b. A hypothesis (H) proposes a possible explanation for a natural phenomenon.
c. An experiment (E) is a procedure that tests the validity of a hypothesis.
d. A hypothesis (H) proposes a possible explanation for a natural phenomenon.

1.47 **a.**

$$2x + 5 = 41$$
$$2x + \cancel{5} - \cancel{5} = 41 - 5$$
$$2x = 36$$
$$\frac{\cancel{2}x}{\cancel{2}} = \frac{36}{2}$$
$$x = 18$$

b.

$$\frac{5x}{3} = 40$$
$$\cancel{3}\left(\frac{5x}{\cancel{3}}\right) = 3(40)$$
$$5x = 120$$
$$\frac{\cancel{5}x}{\cancel{5}} = \frac{120}{5}$$
$$x = 24$$

1.49 **a.** The graph shows the relationship between the solubility of carbon dioxide in water and the temperature.
b. The vertical axis measures the solubility of carbon dioxide in water (g CO_2/100 g water).
c. The values on the vertical axis range from 0 to 0.35 g CO_2/100 g water.
d. As temperature increases, the solubility of carbon dioxide in water decreases.

2

Measurements

Learning Goals

- Write the names and abbreviations for the metric or SI units used in measurements of length, volume, mass, temperature, and time.
- Write a number in scientific notation.
- Identify a number as measured or exact; determine the number of significant figures in a measured number.
- Adjust calculated answers to give the correct number of significant figures.
- Use the numerical values of prefixes to write a metric equality.
- Write a conversion factor for two units that describe the same quantity.
- Use conversion factors to change from one unit to another.
- Calculate the density of a substance; use the density to calculate the mass or volume of a substance.

Chapter Outline

Chapter Opener: Registered Nurse

2.1 Units of Measurement

2.2 Scientific Notation

2.3 Measured Numbers and Significant Figures

2.4 Significant Figures in Calculations

2.5 Prefixes and Equalities

2.6 Writing Conversion Factors

Chemistry Link to Health: Toxicology and Risk–Benefit Assessment

2.7 Problem Solving Using Unit Conversion

2.8 Density

Chemistry Link to the Environment: Density of Crude Oil

Chemistry Link to Health: Bone Density

Key Math Skills

- Using Scientific Notation (2.2)

Core Chemistry Skills

- Counting Significant Figures (2.3)
- Rounding Off (2.4)
- Using Significant Figures in Calculations (2.4)
- Using Prefixes (2.5)
- Writing Conversion Factors from Equalities (2.6)
- Using Conversion Factors (2.7)
- Using Density as a Conversion Factor (2.8)

Answers and Solutions to Text Problems

2.1 **a.** The unit is a meter, which is a unit of length.
b. The unit is a gram, which is a unit of mass.
c. The unit is a liter, which is a unit of volume.
d. The unit is a second, which is a unit of time.
e. The unit is a degree Celsius, which is a unit of temperature.

2.3 **a.** The unit is a milliliter, which is a metric unit.
b. The unit is a kilogram, which is both a metric unit and an SI unit.
c. The unit is an inch, which is neither a metric unit nor an SI unit.
d. The unit is a second, which is both a metric unit and an SI unit.
e. The unit is a degree Celsius, which is a metric unit.

2.5 **a.** The abbreviation for the unit gram is g.
b. The abbreviation for the unit liter is L.
c. The abbreviation for the unit degree Celsius is °C.
d. The abbreviation for the unit pound is lb.
e. The abbreviation for the unit second is s.

2.7 **a.** Move the decimal point four places to the left to give 5.5×10^4 m.
b. Move the decimal point two places to the left to give 4.8×10^2 g.
c. Move the decimal point six places to the right to give 5×10^{-6} cm.
d. Move the decimal point four places to the right to give 1.4×10^{-4} s.
e. Move the decimal point three places to the right to give 7.85×10^{-3} L.
f. Move the decimal point five places to the left to give 6.7×10^5 kg.

2.9 **a.** 7.2×10^3 cm is larger than 4.2×10^3 cm.
b. 3.2×10^{-2} kg, which is also 320×10^{-4} kg, is larger than 4.5×10^{-4} kg.
c. 1×10^4 L or 10 000 L is larger than 1×10^{-4} L or 0.0001 L.
d. 6.8×10^{-2} m or 0.068 m is larger than 0.000 52 m.

2.11 **a.** The standard number is 1.2 times the power of 10^4 or 10 000, which gives 12 000 s.
b. The standard number is 8.25 times the power of 10^{-2} or 0.01, which gives 0.0825 kg.
c. The standard number is 4 times the power of 10^6 or 1 000 000, which gives 4 000 000 g.
d. The standard number is 5.8 times the power of 10^{-3} or 0.001, which gives 0.0058 m^3.

2.13 The estimated digit is the last digit reported in a measurement.
a. In 8.6 m, the estimated digit is the 6 in the tenths place.
b. In 45.25 g, the estimated digit is the 5 in the hundredths place.
c. In 29 °C, the estimated digit is the 9 in the ones place.

2.15 Measured numbers are obtained using some type of measuring device. Exact numbers are numbers obtained by counting items or using a definition that compares two units in the same measuring system.
a. The value 67.5 kg is a measured number; measurement of mass requires a measuring device.
b. The value 8 apples is obtained by counting, making it an exact, not a measured, number.
c. The values in the metric definition 1 m = 1000 mm are exact, not measured, numbers.
d. The value 1720 km is a measured number; measurement of distance requires a measuring device.

2.17 Measured numbers are obtained using some kind of measuring tool. Exact numbers are numbers obtained by counting items or using a definition that compares two units in the same measuring system.
a. 6 oz of hamburger meat is a measured number (3 hamburgers is a counted/exact number).
b. Neither are measured numbers (both 1 table and 4 chairs are counted/exact numbers).
c. Both 0.75 lb of grapes and 350 g of butter are measured numbers.
d. Neither are measured numbers (the values in a definition are exact numbers).

2.19 **a.** None; zeros at the beginning of a decimal number are not significant.
b. The zero between nonzero digits is significant.
c. The zeros after the first nonzero digit in a decimal number are significant.
d. The zero in the coefficient of a number written in scientific notation is significant.
e. None; zeros used as placeholders in a large number without a decimal point are not significant.

2.21 **a.** All five numbers are significant (5 SFs).
b. Only the two nonzero numbers are significant (2 SFs); the preceding zeros are placeholders.
c. Only the two nonzero numbers are significant (2 SFs); the zeros that follow are placeholders.
d. All three numbers in the coefficient of a number written in scientific notation are significant (3 SFs).
e. All four numbers to the right of the decimal point, including the last zero, in a decimal number are significant (4 SFs).
f. All three numbers including the zero at the end of a decimal number are significant (3 SFs).

2.23 **a.** 11.00 m (4 SFs) contains more significant figures than does 11.0 m (3 SFs).
b. 405.0 K (4 SFs) contains more significant figures than does 405 K (3 SFs).
c. 0.0120 s (3 SFs) contains more significant figures than does 12 000 s (2 SFs).
d. 2.50×10^{-2} mL (3 SFs) contains more significant figures than does 2500 mL (2 SFs).

2.25 **a.** 5000 L is the same as 5×1000 L, which is written in scientific notation as 5.0×10^3 L with two significant figures.
b. 30 000 g is the same as $3 \times 10\,000$ g, which is written in scientific notation as 3.0×10^4 g with two significant figures.
c. 100 000 m is the same as $1 \times 100\,000$ m, which is written in scientific notation as 1.0×10^5 m with two significant figures.
d. 0.000 25 cm is the same as $2.5 \times \frac{1}{10\,000}$ cm, which is written in scientific notation as 2.5×10^{-4} cm.

2.27 The significant figures in the measurements used in a calculation are usually fewer than the number of figures in the calculator display, which requires rounding off.

2.29 **a.** 1.85 kg; the last digit is dropped since it is 4 or less.
b. 88.0 L; since the fourth digit is 4 or less, the last three digits are dropped.
c. 0.004 74 cm; since the fourth significant digit (the first digit to be dropped) is 5 or greater, the last retained digit is increased by 1 when the last four digits are dropped.
d. 8810 m; since the fourth significant digit (the first digit to be dropped) is 5 or greater, the last retained digit is increased by 1 when the last digit is dropped (a nonsignificant zero is added at the end as a placeholder).
e. 1.83×10^3 s; since the fourth digit is 4 or less, the last two digits are dropped. The $\times 10^3$ is retained so that the magnitude of the answer is not changed.

2.31 **a.** To round off 56.855 m to three significant figures, drop the final digits 55 and increase the last retained digit by 1 to give 56.9 m.
b. To round off 0.002 282 g to three significant figures, drop the final digit 2 to give 0.002 28 g.
c. To round off 11 527 s to three significant figures, drop the final digits 27 and add two zeros as placeholders to give 11 500 s (1.15×10^4 s).
d. To express 8.1 L to three significant figures, add a significant zero to give 8.10 L.

2.33 **a.** $45.7 \times 0.034 = 1.6$ Two significant figures are allowed since 0.034 has 2 SFs.
b. $0.002\,78 \times 5 = 0.01$ One significant figure is allowed since 5 has 1 SF.
c. $\frac{34.56}{1.25} = 27.6$ Three significant figures are allowed since 1.25 has 3 SFs.

d. $\dfrac{(0.2465)(25)}{1.78} = 3.5$ Two significant figures are allowed since 25 has 2 SFs.

e. $(2.8 \times 10^4)(5.05 \times 10^{-6}) = 0.14$ or 1.4×10^{-1} Two significant figures are allowed since 2.8×10^4 has 2 SFs.

f. $\dfrac{(3.45 \times 10^{-2})(1.8 \times 10^5)}{(8 \times 10^3)} = 0.8$ or 8×10^{-1} One significant figure is allowed since 8×10^3 has 1 SF.

2.35 **a.** 45.48 cm + 8.057 cm = 53.54 cm Two decimal places are allowed since 45.48 cm has two decimal places.
b. 23.45 g + 104.1 g + 0.025 g = 127.6 g One decimal place is allowed since 104.1 g has one decimal place.
c. 145.675 mL − 24.2 mL = 121.5 mL One decimal place is allowed since 24.2 mL has one decimal place.
d. 1.08 L − 0.585 L = 0.50 L Two decimal places are allowed since 1.08 L has two decimal places.

2.37 The km/h markings indicate how many kilometers (how much distance) will be traversed in 1 hour's time if the speed is held constant. The mph (mi/h) markings indicate the same distance traversed *but measured in miles* during the 1 hour of travel.

2.39 **a.** mg **b.** dL **c.** km **d.** fg

2.41 **a.** centiliter **b.** kilogram **c.** millisecond **d.** gigameter

2.43 **a.** 0.01 **b.** 1 000 000 000 000 (or 1×10^{12})
c. 0.001 (or 1×10^{-3}) **d.** 0.1

2.45 **a.** 1 decigram **b.** 1 microgram
c. 1 kilogram **d.** 1 centigram

2.47 **a.** 1 m = 100 cm **b.** 1 nm = 1×10^{-9} m
c. 1 mm = 0.001 m **d.** 1 L = 1000 mL

2.49 **a.** kilogram, since 10^3 g is greater than 10^{-3} g.
b. milliliter, since 10^{-3} L is greater than 10^{-6} L.
c. cm, since 10^{-2} m is greater than 10^{-12} m.
d. kL, since 10^3 L is greater than 10^{-1} L.

2.51 A conversion factor can be inverted to give a second conversion factor: $\dfrac{1\text{ m}}{100\text{ cm}}$ and $\dfrac{100\text{ cm}}{1\text{ m}}$

2.53 **a.** 1 m = 100 cm; $\dfrac{1\text{ m}}{100\text{ cm}}$ and $\dfrac{100\text{ cm}}{1\text{ m}}$

b. 1 g = 1×10^9 ng; $\dfrac{1\text{ g}}{1 \times 10^9\text{ ng}}$ and $\dfrac{1 \times 10^9\text{ ng}}{1\text{ g}}$

c. 1 kL = 1000 L; $\dfrac{1\text{ kL}}{1000\text{ L}}$ and $\dfrac{1000\text{ L}}{1\text{ kL}}$

d. 1 s = 1000 ms; $\dfrac{1\text{ s}}{1000\text{ ms}}$ and $\dfrac{1000\text{ ms}}{1\text{ s}}$

e. $(1\text{ m})^3 = (100\text{ cm})^3$; $\dfrac{(1\text{ m})^3}{(100\text{ cm})^3}$ and $\dfrac{(100\text{ cm})^3}{(1\text{ m})^3}$

2.55 **a.** 1 yd = 3 ft; $\dfrac{3\text{ ft}}{1\text{ yd}}$ and $\dfrac{1\text{ yd}}{3\text{ ft}}$; the numbers 1 and 3 are both exact (U.S. definition).

b. 1 kg = 2.205 lb; $\dfrac{2.205\text{ lb}}{1\text{ kg}}$ and $\dfrac{1\text{ kg}}{2.205\text{ lb}}$; the number 1 is exact; the number 2.205 is measured and has 4 SFs.

c. 1 min = 60 s; $\frac{60\text{ s}}{1\text{ min}}$ and $\frac{1\text{ min}}{60\text{ s}}$; the numbers 1 and 60 are both exact (metric definition).

d. 1 gal of gasoline = 27 mi; $\frac{27\text{ mi}}{1\text{ gal gasoline}}$ and $\frac{1\text{ gal gasoline}}{27\text{ mi}}$; the number 1 is exact; the number 27 is measured and has 2 SFs.

e. 100 g of crust = 48 g of silicon; $\frac{48\text{ g silicon}}{100\text{ g crust}}$ and $\frac{100\text{ g crust}}{48\text{ g silicon}}$; the number 100 is exact; the number 48 is measured has 2 SFs.

2.57 **a.** 1 s = 3.5 m; $\frac{1\text{ s}}{3.5\text{ m}}$ and $\frac{3.5\text{ m}}{1\text{ s}}$; the number 1 is exact; the number 3.5 is measured and has 2 SFs.

b. 1 mL = 0.74 g; $\frac{1\text{ mL}}{0.74\text{ g}}$ and $\frac{0.74\text{ g}}{1\text{ mL}}$; the number 1 is exact; the number 0.74 is measured and has 2 SFs.

c. 1 gal of gasoline = 46.0 km; $\frac{1\text{ gal gasoline}}{46.0\text{ km}}$ and $\frac{46.0\text{ km}}{1\text{ gal gasoline}}$; the number 1 is exact; the number 46.0 is measured and has 3 SFs.

d. 100 g of sterling = 93 g of silver; $\frac{93\text{ g silver}}{100\text{ g sterling}}$ and $\frac{100\text{ g sterling}}{93\text{ g silver}}$; the number 100 is exact; the number 93 is measured and has 2 SFs.

e. 1 kg of plums = 29 μg of pesticide; $\frac{1\text{ kg plums}}{29\ \mu\text{g pesticide}}$ and $\frac{29\ \mu\text{g pesticide}}{1\text{ kg plums}}$; the number 1 is exact; the number 29 is measured and has 2 SFs.

2.59 When using a conversion factor, you are trying to cancel existing units and arrive at a new (needed) unit. The conversion factor must be properly oriented so that the unit in the denominator cancels the preceding unit in the numerator.

2.61 **a.** **Given** 44.2 mL **Need** liters

Plan mL → L $\frac{1\text{ L}}{1000\text{ mL}}$

Set Up $44.2\ \cancel{\text{mL}} \times \frac{1\text{ L}}{1000\ \cancel{\text{mL}}} = 0.0442\text{ L (3 SFs)}$

b. **Given** 8.65 m **Need** nanometers

Plan m → nm $\frac{1 \times 10^9\text{ nm}}{1\text{ m}}$

Set Up $8.65\ \cancel{\text{m}} \times \frac{1 \times 10^9\text{ nm}}{1\ \cancel{\text{m}}} = 8.65 \times 10^9\text{ nm (3 SFs)}$

c. **Given** 5.2×10^8 g **Need** megagrams

Plan g → Mg $\frac{1\text{ Mg}}{1 \times 10^6\text{ g}}$

Set Up $5.2 \times 10^8\ \cancel{\text{g}} \times \frac{1\text{ Mg}}{1 \times 10^6\ \cancel{\text{g}}} = 5.2 \times 10^2\text{ Mg (2 SFs)}$

d. **Given** 0.72 ks **Need** milliseconds

Plan ks → s → ms $\frac{1000\text{ s}}{1\text{ ks}}$ $\frac{1000\text{ ms}}{1\text{ s}}$

Set Up $0.72\ \cancel{\text{ks}} \times \frac{1000\ \cancel{\text{s}}}{1\ \cancel{\text{ks}}} \times \frac{1000\text{ ms}}{1\ \cancel{\text{s}}} = 7.2 \times 10^5\text{ ms (2 SFs)}$

2.63 **a.** **Given** 3.428 lb **Need** kilograms

Plan lb → kg $\frac{1\text{ kg}}{2.205\text{ lb}}$

Set Up $3.428\text{ lb} \times \frac{1\text{ kg}}{2.205\text{ lb}} = 1.555\text{ kg (4 SFs)}$

b. **Given** 1.6 m **Need** inches

Plan m → in. $\frac{39.37\text{ in.}}{1\text{ m}}$

Set Up $1.6\text{ m} \times \frac{39.37\text{ in.}}{1\text{ m}} = 63\text{ in. (2 SFs)}$

c. **Given** 4.2 L **Need** quarts

Plan L → qt $\frac{1.057\text{ qt}}{1\text{ L}}$

Set Up $4.2\text{ L} \times \frac{1.057\text{ qt}}{1\text{ L}} = 4.4\text{ qt (2 SFs)}$

d. **Given** 0.672 ft **Need** millimeters

Plan ft → in. → cm → mm $\frac{12\text{ in.}}{1\text{ ft}}$ $\frac{2.54\text{ cm}}{1\text{ in.}}$ $\frac{10\text{ mm}}{1\text{ cm}}$

Set Up $0.672\text{ ft} \times \frac{12\text{ in.}}{1\text{ ft}} \times \frac{2.54\text{ cm}}{1\text{ in.}} \times \frac{10\text{ mm}}{1\text{ cm}} = 205\text{ mm (3 SFs)}$

2.65 **a.** **Given** 175 cm **Need** meters

Plan cm → m $\frac{1\text{ m}}{100\text{ cm}}$

Set Up $175\text{ cm} \times \frac{1\text{ m}}{100\text{ cm}} = 1.75\text{ m (3 SFs)}$

b. **Given** 5500 mL **Need** liters

Plan mL → L $\frac{1\text{ L}}{1000\text{ mL}}$

Set Up $5500\text{ mL} \times \frac{1\text{ L}}{1000\text{ mL}} = 5.5\text{ L (2 SFs)}$

c. **Given** 0.0055 kg **Need** grams

Plan kg → g $\frac{1000\text{ g}}{1\text{ kg}}$

Set Up $0.0055\text{ kg} \times \frac{1000\text{ g}}{1\text{ kg}} = 5.5\text{ g (2 SFs)}$

d. **Given** 350 cm^3 **Need** cubic meters

Plan $\text{cm}^3 \rightarrow \text{m}^3$ $\frac{(1\text{ m})^3}{(100\text{ cm})^3}$

Set Up $350\text{ cm}^3 \times \frac{(1\text{ m})^3}{(100\text{ cm})^3} = 3.5 \times 10^{-4}\text{ m}^3\text{ (2 SFs)}$

2.67 **a.** **Given** 0.750 qt **Need** milliliters

Plan qt → mL $\frac{946.3\text{ mL}}{1\text{ qt}}$

Set Up $0.750\text{ qt} \times \frac{946.3\text{ mL}}{1\text{ qt}} = 710.\text{ mL (3 SFs)}$

b. Given 11.8 stones **Need** kilograms

Plan stones → lb → kg $\frac{14.0\text{ lb}}{1\text{ stone}}$ $\frac{1\text{ kg}}{2.205\text{ lb}}$

Set Up $11.8\ \cancel{\text{stones}} \times \frac{14.0\ \cancel{\text{lb}}}{1\ \cancel{\text{stone}}} \times \frac{1\text{ kg}}{2.205\ \cancel{\text{lb}}} = 74.9\text{ kg (3 SFs)}$

c. Given 19.5 in. **Need** millimeters

Plan in. → cm → mm $\frac{2.54\text{ cm}}{1\text{ in.}}$ $\frac{10\text{ mm}}{1\text{ cm}}$

Set Up $19.5\ \cancel{\text{in.}} \times \frac{2.54\ \cancel{\text{cm}}}{1\ \cancel{\text{in.}}} \times \frac{10\text{ mm}}{1\ \cancel{\text{cm}}} = 495\text{ mm (3 SFs)}$

d. Given 0.50 μm **Need** inches

Plan μm → m → cm → in. $\frac{1\text{ m}}{10^6\ \mu\text{m}}$ $\frac{100\text{ cm}}{1\text{ m}}$ $\frac{1\text{ in.}}{2.54\text{ cm}}$

Set Up $0.50\ \cancel{\mu\text{m}} \times \frac{1\ \cancel{\text{m}}}{10^6\ \cancel{\mu\text{m}}} \times \frac{100\ \cancel{\text{cm}}}{1\ \cancel{\text{m}}} \times \frac{1\text{ in.}}{2.54\ \cancel{\text{cm}}} = 2.0 \times 10^{-5}\text{ in. (2 SFs)}$

2.69 a. Given 78.0 ft (length) **Need** meters

Plan ft → in. → cm → m $\frac{12\text{ in.}}{1\text{ ft}}$ $\frac{2.54\text{ cm}}{1\text{ in.}}$ $\frac{1\text{ m}}{100\text{ cm}}$

Set Up $78.0\ \cancel{\text{ft}} \times \frac{12\ \cancel{\text{in.}}}{1\ \cancel{\text{ft}}} \times \frac{2.54\ \cancel{\text{cm}}}{1\ \cancel{\text{in.}}} \times \frac{1\text{ m}}{100\ \cancel{\text{cm}}} = 23.8\text{ m (length) (3 SFs)}$

b. Given 27.0 ft (width), 23.8 m (length) **Need** square meters

Plan ft → in. → cm → m, then length (m), width (m) → m^2

$\frac{12\text{ in.}}{1\text{ ft}}$ $\frac{2.54\text{ cm}}{1\text{ in.}}$ $\frac{1\text{ m}}{100\text{ cm}}$ *Area = length × width*

Set Up $27.0\ \cancel{\text{ft}} \times \frac{12\ \cancel{\text{in.}}}{1\ \cancel{\text{ft}}} \times \frac{2.54\ \cancel{\text{cm}}}{1\ \cancel{\text{in.}}} \times \frac{1\text{ m}}{100\ \cancel{\text{cm}}} = 8.23\text{ m (width) (3 SFs)}$

$\therefore$ Area $= 23.8\text{ m} \times 8.23\text{ m} = 196\text{ m}^2$ (3 SFs)

c. Given 23.8 m (length), 185 km/h (speed) **Need** seconds

Plan m → km → h → min → s $\frac{1\text{ km}}{1000\text{ m}}$ $\frac{1\text{ h}}{185\text{ km}}$ $\frac{60\text{ min}}{1\text{ h}}$ $\frac{60\text{ s}}{1\text{ min}}$

Set Up $23.8\ \cancel{\text{m}} \times \frac{1\ \cancel{\text{km}}}{1000\ \cancel{\text{m}}} \times \frac{1\ \cancel{\text{h}}}{185\ \cancel{\text{km}}} \times \frac{60\ \cancel{\text{min}}}{1\ \cancel{\text{h}}} \times \frac{60\text{ s}}{1\ \cancel{\text{min}}} = 0.463\text{ s (3 SFs)}$

d. Given 78.0 ft (length), 27.0 ft (width) **Need** liters of paint

Plan use length (ft), width (ft) → ft^2, then ft^2 → gal of paint → qt of paint → L of paint

$\frac{1\text{ gal}}{150\text{ ft}^2}$ $\frac{4\text{ qt}}{1\text{ gal}}$ $\frac{1\text{ L}}{1.057\text{ qt}}$

Set Up Area $= 78.0\text{ ft} \times 27.0\text{ ft} = 2100\text{ ft}^2$ (3 SFs)

$2100\ \cancel{\text{ft}^2} \times \frac{1\ \cancel{\text{gal}}}{150\ \cancel{\text{ft}^2}} \times \frac{4\ \cancel{\text{qt}}}{1\ \cancel{\text{gal}}} \times \frac{1\text{ L}}{1.057\ \cancel{\text{qt}}} = 53\text{ L of paint (2 SFs)}$

2.71 Each of the following requires a percent factor from the problem information.

a. Given 325 g of crust **Need** grams of oxygen

Plan g of crust → g of oxygen

(percent equality: 100 g of crust = 46.7 g of oxygen) $\frac{46.7\text{ g oxygen}}{100\text{ g crust}}$

Set Up $325\text{ g}\ \cancel{\text{crust}} \times \frac{46.7\text{ g oxygen}}{100\text{ g}\ \cancel{\text{crust}}} = 152\text{ g of oxygen (3 SFs)}$

b. Given 1.25 g of crust **Need** grams of magnesium
Plan g of crust → g of magnesium

(percent equality: 100 g of crust = 2.1 g of magnesium) $\frac{2.1\text{ g magnesium}}{100\text{ g crust}}$

Set Up $1.25\ \cancel{\text{g crust}} \times \frac{2.1\text{ g magnesium}}{100\ \cancel{\text{g crust}}} = 0.026\text{ g of magnesium (2 SFs)}$

c. Given 10.0 oz of fertilizer **Need** grams of nitrogen
Plan oz of fertilizer → lb of fertilizer → g of fertilizer → g of nitrogen
(percent equality: 100 g of fertilizer = 15 g of nitrogen)
$\frac{1\text{ lb}}{16\text{ oz}}$ $\frac{453.6\text{ g}}{1\text{ lb}}$ $\frac{15\text{ g nitrogen}}{100\text{ g fertilizer}}$

Set Up $10.0\ \cancel{\text{oz fertilizer}} \times \frac{1\ \cancel{\text{lb}}}{16\ \cancel{\text{oz}}} \times \frac{453.6\ \cancel{\text{g}}}{1\ \cancel{\text{lb}}} \times \frac{15\text{ g nitrogen}}{100\ \cancel{\text{g fertilizer}}}$
$= 43\text{ g of nitrogen (2 SFs)}$

d. Given 5.0 kg of pecans **Need** pounds of chocolate bars
Plan kg of pecans → kg of bars → lb of bars

(percent equality: 100 kg of bars = 22.0 kg of pecans) $\frac{100\text{ kg choc. bars}}{22.0\text{ kg pecans}}$ $\frac{2.205\text{ lb}}{1\text{ kg}}$

Set Up $5.0\ \cancel{\text{kg pecans}} \times \frac{100\ \cancel{\text{kg}}\text{ choc. bars}}{22.0\ \cancel{\text{kg pecans}}} \times \frac{2.205\text{ lb}}{1\ \cancel{\text{kg}}}$
$= 50.\text{ lb of chocolate bars (2 SFs)}$

2.73 Because the density of aluminum is 2.70 g/cm^3, silver is 10.5 g/cm^3, and lead is 11.3 g/cm^3, we can identify the unknown metal by calculating its density as follows:

$$\text{Density} = \frac{\text{mass of metal}}{\text{volume of metal}} = \frac{217\text{ g}}{19.2\text{ cm}^3} = 11.3\text{ g/cm}^3\text{ (3 SFs)}$$

∴ the metal is lead.

2.75 Density is the mass of a substance divided by its volume. $\text{Density} = \frac{\text{mass (grams)}}{\text{volume (mL)}}$

The densities of solids and liquids are usually stated in g/mL or g/cm^3, so in some problems the units will need to be converted.

a. $\text{Density} = \frac{\text{mass (grams)}}{\text{volume (mL)}} = \frac{24.0\text{ g}}{20.0\text{ mL}} = 1.20\text{ g/mL (3 SFs)}$

b. Given 0.250 lb of butter, 130.3 mL **Need** density (g/mL)

Plan lb → g, then calculate density $\frac{453.6\text{ g}}{1\text{ lb}}$

Set Up $0.250\ \cancel{\text{lb}} \times \frac{453.6\text{ g}}{1\ \cancel{\text{lb}}} = 113\text{ g (3 SFs)}$

$\therefore \text{Density} = \frac{\text{mass}}{\text{volume}} = \frac{113\text{ g}}{130.3\text{ mL}} = 0.867\text{ g/mL (3 SFs)}$

c. Given 12.00 mL initial volume, 13.45 mL final volume, 4.50 g **Need** density (g/mL)
Plan calculate volume by difference, then calculate density
Set Up volume of gem: 13.45 mL total − 12.00 mL water = 1.45 mL

$\therefore \text{Density} = \frac{\text{mass}}{\text{volume}} = \frac{4.50\text{ g}}{1.45\text{ mL}} = 3.10\text{ g/mL (3 SFs)}$

2.77 **a. Given** 514.1 g, 114 cm^3 **Need** density (g/mL)

Plan convert volume $cm^3 \rightarrow$ mL, then calculate the density

Set Up $114\ \cancel{cm^3} \times \dfrac{1\ mL}{1\ \cancel{cm^3}} = 114\ mL$

$\therefore \text{Density} = \dfrac{\text{mass}}{\text{volume}} = \dfrac{514.1\ g}{114\ mL} = 4.51\ g/mL\ (3\ SFs)$

b. Given 0.100 pt, 115.25 g initial, 182.48 g final **Need** density (g/mL)

Plan pt $\rightarrow$ qt $\rightarrow$ L $\rightarrow$ mL, then calculate mass by difference

$\dfrac{1\ qt}{2\ pt}$ $\dfrac{1\ L}{1.057\ qt}$ $\dfrac{1000\ mL}{1\ L}$, then calculate the density

Set Up $0.100\ \cancel{pt} \times \dfrac{1\ \cancel{qt}}{2\ \cancel{pt}} \times \dfrac{1\ \cancel{L}}{1.057\ \cancel{qt}} \times \dfrac{1000\ mL}{1\ \cancel{L}} = 47.3\ mL\ (3\ SFs)$

mass of syrup = 182.48 g − 115.25 g = 67.23 g

$\therefore \text{Density} = \dfrac{\text{mass}}{\text{volume}} = \dfrac{67.23\ g}{47.3\ mL} = 1.42\ g/mL\ (3\ SFs)$

c. Given 8.51 kg, 3.15 L **Need** density (g/mL)

Plan kg $\rightarrow$ g and L $\rightarrow$ mL, then calculate the density $\dfrac{1000\ g}{1\ kg}$ $\dfrac{1000\ mL}{1\ L}$

Set Up $8.51\ \cancel{kg} \times \dfrac{1000\ g}{1\ \cancel{kg}} = 8510\ g\ (3\ SFs)$ and

$3.15\ \cancel{L} \times \dfrac{1000\ mL}{1\ \cancel{L}} = 3150\ mL\ (3\ SFs)$

$\therefore \text{Density} = \dfrac{\text{mass}}{\text{volume}} = \dfrac{8510\ g}{3150\ mL} = 2.70\ g/mL\ (3\ SFs)$

2.79 In these problems, the density is used as a conversion factor.

a. Given 1.50 kg of ethanol **Need** liters of ethanol

Plan kg $\rightarrow$ g $\rightarrow$ mL $\rightarrow$ L $\dfrac{1000\ g}{1\ kg}$ $\dfrac{1\ mL}{0.785\ g}$ $\dfrac{1\ L}{1000\ mL}$

Set Up $1.50\ \cancel{kg\ alcohol} \times \dfrac{1000\ \cancel{g}}{1\ \cancel{kg\ alcohol}} \times \dfrac{1\ \cancel{mL}}{0.785\ \cancel{g}} \times \dfrac{1\ L}{1000\ \cancel{mL}}$

= 1.91 L of ethanol (3 SFs)

b. Given 6.5 mL of mercury **Need** grams of mercury

Plan mL $\rightarrow$ g $\dfrac{13.6\ g}{1\ mL}$

Set Up $6.5\ \cancel{mL} \times \dfrac{13.6\ g}{1\ \cancel{mL}} = 88\ g$ of mercury (2 SFs)

c. Given 225 cm^3 of silver **Need** ounces of silver

Plan mL $\rightarrow$ g $\rightarrow$ lb $\rightarrow$ oz $\dfrac{10.5\ g}{1\ cm^3}$ $\dfrac{1\ lb}{453.6\ g}$ $\dfrac{16\ oz}{1\ lb}$

Set Up $225\ \cancel{cm^3} \times \dfrac{10.5\ \cancel{g}}{1\ \cancel{cm^3}} \times \dfrac{1\ \cancel{lb}}{453.6\ \cancel{g}} \times \dfrac{16\ oz}{1\ \cancel{lb}} = 83.3$ oz of silver (3SFs)

2.81 **a. Given** 74.1 cm^3 of copper **Need** grams of copper

Plan $cm^3 \rightarrow$ g $\dfrac{8.92\ g}{1\ cm^3}$

Set Up $74.1\ \cancel{cm^3} \times \dfrac{8.92\ g}{1\ \cancel{cm^3}} = 661$ g of copper (3 SFs)

b. Given 12.0 gal of gasoline **Need** kilograms of gasoline

Plan gal → qt → mL → g → kg $\frac{4\text{ qt}}{1\text{ gal}}$ $\frac{946.3\text{ mL}}{1\text{ qt}}$ $\frac{0.74\text{ g}}{1\text{ mL}}$ $\frac{1\text{ kg}}{1000\text{ g}}$

Set Up $12.0\text{ gal} \times \frac{4\text{ qt}}{1\text{ gal}} \times \frac{946.3\text{ mL}}{1\text{ qt}} \times \frac{0.74\text{ g}}{1\text{ mL}} \times \frac{1\text{ kg}}{1000\text{ g}} = 34\text{ kg of gasoline (2 SFs)}$

c. Given 27 g of ice **Need** cubic centimeters of ice

Plan $\text{g} \rightarrow \text{cm}^3$ $\frac{1\text{ cm}^3}{0.92\text{ g}}$

Set Up $27\text{ g} \times \frac{1\text{ cm}^3}{0.92\text{ g}} = 29\text{ cm}^3\text{ (2 SFs)}$

2.83 **a.** The number of legs is a counted number; it is exact.
b. The height is measured with a ruler or tape measure; it is a measured number.
c. The number of chairs is a counted number; it is exact.
d. The area is measured with a ruler or tape measure; it is a measured number.

2.85 **a.** $\text{length} = 38.4\text{ in.} \times \frac{2.54\text{ cm}}{1\text{ in.}} = 97.5\text{ cm (3 SFs)}$

b. $\text{width} = 24.2\text{ in.} \times \frac{2.54\text{ cm}}{1\text{ in.}} = 61.5\text{ cm (3 SFs)}$

c. There are three significant figures in the length measurement.
d. $\text{Area} = \text{length} \times \text{width} = 97.5\text{ cm} \times 61.5\text{ cm} = 6.00 \times 10^3\text{ cm}^2\text{ (3 SFs)}$

2.87 **Given** 18.5 mL initial volume, 23.1 mL final volume, 8.24 g **Need** density (g/mL)
Plan calculate volume by difference, then calculate density
Set Up The volume of the object is 23.1 mL − 18.5 mL = 4.6 mL

$\therefore \text{Density} = \frac{\text{mass}}{\text{volume}} = \frac{8.24\text{ g}}{4.6\text{ mL}} = 1.8\text{ g/mL (2 SFs)}$

2.89 The liquid with the highest density will be at the bottom of the cylinder; the liquid with the lowest density will be at the top of the cylinder:
A is vegetable oil (D = 0.92 g/mL), **B** is water (D = 1.00 g/mL),
C is mercury (D = 13.6 g/mL).

2.91 **a.** To round off 0.000 012 58 L to three significant figures, drop the final digit 8 and increase the last retained digit by 1 to give 0.000 012 6 L or 1.26×10^{-5} L.
b. To round off 3.528×10^2 kg to three significant figures, drop the final digit 8 and increase the last retained digit by 1 to give 3.53×10^2 kg.
c. To round off 125 111 m^3 to three significant figures, drop the final digits 111 and add three zeros as placeholders to give 125 000 m^3 (or $1.25 \times 10^5\text{ m}^3$).
d. To express 58.703 m to three significant figures, drop the final digits 03 to give 58.7 m.

2.93 The total mass is the sum of the individual components. 137.25 g + 84 g + 43.7 g = 265 g
No places to the right of the decimal point are allowed since the mass of the fudge sauce (84 g) has no digits to the right of the decimal point.

2.95 This problem requires several conversion factors. Let's take a look first at a possible unit plan. When you write out the unit plan, be sure you know a conversion factor you can use for each step.
Given 7500 ft **Need** minutes

Plan ft → in. → cm → m → min $\frac{12\text{ in.}}{1\text{ ft}}$ $\frac{2.54\text{ cm}}{1\text{ in.}}$ $\frac{1\text{ m}}{100\text{ cm}}$ $\frac{1\text{ min}}{55.0\text{ m}}$

Set Up $7500\text{ ft} \times \frac{12\text{ in.}}{1\text{ ft}} \times \frac{2.54\text{ cm}}{1\text{ in.}} \times \frac{1\text{ m}}{100\text{ cm}} \times \frac{1\text{ min}}{55.0\text{ m}} = 42\text{ min (2 SFs)}$

2.97 Given 4.0 lb of onions **Need** number of onions

Plan lb → g → number of onions $\frac{453.6\text{ g}}{1\text{ lb}}$ $\frac{1\text{ onion}}{115\text{ g}}$

Set Up $4.0\ \cancel{\text{lb onions}} \times \frac{453.6\ \cancel{\text{g}}}{1\ \cancel{\text{lb}}} \times \frac{1\text{ onion}}{115\ \cancel{\text{g}}} = 16\text{ onions (2 SFs)}$

2.99 Given 0.45 lb **Need** cents

Plan lb → kg → pesos → dollars → cents $\frac{1\text{ kg}}{2.205\text{ lb}}$ $\frac{48\text{ pesos}}{1\text{ kg}}$ $\frac{1\text{ dollar}}{13.0\text{ pesos}}$ $\frac{100\text{ cents}}{1\text{ dollar}}$

Set Up $0.45\ \cancel{\text{lb}} \times \frac{1\ \cancel{\text{kg}}}{2.205\ \cancel{\text{lb}}} \times \frac{48\ \cancel{\text{pesos}}}{1\ \cancel{\text{kg}}} \times \frac{1\ \cancel{\text{dollar}}}{13.0\ \cancel{\text{pesos}}} \times \frac{100\text{ cents}}{1\ \cancel{\text{dollar}}} = 75\text{ cents (2 SFs)}$

2.101 Given 65 kg of body mass **Need** pounds of fat

Plan kg of body mass → kg of fat → lb of fat

(percent equality: 100 kg of body mass = 3.0 kg of fat)

$\frac{3.0\text{ kg fat}}{100\text{ kg body mass}}$ $\frac{2.205\text{ lb fat}}{1\text{ kg fat}}$

Set Up $65\ \cancel{\text{kg body mass}} \times \frac{3.0\ \cancel{\text{kg fat}}}{100\ \cancel{\text{kg body mass}}} \times \frac{2.205\text{ lb fat}}{1\ \cancel{\text{kg fat}}} = 4.3\text{ lb of fat (2 SFs)}$

2.103 Given 27.0 cm^3 of sterling silver **Need** ounces of pure silver

Plan cm^3 of sterling → g of sterling → g of silver → lb → oz of silver

(percent equality: 100 g of sterling = 92.5 g of silver)

$\frac{10.3\text{ g sterling}}{1\text{ cm}^3\text{ sterling}}$ $\frac{92.5\text{ g silver}}{100\text{ g sterling}}$ $\frac{1\text{ lb}}{453.6\text{ g}}$ $\frac{16\text{ oz}}{1\text{ lb}}$

Set Up $27.0\ \cancel{\text{cm}^3\text{ sterling}} \times \frac{10.3\ \cancel{\text{g sterling}}}{1\ \cancel{\text{cm}^3\text{ sterling}}} \times \frac{92.5\ \cancel{\text{g silver}}}{100\ \cancel{\text{g sterling}}} \times \frac{1\ \cancel{\text{lb}}}{453.6\ \cancel{\text{g}}} \times \frac{16\text{ oz}}{1\ \cancel{\text{lb}}}$

$= 9.07\text{ oz of pure silver (3 SFs)}$

2.105 Given 1254 aluminum cans **Need** liters of aluminum

Plan number of cans → lb → g → mL → L $\frac{1\text{ lb}}{22\text{ cans}}$ $\frac{453.6\text{ g}}{1\text{ lb}}$ $\frac{1\text{ mL}}{2.70\text{ g}}$ $\frac{1\text{ L}}{1000\text{ mL}}$

Set Up $1254\ \cancel{\text{cans}} \times \frac{1\ \cancel{\text{lb}}}{22\ \cancel{\text{cans}}} \times \frac{453.6\ \cancel{\text{g}}}{1\ \cancel{\text{lb}}} \times \frac{1\ \cancel{\text{mL}}}{2.70\ \cancel{\text{g}}} \times \frac{1\text{ L}}{1000\ \cancel{\text{mL}}}$

$= 9.58\text{ L of aluminum (3 SFs)}$

2.107 Given 215 mL initial, 285 mL final **Need** grams of lead

Plan calculate the volume by difference and mL → g $\frac{11.3\text{ g}}{1\text{ mL}}$

Set Up The difference between the initial volume of the water and its volume with the lead object will give us the volume of the lead object: 285 mL total − 215 mL water = 70. mL lead

then $70.\ \cancel{\text{mL lead}} \times \frac{11.3\text{ g lead}}{1\ \cancel{\text{mL lead}}} = 790\text{ g of lead (2 SFs)}$

2.109 Because the balance can measure mass to 0.001 g, the mass should be reported to 0.001 g. You should record the mass of the object as 32.075 g.

2.111 Given 325 tubes **Need** kilograms of benzyl salicylate

Plan tubes → oz of sunscreen → lb of sunscreen → kg of sunscreen → kg of benzyl salicylate

$\frac{4\text{ oz sunscreen}}{1\text{ tube}}$ $\frac{1\text{ lb}}{16\text{ oz}}$ $\frac{1\text{ kg}}{2.205\text{ lb}}$ $\frac{2.50\text{ kg benzyl salicylate}}{100\text{ kg sunscreen}}$

Set Up $325 \cancel{\text{tubes}} \times \frac{4.0 \cancel{\text{oz sunscreen}}}{1 \cancel{\text{tube}}} \times \frac{1 \cancel{\text{lb}}}{16 \cancel{\text{oz}}} \times \frac{1 \cancel{\text{kg}}}{2.205 \cancel{\text{lb}}} \times \frac{2.50 \text{ kg benzyl salicylate}}{100 \cancel{\text{kg sunscreen}}}$
$= 0.92$ kg of benzyl salicylate (2 SFs)

2.113 Given 66.7 yd long, 12 in. wide, 0.000 30 in. thick **Need** grams of aluminum

Plan Length: yd → ft → in. → cm $\frac{3 \text{ ft}}{1 \text{ yd}} \quad \frac{12 \text{ in.}}{1 \text{ ft}} \quad \frac{2.54 \text{ cm}}{1 \text{ in.}}$ and

width and thickness: in. → cm $\frac{2.54 \text{ cm}}{1 \text{ in.}}$

then calculate volume: cm^3 → g $\frac{2.70 \text{ g}}{1 \text{ cm}^3}$ $V = l \times w \times h$

Set Up Length: $66.7 \cancel{\text{yd}} \times \frac{3 \cancel{\text{ft}}}{1 \cancel{\text{yd}}} \times \frac{12 \cancel{\text{in.}}}{1 \cancel{\text{ft}}} \times \frac{2.54 \text{ cm}}{1 \cancel{\text{in.}}} = 6100 \text{ cm (2 SFs)}$

Width: $12 \cancel{\text{in.}} \times \frac{2.54 \text{ cm}}{1 \cancel{\text{in.}}} = 30. \text{ cm (2 SFs)}$

Thickness: $0.000\,30 \cancel{\text{in.}} \times \frac{2.54 \text{ cm}}{1 \cancel{\text{in.}}} = 0.000\,76 \text{ cm (2 SFs)}$

Volume $= 6100 \text{ cm} \times 30. \text{ cm} \times 0.000\,76 \text{ cm} = 140 \text{ cm}^3$, then

$140 \cancel{\text{cm}^3} \times \frac{2.70 \text{ g}}{1 \cancel{\text{cm}^3}} = 3.8 \times 10^2$ g of aluminum (2 SFs)

2.115 a. Given 180 lb of body mass **Need** cups of coffee

Plan lb of body mass → kg of body mass → mg of caffeine → fl oz of coffee → cups of coffee

$\frac{1 \text{ kg body mass}}{2.205 \text{ lb body mass}} \quad \frac{192 \text{ mg caffeine}}{1 \text{ kg body mass}} \quad \frac{6 \text{ fl oz coffee}}{100. \text{ mg caffeine}} \quad \frac{1 \text{ cup coffee}}{12 \text{ fl oz coffee}}$

Set Up $180 \cancel{\text{lb body mass}} \times \frac{1 \cancel{\text{kg body mass}}}{2.205 \cancel{\text{lb body mass}}} \times \frac{192 \text{ mg caffeine}}{1 \cancel{\text{kg body mass}}}$
$= 1.567 \times 10^4$ mg of caffeine (2 SFs allowed)

$1.567 \times 10^4 \cancel{\text{mg caffeine}} \times \frac{6 \cancel{\text{fl oz coffee}}}{100. \cancel{\text{mg caffeine}}} \times \frac{1 \text{ cup coffee}}{12 \cancel{\text{fl oz coffee}}}$
$= 78$ cups of coffee (2 SFs)

b. Given mg of caffeine from part **a** **Need** cans of cola

Plan mg of caffeine → cans of cola $\frac{1 \text{ can cola}}{50. \text{ mg caffeine}}$

Set Up $1.567 \times 10^4 \cancel{\text{mg caffeine}} \times \frac{1 \text{ can cola}}{50. \cancel{\text{mg caffeine}}} = 310$ cans of cola (2 SFs)

c. Given mg of caffeine from part **a** **Need** number of No-Doz tablets

Plan mg of caffeine → No-Doz tablets $\frac{1 \text{ No-Doz tablet}}{100. \text{ mg caffeine}}$

Set Up $1.567 \times 10^4 \cancel{\text{mg caffeine}} \times \frac{1 \text{ No-Doz tablet}}{100. \cancel{\text{mg caffeine}}} = 160$ No-Doz tablets (2 SFs)

2.117 Given 1.00 L of gasoline **Need** cm^3 of olive oil

Plan L of gasoline → mL of gasoline → g of gasoline → g of olive oil → cm^3 of olive oil

$\frac{1000 \text{ mL gasoline}}{1 \text{ L gasoline}} \quad \frac{0.74 \text{ g gasoline}}{1 \text{ mL gasoline}} \quad \frac{1 \text{ g olive oil}}{1 \text{ g gasoline}} \quad \frac{1 \text{ cm}^3 \text{ olive oil}}{0.92 \text{ g olive oil}}$

Set Up $1.00 \text{ L gasoline} \times \frac{1000 \text{ mL}}{1 \text{ L}} \times \frac{0.74 \text{ g gasoline}}{1 \text{ mL gasoline}} \times \frac{1 \text{ g olive oil}}{1 \text{ g gasoline}} \times \frac{1 \text{ cm}^3 \text{ olive oil}}{0.92 \text{ g olive oil}}$
$= 800 \text{ cm}^3$ $(8.0 \times 10^2 \text{ cm}^3)$ of olive oil (2 SFs)

2.119 Given 155 mL initially, 15.0 g of iron, 20.0 g of lead **Need** final volume (mL)
Plan for both iron and lead g → mL, then calculate the final volume using addition
$\frac{1 \text{ cm}^3 \text{ iron}}{7.86 \text{ g iron}}$ $\frac{1 \text{ cm}^3 \text{ lead}}{11.3 \text{ g lead}}$

Set Up $15.0 \text{ g iron} \times \frac{1 \text{ cm}^3 \text{ iron}}{7.86 \text{ g iron}} \times \frac{1 \text{ mL}}{1 \text{ cm}^3} = 1.91 \text{ mL of iron (3 SFs)}$

$20.0 \text{ g lead} \times \frac{1 \text{ cm}^3 \text{ lead}}{11.3 \text{ g lead}} \times \frac{1 \text{ mL}}{1 \text{ cm}^3} = 1.77 \text{ mL of lead (3 SFs)}$

∴ Final volume = 155 mL water + 1.91 mL iron + 1.77 mL lead = 159 mL
(no places to the right of the decimal point are allowed)

3

Matter and Energy

Learning Goals

- Classify examples of matter as pure substances or mixtures.
- Identify the states and the physical and chemical properties of matter.
- Given a temperature, calculate a corresponding value on another temperature scale.
- Identify energy as potential or kinetic; convert between units of energy.
- Use specific heat to calculate heat loss or gain, temperature change, or mass of a sample.
- Use the energy values to calculate the kilojoules (kJ) or kilocalories (kcal) in a food.

Chapter Outline

Chapter Opener: Dietitian

3.1 Classification of Matter

Chemistry Link to Health: Breathing Mixtures for Scuba

3.2 States and Properties of Matter

3.3 Temperature

Chemistry Link to Health: Variation in Body Temperature

3.4 Energy

3.5 Specific Heat

Chemistry Link to the Environment: Carbon Dioxide and Global Warming

3.6 Energy and Nutrition

Chemistry Link to Health: Losing and Gaining Weight

Core Chemistry Skills

- Classifying Matter (3.1)
- Identifying Physical and Chemical Changes (3.2)
- Converting between Temperature Scales (3.3)
- Using Energy Units (3.4)
- Calculating Specific Heat (3.5)
- Using the Heat Equation (3.5)

Answers and Solutions to Text Problems

3.1 A *pure substance* is matter that has a fixed or definite composition: either elements or compounds. In a *mixture*, two or more substances are physically mixed but not chemically combined.

a. Baking soda is composed of one type of matter ($NaHCO_3$), which makes it a pure substance.

b. A blueberry muffin is composed of several substances mixed together, which makes it a mixture.

c. Ice is composed of one type of matter (H_2O molecules), which makes it a pure substance.

d. Zinc is composed of one type of matter (Zn atoms), which makes it a pure substance.

e. Trimix is a physical mixture of oxygen, nitrogen, and helium gases, which makes it a mixture.

3.3 *Elements* are the simplest type of pure substance, containing only one type of atom. *Compounds* contain two or more elements chemically combined in a specific proportion.
a. A silicon (Si) chip is an element since it contains only one type of atom (Si).
b. Hydrogen peroxide (H_2O_2) is a compound since it contains two elements (H, O) chemically combined.
c. Oxygen (O_2) is an element since it contains only one type of atom (O).
d. Rust (Fe_2O_3) is a compound since it contains two elements (Fe, O) chemically combined.
e. Methane (CH_4) in natural gas is a compound since it contains two elements (C, H) chemically combined.

3.5 A *homogeneous mixture* has a uniform composition; a *heterogeneous mixture* does not have a uniform composition throughout the mixture.
a. Vegetable soup is a heterogeneous mixture since it has chunks of vegetables.
b. Seawater is a homogeneous mixture since it has a uniform composition.
c. Tea is a homogeneous mixture since it has a uniform composition.
d. Tea with ice and lemon slices is a heterogeneous mixture since it has chunks of ice and lemon.
e. Fruit salad is a heterogeneous mixture since it has chunks of fruit.

3.7 **a.** A gas has no definite volume or shape.
b. In a gas, the particles do not interact with each other.
c. In a solid, the particles are held in a rigid structure.

3.9 A *physical property* is a characteristic of the substance such as color, shape, odor, luster, size, melting point, and density. A *chemical property* is a characteristic that indicates the ability of a substance to change into a new substance.
a. Color and physical state are physical properties.
b. The ability to react with oxygen is a chemical property.
c. The freezing point of a substance is a physical property.
d. Milk souring describes chemical reactions and is thus a chemical property.
e. Burning butane gas in oxygen forms new substances, which makes it a chemical property.

3.11 When matter undergoes a *physical change*, its state or appearance changes, but its composition remains the same. When a *chemical change* occurs, the original substance is converted into a new substance, which has different physical and chemical properties.
a. Water vapor condensing is a physical change since the physical form of the water changes, but the composition of the substance does not.
b. Cesium metal reacting is a chemical change since new substances form.
c. Gold melting is a physical change since the physical state changes, but not the composition of the substance.
d. Cutting a puzzle results in a physical change since the size and shape change, but not the composition of the substance.
e. Grating cheese results in a physical change since the size and shape change, but not the composition of the substance.

3.13 **a.** The high reactivity of fluorine is a chemical property since it allows for the formation of new substances.
b. The physical state of fluorine is a physical property.
c. The color of fluorine is a physical property.
d. The reactivity of fluorine with hydrogen is a chemical property since it allows for the formation of new substances.
e. The melting point of fluorine is a physical property.

3.15 The Fahrenheit temperature scale is still used in the United States. A normal body temperature is 98.6 °F on this scale. To convert her 99.8 °F temperature to the equivalent reading on the Celsius scale, the following calculation must be performed:

$$T_C = \frac{(99.8 - 32)}{1.8} = \frac{67.8}{1.8} = 37.7\ °C\ (3\ SFs)\ (1.8 \text{ and } 32 \text{ are exact numbers})$$

Because a normal body temperature is 37.0 on the Celsius scale, her temperature of 37.7 °C would indicate a mild fever.

3.17 To convert Celsius to Fahrenheit: $T_F = 1.8(T_C) + 32$

To convert Fahrenheit to Celsius: $T_C = \frac{(T_F - 32)}{1.8}$ (1.8 and 32 are exact numbers)

To convert Celsius to Kelvin: $T_K = T_C + 273$

To convert Kelvin to Celsius: $T_C = T_K - 273$

a. $T_F = 1.8(T_C) + 32 = 1.8(37.0) + 32 = 66.6 + 32 = 98.6\ °F$

b. $T_C = \frac{(T_F - 32)}{1.8} = \frac{(65.3 - 32)}{1.8} = \frac{33.3}{1.8} = 18.5\ °C$

c. $T_K = T_C + 273 = -27 + 273 = 246\ K$

d. $T_C = T_K - 273 = 224 - 273 = -49\ °C$

e. $T_C = \frac{(T_F - 32)}{1.8} = \frac{(15 - 32)}{1.8} = \frac{-17}{1.8} = -9.4\ °C$

3.19 **a.** $T_C = \frac{(T_F - 32)}{1.8} = \frac{(106 - 32)}{1.8} = \frac{74}{1.8} = 41\ °C$

b. $T_C = \frac{(T_F - 32)}{1.8} = \frac{(103 - 32)}{1.8} = \frac{71}{1.8} = 39\ °C$

No, there is no need to phone the doctor. The child's temperature is less than 40.0 °C.

3.21 When the car is at the top of the ramp, it has its maximum potential energy. As it descends, potential energy is converted into kinetic energy. At the bottom, all of its energy is kinetic.

3.23 **a.** Potential energy is stored in the water at the top of the waterfall.
b. Kinetic energy is displayed as the kicked ball moves.
c. Potential energy is stored in the chemical bonds in the coal.
d. Potential energy is stored when the skier is at the top of the hill.

3.25 **a. Given** 3.0 h, 270 kJ/h **Need** joules

Plan h → kJ → J $\frac{1000\ J}{1\ kJ}$

Set Up $3.0\ \cancel{h} \times \frac{270\ \cancel{kJ}}{1.0\ \cancel{h}} \times \frac{1000\ J}{1\ \cancel{kJ}} = 8.1 \times 10^5\ J\ (2\ SFs)$

b. Given 3.0 h, 270 kJ/h **Need** kilocalories

Plan h → kJ → J → cal → kcal $\frac{1000\ J}{1\ kJ}$ $\frac{1\ cal}{4.184\ J}$ $\frac{1\ kcal}{1000\ cal}$

Set Up $3.0\ \cancel{h} \times \frac{270\ \cancel{kJ}}{1.0\ \cancel{h}} \times \frac{1000\ \cancel{J}}{1\ \cancel{kJ}} \times \frac{1\ \cancel{cal}}{4.184\ \cancel{J}} \times \frac{1\ kcal}{1000\ \cancel{cal}} = 190\ kcal\ (2\ SFs)$

3.27 **a. Given** 3500 cal **Need** kilocalories

Plan cal → kcal $\frac{1\ kcal}{1000\ cal}$

Set Up $3500\ \cancel{cal} \times \frac{1\ kcal}{1000\ \cancel{cal}} = 3.5\ kcal\ (2\ SFs)$

b. Given 415 J **Need** calories

Plan J → cal $\frac{1\ \text{cal}}{4.184\ \text{J}}$

Set Up $415\ \cancel{\text{J}} \times \frac{1\ \text{cal}}{4.184\ \cancel{\text{J}}} = 99.2\ \text{cal (3 SFs)}$

c. Given 28 cal **Need** joules

Plan cal → J $\frac{4.184\ \text{J}}{1\ \text{cal}}$

Set Up $28\ \cancel{\text{cal}} \times \frac{4.184\ \text{J}}{1\ \cancel{\text{cal}}} = 120\ \text{J (2 SFs)}$

d. Given 4.5 kJ **Need** calories

Plan kJ → J → cal $\frac{1000\ \text{J}}{1\ \text{kJ}}$ $\frac{1\ \text{cal}}{4.184\ \text{J}}$

Set Up $4.5\ \cancel{\text{kJ}} \times \frac{1000\ \cancel{\text{J}}}{1\ \cancel{\text{kJ}}} \times \frac{1\ \text{cal}}{4.184\ \cancel{\text{J}}} = 1100\ \text{cal (2 SFs)}$

3.29 Copper, which has the lowest specific heat of the samples, would reach the highest temperature.

3.31 a. Given heat = 312 J; mass = 13.5 g; ΔT = 83.6 °C − 24.2 °C = 59.4 °C
Need specific heat (J/g °C)

Plan $SH = \frac{\text{heat}}{\text{mass} \times \Delta T}$

Set Up specific heat (SH) $= \frac{312\ \text{J}}{13.5\ \text{g} \times 59.4\ °\text{C}} = \frac{0.389\ \text{J}}{\text{g}\ °\text{C}}$ (3 SFs)

b. Given heat = 345 J; mass = 48.2 g; ΔT = 57.9 °C − 35.0 °C = 22.9 °C
Need specific heat (J/g °C)

Plan $SH = \frac{\text{heat}}{\text{mass} \times \Delta T}$

Set Up specific heat (SH) $= \frac{345\ \text{J}}{48.2\ \text{g} \times 22.9\ °\text{C}} = \frac{0.313\ \text{J}}{\text{g}\ °\text{C}}$ (3 SFs)

3.33 a. Given SH_{water} = 4.184 J/g °C; mass = 25.0 g; ΔT = 25.7 °C − 12.5 °C = 13.2 °C
Need heat (q) in joules
Plan heat (q) = mass × ΔT × SH

Set Up heat $= 25.0\ \cancel{\text{g}} \times 13.2\ \cancel{°\text{C}} \times \frac{4.184\ \text{J}}{\cancel{\text{g}}\ \cancel{°\text{C}}} = 1380\ \text{J (3 SFs)}$

∴ 1380 J of energy are required.

b. Given SH_{copper} = 0.385 J/g °C; mass = 38.0 g; ΔT = 246 °C − 122 °C = 124 °C
Need heat (q) in joules
Plan heat (q) = mass × ΔT × SH

Set Up heat $= 38.0\ \cancel{\text{g}} \times 124\ \cancel{°\text{C}} \times \frac{0.385\ \text{J}}{\cancel{\text{g}}\ \cancel{°\text{C}}} = 1810\ \text{J (3 SFs)}$

∴ 1810 J of energy are required.

3.35 a. Given $SH_{ethanol}$ = 2.46 J/g °C; mass = 15.0 g; ΔT = −42.0 °C − 60.5 °C = −102.5 °C
Need heat (q) in joules
Plan heat (q) = mass × ΔT × SH

Set Up heat $= 15.0\,\cancel{g} \times (-102.5\,\cancel{°C}) \times \dfrac{2.46\text{ J}}{\cancel{g}\,\cancel{°C}} = -3780\text{ J (3 SFs)}$

∴ 3780 J of energy are lost.

b. Given $SH_{iron} = 0.452$ J/g °C; mass = 125 g; $\Delta T = 55\ °C - 118\ °C = -63\ °C$
Need heat (q) in joules
Plan heat (q) = mass × ΔT × SH

Set Up heat $= 125\,\cancel{g} \times (-63\,\cancel{°C}) \times \dfrac{0.452\text{ J}}{\cancel{g}\,\cancel{°C}} = -3600\text{ J (2 SFs)}$

∴ 3600 J of energy are lost.

3.37 a. Given $SH_{gold} = 0.129$ J/g °C; heat = 225 J; $\Delta T = 47.0\ °C - 15.0\ °C = 32.0\ °C$
Need mass in grams

Plan heat (q) = mass × ΔT × SH ∴ mass $= \dfrac{\text{heat}}{\Delta T \times SH}$

Set Up mass $= \dfrac{\text{heat}}{\Delta T \times SH} = \dfrac{225\,\cancel{J}}{32.0\,\cancel{°C} \times \dfrac{0.129\,\cancel{J}}{g\,\cancel{°C}}} = 54.5\text{ g of gold (3 SFs)}$

b. Given $SH_{iron} = 0.452$ J/g °C; heat = −8.40 kJ; $\Delta T = 82.0\ °C - 168.0\ °C = -86.0\ °C$
Need mass in grams

Plan heat (q) = mass × ΔT × SH ∴ mass $= \dfrac{\text{heat}}{\Delta T \times SH}$ $\dfrac{1000\text{ J}}{1\text{ kJ}}$

Set Up mass $= \dfrac{\text{heat}}{\Delta T \times SH} = \dfrac{(-8.40\,\cancel{kJ}) \times \dfrac{1000\,\cancel{J}}{1\,\cancel{kJ}}}{(-86.0\,\cancel{°C}) \times \dfrac{0.452\,\cancel{J}}{g\,\cancel{°C}}} = 216\text{ g of iron (3 SFs)}$

3.39 a. Given $SH_{aluminum} = 0.897$ J/g °C; heat = 8.80 kJ; $\Delta T = 26.8\ °C - 12.5\ °C = 14.3\ °C$
Need mass in grams

Plan heat (q) = mass × ΔT × SH ∴ mass $= \dfrac{\text{heat}}{\Delta T \times SH}$ $\dfrac{1000\text{ J}}{1\text{ kJ}}$

Set Up mass $= \dfrac{\text{heat}}{\Delta T \times SH} = \dfrac{8.80\,\cancel{kJ} \times \dfrac{1000\,\cancel{J}}{1\,\cancel{kJ}}}{14.3\,\cancel{°C} \times \dfrac{0.897\,\cancel{J}}{g\,\cancel{°C}}} = 686\text{ g of aluminum (3 SFs)}$

b. Given $SH_{titanium} = 0.523$ J/g °C; heat = −14 200 J; $\Delta T = 42\ °C - 185\ °C = -143\ °C$
Need mass in grams

Plan heat (q) = mass × ΔT × SH ∴ mass $= \dfrac{\text{heat}}{\Delta T \times SH}$

Set Up mass $= \dfrac{\text{heat}}{\Delta T \times SH} = \dfrac{(-14\,200\,\cancel{J})}{(-143\,\cancel{°C}) \times \dfrac{0.523\,\cancel{J}}{g\,\cancel{°C}}} = 190.\text{ g of titanium (3 SFs)}$

3.41 a. Given $SH_{iron} = 0.452$ J/g °C; heat = 1580 J; mass = 20.0 g
Need ΔT

Plan heat (q) = mass × ΔT × SH ∴ $\Delta T = \dfrac{\text{heat}}{\text{mass} \times SH}$

Set Up $\Delta T = \dfrac{\text{heat}}{\text{mass} \times SH} = \dfrac{1580\,\cancel{J}}{20.0\,\cancel{g} \times \dfrac{0.452\,\cancel{J}}{\cancel{g}\,°C}} = 175\ °C\text{ (3 SFs)}$

b. Given SH_{water} = 4.184 J/g °C; heat = 7.10 kJ; mass = 150.0 g
Need ΔT

Plan heat (q) = mass × ΔT × SH $\quad \therefore \Delta T = \dfrac{\text{heat}}{\text{mass} \times SH} \quad \dfrac{1000\text{ J}}{1\text{ kJ}}$

Set Up $\Delta T = \dfrac{\text{heat}}{\text{mass} \times SH} = \dfrac{7.10\text{ kJ} \times \dfrac{1000\text{ J}}{1\text{ kJ}}}{150.0\text{ g} \times \dfrac{4.184\text{ J}}{\text{g °C}}} = 11.3\text{ °C (3 SFs)}$

3.43 a. Given SH_{gold} = 0.129 J/g °C; heat = 7680 J; mass = 85.0 g **Need** ΔT

Plan heat (q) = mass × ΔT × SH $\quad \therefore \Delta T = \dfrac{\text{heat}}{\text{mass} \times SH}$

Set Up $\Delta T = \dfrac{\text{heat}}{\text{mass} \times SH} = \dfrac{7680\text{ J}}{85.0\text{ g} \times \dfrac{0.129\text{ J}}{\text{g °C}}} = 700.\text{ °C (3 SFs)}$

b. Given SH_{copper} = 0.385 J/g °C; heat = 6.75 kJ; mass = 50.0 g **Need** ΔT

Plan heat (q) = mass × ΔT × SH $\quad \therefore \Delta T = \dfrac{\text{heat}}{\text{mass} \times SH} \quad \dfrac{1000\text{ J}}{1\text{ kJ}}$

Set Up $\Delta T = \dfrac{\text{heat}}{\text{mass} \times SH} = \dfrac{6.75\text{ kJ} \times \dfrac{1000\text{ J}}{1\text{ kJ}}}{50.0\text{ g} \times \dfrac{0.385\text{ J}}{\text{g °C}}} = 351\text{ °C (3 SFs)}$

3.45 a. Given mass = 505 g of water; ΔT = 35.7 °C − 25.2 °C = 10.5 °C;
SH_{water} = 4.184 J/g °C **Need** energy (q) in kilojoules
Plan heat (q) = mass × ΔT × SH
Set Up $505\text{ g} \times 10.5\text{ °C} \times \dfrac{4.184\text{ J}}{\text{g °C}} \times \dfrac{1\text{ kJ}}{1000\text{ J}} = 22.2\text{ kJ (3 SFs)}$

b. Given mass = 4980 g of water; ΔT = 62.4 °C − 20.6 °C = 41.8 °C;
SH_{water} = 4.184 J/g °C **Need** energy (q) in kilojoules
Plan heat (q) = mass × ΔT × SH
Set Up $4980\text{ g} \times 41.8\text{ °C} \times \dfrac{4.184\text{ J}}{\text{g °C}} \times \dfrac{1\text{ kJ}}{1000\text{ J}} = 871\text{ kJ (3 SFs)}$

3.47 a. Given 1 cup of orange juice that contains 26 g of carbohydrate, 2 g of protein, and no fat
Need total energy (q) in kilojoules and kilocalories

Food Type	Mass		Energy Value		Energy (rounded off to the tens place)
Carbohydrate	26 g	×	$\dfrac{17\text{ kJ (or 4 kcal)}}{1\text{ g}}$	=	440 kJ (or 100 kcal)
Protein	2 g	×	$\dfrac{17\text{ kJ (or 4 kcal)}}{1\text{ g}}$	=	30 kJ (or 10 kcal)
			Total energy content	=	470 kJ (or 110 kcal)

b. Given one apple that provides 72 kcal of energy and contains no fat or protein
Need grams of carbohydrate

Plan kcal → g of carbohydrate $\frac{1 \text{ g of carbohydrate}}{4 \text{ kcal}}$

Set Up $72 \text{ kcal} \times \frac{1 \text{ g carbohydrate}}{4 \text{ kcal}} = 18 \text{ g of carbohydrate (2 SFs)}$

3.49 a. Given 1 tablespoon of vegetable oil that contains 14 g of fat, and no carbohydrate or protein
Need total energy (q) in kilojoules and kilocalories

Food Type	Mass		Energy Value		Energy (rounded off to the tens place)
Fat	14 g	×	$\frac{38 \text{ kJ (or 9 kcal)}}{1 \text{ g}}$	=	530 kJ (or 130 kcal)

b. Given a diet that consists of 68 g of carbohydrate, 9.0 g of fat, and 150 g of protein
Need total energy (q) in kJ and kcal

Food Type	Mass		Energy Value		Energy (rounded off to the tens place)
Carbohydrate	68 g	×	$\frac{17 \text{ kJ (or 4 kcal)}}{1 \text{ g}}$	=	1160 kJ (or 270 kcal)
Fat	9 g	×	$\frac{38 \text{ kJ (or 9 kcal)}}{1 \text{ g}}$	=	340 kJ (or 80 kcal)
Protein	150 g	×	$\frac{17 \text{ kJ (or 4 kcal)}}{1 \text{ g}}$	=	2550 kJ (or 600 kcal)
			Total energy content	=	4050 kJ (or 950 kcal)

3.51 a. The diagram shows two different types of atoms chemically combined in a definite proportion; it represents a compound.
b. The diagram shows two different types of atoms physically mixed, but not chemically combined; it represents a mixture.
c. The diagram contains only one type of atom; it represents an element.

3.53 A *homogeneous mixture* has a uniform composition; a *heterogeneous mixture* does not have a uniform composition.
a. Lemon-flavored water is a homogeneous mixture since it has a uniform composition (as long as there are no lemon pieces).
b. Stuffed mushrooms are a heterogeneous mixture since they include both mushrooms and chunks of filling.
c. Eye drops are a homogeneous mixture since they have a uniform composition.

3.55 $T_C = \frac{(T_F - 32)}{1.8} = \frac{(155 - 32)}{1.8} = \frac{123}{1.8} = 68.3\text{ °C (3 SFs)}$
(1.8 and 32 are exact numbers)

3.57 Given 10.0 cm^3 cubes of gold ($SH_{\text{gold}} = 0.129$ J/g °C), and aluminum ($SH_{\text{aluminum}} =$ 0.897 J/g °C); $\Delta T = 25\text{ °C} - 15\text{ °C} = 10.\text{ °C}$
Need energy in joules and calories

Plan cm^3 → g heat = mass × ΔT × SH $\frac{19.3 \text{ g gold}}{1 \text{ cm}^3}$ $\frac{2.70 \text{ g aluminum}}{1 \text{ cm}^3}$

Set Up <u>for gold:</u> $10.0\ \cancel{cm^3} \times \dfrac{19.3\ \cancel{g}}{1\ \cancel{cm^3}} \times 10.\ \cancel{°C} \times \dfrac{0.129\ J}{\cancel{g}\ \cancel{°C}} = 250\ J\ (2\ SFs)$

$$10.0\ \cancel{cm^3} \times \frac{19.3\ \cancel{g}}{1\ \cancel{cm^3}} \times 10.\ \cancel{°C} \times \frac{0.129\ \cancel{J}}{\cancel{g}\ \cancel{°C}} \times \frac{1\ cal}{4.184\ \cancel{J}} = 60.\ cal\ (2\ SFs)$$

<u>for aluminum:</u> $10.0\ \cancel{cm^3} \times \dfrac{2.70\ \cancel{g}}{1\ \cancel{cm^3}} \times 10.\ \cancel{°C} \times \dfrac{0.897\ J}{\cancel{g}\ \cancel{°C}} = 240\ J\ (2\ SFs)$

$$10.0\ \cancel{cm^3} \times \frac{2.70\ \cancel{g}}{1\ \cancel{cm^3}} \times 10.\ \cancel{°C} \times \frac{0.897\ \cancel{J}}{\cancel{g}\ \cancel{°C}} \times \frac{1\ cal}{4.184\ \cancel{J}} = 58\ cal\ (2SFs)$$

Thus, the energy needed to heat each of the metals is almost the same.

3.59 **a.** **Given** a meal consisting of: cheeseburger: 46 g of carbohydrate, 40. g of fat, 47 g of protein
french fries: 47 g of carbohydrate, 16 g of fat, 4 g of protein
chocolate shake: 76 g of carbohydrate, 10. g of fat, 10. g of protein

Need energy from each food type in kilocalories
Plan total grams of each food type, then g → kcal
Set Up total carbohydrate = 46 g + 47 g + 76 g = 169 g
total fat = 40. g + 16 g + 10. g = 66 g
total protein = 47 g + 4 g + 10. g = 61 g

Food Type	Mass		Energy Value		Energy (rounded off to the tens place)
Carbohydrate	$169\ \cancel{g}$	×	$\dfrac{4\ kcal}{1\ \cancel{g}}$	=	680 kcal
Fat	$66\ \cancel{g}$	×	$\dfrac{9\ kcal}{1\ \cancel{g}}$	=	590 kcal
Protein	$61\ \cancel{g}$	×	$\dfrac{4\ kcal}{1\ \cancel{g}}$	=	240 kcal

b. total energy for the meal = 680 kcal + 590 kcal + 240 kcal = 1510 kcal

c. **Given** 1510 kcal from meal **Need** hours of sleeping to "burn off"

Plan kcal → h $\dfrac{1\ h\ sleeping}{60\ kcal}$

Set Up $1510\ \cancel{kcal} \times \dfrac{1\ h\ sleeping}{60\ \cancel{kcal}} = 25\ h$ of sleeping (2 SFs)

d. **Given** 1510 kcal from meal **Need** hours of running to "burn off"

Plan kcal → h $\dfrac{1\ h\ running}{750\ kcal}$

Set Up $1510\ \cancel{kcal} \times \dfrac{1\ h\ running}{750\ \cancel{kcal}} = 2.0\ h$ of running (2 SFs)

3.61 *Elements* are the simplest type of pure substance, containing only one type of atom. *Compounds* contain two or more elements chemically combined in a definite proportion. In a *mixture*, two or more substances are physically mixed but not chemically combined.

a. Carbon in pencils is an element since it contains only one type of atom (C).
b. Carbon dioxide is a compound since it contains two elements (C, O) chemically combined.
c. Orange juice is composed of several substances physically mixed together (e.g. water, sugar, citric acid), which makes it a mixture.

3.63 A *homogeneous mixture* has a uniform composition; a *heterogeneous mixture* does not have a uniform composition throughout the mixture.

a. A hot fudge sundae is a heterogeneous mixture since it has ice cream, fudge sauce, and perhaps a cherry.
b. Herbal tea is a homogeneous mixture since it has a uniform composition.
c. Vegetable oil is a homogeneous mixture since it has a uniform composition.

3.65 **a.** A vitamin tablet is a solid.
b. Helium in a balloon is a gas.
c. Milk is a liquid.
d. Air is a mixture of gases.
e. Charcoal is a solid.

3.67 A *physical property* is a characteristic of the substance such as color, shape, odor, luster, size, melting point, and density. A *chemical property* is a characteristic that indicates the ability of a substance to change into a new substance.

a. The luster of gold is a physical property.
b. The melting point of gold is a physical property.
c. The ability of gold to conduct electricity is a physical property.
d. The ability of gold to form a new substance with sulfur is a chemical property.

3.69 When matter undergoes a *physical change*, its state or appearance changes, but its composition remains the same. When a *chemical change* occurs, the original substance is converted into a new substance, which has different physical and chemical properties.

a. Plant growth produces new substances, so it is a chemical change.
b. A change of state from solid to liquid is a physical change.
c. Chopping wood into smaller pieces is a physical change.
d. Burning wood, which forms new substances, results in a chemical change.

3.71 **a.** $T_C = \dfrac{(T_F - 32)}{1.8} = \dfrac{(134 - 32)}{1.8} = \dfrac{102}{1.8} = 56.7\ ^\circ\text{C}$

$T_K = T_C + 273 = 56.7 + 273 = 330.\ \text{K}$

b. $T_C = \dfrac{(T_F - 32)}{1.8} = \dfrac{(-69.7 - 32)}{1.8} = \dfrac{-101.7}{1.8} = -56.5\ ^\circ\text{C}$

$T_K = T_C + 273 = -56.5 + 273 = 217\ \text{K}$

3.73 $T_C = \dfrac{(T_F - 32)}{1.8} = \dfrac{(-15 - 32)}{1.8} = \dfrac{-47}{1.8} = -26\ ^\circ\text{C}$

$T_K = T_C + 273 = -26 + 273 = 247\ \text{K}$

3.75 **Given** 1 lb of body fat; 15% (m/m) water in body fat
Need kilocalories to "burn off"
Plan Because each gram of body fat contains 15% water, a person actually loses 85 grams of fat per hundred grams of body fat. (We considered 1 lb of fat as exactly 1 lb.)

lb of body fat → g of body fat → g of fat → kcal

$$\frac{453.6\ \text{g body fat}}{1\ \text{lb body fat}} \quad \frac{85\ \text{g fat}}{100\ \text{g body fat}} \quad \frac{9\ \text{kcal}}{1\ \text{g fat}}$$

Set Up $1\ \cancel{\text{lb body fat}} \times \dfrac{453.6\ \cancel{\text{g body fat}}}{1\ \cancel{\text{lb body fat}}} \times \dfrac{85\ \cancel{\text{g fat}}}{100\ \cancel{\text{g body fat}}} \times \dfrac{9\ \text{kcal}}{1\ \cancel{\text{g fat}}} = 3500\ \text{kcal (2 SFs)}$

3.77 Water has a higher specific heat than sand, which means that a large amount of energy is required to cause a significant temperature change. The lower specific heat of the sand means that even a small amount of energy will cause a significant temperature change, making it uncomfortable to walk on the hot sand.

3.79 **Given** mass = 725 g of water; SH_{water} = 4.184 J/g °C; ΔT = 37 °C − 65 °C = −28 °C
Need heat (q) in kilojoules

Plan heat (q) = mass × ΔT × SH, then J → kJ $\frac{1\text{ kJ}}{1000\text{ J}}$

Set Up $\text{Heat} = 725\text{ g} \times (-28\text{ °C}) \times \frac{4.184\text{ J}}{\text{g °C}} \times \frac{1\text{ kJ}}{1000\text{ J}} = -85\text{ kJ (2 SFs)}$

∴ 85 kJ are lost from the water bottle and are transferred to the muscles.

3.81 **Given** mass = 0.50 g of oil; 18.9 kJ produced
Need energy value (kcal/g)

Plan kJ → kcal $\frac{1000\text{ J}}{1\text{ kJ}}$ $\frac{1\text{ cal}}{4.184\text{ J}}$ $\frac{1\text{ kcal}}{1000\text{ cal}}$

then energy value = $\frac{\text{heat in kcal}}{\text{mass in g}}$

Set Up $18.9\text{ kJ} \times \frac{1000\text{ J}}{1\text{ kJ}} \times \frac{1\text{ cal}}{4.184\text{ J}} \times \frac{1\text{ kcal}}{1000\text{ cal}} = 4.52\text{ kcal (3 SFs)}$

∴ energy value = $\frac{\text{heat produced}}{\text{mass of oil}} = \frac{4.52\text{ kcal}}{0.50\text{ g oil}}$ = 9.0 kcal/g of vegetable oil (2 SFs)

3.83 **Given**

Metal	**Water**
Mass = 25.0 g	mass = 50.0 g
Initial temperature = 98.0 °C	initial temperature = 18.0 °C
Final temperature = 27.4 °C	final temperature = 27.4 °C
SH = ? J/g °C	SH = 4.184 J/g °C

Need specific heat (SH) of metal
Plan Heat lost by metal = heat gained by water
For both, heat (q) = mass × ΔT × SH
Set Up

For water: $q = m \times \Delta T \times SH = 50.0\text{ g} \times (27.4\text{ °C} - 18.0\text{ °C}) \times \frac{4.184\text{ J}}{\text{g °C}} = 1970\text{ J}$

For metal: $SH = \frac{-q}{\Delta T \times \text{mass}} = \frac{(-1970\text{ J})}{(27.4\text{ °C} - 98.0\text{ °C}) \times 25.0\text{ g}} = \frac{1.1\text{ J}}{\text{g °C}}\text{ (2 SFs)}$

3.85 **Given** 0.66 g of olive oil; 370 g of water; SH_{water} = 4.184 J/g °C;
ΔT = 38.8 °C − 22.7 °C = 16.1 °C
Need energy value (in kJ/g and kcal/g)

Plan heat (q) = mass × ΔT × SH then J → kJ $\frac{1\text{ kJ}}{1000\text{ J}}$

Set Up $\text{heat} = 370\text{ g} \times 16.1\text{ °C} \times \frac{4.184\text{ J}}{\text{g °C}} \times \frac{1\text{ kJ}}{1000\text{ J}} = 25\text{ kJ}$

∴ energy value = $\frac{\text{heat produced}}{\text{mass of oil}} = \frac{25\text{ kJ}}{0.66\text{ g oil}}$ = 38 kJ/g of olive oil (2 SFs)

$\text{heat} = 370\text{ g} \times 16.1\text{ °C} \times \frac{1\text{ cal}}{\text{g °C}} \times \frac{1\text{ kcal}}{1000\text{ cal}} = 6.0\text{ kcal}$

∴ energy value = $\frac{\text{heat produced}}{\text{mass of oil}} = \frac{6.0\text{ kcal}}{0.66\text{ g oil}}$ = 9.1 kcal/g of olive oil (2 SFs)

3.87 Given

Copper	Water
Mass = ? g	mass = 50.0 g
Initial temperature = 86.0 °C	initial temperature = 16.0 °C
Final temperature = 24.0 °C	final temperature = 24.0 °C
SH = 0.385 J/g °C	SH = 4.184 J/g °C

Need mass of copper in grams

Plan Heat lost by copper = heat gained by water

For both, heat (q) = mass × ΔT × SH

Set Up

For water: $q = m \times \Delta T \times SH = 50.0\,\text{g} \times (24.0\,°\text{C} - 16.0\,°\text{C}) \times \dfrac{4.184\ \text{J}}{\text{g}\,°\text{C}} = 1670\ \text{J}$

For copper: $\text{mass} = \dfrac{-q}{\Delta T \times SH} = \dfrac{(-1670\ \text{J})}{(24.0\,°\text{C} - 86.0\,°\text{C}) \times \left(\dfrac{0.385\ \text{J}}{\text{g}\,°\text{C}}\right)} = 70.\ \text{g of copper (2 SFs)}$

Answers to Combining Ideas from Chapters 1 to 3

CI.1 **a.** There are five significant figures in the weight of 294.10 troy ounces.

b. $$294.10\ \cancel{\text{troy ounces}} \times \frac{31.1035\ \text{g}}{1\ \cancel{\text{troy ounce}}} = 9147.5\ \text{g}\ (5\ \text{SFs})$$

$$294.10\ \cancel{\text{troy ounces}} \times \frac{31.1035\ \cancel{\text{g}}}{1\ \cancel{\text{troy ounce}}} \times \frac{1\ \text{kg}}{1000\ \cancel{\text{g}}} = 9.1475\ \text{kg}\ (5\ \text{SFs})$$

c. $$9147.5\ \cancel{\text{g}} \times \frac{1\ \text{cm}^3}{19.3\ \cancel{\text{g}}} = 474\ \text{cm}^3\ (3\ \text{SFs})$$

d. Volume = Area × height (i.e. thickness),

$$\therefore \text{Area of foil} = \frac{\text{volume of gold}}{\text{thickness of foil}} = \frac{474\ \cancel{\text{cm}^3}}{0.0035\ \cancel{\text{in.}}} \times \frac{1\ \cancel{\text{in.}}}{2.54\ \cancel{\text{cm}}} \times \left(\frac{1\ \text{m}}{100\ \cancel{\text{cm}}}\right)^2$$

$$= 5.3\ \text{m}^2\ (2\ \text{SFs})$$

e. Melting point of gold = 1064 °C
in degrees Fahrenheit:

$$T_F = 1.8(T_C) + 32 = 1.8(1064) + 32 = 1947\ °\text{F} \text{ (answer to the ones place)}$$

in kelvins:

$$T_K = T_C + 273 = 1064 + 273 = 1337\ \text{K} \text{ (answer to the ones place)}$$

f. **Given** SH_{gold} = 0.129 J/g °C; mass = 9147.5 g of gold;
initial temperature = 63 °F; final temperature = 85 °F

Need heat (q) in kilojoules

Plan heat (q) = mass × ΔT × SH

Set Up convert temperatures to Celsius:

$$\text{initial temperature: } T_C = \frac{T_F - 32}{1.8} = \frac{(63 - 32)}{1.8} = 17\ °\text{C}$$

$$\text{final temperature: } T_C = \frac{T_F - 32}{1.8} = \frac{(85 - 32)}{1.8} = 29\ °\text{C}$$

$$\Delta T = T_f - T_i = 29\ °\text{C} - 17\ °\text{C} = 12\ °\text{C}$$

$$\text{heat} = 9147.5\ \cancel{\text{g}} \times 12\ \cancel{°\text{C}} \times \frac{0.129\ \cancel{\text{J}}}{\cancel{\text{g}}\ \cancel{°\text{C}}} \times \frac{1\ \text{kJ}}{1000\ \cancel{\text{J}}} = 14\ \text{kJ}\ (2\ \text{SFs})$$

CI.3 **a.** The water has its own shape in sample **B**.

b. Water sample **A** is represented by diagram **2**, which shows the particles in a random arrangement but close together.

c. Water sample **B** is represented by diagram **1**, which shows the particles fixed in a definite arrangement.

d. The state of matter indicated in diagram **1** is a solid; in diagram **2**, it is a liquid; and in diagram **3**, it is a gas.

e. The motion of the particles is slowest in diagram **1**.

f. The arrangement of particles is farthest apart in diagram **3**.

g. The particles fill the volume of the container in diagram <u>3</u>.
h. water 45 °C → 0 °C: $\Delta T = 0\,°C - 45\,°C = -45\,°C$

$$\text{heat} = 19\,\cancel{g} \times (-45\,\cancel{°C}) \times \frac{4.184\,\cancel{J}}{\cancel{g}\,\cancel{°C}} \times \frac{1\,kJ}{1000\,\cancel{J}} = -3.6\,kJ\ (2\ SFs)$$

∴ 3.6 kJ of heat is removed.

CI.5 a. $0.250\,\cancel{lb} \times \frac{453.6\,\cancel{g}}{1\,\cancel{lb}} \times \frac{1\,cm^3}{7.86\,\cancel{g}} = 14.4\,cm^3\ (3\ SFs)$

b. $30\,\cancel{nails} \times \frac{14.4\,cm^3}{75\,\cancel{nails}} = 5.76\,cm^3 = 5.76\,mL\ (3\ SFs)$

New water level = 17.6 mL water + 5.76 mL = 23.4 mL (3 SFs)

c. $\Delta T = T_f - T_i = 125\,°C - 16\,°C = 109\,°C$

$$\text{mass of iron nails} = 0.250\,\cancel{lb} \times \frac{453.6\,g}{1\,\cancel{lb}} = 113\,g\ (3\ SFs)$$

$$\text{heat} = \text{mass} \times \Delta T \times SH$$

$$= 0.250\,\cancel{lb} \times \frac{453.6\,\cancel{g}}{1\,\cancel{lb}} \times 109\,\cancel{°C} \times \frac{0.452\,J}{\cancel{g}\,\cancel{°C}} = 5590\,J\ (3\ SFs)$$

d. Given

Iron nails	Water
Mass = 113 g	mass = 325 g
Initial temperature = 55.0 °C	initial temperature = 4.0 °C
Final temperature = ?	final temperature = ?
SH = 0.452 J/g °C	*SH* = 4.184 J/g °C

Need final temperature
Plan heat $(-q)$ lost by nails at 55.0 °C = heat (q) gained by water at 4.0 °C
$-q_{iron} = q_{water}$
Set Up $-(m \times \Delta T \times SH)_{iron} = (m \times \Delta T \times SH)_{water}$

$$-\left(\left(0.250\,\cancel{lb} \times \frac{453.6\,\cancel{g}}{1\,\cancel{lb}}\right) \times (T_f - 55.0)\,\cancel{°C} \times \frac{0.452\,J}{\cancel{g}\,\cancel{°C}}\right) = \left(325\,\cancel{g} \times (T_f - 4.0)\,\cancel{°C} \times \frac{4.184\,J}{\cancel{g}\,\cancel{°C}}\right)$$

$$2820 - 51.3T_f = 1360T_f - 5440$$
$$1411T_f = 8260$$
$$T_f = 5.8\,°C$$

4

Atoms and Elements

Learning Goals

- Given the name of an element, write its correct symbol; from the symbol, write the correct name.
- Use the periodic table to identify the group and the period of an element; identify the element as a metal, a nonmetal, or a metalloid.
- Describe the electrical charge and location in an atom for a proton, a neutron, and an electron.
- Given the atomic number and the mass number of an atom, state the number of protons, neutrons, and electrons.
- Give the number of protons, electrons, and neutrons in one or more isotopes of an element.
- Calculate the atomic mass of an element using the percent abundance and mass of its naturally occurring isotopes.

Chapter Outline

Chapter Opener: Farmer

4.1 Elements and Symbols

Chemistry Link to Industry: Many Forms of Carbon

Chemistry Link to Health: Mercury

4.2 The Periodic Table

Chemistry Link to Health: Elements Essential to Health

4.3 The Atom

4.4 Atomic Number and Mass Number

4.5 Isotopes and Atomic Mass

Core Chemistry Skills

- Counting Protons and Neutrons (4.4)
- Writing Atomic Symbols for Isotopes (4.5)

Answers and Solutions to Text Problems

4.1 **a.** Cu is the symbol for copper.
b. Pt is the symbol for platinum.
c. Ca is the symbol for calcium.
d. Mn is the symbol for manganese.
e. Fe is the symbol for iron.
f. Ba is the symbol for barium.
g. Pb is the symbol for lead.
h. Sr is the symbol for strontium.

4.3 **a.** Carbon is the element with the symbol C.
b. Chlorine is the element with the symbol Cl.

c. Iodine is the element with the symbol I.
d. Mercury is the element with the symbol Hg.
e. Silver is the element with the symbol Ag.
f. Argon is the element with the symbol Ar.
g. Boron is the element with the symbol B.
h. Nickel is the element with the symbol Ni.

4.5 a. Sodium (Na) and chlorine (Cl) are in NaCl.
b. Calcium (Ca), sulfur (S), and oxygen (O) are in $CaSO_4$.
c. Carbon (C), hydrogen (H), chlorine (Cl), nitrogen (N), and oxygen (O) are in $C_{15}H_{22}ClNO_2$.
d. Barium (Ba), chromium (Cr), and oxygen (O) are in $BaCrO_4$.

4.7 a. C, N, and O are in Period 2.
b. He is the element at the top of Group 8A (18).
c. The alkali metals are the elements in Group 1A (1).
d. Period 2 is the horizontal row of elements that ends with neon (Ne).

4.9 a. Ca is an alkaline earth metal.
b. Fe is a transition element.
c. Xe is a noble gas.
d. K is an alkali metal.
e. Cl is a halogen.

4.11 a. C is Group 4A (14), Period 2.
b. He is the noble gas in Period 1.
c. Na is the alkali metal in Period 3.
d. Ca is Group 2A (2), Period 4.
e. Al is Group 3A (13), Period 3.

4.13 On the periodic table, *metals* are located to the left of the heavy zigzag line, *nonmetals* are located to the right, and *metalloids* (B, Si, Ge, As, Sb, Te, Po, and At) are located along the line.
a. Ca is a metal.
b. S is a nonmetal.
c. Metals are shiny.
d. An element that is a gas at room temperature is a nonmetal.
e. Group 8A (18) elements are nonmetals.
f. Br is a nonmetal.
g. B is a metalloid.
h. Ag is a metal.

4.15 a. The electron has the smallest mass.
b. The proton has a 1+ charge.
c. The electron is found outside the nucleus.
d. The neutron is electrically neutral.

4.17 Rutherford determined that positively charged particles called protons are located in a very small, dense region of the atom called the nucleus.

4.19 a. True
b. True
c. True
d. False; since a neutron has no charge, it is not attracted to an electron (a proton is attracted to an electron).

4.21 In the process of brushing your hair, strands of hair become charged with like charges that repel each other.

4.23 **a.** The atomic number is the same as the number of protons in an atom.
b. Both are needed since the number of neutrons is the (mass number) − (atomic number).
c. The mass number is the number of particles (protons + neutrons) in the nucleus.
d. The atomic number is the same as the number of electrons in a neutral atom.

4.25 The atomic number defines the element and is found above the symbol of the element in the periodic table.
a. Lithium, Li, has an atomic number of 3.
b. Fluorine, F, has an atomic number of 9.
c. Calcium, Ca, has an atomic number of 20.
d. Zinc, Zn, has an atomic number of 30.
e. Neon, Ne, has an atomic number of 10.
f. Silicon, Si, has an atomic number of 14.
g. Iodine, I, has an atomic number of 53.
h. Oxygen, O, has an atomic number of 8.

4.27 The atomic number gives the number of protons in the nucleus of an atom. Since atoms are neutral, the atomic number also gives the number of electrons in the neutral atom.
a. There are 18 protons and 18 electrons in a neutral argon atom.
b. There are 25 protons and 25 electrons in a neutral manganese atom.
c. There are 53 protons and 53 electrons in a neutral iodine atom.
d. There are 48 protons and 48 electrons in a neutral cadmium atom.

4.29 The atomic number is the same as the number of protons in an atom and the number of electrons in a neutral atom; the atomic number defines the element. The number of neutrons is the (mass number) − (atomic number).

Name of the Element	Symbol	Atomic Number	Mass Number	Number of Protons	Number of Neutrons	Number of Electrons
Aluminum	Al	13	27	13	27 − 13 = 14	13
Magnesium	Mg	12	12 + 12 = 24	12	12	12
Potassium	K	19	19 + 20 = 39	19	20	19
Sulfur	S	16	16 + 15 = 31	16	15	16
Iron	Fe	26	56	26	56 − 26 = 30	26

4.31 **a.** Since the atomic number of strontium is 38, every Sr atom has 38 protons. An atom of strontium (mass number 89) has 51 neutrons ($89 - 38 = 51\ n$). Neutral atoms have the same number of protons and electrons. Therefore, 38 protons, 51 neutrons, 38 electrons.
b. Since the atomic number of chromium is 24, every Cr atom has 24 protons. An atom of chromium (mass number 52) has 28 neutrons ($52 - 24 = 28\ n$). Neutral atoms have the same number of protons and electrons. Therefore, 24 protons, 28 neutrons, 24 electrons.
c. Since the atomic number of sulfur is 16, every S atom has 16 protons. An atom of sulfur (mass number 34) has 18 neutrons ($34 - 16 = 18\ n$). Neutral atoms have the same number of protons and electrons. Therefore, 16 protons, 18 neutrons, 16 electrons.
d. Since the atomic number of bromine is 35, every Br atom has 35 protons. An atom of bromine (mass number 81) has 46 neutrons ($81 - 35 = 46\ n$). Neutral atoms have the same number of protons and electrons. Therefore, 35 protons, 46 neutrons, 35 electrons.

4.33 **a.** Since the number of protons is 15, the atomic number is 15 and the element symbol is P. The mass number is the sum of the number of protons and the number of neutrons, $15 + 16 = 31$. The atomic symbol for this isotope is $^{31}_{15}P$.

b. Since the number of protons is 35, the atomic number is 35 and the element symbol is Br. The mass number is the sum of the number of protons and the number of neutrons, $35 + 45 = 80$. The atomic symbol for this isotope is $^{80}_{35}Br$.

c. Since the number of electrons is 50, there must be 50 protons in a neutral atom. Since the number of protons is 50, the atomic number is 50, and the element symbol is Sn. The mass number is the sum of the number of protons and the number of neutrons, $50 + 72 = 122$. The atomic symbol for this isotope is $^{122}_{50}Sn$.

d. Since the element is chlorine, the element symbol is Cl, the atomic number is 17, and the number of protons is 17. The mass number is the sum of the number of protons and the number of neutrons, $17 + 18 = 35$. The atomic symbol for this isotope is $^{35}_{17}Cl$.

e. Since the element is mercury, the element symbol is Hg, the atomic number is 80, and the number of protons is 80. The mass number is the sum of the number of protons and the number of neutrons, $80 + 122 = 202$. The atomic symbol for this isotope is $^{202}_{80}Hg$.

4.35 **a.** Since the element is argon, the element symbol is Ar, the atomic number is 18, and the number of protons is 18. The atomic symbols for the isotopes with mass numbers of 36, 38, and 40, are $^{36}_{18}Ar$, $^{38}_{18}Ar$, and $^{40}_{18}Ar$, respectively.

b. They all have the same atomic number (the same number of protons and electrons).

c. They have different numbers of neutrons, which gives them different mass numbers.

d. The atomic mass of argon listed on the periodic table is the weighted average atomic mass of all the naturally occurring isotopes of argon.

e. The isotope Ar-40 ($^{40}_{18}Ar$) is the most prevalent in a sample of argon because its mass is closest to the average atomic mass of argon listed on the periodic table (39.95 amu).

4.37 The mass of an isotope is the mass of an individual atom. The atomic mass is the weighted average of all the naturally occurring isotopes of that element.

4.39 Since the atomic mass of copper (63.55 amu) is closer to 63 amu, there are more atoms of $^{63}_{29}Cu$ in a sample of copper.

4.41 Since the atomic mass of neon (20.18 amu) is closest to 20 amu, the most prevalent isotope of neon is $^{20}_{10}Ne$.

4.43

$$^{69}_{31}\text{Ga} \quad 68.93 \times \frac{60.11}{100} = 41.43 \text{ amu}$$

$$^{71}_{31}\text{Ga} \quad 70.92 \times \frac{39.89}{100} = 28.29 \text{ amu}$$

$$\text{Atomic mass of Ga} = 69.72 \text{ amu (4 SFs)}$$

4.45 **a.** False; all atoms of a given element are identical to one another but different from atoms of other elements.

b. True

c. True

d. False; atoms are never created or destroyed during a chemical reaction.

4.47 **a.** The atomic mass is the weighted average of the masses of all of the naturally occurring isotopes of the element. Isotope masses are based on the masses of all subatomic particles in the atom: protons (1), neutrons (2), and electrons (3), although almost all of that mass comes from the protons (1) and neutrons (2).

b. The number of protons (1) is the atomic number.

c. The protons (1) are positively charged.

d. The electrons (3) are negatively charged.

e. The number of neutrons (2) is the (mass number) − (atomic number).

4.49 **a.** ${}^{16}_{8}X$, ${}^{17}_{8}X$, and ${}^{18}_{8}X$ all have an atomic number of 8, so all have eight protons.
b. ${}^{16}_{8}X$, ${}^{17}_{8}X$, and ${}^{18}_{8}X$ all have an atomic number of 8, so all are isotopes of oxygen.
c. ${}^{16}_{8}X$ and ${}^{16}_{9}X$ have mass numbers of 16, whereas ${}^{18}_{8}X$ and ${}^{18}_{10}X$ have mass numbers of 18.
d. ${}^{16}_{8}X$ $(16 - 8 = 8\,n)$ and ${}^{18}_{10}X$ $(18 - 10 = 8\,n)$ both have eight neutrons.

4.51 **a.** ${}^{37}_{17}Cl$ and ${}^{38}_{18}Ar$ both have 20 neutrons $(37 - 17 = 20\,n; 38 - 18 = 20\,n)$.
b. Since both ${}^{36}_{16}S$ and ${}^{34}_{16}S$ have an atomic number of 16, they will both have 16 protons in the nucleus. Since the number of protons is 16, there must be 16 electrons in the neutral atom for both. They have different numbers of neutrons $(36 - 16 = 20\,n; 34 - 16 = 18\,n)$.
c. ${}^{79}_{34}Se$ (34 protons, 79 − 34 = 45 neutrons, 34 electrons) and ${}^{81}_{35}Br$ (35 protons, 81 − 35 = 46 neutrons, 35 electrons) do not share common numbers of protons, neutrons, or electrons.
d. ${}^{40}_{18}Ar$ and ${}^{39}_{17}Cl$ both have 22 neutrons $(40 - 18 = 22\,n; 39 - 17 = 22\,n)$.

4.53 **a.** The diagram shows 4 protons and 5 neutrons in the nucleus of the element. Since the number of protons is 4, the atomic number is 4, and the element symbol is Be. The mass number is the sum of the number of protons and the number of neutrons, 4 + 5 = 9. The atomic symbol is ${}^{9}_{4}Be$.
b. The diagram shows 5 protons and 6 neutrons in the nucleus of the element. Since the number of protons is 5, the atomic number is 5, and the element symbol is B. The mass number is the sum of the number of protons and the number of neutrons, 5 + 6 = 11. The atomic symbol is ${}^{11}_{5}B$.
c. The diagram shows 6 protons and 7 neutrons in the nucleus of the element. Since the number of protons is 6, the atomic number is 6, and the element symbol is C. The mass number is the sum of the number of protons and the number of neutrons, 6 + 7 = 13. The atomic symbol is ${}^{13}_{6}C$.
d. The diagram shows 5 protons and 5 neutrons in the nucleus of the element. Since the number of protons is 5, the atomic number is 5, and the element symbol is B. The mass number is the sum of the number of protons and the number of neutrons, 5 + 5 = 10. The atomic symbol is ${}^{10}_{5}B$.
e. The diagram shows 6 protons and 6 neutrons in the nucleus of the element. Since the number of protons is 6, the atomic number is 6, and the element symbol is C. The mass number is the sum of the number of protons and the number of neutrons, 6 + 6 = 12. The atomic symbol is ${}^{12}_{6}C$. Both **b** (${}^{11}_{5}B$) and **d** (${}^{10}_{5}B$) are isotopes of boron; **c** (${}^{13}_{6}C$) and **e** (${}^{12}_{6}C$) are isotopes of carbon.

4.55 **a.** The lightest alkali metal (Group 1A (1)) is lithium, which has an atomic number of 3 and the symbol Li.
b. The heaviest noble gas (Group 8A (18)) is radon, which has an atomic number of 86 and the symbol Rn.
c. The alkaline earth metal (Group 2A (2)) in Period 3 is magnesium, which has an atomic mass of 24.31 amu and the symbol Mg.
d. The halogen (Group 7A (17)) with the fewest electrons is fluorine, which has an atomic mass of 19.00 amu and the symbol F.

4.57 The first letter of a symbol is a capital, but a second letter is lowercase. The symbol for cobalt is Co, but the symbols in CO are for the elements carbon and oxygen.

4.59 **a.** Br, bromine is in Group 7A (17), Period 4.
b. Argon is in Group 8A (18), Period 3.
c. Lithium is in Group 1A (1), Period 2.
d. Radium is in Group 2A (2), Period 7.

4.61 On the periodic table, *metals* are located to the left of the heavy zigzag line, *nonmetals* are located to the right of the line, and *metalloids* (B, Si, Ge, As, Sb, Te, Po, and At) are located along the line.
a. Cu is a metal.
b. Se is a nonmetal.
c. As is a metalloid.
d. Cr is a metal.

4.63 **a.** False; the proton is a positively charged particle.
b. False; the neutron has about the same mass as a proton.
c. True
d. False; the nucleus is the small, dense core of the atom.
e. True

4.65 **a.** The atomic number gives the number of <u>protons</u> in the nucleus.
b. In an atom, the number of electrons is equal to the number of <u>protons</u>.
c. Sodium and potassium are examples of elements called <u>alkali metals</u>.

4.67 The atomic number defines the element and is found above the symbol of the element in the periodic table.
a. Nickel, Ni, has an atomic number of 28.
b. Barium, Ba, has an atomic number of 56.
c. Radium, Ra, has an atomic number of 88.
d. Arsenic, As, has an atomic number of 33.
e. Tin, Sn, has an atomic number of 50.
f. Cesium, Cs, has an atomic number of 55.

4.69 The atomic number gives the number of protons in the nucleus of an atom. Since atoms are neutral, the atomic number also gives the number of electrons in the neutral atom.
a. There are 25 protons and 25 electrons in a neutral manganese (Mn) atom.
b. There are 15 protons and 15 electrons in a neutral phosphorus atom.
c. There are 38 protons and 38 electrons in a neutral strontium (Sr) atom.
d. There are 27 protons and 27 electrons in a neutral cobalt (Co) atom.
e. There are 92 protons and 92 electrons in a neutral uranium atom.

4.71 **a.** Since the atomic number of cadmium is 48, every Cd atom has 48 protons. An atom of cadmium (mass number 114) has 66 neutrons $(114 - 48 = 66\,n)$. Neutral atoms have the same number of protons and electrons. Therefore, 48 protons, 66 neutrons, 48 electrons.
b. Since the atomic number of technetium is 43, every Tc atom has 43 protons. An atom of technetium (mass number 98) has 55 neutrons $(98 - 43 = 55\,n)$. Neutral atoms have the same number of protons and electrons. Therefore, 43 protons, 55 neutrons, 43 electrons.
c. Since the atomic number of gold is 79, every Au atom has 79 protons. An atom of gold (mass number 199) has 120 neutrons $(199 - 79 = 120\,n)$. Neutral atoms have the same number of protons and electrons. Therefore, 79 protons, 120 neutrons, 79 electrons.
d. Since the atomic number of radon is 86, every Rn atom has 86 protons. An atom of radon (mass number 222) has 136 neutrons $(222 - 86 = 136\,n)$. Neutral atoms have the same number of protons and electrons. Therefore, 86 protons, 136 neutrons, 86 electrons.
e. Since the atomic number of xenon is 54, every Xe atom has 54 protons. An atom of xenon (mass number 136) has 82 neutrons $(136 - 54 = 82\,n)$. Neutral atoms have the same number of protons and electrons. Therefore, 54 protons, 82 neutrons, 54 electrons.

4.73 **a.** Since the number of protons is 4, the atomic number is 4, and the element symbol is Be. The mass number is the sum of the number of protons and the number of neutrons, $4 + 5 = 9$. The atomic symbol for this isotope is ${}^{9}_{4}\text{Be}$.

b. Since the number of protons is 12, the atomic number is 12, and the element symbol is Mg. The mass number is the sum of the number of protons and the number of neutrons, $12 + 14 = 26$. The atomic symbol for this isotope is ${}^{26}_{12}\text{Mg}$.

c. Since the element is calcium, the element symbol is Ca, and the atomic number is 20. The mass number is given as 46. The atomic symbol for this isotope is ${}^{46}_{20}\text{Ca}$.

d. Since the number of electrons is 30, there must be 30 protons in a neutral atom. Since the number of protons is 30, the atomic number is 30, and the element symbol is Zn. The mass number is the sum of the number of protons and the number of neutrons, $30 + 40 = 70$. The atomic symbol for this isotope is ${}^{70}_{30}\text{Zn}$.

4.75

Name	Atomic Symbol	Number of Protons	Number of Neutrons	Number of Electrons
Selenium	${}^{80}_{34}\text{Se}$	34	$80 - 34 = 46$	34
Nickel	${}^{28+34}_{28}\text{Ni}$ or ${}^{62}_{28}\text{Ni}$	28	34	28
Magnesium	${}^{12+14}_{12}\text{Mg}$ or ${}^{26}_{12}\text{Mg}$	12	14	12
Radon	${}^{220}_{86}\text{Rn}$	86	$220 - 86 = 134$	86

4.77 **a.** Since the element is gold (Au), the atomic number is 79, and the number of protons is 79. In a neutral atom, the number of electrons is equal to the number of protons, so there will be 79 electrons. The number of neutrons is the (mass number) − (atomic number) $= 197 - 79 = 118\,n$. In Au-197, there are 79 protons, 118 neutrons, 79 electrons.

b. Since the element is gold (Au), the atomic number is 79, and the number of protons is 79. The mass number is the sum of the number of protons and the number of neutrons, $79 + 116 = 195$. The atomic symbol for this isotope is ${}^{195}_{79}\text{Au}$.

c. Since the atomic number is 78, the number of protons is 78, and the element symbol is Pt. The mass number is the sum of the number of protons and the number of neutrons, $78 + 116 = 194$. The atomic symbol for this isotope is ${}^{194}_{78}\text{Pt}$.

4.79 **a.** Since the element is lead (Pb), the atomic number is 82, and the number of protons is 82. In a neutral atom, the number of electrons is equal to the number of protons, so there will be 82 electrons. The number of neutrons is the (mass number) − (atomic number) $= 208 - 82 = 126\,n$. In ${}^{208}_{82}\text{Pb}$, there are 82 protons, 126 neutrons, 82 electrons.

b. Since the element is lead (Pb), the atomic number is 82, and the number of protons is 82. The mass number is the sum of the number of protons and the number of neutrons, $82 + 132 = 214$. The atomic symbol for this isotope is ${}^{214}_{82}\text{Pb}$.

c. Since the mass number is 214 (as in part **b**) and the number of neutrons is 131, the number of protons is the (mass number) − (number of neutrons) $= 214 - 131 = 83\,p$. Since there are 83 protons, the atomic number is 83, and the element symbol is Bi (bismuth). The atomic symbol for this isotope is ${}^{214}_{83}\text{Bi}$.

4.81 $^{204}_{82}Pb$ $204.0 \times \frac{1.40}{100} = 2.86$ amu

$^{206}_{82}Pb$ $206.0 \times \frac{24.10}{100} = 49.65$ amu

$^{207}_{82}Pb$ $207.0 \times \frac{22.10}{100} = 45.75$ amu

$^{208}_{82}Pb$ $208.0 \times \frac{52.40}{100} = 109.0$ amu

Atomic mass of Pb = 207.3 amu (4 SFs)

4.83 $^{113}_{49}In$ $112.9 \times \frac{4.30}{100} = 4.85$ amu

$^{115}_{49}In$ $114.9 \times \frac{95.70}{100} = 110.0$ amu

Atomic mass of In = 114.9 amu (4 SFs)

4.85 **Given** length = 1 in., sodium atoms (3.14×10^{-8} cm diameter)
Need number of Na atoms
Plan in. → cm → number of Na atoms

$$\frac{2.54 \text{ cm}}{1 \text{ in.}} \quad \frac{1 \text{ atom Na}}{3.14 \times 10^{-8} \text{ cm}}$$

Set Up $1 \cancel{\text{in.}} \times \frac{2.54 \cancel{\text{cm}}}{1 \cancel{\text{in.}}} \times \frac{1 \text{ atom Na}}{3.14 \times 10^{-8} \cancel{\text{cm}}} = 8.09 \times 10^{7}$ atoms of Na (3 SFs)

5
Electronic Structure and Periodic Trends

Learning Goals

- Compare the wavelength of radiation with its frequency.
- Explain how atomic spectra correlate with the energy levels in atoms.
- Describe the sublevels and orbitals in atoms.
- Draw the orbital diagram and write the electron configuration for an element.
- Write the electron configuration for an atom using the sublevel blocks on the periodic table.
- Use the electron configurations of elements to explain the trends in periodic properties.

Chapter Outline

Chapter Opener: Engineer

5.1 Electromagnetic Radiation

Chemistry Link to Health: Biological Reactions to UV Light

5.2 Atomic Spectra and Energy Levels

Chemistry Link to the Environment: Energy-Saving Fluorescent Bulbs

5.3 Sublevels and Orbitals

5.4 Orbital Diagrams and Electron Configurations

5.5 Electron Configurations and the Periodic Table

5.6 Trends in Periodic Properties

Core Chemistry Skills

- Writing Electron Configurations (5.4)
- Using the Periodic Table to Write Electron Configurations (5.5)
- Identifying Trends in Periodic Properties (5.6)
- Drawing Electron-Dot Symbols (5.6)

Answers and Solutions to Text Problems

5.1 The wavelength of UV light is the distance between crests of the wave of ultraviolet light.

5.3 "White" light has all the colors and wavelengths of the visible spectrum, including red and blue light. Red and blue light are visible light with specific wavelengths.

5.5 **Given** $\nu = 650\text{ kHz}$ **Need** hertz

Plan kHz → Hz $\dfrac{1000\text{ Hz}}{1\text{ kHz}}$

Set Up $650\text{ }\cancel{\text{kHz}} \times \dfrac{1000\text{ Hz}}{1\text{ }\cancel{\text{kHz}}} = 6.5 \times 10^5\text{ Hz (2 SFs)}$

5.7 **Given** $\lambda = 6.3 \times 10^{-5}\text{ cm}$ **Need** meters and nanometers

Plan cm → m $\dfrac{1\text{ m}}{100\text{ cm}}$; cm → m → nm $\dfrac{1\text{ m}}{100\text{ cm}}$ $\dfrac{1 \times 10^9\text{ nm}}{1\text{ m}}$

Set Up $6.3 \times 10^{-5}\ \cancel{\text{cm}} \times \dfrac{1\ \text{m}}{100\ \cancel{\text{cm}}} = 6.3 \times 10^{-7}\ \text{m}\ (2\ \text{SFs})$

$6.3 \times 10^{-5}\ \cancel{\text{cm}} \times \dfrac{1\ \cancel{\text{m}}}{100\ \cancel{\text{cm}}} \times \dfrac{1 \times 10^{9}\ \text{nm}}{1\ \cancel{\text{m}}} = 630\ \text{nm}\ (2\ \text{SFs})$

5.9 Microwaves have longer wavelengths than ultraviolet light or X-rays.

5.11 In order of increasing wavelengths (shortest to longest): X-rays, blue light, microwaves

5.13 In order of increasing frequencies (lowest to highest): TV signal, microwaves, X-rays

5.15 Atomic spectra consist of a series of lines separated by dark sections, indicating that the energy emitted by the elements is not continuous.

5.17 Electrons can jump to higher energy levels when they absorb a photon.

5.19 **a.** A photon of green light has greater energy than a photon of yellow light.
b. A photon of blue light has greater energy than a photon of red light.

5.21 **a.** A 1*s* orbital is spherical.
b. A 2*p* orbital has two lobes.
c. A 5*s* orbital is spherical.

5.23 **a.** **1 and 2**; 1*s* and 2*s* orbitals have the same shape (spherical) and can contain a maximum of two electrons.
b. **3**; 3*s* and 3*p* sublevels are in the same energy level ($n = 3$).
c. **1 and 2**; 3*p* and 4*p* sublevels contain orbitals with the same two-lobed shape and can contain the same maximum number of electrons (six).
d. **1, 2, and 3**; all of the 3*p* orbitals have two lobes, can contain a maximum of two electrons, and are in energy level $n = 3$.

5.25 **a.** There are five orbitals in the 3*d* sublevel.
b. There is one sublevel in the $n = 1$ energy level.
c. There is one orbital in the 6*s* sublevel.
d. There are nine orbitals in the $n = 3$ energy level: one 3*s* orbital, three 3*p* orbitals, and five 3*d* orbitals.

5.27 **a.** Any orbital can hold a maximum of two electrons. Thus, a 2*p* orbital can have a maximum of two electrons.
b. The 3*p* sublevel contains three *p* orbitals, each of which can hold a maximum of two electrons, which gives a maximum of six electrons in the 3*p* sublevel.
c. Using $2n^2$, the calculation for the maximum number of electrons in the $n = 4$ energy level is $2(4)^2 = 2(16) = 32$ electrons.
d. The 5*d* sublevel contains five *d* orbitals, each of which can hold a maximum of two electrons, which gives a maximum of 10 electrons in the 5*d* sublevel.

5.29 The electron configuration shows the number of electrons in each sublevel of an atom. The abbreviated electron configuration uses the symbol of the preceding noble gas to show completed sublevels.

5.31 Determine the number of electrons and then fill orbitals in the following order: $1s^22s^22p^63s^23p^6$
a. Boron is atomic number 5 and so it has 5 electrons. Fill orbitals as

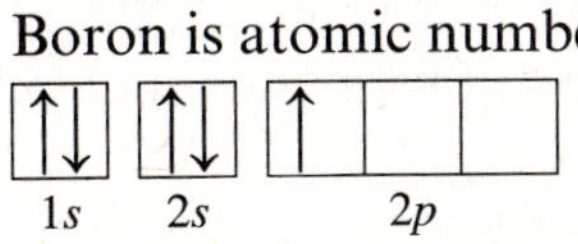

b. Aluminum is atomic number 13 and so it has 13 electrons. Fill orbitals as

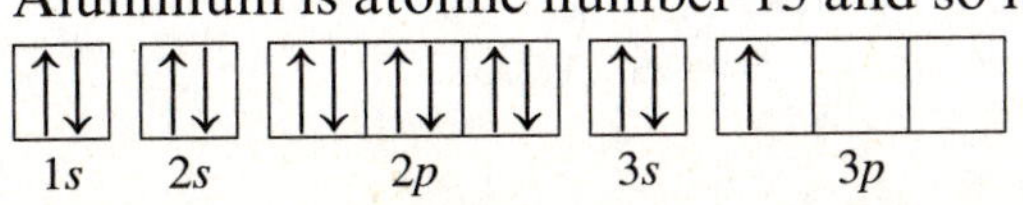

c. Phosphorus is atomic number 15 and so it has 15 electrons. Fill orbitals as

↑↓	↑↓	↑↓ ↑↓ ↑↓	↑↓	↑ ↑ ↑
$1s$	$2s$	$2p$	$3s$	$3p$

d. Argon is atomic number 18 and so it has 18 electrons. Fill orbitals as

↑↓	↑↓	↑↓ ↑↓ ↑↓	↑↓	↑↓ ↑↓ ↑↓
$1s$	$2s$	$2p$	$3s$	$3p$

5.33 **a.** B $1s^2 2s^2 2p^1$
b. Na $1s^2 2s^2 2p^6 3s^1$
c. Li $1s^2 2s^1$
d. Mg $1s^2 2s^2 2p^6 3s^2$

5.35 **a.** C $[\mathrm{He}]2s^2 2p^2$
b. Si $[\mathrm{Ne}]3s^2 3p^2$
c. P $[\mathrm{Ne}]3s^2 3p^3$
d. F $[\mathrm{He}]2s^2 2p^5$

5.37 **a.** Li is the element with one electron in the $2s$ sublevel.
b. S is the element with two electrons in the $3s$ and four electrons in the $3p$ sublevels.
c. Si is the element with two electrons in the $3s$ and two electrons in the $3p$ sublevels.
d. F is the element with two electrons in the $2s$ and five electrons in the $2p$ sublevels.

5.39 **a.** Al has three electrons in the energy level $n = 3$, $3s^2 3p^1$.
b. C has two $2p$ electrons.
c. Ar completes the $3p$ sublevel, $3p^6$.
d. Be completes the $2s$ sublevel, $2s^2$.

5.41 On the periodic table, the *s* sublevel block is on the left, the *p* sublevel block is on the right, and the *d* sublevel block is in the center between the *s* and *p* blocks.
a. As $1s^2 2s^2 2p^6 3s^2 3p^6 4s^2 3d^{10} 4p^3$
b. Fe $1s^2 2s^2 2p^6 3s^2 3p^6 4s^2 3d^6$
c. In $1s^2 2s^2 2p^6 3s^2 3p^6 4s^2 3d^{10} 4p^6 5s^2 4d^{10} 5p^1$
d. I $1s^2 2s^2 2p^6 3s^2 3p^6 4s^2 3d^{10} 4p^6 5s^2 4d^{10} 5p^5$

5.43 On the periodic table, the *s* sublevel block is on the left, the *p* sublevel block is on the right, and the *d* sublevel block is in the center between the *s* and *p* blocks. The abbreviated electron configuration consists of the symbol of the preceding noble gas, followed by the electron configuration in the next period.
a. Ti $[\mathrm{Ar}]4s^2 3d^2$
b. Br $[\mathrm{Ar}]4s^2 3d^{10} 4p^5$
c. Ba $[\mathrm{Xe}]6s^2$
d. Pb $[\mathrm{Xe}]6s^2 4f^{14} 5d^{10} 6p^2$

5.45 Use the final sublevel notation in the electron configuration to locate the element.
a. P (ends in $3p^3$)
b. Co (ends in $3d^7$)
c. Zn (ends in $3d^{10}$)
d. Bi (ends in $6p^3$)

5.47 **a.** Ga has three electrons in energy level $n = 4$; two are in the $4s$ block, and one is in the $4p$ block.
b. N is the third element in the $2p$ block; it has three $2p$ electrons.
c. Xe is the final element (6 electrons) in the $5p$ block.
d. Zr is the second element in the $4d$ block; it has two $4d$ electrons.

5.49 **a.** Zn is the tenth element in the 3*d* block; it has 10 3*d* electrons.
b. Na has one electron in the 3*s* block; the 2*p* block in Na is complete with six electrons.
c. As is the third element in the 4*p* block; it has three 4*p* electrons.
d. Rb is the first element in the 5*s* block; Rb has one 5*s* electron.

5.51 The group numbers 1A–8A indicate the number of valence (outer) electrons for the elements in each vertical column.

5.53 **a.** An element with two valence electrons is in Group 2A (2).
b. An element with five valence electrons is in Group 5A (15).
c. An element with two valence electrons and five electrons in the *d* block is in Group 7B (7).
d. An element with six valence electrons is in Group 6A (16).

5.55 **a.** Alkali metals, which are in Group 1A (1), have a valence electron configuration ns^1.
b. Elements in Group 4A (14) have a valence electron configuration ns^2np^2.
c. Elements in Group 7A(17), have a valence electron configuration ns^2np^5.
d. Elements in Group 5A (15) have a valence electron configuration ns^2np^3.

5.57 **a.** Aluminum in Group 3A (13) has three valence electrons.
b. Any element in Group 5A (15) has five valence electrons.
c. Barium in Group 2A (2) has two valence electrons.
d. Each halogen in Group 7A (17) has seven valence electrons.

5.59 The number of dots is equal to the number of valence electrons, as indicated by the group number.
a. Sulfur is in Group 6A (16); $\cdot\overset{\cdot\cdot}{\underset{\cdot}{\text{S}}}:$
b. Nitrogen is in Group 5A (15); $\cdot\overset{\cdot\cdot}{\underset{\cdot}{\text{N}}}\cdot$
c. Calcium is in Group 2A (2); $\dot{\text{C}}\text{a}\cdot$
d. Sodium is in Group 1A (1); $\text{Na}\cdot$
e. Gallium is in Group 3A (13); $\cdot\dot{\text{G}}\text{a}\cdot$

5.61 The atomic size of representative elements decreases going across a period from Group 1A to 8A and increases going down a group.
a. In Period 3, Na, which is on the left, is larger than Cl.
b. In Group 1A (1), Rb, which is farther down the group, is larger than Na.
c. In Period 3, Na, which is on the left, is larger than Mg.
d. In Period 5, Rb, which is on the left, is larger than I.

5.63 **a.** The atomic size of representative elements decreases from Group 1A to 8A: Mg, Al, Si.
b. The atomic size of representative elements increases going down a group: I, Br, Cl.
c. The atomic size of representative elements decreases from Group 1A to 8A: Sr, Sb, I.
d. The atomic size of representative elements decreases from Group 1A to 8A: Na, Si, P.

5.65 **a.** In Br, the valence electrons are closer to the nucleus, so Br has a higher ionization energy than I.
b. In Mg, the valence electrons are closer to the nucleus, so Mg has a higher ionization energy than Sr.
c. Attraction for the valence electrons increases going from left to right across a period, giving P a higher ionization energy than Si.
d. Attraction for the valence electrons increases going from left to right across a period, giving Xe a higher ionization energy than I.

5.67 **a.** Br, Cl, F; ionization energy decreases going down a group.
b. Na, Al, Cl; going across a period from left to right, ionization energy increases.
c. Cs, K, Na; ionization energy decreases going down a group.
d. Ca, As, Br; going across a period from left to right, ionization energy increases.

5.69 Ca, Ga, Ge, Br; since metallic character decreases from left to right across a period, Ca, in Group 2A (2), will be the most metallic, followed by Ga, then Ge, and finally Br, in Group 7A (17).

5.71 Sr has a lower ionization energy and is more metallic than Sb.

5.73 Going down Group 6A (16),
a. 1; the ionization energy decreases
b. 2; the atomic size increases
c. 2; the metallic character increases
d. 3; the number of valence electrons remains the same

5.75 In Period 2, an atom of N compared to an atom of Li has a larger (greater)
a. False; N has a smaller atomic size than Li.
b. True
c. True
d. False; N has less metallic character than Li.
e. True

5.77 **a.** C has the longest wavelength.
b. A has the shortest wavelength.
c. A has the highest frequency.
d. C has the lowest frequency.

5.79 **a.** An orbital with two lobes is a *p* orbital.
b. A spherical orbital is an *s* orbital.
c. An orbital with two lobes is a *p* orbital.

5.81 **a.** This orbital diagram is possible. The orbitals up to $3s^2$ are filled with electrons having opposite spins. The element represented would be magnesium.
b. Not possible. The 2*p* sublevel would fill completely before the 3*s*, and only 2 electrons are allowed in an orbital (not 3 as shown in the 3*s*).

5.83 Atomic size increases going down a group. Li is **D** because it would be smallest. Na is **A,** K is **C,** and Rb is **B.**

5.85 A continuous spectrum from white light contains wavelengths of all energies. Atomic spectra are line spectra in which a series of lines corresponds to energy emitted when electrons drop from a higher energy level to a lower level.

5.87 The Pauli exclusion principle states that two electrons in the same orbital must have opposite spins.

5.89 A 4*p* orbital is possible because the $n = 4$ energy level has four sublevels, including a *p* sublevel. A 2*d* orbital is not possible because the $n = 2$ energy level has only *s* and *p* sublevels. There are no 3*f* orbitals because only *s*, *p*, and *d* sublevels are allowed for the $n = 3$ energy level. A 5*f* sublevel is possible in the $n = 5$ energy level because five sublevels are allowed, including 5*f*.

5.91 **a.** The 3*p* sublevel starts to fill after completion of the 3*s* sublevel.
b. The 5*s* sublevel starts to fill after completion of the 4*p* sublevel.
c. The 4*p* sublevel starts to fill after completion of the 3*d* sublevel.
d. The 4*s* sublevel starts to fill after completion of the 3*p* sublevel.

5.93 **a.** Iron is the sixth element in the 3*d* block; iron has six 3*d* electrons.
b. Barium has a completely filled 5*p* sublevel, which is six 5*p* electrons.
c. Iodine has a completely filled 4*d* sublevel, which is 10 4*d* electrons.
d. Radium has a filled 7*s* sublevel, which is two 7*s* electrons.

5.95 Ca, Sr, and Ba all have two valence electrons, ns^2, which places them in Group 2A (2), the alkaline earth metals.

5.97 **a.** X is a metal; Y and Z are nonmetals.
b. X has the largest atomic size.
c. Y has the highest ionization energy.
d. Y has the smallest atomic size.

5.99 **a.** Phosphorus in Group 5A (15) has an electron configuration that ends with $3s^23p^3$.
b. Lithium is the alkali metal that is highest in Group 1A (1) and has the smallest atomic size (H in Group 1A (1) is a nonmetal).
c. Cadmium in Period 5 has a complete 4*d* sublevel with 10 electrons.
d. Nitrogen at the top of Group 5A (15) has the highest ionization energy in that group.
e. Sodium, the first element in Period 3, has the largest atomic size of that period.

5.101 Calcium has a greater number of protons than K; this greater nuclear charge makes it more difficult to remove an electron from Ca. The least tightly bound (valence) electron in Ca is farther from the nucleus than in Mg, and less energy is needed to remove it.

5.103 **a.** Since metallic character increases going down a group, Sb will be more metallic than As.
b. Since metallic character decreases going across a period from left to right, Sn will be more metallic than Sb.
c. Since metallic character decreases going across a period from left to right, P will be more metallic than Cl.
d. Since metallic character increases going down a group and decreases going across a period from left to right, P will be more metallic than O.

5.105 **a.** Na on the far left of the heavy zigzag line on the periodic table is a metal.
b. Na at the beginning of Period 3 has the largest atomic size.
c. F at the top of Group 7A (17) and to the far right in Period 2 has the highest ionization energy.
d. Na has the lowest ionization energy and loses an electron most easily.
e. Cl is found in Period 3 in Group 7A.
f. F at the top of Group 7A (17) and to the far right in Period 2 has the least metallic character.

5.107 **a.** Si [Ne]$3s^23p^2$; Group 4A (14) **b.** Se [Ar]$4s^23d^{10}4p^4$; Group 6A (16)
c. Mn [Ar]$4s^23d^5$; Group 7B (7) **d.** Sb [Kr]$5s^24d^{10}5p^3$; Group 5A (15)

5.109 **a.** Atomic size increases going down a group, which gives O the smallest atomic size in Group 6A (16).
b. Atomic size decreases going from left to right across a period, which gives Ar the smallest atomic size in Period 3.
c. Ionization energy decreases going down a group, which gives B the highest ionization energy in Group 3A (13).
d. Ionization energy increases going from left to right across a period, which gives Na the lowest ionization energy in Period 3.
e. Ru in Period 5 has six electrons in the 4*d* sublevel.

5.111 Ultraviolet light and microwaves both travel at the same speed: 3.00×10^8 m/s. The wavelength of ultraviolet light is shorter than the wavelength of microwaves, and the frequency of ultraviolet light is higher than the frequency of microwaves.

5.113 The series of lines separated by dark sections in atomic spectra indicate that the energy emitted by the elements is not continuous and that electrons are moving between discrete energy levels.

5.115 An energy level contains all the electrons with similar energy. A sublevel contains electrons with the same energy, while an orbital is the region around the nucleus where electrons of a certain energy are most likely to be found.

5.117 S has a larger atomic size than Cl; Cl is larger than F: S > Cl > F. F has a higher ionization energy than Cl; Cl has a higher ionization energy than S: F > Cl > S.

6

Ionic and Molecular Compounds

Learning Goals

- Write the symbols for the simple ions of the representative elements.
- Using charge balance, write the correct formula for an ionic compound.
- Given the formula of an ionic compound, write the correct name; given the name of an ionic compound, write the correct formula.
- Write the name and formula for an ionic compound containing a polyatomic ion.
- Given the formula of a molecular compound, write its correct name; given the name of a molecular compound, write its formula.
- Identify the properties characteristic of organic and inorganic compounds.
- Write the IUPAC name and draw the condensed structural formula for an alkane.

Chapter Outline

Core Chemistry Skills

- Writing Positive and Negative Ions (6.1)
- Writing Ionic Formulas (6.2, 6.4)
- Naming Ionic Compounds (6.3, 6.4)
- Writing the Names and Formulas for Molecular Compounds (6.5)
- Naming and Drawing Alkanes (6.7)

Answers and Solutions to Text Problems

6.1 Atoms with one, two, or three valence electrons will lose those electrons to acquire a noble gas electron configuration.

a. Li loses 1 e^-.
b. Ca loses 2 e^-.
c. Ga loses 3 e^-.
d. Cs loses 1 e^-.

6.3 Atoms form ions by losing or gaining electrons to achieve the same electron configuration as the nearest noble gas.

a. Li^+ has an electron configuration $1s^2$.
b. Mg^{2+} has an electron configuration $1s^22s^22p^6$.
c. Ga^{3+} has an electron configuration $1s^22s^22p^63s^23p^63d^{10}$.
d. P^{3-} has an electron configuration $1s^22s^22p^63s^23p^6$.

6.5 Atoms form ions by losing or gaining valence electrons to achieve the same electron configuration as the nearest noble gas. Elements in Groups 1A (1), 2A (2), and 3A (13) lose valence electrons, whereas elements in Groups 5A (15), 6A (16), and 7A (17) gain valence electrons to complete octets.

a. Sr loses 2 e^-.
b. P gains 3 e^-.
c. Elements in Group 7A (17) gain 1 e^-.
d. Na loses 1 e^-.

6.7 **a.** The element with 3 protons is lithium. In a lithium ion with 2 electrons, the ionic charge would be 1+, (3+) + (2−) = 1+. The lithium ion is written as Li^+.
b. The element with 53 protons is iodine. In an iodine ion with 54 electrons, the ionic charge would be 1−, (53+) + (54−) = 1−. The iodide ion is written as I^-.
c. The element with 38 protons is strontium. In a strontium ion with 36 electrons, the ionic charge would be 2+, (38+) + (36−) = 2+. The strontium ion is written as Sr^{2+}.
d. The element with 26 protons is iron. In an iron ion with 23 electrons, the ionic charge would be 3+, (26+) + (23−) = 3+. This iron ion is written as Fe^{3+}.

6.9 **a.** O has an atomic number of 8, which means it has 8 protons. In a neutral atom, the number of electrons equals the number of protons. An oxygen ion with a charge of 2− has gained 2 e^- to have 8 + 2 = 10 electrons. ∴ in an O^{2-} ion, there are 8 protons and 10 electrons.
b. K has an atomic number of 19, which means it has 19 protons. In a neutral atom, the number of electrons equals the number of protons. A potassium ion with a charge of 1+ has lost 1 e^- to have 19 − 1 = 18 electrons. ∴ in a K^+ ion, there are 19 protons and 18 electrons.
c. Br has an atomic number of 35, which means it has 35 protons. In a neutral atom, the number of electrons equals the number of protons. A bromine ion with a charge of 1− has gained 1 e^- to have 35 + 1 = 36 electrons. ∴ in a Br^- ion, there are 35 protons and 36 electrons.
d. S has an atomic number of 16, which means it has 16 protons. In a neutral atom, the number of electrons equals the number of protons. A sulfur ion with a charge of 2− has gained 2 e^- to have 16 + 2 = 18 electrons. ∴ in a S^{2-} ion, there are 16 protons and 18 electrons.

6.11 **a.** Chlorine in Group 7A (17) gains one electron to form chloride ion, Cl^-.
b. Barium in Group 2A (2) loses two electrons to form barium ion, Ba^{2+}.
c. Aluminum in Group 3A (13) loses three electrons to form aluminum ion, Al^{3+}.
d. Selenium in Group 6A (16) gains two electrons to form selenide ion, Se^{2-}.

6.13 **a** (Li and Cl) and **c** (K and O) will form ionic compounds.

6.15 **a.** Na^+ and $O^{2-} \rightarrow Na_2O$
b. Al^{3+} and $Br^- \rightarrow AlBr_3$
c. Ba^{2+} and $N^{3-} \rightarrow Ba_3N_2$
d. Mg^{2+} and $F^- \rightarrow MgF_2$
e. Al^{3+} and $S^{2-} \rightarrow Al_2S_3$

6.17 **a.** Ions: K^+ and $S^{2-} \rightarrow K_2S$ Check: 2(1+) + 1(2−) = 0
b. Ions: Na^+ and $N^{3-} \rightarrow Na_3N$ Check: 3(1+) + 1(3−) = 0
c. Ions: Al^{3+} and $I^- \rightarrow AlI_3$ Check: 1(3+) + 3(1−) = 0
d. Ions: Ga^{3+} and $O^{2-} \rightarrow Ga_2O_3$ Check: 2(3+) + 3(2−) = 0

6.19 **a.** aluminum oxide **b.** cesium sulfide
c. sodium oxide **d.** magnesium phosphide
e. potassium iodide **f.** barium fluoride

6.21 Most of the transition elements form more than one positive ion. The specific ion is indicated in a name by writing a Roman numeral that is the same as the ionic charge. For example, iron forms Fe^{2+} and Fe^{3+} ions, which are named iron(II) and iron(III), respectively.

6.23 **a.** iron(II) **b.** copper(II)
c. zinc **d.** lead(IV)
e. chromium(III) **f.** manganese(II)

6.25 **a.** Ions: Sn^{2+} and Cl^- → tin(II) chloride
b. Ions: Fe^{2+} and O^{2-} → iron(II) oxide
c. Ions: Cu^+ and S^{2-} → copper(I) sulfide
d. Ions: Cu^{2+} and S^{2-} → copper(II) sulfide
e. Ions: Cd^{2+} and Br^- → cadmium bromide
f. Ions: Hg^{2+} and Cl^- → mercury(II) chloride

6.27 **a.** Au^{3+} **b.** Cr^{2+}
c. Pb^{4+} **d.** Sn^{2+}

6.29 **a.** Ions: Mg^{2+} and Cl^- → $MgCl_2$ **b.** Ions: Na^+ and S^{2-} → Na_2S
c. Ions: Cu^+ and O^{2-} → Cu_2O **d.** Ions: Zn^{2+} and P^{3-} → Zn_3P_2
e. Ions: Au^{3+} and N^{3-} → AuN **f.** Ions: Co^{3+} and F^- → CoF_3

6.31 **a.** Ions: Co^{3+} and Cl^- → $CoCl_3$ **b.** Ions: Pb^{4+} and O^{2-} → PbO_2
c. Ions: Ag^+ and I^- → AgI **d.** Ions: Ca^{2+} and N^{3-} → Ca_3N_2
e. Ions: Cu^+ and P^{3-} → Cu_3P **f.** Ions: Cr^{2+} and Cl^- → $CrCl_2$

6.33 **a.** HCO_3^- **b.** NH_4^+
c. PO_4^{3-} **d.** HSO_4^-
e. ClO^-

6.35 **a.** sulfate **b.** carbonate
c. phosphate **d.** nitrate
e. perchlorate

6.37

	NO_2^-	CO_3^{2-}	HSO_4^-	PO_4^{3-}
Li^+	$LiNO_2$ Lithium nitrite	Li_2CO_3 Lithium carbonate	$LiHSO_4$ Lithium hydrogen sulfate	Li_3PO_4 Lithium phosphate
Cu^{2+}	$Cu(NO_2)_2$ Copper(II) nitrite	$CuCO_3$ Copper(II) carbonate	$Cu(HSO_4)_2$ Copper(II) hydrogen sulfate	$Cu_3(PO_4)_2$ Copper(II) phosphate
Ba^{2+}	$Ba(NO_2)_2$ Barium nitrite	$BaCO_3$ Barium carbonate	$Ba(HSO_4)_2$ Barium hydrogen sulfate	$Ba_3(PO_4)_2$ Barium phosphate

6.39 **a.** CO_3^{2-}, sodium carbonate **b.** NH_4^+, ammonium chloride
c. PO_3^{3-}, sodium phosphite **d.** NO_2^-, manganese(II) nitrite
e. SO_3^{2-}, iron(II) sulfite **f.** $C_2H_3O_2^-$, potassium acetate

6.41 **a.** Ions: Ba^{2+} and OH^- → $Ba(OH)_2$ **b.** Ions: Na^+ and HSO_3^- → $NaHSO_3$
c. Ions: Fe^{2+} and NO_3^- → $Fe(NO_3)_2$ **d.** Ions: Zn^{2+} and PO_4^{3-} → $Zn_3(PO_4)_2$
e. Ions: Fe^{3+} and CO_3^{2-} → $Fe_2(CO_3)_3$

6.43 **a.** Ions: Al^{3+} and SO_4^{2-} → aluminum sulfate
b. Ions: Mg^{2+} and PO_4^{3-} → magnesium phosphate
c. Ions: Cr^{3+} and O^{2-} → chromium(III) oxide
d. Ions: Na^{+} and PO_4^{3-} → sodium phosphate
e. Ions: NH_4^{+} and SO_4^{2-} → ammonium sulfate
f. Ions: Bi^{3+} and S^{2-} → bismuth(III) sulfide

6.45 The nonmetallic elements that are not noble gases are most likely to form molecular compounds.

6.47 **a.** Each H atom has one valence electron, for a total of two valence electrons in the molecule. In H_2, there is one bonding pair and no (0) lone pairs.
b. The Br atom has seven valence electrons and the H atom has one valence electron, for a total of eight valence electrons in the molecule. The Br atom achieves an octet by sharing a valence electron with one H atom to give eight valence electrons; the H atom has two valence electrons. The molecule has one bonding pair and there are three lone pairs on the Br atom.
c. Each Br atom has seven valence electrons, for a total of 14 valence electrons in the molecule. Each Br atom achieves an octet by sharing one valence electron with the other Br atom. There is one bonding pair between the Br atoms, and six lone pairs (three lone pairs on each Br atom).

6.49 When naming molecular compounds, prefixes are used to indicate the number of each atom as shown in the subscripts of the formula. The first nonmetal is named by its elemental name; the second nonmetal is named by using its elemental name with the ending changed to *ide*.
a. 1 P and 3 Br → phosphorus tribromide
b. 1 C and 4 Br → carbon tetrabromide
c. 1 Si and 2 O → silicon dioxide
d. 1 H and 1 F → hydrogen fluoride
e. 1 N and 3 I → nitrogen triiodide

6.51 When naming molecular compounds, prefixes are used to indicate the number of each atom as shown in the subscripts of the formula. The first nonmetal is named by its elemental name; the second nonmetal is named by using its elemental name with the ending changed to *ide*.
a. 2 N and 3 O → dinitrogen trioxide
b. 1 N and 3 Cl → nitrogen trichloride
c. 1 Si and 4 Br → silicon tetrabromide
d. 1 P and 5 Cl → phosphorus pentachloride
e. 2 N and 3 S → dinitrogen trisulfide

6.53 **a.** 1 C and 4 Cl → CCl_4
b. 1 C and 1 O → CO
c. 1 P and 3 Cl → PCl_3
d. 2 N and 4 O → N_2O_4

6.55 **a.** 1 O and 2 F → OF_2
b. 1 B and 3 F → BF_3
c. 2 N and 3 O → N_2O_3
d. 1 S and 6 F → SF_6

6.57 **a.** Ions: Al^{3+} and Cl^{-} → aluminum chloride
b. 1 B and 3 O → diboron trioxide
c. 2 N and 4 O → dinitrogen tetroxide
d. Ions: Sn^{2+} and NO_3^{-} → tin(II) nitrate
e. Ions: Cu^{2+} and ClO_2^{-} → copper(II) chlorite

6.59 Organic compounds contain C and H and sometimes O, N, or a halogen atom. Inorganic compounds usually contain elements other than C and H.
a. inorganic
b. organic
c. organic
d. inorganic
e. inorganic
f. organic

6.61 **a.** Inorganic compounds are usually soluble in water.
b. Organic compounds have lower boiling points than most inorganic compounds.
c. Organic compounds contain carbon and hydrogen.
d. Inorganic compounds contain ionic bonds.

6.63 **a.** Ethane; organic compounds have low boiling points.
b. Ethane; organic compounds burn vigorously in air.
c. Sodium bromide; inorganic compounds have high melting points.
d. Sodium bromide; inorganic compounds (ionic) typically dissolve in water.

6.65 **a.** Pentane has a carbon chain of five carbon atoms.
b. Ethane has a carbon chain of two carbon atoms.
c. Hexane has a carbon chain of six carbon atoms.

6.67 In the condensed structural formula for an alkane, each carbon atom and its attached hydrogen atoms are written as a group.
a. CH_4
b. $CH_3—CH_3$
c. $CH_3—CH_2—CH_2—CH_2—CH_3$

6.69 **a.** C_7H_{16}
b. $CH_3—CH_2—CH_2—CH_2—CH_2—CH_2—CH_3$ 
c. Heptane is a liquid at room temperature since it has a boiling point of 98 °C.
d. Heptane contains only nonpolar C—H bonds, which makes it insoluble in water.
e. Since the density of heptane (0.68 g/mL) is less than that of water, heptane will float on water.

6.71 **a.** By losing two valence electrons from the fourth energy level, calcium achieves an octet in the third energy level.
b. The calcium ion, Ca^{2+}, has an electron configuration $1s^22s^22p^63s^23p^6$, which is the same as argon (Ar).
c. Group 1A (1) and 2A (2) elements acquire octets by losing electrons when they form compounds. Group 8A (18) elements already have a stable octet of valence electrons (or two electrons for helium), so they are not normally found in compounds.

6.73 **a.** P^{3-} ion **b.** O atom
c. Zn^{2+} ion **d.** Fe^{3+} ion

6.75 **a.** With 35 protons, the element is bromine (Br). Since there are 36 electrons, it is the bromide ion, Br^-.
b. With 47 protons, the element is silver (Ag). Since there are 46 electrons, it is the silver ion, Ag^+.
c. With 50 protons, the element is tin (Sn). Since there are 46 electrons, it is the tin ion, Sn^{4+}.

6.77 **a.** X is in Group 1A (1); Y is in Group 6A (16) **b.** ionic
c. X^+, Y^{2-} **d.** X_2Y
e. Ions: X^+ and $Cl^- \rightarrow XCl$ **f.** Ions: Na^+ and $Y^{2-} \rightarrow Na_2Y$
g. ionic

6.79

Period	Electron-Dot Symbols	Elements	Formula of Compound	Name of Compound
2	X· and ·Ÿ· (5 dots)	Li and N	Li_3N	Lithium nitride
4	Ẋ· and :Ÿ· (7 dots)	Ca and Br	$CaBr_2$	Calcium bromide
4	·Ẋ· and :Ÿ· (6 dots)	Ga and Se	Ga_2Se_3	Gallium selenide

6.81

Electron Configurations		Symbols of Ions		Formula of Compound	Name of Compound
Metal	Nonmetal	Cation	Anion		
$1s^22s^22p^63s^2$	$1s^22s^22p^3$	Mg^{2+}	N^{3-}	Mg_3N_2	Magnesium nitride
$1s^22s^22p^63s^23p^64s^1$	$1s^22s^22p^4$	K^+	O^{2-}	K_2O	Potassium oxide
$1s^22s^22p^63s^23p^1$	$1s^22s^22p^5$	Al^{3+}	F^-	AlF_3	Aluminum fluoride

6.83 **a.** Butane; organic compounds have low melting points.
b. Butane; organic compounds burn vigorously in air.
c. Potassium chloride; inorganic compounds have high melting points.
d. Potassium chloride; inorganic compounds (ionic) contain ionic bonds.
e. Butane; organic compounds are more likely to be gases at room temperature.

6.85 **a.** Organic, molecular, heptane
b. Inorganic, molecular, 1 N and 2 O $\rightarrow$ nitrogen dioxide
c. Inorganic, ionic, ions are K^+ and PO_4^{3-} $\rightarrow$ potassium phosphate
d. Organic, molecular, ethane

6.87 **a.** An element that forms an ion with a 2+ charge would be in Group 2A (2).
b. The electron-dot symbol for an element in Group 2A (2) is $\dot{X}\cdot$
c. Be is the Group 2A (2) element in Period 2.
d. Ions: X^{2+} and $N^{3-} \rightarrow X_3N_2$

6.89 **a.** Tin(IV) is Sn^{4+}.
b. The Sn^{4+} ion has 50 protons and $50 - 4 = 46$ electrons.
c. Ions: Sn^{4+} and $O^{2-} \rightarrow SnO_2$
d. Ions: Sn^{4+} and $PO_4^{3-} \rightarrow Sn_3(PO_4)_4$

6.91 **a.** X as a X^{3+} ion would be in Group 3A (13).
b. X as a X^{2-} ion would be in Group 6A (16).
c. X as a X^{2+} ion would be in Group 2A (2).

6.93 **a.** Ions: Fe^{3+} and $Br^- \rightarrow$ iron(III) bromide
b. Ions: Ca^{2+} and $PO_4^{3-} \rightarrow$ calcium phosphate
c. Ions: Al^{3+} and $CO_3^{2-} \rightarrow$ aluminum carbonate
d. Ions: Pb^{4+} and $Cl^- \rightarrow$ lead(IV) chloride
e. Ions: Mg^{2+} and $ClO_3^- \rightarrow$ magnesium chlorate
f. Ions: Sn^{2+} and $SO_4^{2-} \rightarrow$ tin(II) sulfate
g. Ions: Cu^{2+} and $S^{2-} \rightarrow$ copper(II) sulfide

6.95 **a.** Ions: Cu^+ and $N^{3-} \rightarrow Cu_3N$
b. Ions: K^+ and $HSO_3^- \rightarrow KHSO_3$
c. Ions: Pb^{4+} and $S^{2-} \rightarrow PbS_2$
d. Ions: Au^{3+} and $CO_3^{2-} \rightarrow Au_2(CO_3)_3$
e. Ions: Zn^{2+} and $ClO_4^- \rightarrow Zn(ClO_4)_2$

6.97 **a.** Ions: Ni^{3+} and $Cl^- \rightarrow NiCl_3$
b. Ions: Pb^{4+} and $O^{2-} \rightarrow PbO_2$
c. Ions: Ag^+ and $Br^- \rightarrow AgBr$
d. Ions: Ca^{2+} and $N^{3-} \rightarrow Ca_3N_2$
e. Ions: Cu^+ and $P^{3-} \rightarrow Cu_3P$
f. Ions: Cr^{2+} and $S^{2-} \rightarrow CrS$

6.99 **a.** Ions: Mg^{2+} and $O^{2-} \rightarrow$ magnesium oxide
b. Ions: Cr^{3+} and $HCO_3^- \rightarrow$ chromium(III) hydrogen carbonate or chromium(III) bicarbonate
c. Ions: Mn^{3+} and $CrO_4^{2-} \rightarrow$ manganese(III) chromate

6.101 **a.** 1 N and 3 Cl $\rightarrow$ nitrogen trichloride
b. 2 N and 3 S $\rightarrow$ dinitrogen trisulfide
c. 2 N and 1 O $\rightarrow$ dinitrogen oxide
d. 1 I and 1 F $\rightarrow$ iodine fluoride
e. 1 P and 5 F $\rightarrow$ phosphorus pentafluoride
f. 2 P and 5 O $\rightarrow$ diphosphorus pentoxide

6.103 **a.** 1 I and 7 F $\rightarrow IF_7$
b. 1 B and 3 F $\rightarrow BF_3$
c. 1 H and 1 I $\rightarrow$ HI
d. 1 S and 2 Cl $\rightarrow SCl_2$

6.105 **a.** Ionic, ions are Co^{3+} and $Cl^- \rightarrow$ cobalt(III) chloride
b. Ionic, ions are Na^+ and $SO_4^{2-} \rightarrow$ sodium sulfate
c. Molecular, 2 N and 3 O $\rightarrow$ dinitrogen trioxide
d. Ionic, ions are Li^+ and $Br^- \rightarrow$ lithium bromide
e. Molecular, 1 I and 3 Cl $\rightarrow$ iodine trichloride
f. Molecular, 1 C and 4 F $\rightarrow$ carbon tetrafluoride

6.107 **a.** Ions: Sn^{2+} and $CO_3^{2-} \rightarrow SnCO_3$
b. Ions: Li^+ and $P^{3-} \rightarrow Li_3P$
c. Molecular, 1 Si and 4 Cl $\rightarrow SiCl_4$
d. Ions: Cr^{2+} and $S^{2-} \rightarrow CrS$
e. Molecular, 1 C and 2 O $\rightarrow CO_2$
f. Ions: Ca^{2+} and $Br^- \rightarrow CaBr_2$
g. Pentane is an alkane with five carbon atoms; its formula can be written as C_5H_{12} or $CH_3—CH_2—CH_2—CH_2—CH_3$.

6.109

Formula of Compound	Type of Compound	Name of Compound
$FeSO_4$	Ionic	Iron(II) sulfate
SiO_2	Molecular	Silicon dioxide
NH_4NO_3	Ionic	Ammonium nitrate
$Al_2(SO_4)_3$	Ionic	Aluminum sulfate
Co_2S_3	Ionic	Cobalt(III) sulfide

6.111 Compounds with a metal and nonmetal are classified as ionic; compounds with two nonmetals are molecular.
a. Inorganic, ionic, ions are Li^+ and $O^{2-} \rightarrow$ lithium oxide
b. Organic, molecular, propane
c. Inorganic, molecular, 1 Si and 4 F $\rightarrow$ silicon tetrafluoride
d. Organic, molecular, methane
e. Inorganic, ionic, ions are Mg^{2+} and $F^- \rightarrow$ magnesium fluoride
f. Organic, molecular, butane

6.113 Ions: Na^+ and $S^{2-} \rightarrow Na_2S$

$$4.8 \times 10^{22} \cancel{\text{Na}^+ \text{ions}} \times \frac{1\ \text{S}^{2-}\ \text{ions}}{2\ \cancel{\text{Na}^+\ \text{ions}}} = 2.4 \times 10^{22}\ \text{S}^{2-}\ \text{ions (2 SFs)}$$

6.115 In the condensed structural formula for an alkane, each carbon atom and its attached hydrogen atoms are written as a group.
a. $CH_3—CH_2—CH_3$
b. $CH_3—CH_2—CH_2—CH_2—CH_2—CH_3$

7

Chemical Quantities

Learning Goals

- Use Avogadro's number to determine the number of particles in a given number of moles.
- Given the chemical formula of a substance, calculate its molar mass.
- Given the number of moles of a substance, calculate the mass in grams; given the mass, calculate the number of moles.
- Given the formula of a compound, calculate the mass percent composition; from the mass percent composition, determine the empirical formula of a compound.
- Given a quantity in moles of reactant or product, use a mole–mole factor from the balanced equation to calculate the moles of another substance in the reaction.
- Determine the molecular formula of a substance from the empirical formula and molar mass.

Chapter Outline

Chapter Opener: Veterinarian

7.1 **The Mole**

7.2 **Molar Mass**

7.3 **Calculations Using Molar Mass**

7.4 **Mass Percent Composition and Empirical Formulas**

Chemistry Link to the Environment: Fertilizers

7.5 **Molecular Formulas**

Core Chemistry Skills

- Converting Particles to Moles (7.1)
- Calculating Molar Mass (7.2)
- Using Molar Mass as a Conversion Factor (7.3)
- Calculating Mass Percent Composition (7.4)
- Calculating an Empirical Formula (7.4)
- Calculating a Molecular Formula (7.5)

Answers and Solutions to Text Problems

7.1 **a.** $0.200 \text{ mol Ag} \times \dfrac{6.022 \times 10^{23} \text{ atoms Ag}}{1 \text{ mol Ag}} = 1.20 \times 10^{23}$ atoms of Ag (3 SFs)

b. $0.750 \text{ mol } C_3H_8O \times \dfrac{6.022 \times 10^{23} \text{ molecules } C_3H_8O}{1 \text{ mol } C_3H_8O} = 4.52 \times 10^{23}$ molecules of C_3H_8O (3 SFs)

c. $1.25 \text{ mol Cr} \times \dfrac{6.022 \times 10^{23} \text{ atoms Cr}}{1 \text{ mol Cr}} = 7.53 \times 10^{23}$ atoms of Cr (3 SFs)

7.3 **a.** $1.0\ \cancel{\text{mol } C_{20}H_{24}N_2O_2} \times \dfrac{24 \text{ mol H}}{1\ \cancel{\text{mol } C_{20}H_{24}N_2O_2}} = 24 \text{ mol of H (2 SFs)}$

b. $5.0\ \cancel{\text{mol } C_{20}H_{24}N_2O_2} \times \dfrac{20 \text{ mol C}}{1\ \cancel{\text{mol } C_{20}H_{24}N_2O_2}} = 1.0 \times 10^2 \text{ mol of C (2 SFs)}$

c. $0.020\ \cancel{\text{mol } C_{20}H_{24}N_2O_2} \times \dfrac{2 \text{ mol N}}{1\ \cancel{\text{mol } C_{20}H_{24}N_2O_2}} = 0.040 \text{ mol of N (2 SFs)}$

7.5 **a.** $0.500\ \cancel{\text{mol C}} \times \dfrac{6.022 \times 10^{23} \text{ atoms C}}{1\ \cancel{\text{mol C}}} = 3.01 \times 10^{23} \text{ atoms of C (3 SFs)}$

b. $1.28\ \cancel{\text{mol } SO_2} \times \dfrac{6.022 \times 10^{23} \text{ molecules } SO_2}{1\ \cancel{\text{mol } SO_2}} = 7.71 \times 10^{23} \text{ molecules of } SO_2 \text{ (3 SFs)}$

c. $5.22 \times 10^{22}\ \cancel{\text{atoms Fe}} \times \dfrac{1 \text{ mol Fe}}{6.022 \times 10^{23}\ \cancel{\text{atoms Fe}}} = 0.0867 \text{ mol of Fe (3 SFs)}$

7.7 1 mol of H_3PO_4 molecules contains 3 mol of H atoms, 1 mol of P atoms, and 4 mol of O atoms.

a. $2.00\ \cancel{\text{mol } H_3PO_4} \times \dfrac{3 \text{ mol H}}{1\ \cancel{\text{mol } H_3PO_4}} = 6.00 \text{ mol of H (3 SFs)}$

b. $2.00\ \cancel{\text{mol } H_3PO_4} \times \dfrac{4 \text{ mol O}}{1\ \cancel{\text{mol } H_3PO_4}} = 8.00 \text{ mol of O (3 SFs)}$

c. $2.00\ \cancel{\text{mol } H_3PO_4} \times \dfrac{1\ \cancel{\text{mol P}}}{1\ \cancel{\text{mol } H_3PO_4}} \times \dfrac{6.022 \times 10^{23} \text{ atoms P}}{1\ \cancel{\text{mol P}}} = 1.20 \times 10^{24} \text{ atoms of P (3 SFs)}$

d. $2.00\ \cancel{\text{mol } H_3PO_4} \times \dfrac{4\ \cancel{\text{mol O}}}{1\ \cancel{\text{mol } H_3PO_4}} \times \dfrac{6.022 \times 10^{23} \text{ atoms O}}{1\ \cancel{\text{mol O}}} = 4.82 \times 10^{24} \text{ atoms of O (3 SFs)}$

7.9 **a.** $1\ \cancel{\text{mol Na}} \times \dfrac{22.99 \text{ g Na}}{1\ \cancel{\text{mol Na}}} = 22.99 \text{ g of Na}$

$1\ \cancel{\text{mol Cl}} \times \dfrac{35.45 \text{ g Cl}}{1\ \cancel{\text{mol Cl}}} = 35.45 \text{ g of Cl}$

1 mol of Na = 22.99 g of Na

1 mol of Cl = 35.45 g of Cl

$\therefore$ Molar mass of NaCl = 58.44 g

b. $2\ \cancel{\text{mol Fe}} \times \dfrac{55.85 \text{ g Fe}}{1\ \cancel{\text{mol Fe}}} = 111.7 \text{ g of Fe}$

$3\ \cancel{\text{mol O}} \times \dfrac{16.00 \text{ g O}}{1\ \cancel{\text{mol O}}} = 48.00 \text{ g of O}$

2 mol of Fe = 111.7 g of Fe

3 mol of O = 48.00 g of O

$\therefore$ Molar mass of Fe_2O_3 = 159.7 g

c. $19\ \cancel{\text{mol C}} \times \dfrac{12.01 \text{ g C}}{1\ \cancel{\text{mol C}}} = 228.2 \text{ g of C}$

$20\ \cancel{\text{mol H}} \times \dfrac{1.008 \text{ g H}}{1\ \cancel{\text{mol H}}} = 20.16 \text{ g of H}$

$1\ \cancel{\text{mol F}} \times \dfrac{19.00 \text{ g F}}{1\ \cancel{\text{mol F}}} = 19.00 \text{ g of F}$

$$1\ \cancel{\text{mol N}} \times \frac{14.01\ \text{g N}}{1\ \cancel{\text{mol N}}} = 14.01\ \text{g of N}$$

$$3\ \cancel{\text{mol O}} \times \frac{16.00\ \text{g O}}{1\ \cancel{\text{mol O}}} = 48.00\ \text{g of O}$$

19 mol of C	=	228.2 g of C
20 mol of H	=	20.16 g of H
1 mol of F	=	19.00 g of F
1 mol of N	=	14.01 g of N
3 mol of O	=	48.00 g of O
∴ Molar mass of $C_{19}H_{20}FNO_3$	=	329.4 g

7.11 **a.** $$2\ \cancel{\text{mol Al}} \times \frac{26.98\ \text{g Al}}{1\ \cancel{\text{mol Al}}} = 53.96\ \text{g of Al}$$

$$3\ \cancel{\text{mol S}} \times \frac{32.07\ \text{g S}}{1\ \cancel{\text{mol S}}} = 96.21\ \text{g of S}$$

$$12\ \cancel{\text{mol O}} \times \frac{16.00\ \text{g O}}{1\ \cancel{\text{mol O}}} = 192.0\ \text{g of O}$$

2 mol of Al	=	53.96 g of Al
3 mol of S	=	96.21 g of S
12 mol of O	=	192.0 g of O
∴ Molar mass of $Al_2(SO_4)_3$	=	342.2 g

b. $$1\ \cancel{\text{mol K}} \times \frac{39.10\ \text{g K}}{1\ \cancel{\text{mol K}}} = 39.10\ \text{g of K}$$

$$4\ \cancel{\text{mol C}} \times \frac{12.01\ \text{g C}}{1\ \cancel{\text{mol C}}} = 48.04\ \text{g of C}$$

$$5\ \cancel{\text{mol H}} \times \frac{1.008\ \text{g H}}{1\ \cancel{\text{mol H}}} = 5.040\ \text{g of H}$$

$$6\ \cancel{\text{mol O}} \times \frac{16.00\ \text{g O}}{1\ \cancel{\text{mol O}}} = 96.00\ \text{g of O}$$

1 mol of K	=	39.10 g of K
4 mol of C	=	48.04 g of C
5 mol of H	=	5.040 g of H
6 mol of O	=	96.00 g of O
∴ Molar mass of $KC_4H_5O_6$	=	188.18 g

c. $$16\ \cancel{\text{mol C}} \times \frac{12.01\ \text{g C}}{1\ \cancel{\text{mol C}}} = 192.2\ \text{g of C}$$

$$19\ \cancel{\text{mol H}} \times \frac{1.008\ \text{g H}}{1\ \cancel{\text{mol H}}} = 19.15\ \text{g of H}$$

$$3\ \cancel{\text{mol N}} \times \frac{14.01\ \text{g N}}{1\ \cancel{\text{mol N}}} = 42.03\ \text{g of N}$$

$$5\ \cancel{\text{mol O}} \times \frac{16.00\ \text{g O}}{1\ \cancel{\text{mol O}}} = 80.00\ \text{g of O}$$

$$1\ \cancel{\text{mol S}} \times \frac{32.07\ \text{g S}}{1\ \cancel{\text{mol S}}} = 32.07\ \text{g of S}$$

$$\begin{aligned}
&16 \text{ mol of C} &&= 192.2 \text{ g of C}\\
&19 \text{ mol of H} &&= 19.15 \text{ g of H}\\
&3 \text{ mol of N} &&= 42.03 \text{ g of N}\\
&5 \text{ mol of O} &&= 80.00 \text{ g of O}\\
&1 \text{ mol of S} &&= 32.07 \text{ g of S}\\
&\therefore \text{ Molar mass of } C_{16}H_{19}N_3O_5S &&= 365.5 \text{ g}
\end{aligned}$$

7.13 **a.** $2 \cancel{\text{mol Cl}} \times \dfrac{35.45 \text{ g Cl}}{1 \cancel{\text{mol Cl}}} = 70.90 \text{ g of Cl}$

$\therefore$ Molar mass of Cl_2 = 70.90 g

b. $3 \cancel{\text{mol C}} \times \dfrac{12.01 \text{ g C}}{1 \cancel{\text{mol C}}} = 36.03 \text{ g of C}$

$6 \cancel{\text{mol H}} \times \dfrac{1.008 \text{ g H}}{1 \cancel{\text{mol H}}} = 6.048 \text{ g of H}$

$3 \cancel{\text{mol O}} \times \dfrac{16.00 \text{ g O}}{1 \cancel{\text{mol O}}} = 48.00 \text{ g of O}$

$$\begin{aligned}
&3 \text{ mol of C} &&= 36.03 \text{ g of C}\\
&6 \text{ mol of H} &&= 6.048 \text{ g of H}\\
&3 \text{ mol of O} &&= 48.00 \text{ g of O}\\
&\therefore \text{ Molar mass of } C_3H_6O_3 &&= 90.08 \text{ g}
\end{aligned}$$

c. $3 \cancel{\text{mol Mg}} \times \dfrac{24.31 \text{ g Mg}}{1 \cancel{\text{mol Mg}}} = 72.93 \text{ g of Mg}$

$2 \cancel{\text{mol P}} \times \dfrac{30.97 \text{ g P}}{1 \cancel{\text{mol P}}} = 61.94 \text{ g of P}$

$8 \cancel{\text{mol O}} \times \dfrac{16.00 \text{ g O}}{1 \cancel{\text{mol O}}} = 128.0 \text{ g of O}$

$$\begin{aligned}
&3 \text{ mol of Mg} &&= 72.93 \text{ g of Mg}\\
&2 \text{ mol of P} &&= 61.94 \text{ g of P}\\
&8 \text{ mol of O} &&= 128.0 \text{ g of O}\\
&\therefore \text{ Molar mass of } Mg_3(PO_4)_2 &&= 262.9 \text{ g}
\end{aligned}$$

7.15 **a.** $1 \cancel{\text{mol Al}} \times \dfrac{26.98 \text{ g Al}}{1 \cancel{\text{mol Al}}} = 26.98 \text{ g of Al}$

$3 \cancel{\text{mol F}} \times \dfrac{19.00 \text{ g F}}{1 \cancel{\text{mol F}}} = 57.00 \text{ g of F}$

$$\begin{aligned}
&1 \text{ mol of Al} &&= 26.98 \text{ g of Al}\\
&3 \text{ mol of F} &&= 57.00 \text{ g of F}\\
&\therefore \text{ Molar mass of } AlF_3 &&= 83.98 \text{ g}
\end{aligned}$$

b. $2 \cancel{\text{mol C}} \times \dfrac{12.01 \text{ g C}}{1 \cancel{\text{mol C}}} = 24.02 \text{ g of C}$

$4 \cancel{\text{mol H}} \times \dfrac{1.008 \text{ g H}}{1 \cancel{\text{mol H}}} = 4.032 \text{ g of H}$

$2 \cancel{\text{mol Cl}} \times \dfrac{35.45 \text{ g Cl}}{1 \cancel{\text{mol Cl}}} = 70.90 \text{ g of Cl}$

$$\begin{aligned} &\text{2 mol of C} &&= 24.02 \text{ g of C} \\ &\text{4 mol of H} &&= 4.032 \text{ g of H} \\ &\text{2 mol of Cl} &&= 70.90 \text{ g of Cl} \\ &\therefore \text{Molar mass of } C_2H_4Cl_2 &&= 98.95 \text{ g} \end{aligned}$$

c. $1 \cancel{\text{mol Sn}} \times \frac{118.7 \text{ g Sn}}{1 \cancel{\text{mol Sn}}} = 118.7 \text{ g of Sn}$

$2 \cancel{\text{mol F}} \times \frac{19.00 \text{ g F}}{1 \cancel{\text{mol F}}} = 38.00 \text{ g of F}$

$$\begin{aligned} &\text{1 mol of Sn} &&= 118.7 \text{ g of Sn} \\ &\text{2 mol of F} &&= 38.00 \text{ g of F} \\ &\therefore \text{Molar mass of } SnF_2 &&= 156.7 \text{ g} \end{aligned}$$

7.17 **a.** $1.50 \cancel{\text{mol Na}} \times \frac{22.99 \text{ g Na}}{1 \cancel{\text{mol Na}}} = 34.5 \text{ g of Na (3 SFs)}$

b. $2.80 \cancel{\text{mol Ca}} \times \frac{40.08 \text{ g Ca}}{1 \cancel{\text{mol Ca}}} = 112 \text{ g of Ca (3 SFs)}$

c. Molar mass of CO_2

$= 1 \text{ mol of C} + 2 \text{ mol of O} = 12.01 \text{ g} + 2(16.00 \text{ g}) = 44.01 \text{ g}$

$0.125 \cancel{\text{mol } CO_2} \times \frac{44.01 \text{ g } CO_2}{1 \cancel{\text{mol } CO_2}} = 5.50 \text{ g of } CO_2 \text{ (3 SFs)}$

d. Molar mass of Na_2CO_3

$= 2 \text{ mol of Na} + 1 \text{ mol of C} + 3 \text{ mol of O} = 2(22.99 \text{ g}) + 12.01 \text{ g} + 3(16.00 \text{ g})$
$= 105.99 \text{ g}$

$0.0485 \cancel{\text{mol } Na_2CO_3} \times \frac{105.99 \text{ g } Na_2CO_3}{1 \cancel{\text{mol } Na_2CO_3}} = 5.14 \text{ g of } Na_2CO_3 \text{ (3 SFs)}$

e. Molar mass of PCl_3

$= 1 \text{ mol of P} + 3 \text{ mol of Cl} = 30.97 \text{ g} + 3(35.45 \text{ g}) = 137.32 \text{ g}$

$7.14 \times 10^2 \cancel{\text{mol } PCl_3} \times \frac{137.32 \text{ g } PCl_3}{1 \cancel{\text{mol } PCl_3}} = 9.80 \times 10^4 \text{ g of } PCl_3 \text{ (3 SFs)}$

7.19 **a.** $0.150 \cancel{\text{mol Ne}} \times \frac{20.18 \text{ g Ne}}{1 \cancel{\text{mol Ne}}} = 3.03 \text{ g of Ne}$

b. Molar mass of I_2

$= 2 \text{ mol of I} = 2(126.9 \text{ g}) = 253.8 \text{ g}$

$0.150 \cancel{\text{mol } I_2} \times \frac{253.8 \text{ g } I_2}{1 \cancel{\text{mol } I_2}} = 38.1 \text{ g of } I_2 \text{ (3 SFs)}$

c. Molar mass of Na_2O

$= 2 \text{ mol of Na} + 1 \text{ mol of O} = 2(22.99 \text{ g}) + 16.00 \text{ g} = 61.98 \text{ g}$

$0.150 \cancel{\text{mol } Na_2O} \times \frac{61.98 \text{ g } Na_2O}{1 \cancel{\text{mol } Na_2O}} = 9.30 \text{ g of } Na_2O \text{ (3 SFs)}$

d. Molar mass of $Ca(NO_3)_2$

$= 1 \text{ mol of Ca} + 2 \text{ mol of N} + 6 \text{ mol of O} = 40.08 \text{ g} + 2(14.01 \text{ g}) + 6(16.00 \text{ g}) = 164.10 \text{ g}$

$0.150 \cancel{\text{mol } Ca(NO_3)_2} \times \frac{164.10 \text{ g } Ca(NO_3)_2}{1 \cancel{\text{mol } Ca(NO_3)_2}} = 24.6 \text{ g of } Ca(NO_3)_2 \text{ (3 SFs)}$

e. Molar mass of C_6H_{14}
$= 6$ mol of C + 14 mol of H $= 6(12.01\text{ g}) + 14(1.008\text{ g}) = 86.17\text{ g}$

$$0.150\ \cancel{\text{mol } C_6H_{14}} \times \frac{86.17\text{ g } C_6H_{14}}{1\ \cancel{\text{mol } C_6H_{14}}} = 12.9\text{ g of } C_6H_{14}\ (3\text{ SFs})$$

7.21 **a.** $$82.0\text{ g}\ \cancel{\text{Ag}} \times \frac{1\text{ mol Ag}}{107.9\text{ g}\ \cancel{\text{Ag}}} = 0.760\text{ mol of Ag } (3\text{ SFs})$$

b. $$0.288\text{ g}\ \cancel{\text{C}} \times \frac{1\text{ mol C}}{12.01\text{ g}\ \cancel{\text{C}}} = 0.0240\text{ mol of C } (3\text{ SFs})$$

c. Molar mass of NH_3
$= 1$ mol of N + 3 mol of H $= 1(14.01\text{ g}) + 3(1.008\text{ g}) = 17.03\text{ g}$

$$15.0\text{ g}\ \cancel{NH_3} \times \frac{1\text{ mol } NH_3}{17.03\text{ g}\ \cancel{NH_3}} = 0.881\text{ mol of } NH_3\ (3\text{ SFs})$$

d. Molar mass of C_3H_8
$= 3$ mol of C + 8 mol of H $= 3(12.01\text{ g}) + 8(1.008\text{ g}) = 44.09\text{ g}$

$$7.25\text{ g}\ \cancel{C_3H_8} \times \frac{1\text{ mol } C_3H_8}{44.09\text{ g}\ \cancel{C_3H_8}} = 0.164\text{ mol of } C_3H_8\ (3\text{ SFs})$$

e. Molar mass of Fe_2O_3
$= 2$ mol of Fe + 3 mol of O $= 2(55.85\text{ g}) + 3(16.00\text{ g}) = 159.7\text{ g}$

$$245\text{ g}\ \cancel{Fe_2O_3} \times \frac{1\text{ mol } Fe_2O_3}{159.7\text{ g}\ \cancel{Fe_2O_3}} = 1.53\text{ mol of } Fe_2O_3\ (3\text{ SFs})$$

7.23 **a.** $$25.0\text{ g}\ \cancel{\text{He}} \times \frac{1\text{ mol He}}{4.003\text{ g}\ \cancel{\text{He}}} = 6.25\text{ mol of He } (3\text{ SFs})$$

b. Molar mass of O_2
$= 2$ mol of O $= 2(16.00\text{ g}) = 32.00\text{ g}$

$$25.0\text{ g}\ \cancel{O_2} \times \frac{1\text{ mol } O_2}{32.00\text{ g}\ \cancel{O_2}} = 0.781\text{ mol of } O_2\ (3\text{ SFs})$$

c. Molar mass of $Al(OH)_3$
$= 1$ mol of Al + 3 mol of O + 3 mol of H $= 26.98\text{ g} + 3(16.00\text{ g}) + 3(1.008\text{ g}) = 78.00\text{ g}$

$$25.0\text{ g}\ \cancel{Al(OH)_3} \times \frac{1\text{ mol } Al(OH)_3}{78.00\text{ g}\ \cancel{Al(OH)_3}} = 0.321\text{ mol of } Al(OH)_3\ (3\text{ SFs})$$

d. Molar mass of Ga_2S_3
$= 2$ mol of Ga + 3 mol of S $= 2(69.72\text{ g}) + 3(32.07\text{ g}) = 235.7\text{ g}$

$$25.0\text{ g}\ \cancel{Ga_2S_3} \times \frac{1\text{ mol } Ga_2S_3}{235.7\text{ g}\ \cancel{Ga_2S_3}} = 0.106\text{ mol of } Ga_2S_3\ (3\text{ SFs})$$

e. Molar mass of C_4H_{10}
$= 4$ mol of C + 10 mol of H $= 4(12.01\text{ g}) + 10(1.008\text{ g}) = 58.12\text{ g}$

$$25.0\text{ g}\ \cancel{C_4H_{10}} \times \frac{1\text{ mol } C_4H_{10}}{58.12\text{ g}\ \cancel{C_4H_{10}}} = 0.430\text{ mol of } C_4H_{10}\ (3\text{ SFs})$$

7.25 **a.** $25.0\ \cancel{g\ C} \times \frac{1\ \cancel{mol\ C}}{12.01\ \cancel{g\ C}} \times \frac{6.022 \times 10^{23}\ \text{atoms C}}{1\ \cancel{mol\ C}} = 1.25 \times 10^{24}\ \text{atoms of C}$

b. $0.688\ \cancel{mol\ CO_2} \times \frac{1\ \cancel{mol\ C}}{1\ \cancel{mol\ CO_2}} \times \frac{6.022 \times 10^{23}\ \text{atoms C}}{1\ \cancel{mol\ C}} = 4.14 \times 10^{23}\ \text{atoms of C}$

c. Molar mass of C_3H_8

$= 3\ \text{mol of C} + 8\ \text{mol of H} = 3(12.01\ g) + 8(1.008\ g) = 44.09\ g$

$275\ \cancel{g\ C_3H_8} \times \frac{1\ \cancel{mol\ C_3H_8}}{44.09\ \cancel{g\ C_3H_8}} \times \frac{3\ \cancel{mol\ C}}{1\ \cancel{mol\ C_3H_8}} \times \frac{6.022 \times 10^{23}\ \text{atoms C}}{1\ \cancel{mol\ C}} = 1.13 \times 10^{25}\ \text{atoms of C}$

d. $1.84\ \cancel{mol\ C_2H_6O} \times \frac{2\ \cancel{mol\ C}}{1\ \cancel{mol\ C_2H_6O}} \times \frac{6.022 \times 10^{23}\ \text{atoms C}}{1\ \cancel{mol\ C}} = 2.22 \times 10^{24}\ \text{atoms of C}$

e. $7.5 \times 10^{24}\ \cancel{molecules\ CH_4} \times \frac{1\ \text{atom C}}{1\ \cancel{molecule\ CH_4}} = 7.5 \times 10^{24}\ \text{atoms of C (2 SFs)}$

7.27 **a.** Molar mass of C_3H_8

$= 3\ \text{mol of C} + 8\ \text{mol of H} = 3(12.01\ g) + 8(1.008\ g) = 44.09\ g$

$1.50\ \cancel{mol\ C_3H_8} \times \frac{44.09\ g\ C_3H_8}{1\ \cancel{mol\ C_3H_8}} = 66.1\ g\ \text{of}\ C_3H_8\ \text{(3 SFs)}$

b. $34.0\ \cancel{g\ C_3H_8} \times \frac{1\ mol\ C_3H_8}{44.09\ \cancel{g\ C_3H_8}} = 0.771\ \text{mol of}\ C_3H_8\ \text{(3 SFs)}$

c. $34.0\ \cancel{g\ C_3H_8} \times \frac{1\ \cancel{mol\ C_3H_8}}{44.09\ \cancel{g\ C_3H_8}} \times \frac{3\ \cancel{mol\ C}}{1\ \cancel{mol\ C_3H_8}} \times \frac{12.01\ g\ C}{1\ \cancel{mol\ C}} = 27.8\ g\ \text{of C (3 SFs)}$

d. $0.254\ \cancel{g\ C_3H_8} \times \frac{1\ \cancel{mol\ C_3H_8}}{44.09\ \cancel{g\ C_3H_8}} \times \frac{8\ \cancel{mol\ H}}{1\ \cancel{mol\ C_3H_8}} \times \frac{6.022 \times 10^{23}\ \text{atoms H}}{1\ \cancel{mol\ H}}$

$= 2.78 \times 10^{22}\ \text{atoms of H (3 SFs)}$

7.29 Mass percent of an element $= \frac{\text{mass of an element}}{\text{total mass of the compound}} \times 100\%$

a. Mass of compound = 4.68 g Si + 5.32 g O = 10.00 g of compound

Mass % Si $= \frac{4.68\ g\ Si}{10.00\ g\ compound} \times 100\% = 46.8\%\ Si\ \text{(3 SFs)}$

Mass % O $= \frac{5.32\ g\ O}{10.00\ g\ compound} \times 100\% = 53.2\%\ O\ \text{(3 SFs)}$

b. Mass of compound = 5.72 g C + 1.28 g H = 7.00 g of compound

Mass % C $= \frac{5.72\ g\ C}{7.00\ g\ compound} \times 100\% = 81.7\%\ C\ \text{(3 SFs)}$

Mass % H $= \frac{1.28\ g\ H}{7.00\ g\ compound} \times 100\% = 18.3\%\ H\ \text{(3 SFs)}$

c. Mass of compound = 16.1 g Na + 22.5 g S + 11.3 g O = 49.9 g of compound

Mass % Na $= \frac{16.1\ g\ Na}{49.9\ g\ compound} \times 100\% = 32.3\%\ Na\ \text{(3 SFs)}$

$$\text{Mass \% S} = \frac{22.5 \text{ g S}}{49.9 \text{ g compound}} \times 100\% = 45.1\% \text{ S (3 SFs)}$$

$$\text{Mass \% O} = \frac{11.3 \text{ g O}}{49.9 \text{ g compound}} \times 100\% = 22.6\% \text{ O (3 SFs)}$$

d. Mass of compound = 6.22 g C + 1.04 g H + 4.14 g O = 11.4 g of compound

$$\text{Mass \% C} = \frac{6.22 \text{ g C}}{11.4 \text{ g compound}} \times 100\% = 54.6\% \text{ C (3 SFs)}$$

$$\text{Mass \% H} = \frac{1.04 \text{ g H}}{11.4 \text{ g compound}} \times 100\% = 9.12\% \text{ H (3 SFs)}$$

$$\text{Mass \% O} = \frac{4.14 \text{ g O}}{11.4 \text{ g compound}} \times 100\% = 36.3\% \text{ O (3 SFs)}$$

7.31 a. $1 \cancel{\text{mol Mg}} \times \frac{24.31 \text{ g Mg}}{1 \cancel{\text{mol Mg}}} = 24.31 \text{ g of Mg}$

$2 \cancel{\text{mol F}} \times \frac{19.00 \text{ g F}}{1 \cancel{\text{mol F}}} = 38.00 \text{ g of F}$

$\therefore$ Molar mass of MgF_2 = 62.31 g

$$\text{Mass \% Mg} = \frac{24.31 \text{ g Mg}}{62.31 \text{ g } MgF_2} \times 100\% = 39.01\% \text{ Mg (4 SFs)}$$

$$\text{Mass \% F} = \frac{38.00 \text{ g F}}{62.31 \text{ g } MgF_2} \times 100\% = 60.99\% \text{ F (4 SFs)}$$

b. $1 \cancel{\text{mol Ca}} \times \frac{40.08 \text{ g Ca}}{1 \cancel{\text{mol Ca}}} = 40.08 \text{ g of Ca}$

$2 \cancel{\text{mol O}} \times \frac{16.00 \text{ g O}}{1 \cancel{\text{mol O}}} = 32.00 \text{ g of O}$

$2 \cancel{\text{mol H}} \times \frac{1.008 \text{ g H}}{1 \cancel{\text{mol H}}} = 2.016 \text{ g of H}$

$\therefore$ Molar mass of $Ca(OH)_2$ = 74.10 g

$$\text{Mass \% Ca} = \frac{40.08 \text{ g Ca}}{74.10 \text{ g } Ca(OH)_2} \times 100\% = 54.09\% \text{ Ca (4 SFs)}$$

$$\text{Mass \% O} = \frac{32.00 \text{ g O}}{74.10 \text{ g } Ca(OH)_2} \times 100\% = 43.18\% \text{ O (4 SFs)}$$

$$\text{Mass \% H} = \frac{2.016 \text{ g H}}{74.10 \text{ g } Ca(OH)_2} \times 100\% = 2.72\% \text{ H (rounded off to hundredths)}$$

c. $4 \cancel{\text{mol C}} \times \frac{12.01 \text{ g C}}{1 \cancel{\text{mol C}}} = 48.04 \text{ g of C}$

$8 \cancel{\text{mol H}} \times \frac{1.008 \text{ g H}}{1 \cancel{\text{mol H}}} = 8.064 \text{ g of H}$

$4 \cancel{\text{mol O}} \times \frac{16.00 \text{ g O}}{1 \cancel{\text{mol O}}} = 64.00 \text{ g of O}$

$\therefore$ Molar mass of $C_4H_8O_4$ = 120.10 g

$$\text{Mass \% C} = \frac{48.04 \text{ g C}}{120.10 \text{ g } C_4H_8O_4} \times 100\% = 40.00\% \text{ C (4 SFs)}$$

$$\text{Mass \% H} = \frac{8.064 \text{ g H}}{120.10 \text{ g } C_4H_8O_4} \times 100\% = 6.71\% \text{ H (rounded off to hundredths)}$$

$$\text{Mass \% O} = \frac{64.00 \text{ g O}}{120.10 \text{ g } C_4H_8O_4} \times 100\% = 53.29\% \text{ O (4 SFs)}$$

d. $3 \cancel{\text{mol N}} \times \frac{14.01 \text{ g N}}{1 \cancel{\text{mol N}}} = 42.03 \text{ g of N}$

$12 \cancel{\text{mol H}} \times \frac{1.008 \text{ g H}}{1 \cancel{\text{mol H}}} = 12.10 \text{ g of H}$

$1 \cancel{\text{mol P}} \times \frac{30.97 \text{ g P}}{1 \cancel{\text{mol P}}} = 30.97 \text{ g of P}$

$4 \cancel{\text{mol O}} \times \frac{16.00 \text{ g O}}{1 \cancel{\text{mol O}}} = \underline{64.00 \text{ g of O}}$

$\therefore$ Molar mass of $(NH_4)_3PO_4 = 149.10$ g

$$\text{Mass \% N} = \frac{42.03 \text{ g N}}{149.10 \text{ g } (NH_4)_3PO_4} \times 100\% = 28.19\% \text{ N (4 SFs)}$$

$$\text{Mass \% H} = \frac{12.10 \text{ g H}}{149.10 \text{ g } (NH_4)_3PO_4} \times 100\% = 8.12\% \text{ H (rounded off to hundredths)}$$

$$\text{Mass \% P} = \frac{30.97 \text{ g P}}{149.10 \text{ g } (NH_4)_3PO_4} \times 100\% = 20.77\% \text{ P (4 SFs)}$$

$$\text{Mass \% O} = \frac{64.00 \text{ g O}}{149.10 \text{ g } (NH_4)_3PO_4} \times 100\% = 42.92\% \text{ O (4 SFs)}$$

e. $17 \cancel{\text{mol C}} \times \frac{12.01 \text{ g C}}{1 \cancel{\text{mol C}}} = 204.2 \text{ g of C}$

$19 \cancel{\text{mol H}} \times \frac{1.008 \text{ g H}}{1 \cancel{\text{mol H}}} = 19.15 \text{ g of H}$

$1 \cancel{\text{mol N}} \times \frac{14.01 \text{ g N}}{1 \cancel{\text{mol N}}} = 14.01 \text{ g of N}$

$3 \cancel{\text{mol O}} \times \frac{16.00 \text{ g O}}{1 \cancel{\text{mol O}}} = \underline{48.00 \text{ g of O}}$

$\therefore$ Molar mass of $C_{17}H_{19}NO_3 = 285.4$ g

$$\text{Mass \% C} = \frac{204.2 \text{ g C}}{285.4 \text{ g } C_{17}H_{19}NO_3} \times 100\% = 71.55\% \text{ C (4 SFs)}$$

$$\text{Mass \% H} = \frac{19.15 \text{ g H}}{285.4 \text{ g } C_{17}H_{19}NO_3} \times 100\% = 6.71\% \text{ H (rounded off to hundredths)}$$

$$\text{Mass \% N} = \frac{14.01 \text{ g N}}{285.4 \text{ g } C_{17}H_{19}NO_3} \times 100\% = 4.91\% \text{ N (rounded off to hundredths)}$$

$$\text{Mass \% O} = \frac{48.00 \text{ g O}}{285.4 \text{ g } C_{17}H_{19}NO_3} \times 100\% = 16.82\% \text{ O (4 SFs)}$$

7.33 **a.** $2\ \cancel{\text{mol N}} \times \dfrac{14.01\text{ g N}}{1\ \cancel{\text{mol N}}} = 28.02\text{ g of N}$

$5\ \cancel{\text{mol O}} \times \dfrac{16.00\text{ g O}}{1\ \cancel{\text{mol O}}} = 80.00\text{ g of O}$

$\therefore$ Molar mass of $N_2O_5 = 108.02\text{ g}$

$$\text{Mass \% N} = \frac{28.02\text{ g N}}{108.02\text{ g } N_2O_5} \times 100\% = 25.94\%\text{ N (4 SFs)}$$

b. $1\ \cancel{\text{mol N}} \times \dfrac{14.01\text{ g N}}{1\ \cancel{\text{mol N}}} = 14.01\text{ g of N}$

$4\ \cancel{\text{mol H}} \times \dfrac{1.008\text{ g H}}{1\ \cancel{\text{mol H}}} = 4.032\text{ g of H}$

$1\ \cancel{\text{mol Cl}} \times \dfrac{35.45\text{ g Cl}}{1\ \cancel{\text{mol Cl}}} = 35.45\text{ g of Cl}$

$\therefore$ Molar mass of $NH_4Cl = 53.49\text{ g}$

$$\text{Mass \% N} = \frac{14.01\text{ g N}}{53.49\text{ g } NH_4Cl} \times 100\% = 26.19\%\text{ N (4 SFs)}$$

c. $2\ \cancel{\text{mol C}} \times \dfrac{12.01\text{ g C}}{1\ \cancel{\text{mol C}}} = 24.02\text{ g of C}$

$8\ \cancel{\text{mol H}} \times \dfrac{1.008\text{ g H}}{1\ \cancel{\text{mol H}}} = 8.064\text{ g of H}$

$2\ \cancel{\text{mol N}} \times \dfrac{14.01\text{ g N}}{1\ \cancel{\text{mol N}}} = 28.02\text{ g of N}$

$\therefore$ Molar mass of $C_2H_8N_2 = 60.10\text{ g}$

$$\text{Mass \% N} = \frac{28.02\text{ g N}}{60.10\text{ g } C_2H_8N_2} \times 100\% = 46.62\%\text{ N (4 SFs)}$$

d. $9\ \cancel{\text{mol C}} \times \dfrac{12.01\text{ g C}}{1\ \cancel{\text{mol C}}} = 108.1\text{ g of C}$

$15\ \cancel{\text{mol H}} \times \dfrac{1.008\text{ g H}}{1\ \cancel{\text{mol H}}} = 15.12\text{ g of H}$

$5\ \cancel{\text{mol N}} \times \dfrac{14.01\text{ g N}}{1\ \cancel{\text{mol N}}} = 70.05\text{ g of N}$

$1\ \cancel{\text{mol O}} \times \dfrac{16.00\text{ g O}}{1\ \cancel{\text{mol O}}} = 16.00\text{ g of O}$

$\therefore$ Molar mass of $C_9H_{15}N_5O = 209.3\text{ g}$

$$\text{Mass \% N} = \frac{70.05\text{ g N}}{209.3\text{ g } C_9H_{15}N_5O} \times 100\% = 33.47\%\text{ N (4 SFs)}$$

e. $14 \text{ mol C} \times \frac{12.01 \text{ g C}}{1 \text{ mol C}} = 168.1 \text{ g of C}$

$22 \text{ mol H} \times \frac{1.008 \text{ g H}}{1 \text{ mol H}} = 22.18 \text{ g of H}$

$2 \text{ mol N} \times \frac{14.01 \text{ g N}}{1 \text{ mol N}} = 28.02 \text{ g of N}$

$1 \text{ mol O} \times \frac{16.00 \text{ g O}}{1 \text{ mol O}} = 16.00 \text{ g of O}$

$\therefore$ Molar mass of $C_{14}H_{22}N_2O = 234.3 \text{ g}$

$\text{Mass \% N} = \frac{28.02 \text{ g N}}{234.3 \text{ g } C_{14}H_{22}N_2O} \times 100\% = 11.96\% \text{ N (4 SFs)}$

7.35 **a.** $3.57 \text{ g N} \times \frac{1 \text{ mol N}}{14.01 \text{ g N}} = 0.255 \text{ mol of N}$

$2.04 \text{ g O} \times \frac{1 \text{ mol O}}{16.00 \text{ g O}} = 0.128 \text{ mol of O}$ (smaller number of moles)

$\frac{0.255 \text{ mol N}}{0.128} = 1.99 \text{ mol N}$ $\qquad \frac{0.128 \text{ mol O}}{0.128} = 1.00 \text{ mol of O}$

$\therefore$ Empirical formula $= N_{1.99}O_{1.00} \rightarrow N_2O$

b. $7.00 \text{ g C} \times \frac{1 \text{ mol C}}{12.01 \text{ g C}} = 0.583 \text{ mol of C}$ (smaller number of moles)

$1.75 \text{ g H} \times \frac{1 \text{ mol H}}{1.008 \text{ g H}} = 1.74 \text{ mol of H}$

$\frac{0.583 \text{ mol C}}{0.583} = 1.00 \text{ mol of C}$ $\qquad \frac{1.74 \text{ mol H}}{0.583} = 2.98 \text{ mol of H}$

$\therefore$ Empirical formula $= C_{1.00}H_{2.98} \rightarrow CH_3$

c. $0.175 \text{ g H} \times \frac{1 \text{ mol H}}{1.008 \text{ g H}} = 0.174 \text{ mol of H}$ (smallest number of moles)

$2.44 \text{ g N} \times \frac{1 \text{ mol N}}{14.01 \text{ g N}} = 0.174 \text{ mol of N}$

$8.38 \text{ g O} \times \frac{1 \text{ mol O}}{16.00 \text{ g O}} = 0.524 \text{ mol of O}$

$\frac{0.174 \text{ mol H}}{0.174} = 1.00 \text{ mol of H}$ $\qquad \frac{0.174 \text{ mol N}}{0.174} = 1.00 \text{ mol of N}$

$\frac{0.524 \text{ mol O}}{0.174} = 3.01 \text{ mol of O}$

$\therefore$ Empirical formula $= H_{1.00}N_{1.00}O_{3.01} \rightarrow HNO_3$

d. $2.06\ \text{g Ca} \times \dfrac{1\ \text{mol Ca}}{40.08\ \text{g Ca}} = 0.0514\ \text{mol of Ca}$

$2.66\ \text{g Cr} \times \dfrac{1\ \text{mol Cr}}{52.00\ \text{g Cr}} = 0.0512\ \text{mol of Cr (smallest number of moles)}$

$3.28\ \text{g O} \times \dfrac{1\ \text{mol O}}{16.00\ \text{g O}} = 0.205\ \text{mol of O}$

$\dfrac{0.0514\ \text{mol Ca}}{0.0512} = 1.00\ \text{mol of Ca}$ $\qquad \dfrac{0.0512\ \text{mol Cr}}{0.0512} = 1.00\ \text{mol of Cr}$

$\dfrac{0.205\ \text{mol O}}{0.0512} = 4.00\ \text{mol of O}$

$\therefore$ Empirical formula $= Ca_{1.00}Cr_{1.00}O_{4.00} \rightarrow CaCrO_4$

7.37 **a.** In exactly 100. g of compound, there are 70.9 g of K and 29.1 g of S.

$70.9\ \text{g K} \times \dfrac{1\ \text{mol K}}{39.10\ \text{g K}} = 1.81\ \text{mol of K}$

$29.1\ \text{g S} \times \dfrac{1\ \text{mol S}}{32.07\ \text{g S}} = 0.907\ \text{mol of S (smaller number of moles)}$

$\dfrac{1.81\ \text{mol K}}{0.907} = 2.00\ \text{mol of K}$ $\qquad \dfrac{0.907\ \text{mol S}}{0.907} = 1.00\ \text{mol of S}$

$\therefore$ Empirical formula $= K_{2.00}S_{1.00} \rightarrow K_2S$

b. In exactly 100. g of compound, there are 55.0 g of Ga and 45.0 g of F.

$55.0\ \text{g Ga} \times \dfrac{1\ \text{mol Ga}}{69.72\ \text{g Ga}} = 0.789\ \text{mol of Ga (smaller number of moles)}$

$45.0\ \text{g F} \times \dfrac{1\ \text{mol F}}{19.00\ \text{g F}} = 2.37\ \text{mol of F}$

$\dfrac{0.789\ \text{mol Ga}}{0.789} = 1.00\ \text{mol of Ga}$ $\qquad \dfrac{2.37\ \text{mol F}}{0.789} = 3.00\ \text{mol of F}$

$\therefore$ Empirical formula $= Ga_{1.00}F_{3.00} \rightarrow GaF_3$

c. In exactly 100. g of compound, there are 31.0 g of B and 69.0 g of O.

$31.0\ \text{g B} \times \dfrac{1\ \text{mol B}}{10.81\ \text{g B}} = 2.87\ \text{mol of B (smaller number of moles)}$

$69.0\ \text{g O} \times \dfrac{1\ \text{mol O}}{16.00\ \text{g O}} = 4.31\ \text{mol of O}$

$\dfrac{2.87\ \text{mol B}}{2.87} = 1.00\ \text{mol of B}$ $\qquad \dfrac{4.31\ \text{mol O}}{2.87} = 1.50\ \text{mol of O}$

$\therefore$ Empirical formula $= B_{1.00}O_{1.50} \rightarrow B_{(1.00\times 2)}O_{(1.50\times 2)} = B_{2.00}O_{3.00} \rightarrow B_2O_3$

d. In exactly 100. g of compound, there are 18.8 g of Li, 16.3 g of C, and 64.9 g of O.

$$18.8\ g\,\cancel{Li} \times \frac{1\ mol\ Li}{6.941\ g\,\cancel{Li}} = 2.71\ mol\ of\ Li$$

$$16.3\ g\,\cancel{C} \times \frac{1\ mol\ C}{12.01\ g\,\cancel{C}} = 1.36\ mol\ of\ C\ (smallest\ number\ of\ moles)$$

$$64.9\ g\,\cancel{O} \times \frac{1\ mol\ O}{16.00\ g\,\cancel{O}} = 4.06\ mol\ of\ O$$

$$\frac{2.71\ mol\ Li}{1.36} = 1.99\ mol\ of\ Li \qquad \frac{1.36\ mol\ C}{1.36} = 1.00\ mol\ of\ C$$

$$\frac{4.06\ mol\ O}{1.36} = 2.99\ mol\ of\ O$$

$\therefore$ Empirical formula $= Li_{1.99}C_{1.00}O_{2.99} \rightarrow Li_2CO_3$

e. In exactly 100. g of compound, there are 51.7 g of C, 6.95 g of H, and 41.3 g of O.

$$51.7\ g\,\cancel{C} \times \frac{1\ mol\ C}{12.01\ g\,\cancel{C}} = 4.30\ mol\ of\ C$$

$$6.95\ g\,\cancel{H} \times \frac{1\ mol\ H}{1.008\ g\,\cancel{H}} = 6.89\ mol\ of\ H$$

$$41.3\ g\,\cancel{O} \times \frac{1\ mol\ O}{16.00\ g\,\cancel{O}} = 2.58\ mol\ of\ O\ (smallest\ number\ of\ moles)$$

$$\frac{4.30\ mol\ C}{2.58} = 1.67\ mol\ of\ C \qquad \frac{6.89\ mol\ H}{2.58} = 2.67\ mol\ of\ H$$

$$\frac{2.58\ mol\ O}{2.58} = 1.00\ mol\ of\ O$$

$\therefore$ Empirical formula $= C_{1.67}H_{2.67}O_{1.00} \rightarrow C_{(1.67\times3)}H_{(2.67\times3)}O_{(1.00\times3)}$

$= C_{5.01}H_{8.01}O_{3.00} \rightarrow C_5H_8O_3$

7.39 **a.** $H_2O_2 \rightarrow H_{(2\div2)}O_{(2\div2)} = HO$ (empirical formula)
b. $C_{18}H_{12} \rightarrow C_{(18\div6)}H_{(12\div6)} = C_3H_2$ (empirical formula)
c. $C_{10}H_{16}O_2 \rightarrow C_{(10\div2)}H_{(16\div2)}O_{(2\div2)} = C_5H_8O$ (empirical formula)
d. $C_9H_{18}N_6 \rightarrow C_{(9\div3)}H_{(18\div3)}N_{(6\div3)} = C_3H_6N_2$ (empirical formula)
e. $C_2H_4N_2O_2 \rightarrow C_{(2\div2)}H_{(4\div2)}N_{(2\div2)}O_{(2\div2)} = CH_2NO$ (empirical formula)

7.41 Empirical formula mass of $CH_2O = 12.01 + 2(1.008) + 16.00 = 30.03$ g

$$\text{Small integer} = \frac{\text{molar mass of fructose}}{\text{empirical formula mass of } CH_2O} = \frac{180\ \cancel{g}}{30.03\ \cancel{g}} = 6$$

$\therefore$ Molecular formula of fructose $= C_{(1\times6)}H_{(2\times6)}O_{(1\times6)} = C_6H_{12}O_6$

7.43 Empirical formula mass of CH $= 12.01 + 1.008 = 13.02$ g

For benzene: $$\text{Small integer} = \frac{\text{molar mass of benzene}}{\text{empirical formula mass of CH}} = \frac{78\ \cancel{g}}{13.02\ \cancel{g}} = 6$$

$\therefore$ Molecular formula of benzene $= C_{(1\times6)}H_{(1\times6)} = C_6H_6$

For acetylene: $$\text{Small integer} = \frac{\text{molar mass of acetylene}}{\text{empirical formula mass of CH}} = \frac{26\ \cancel{g}}{13.02\ \cancel{g}} = 2$$

$\therefore$ Molecular formula of acetylene $= C_{(1\times2)}H_{(1\times2)} = C_2H_2$

7.45 In exactly 100. g of mevalonic acid, there are 48.64 g of C, 8.16 g of H, and 43.20 g of O.

$$48.64 \text{ g C} \times \frac{1 \text{ mol C}}{12.01 \text{ g C}} = 4.050 \text{ mol of C}$$

$$8.16 \text{ g H} \times \frac{1 \text{ mol H}}{1.008 \text{ g H}} = 8.10 \text{ mol of H}$$

$$43.20 \text{ g O} \times \frac{1 \text{ mol O}}{16.00 \text{ g O}} = 2.70 \text{ mol of O (smallest number of moles)}$$

$$\frac{4.050 \text{ mol C}}{2.70} = 1.50 \text{ mol of C} \qquad \frac{8.10 \text{ mol H}}{2.70} = 3.00 \text{ mol of H}$$

$$\frac{2.70 \text{ mol O}}{2.70} = 1.00 \text{ mol of O}$$

$$\therefore \text{ Empirical formula} = C_{1.50}H_{3.00}O_{1.00} \rightarrow C_{(1.50\times2)}H_{(3.00\times2)}O_{(1.00\times2)}$$
$$= C_{3.00}H_{6.00}O_{2.00} \rightarrow C_3H_6O_2$$

Empirical formula mass of $C_3H_6O_2$ = 3(12.01) + 6(1.008) + 2(16.00) = 74.08 g

$$\text{Small integer} = \frac{\text{molar mass of mevalonic acid}}{\text{empirical formula mass of } C_3H_6O_2} = \frac{148 \text{ g}}{74.08 \text{ g}} = 2$$

$$\therefore \text{ Molecular formula of mevalonic acid} = C_{(3\times2)}H_{(6\times2)}O_{(2\times2)} = C_6H_{12}O_4$$

7.47 In exactly 100. g of vanillic acid, there are 57.14 g of C, 4.80 g of H, and 38.06 g of O.

$$57.14 \text{ g C} \times \frac{1 \text{ mol C}}{12.01 \text{ g C}} = 4.758 \text{ mol of C}$$

$$4.80 \text{ g H} \times \frac{1 \text{ mol H}}{1.008 \text{ g H}} = 4.76 \text{ mol of H}$$

$$38.06 \text{ g O} \times \frac{1 \text{ mol O}}{16.00 \text{ g O}} = 2.379 \text{ mol of O (smallest number of moles)}$$

$$\frac{4.758 \text{ mol C}}{2.379} = 2.000 \text{ mol of C} \qquad \frac{4.76 \text{ mol H}}{2.379} = 2.00 \text{ mol of H}$$

$$\frac{2.379 \text{ mol O}}{2.379} = 1.000 \text{ mol of O}$$

$$\therefore \text{ Empirical formula} = C_{2.000}H_{2.00}O_{1.000} \rightarrow C_2H_2O$$

Empirical formula mass of C_2H_2O = 2(12.01) + 2(1.008) + 16.00 = 42.04 g

$$\text{Small integer} = \frac{\text{molar mass of vanillic acid}}{\text{empirical formula mass of } C_2H_2O} = \frac{168 \text{ g}}{42.04 \text{ g}} = 4$$

$$\therefore \text{ Molecular formula of vanillic acid} = C_{(2\times4)}H_{(2\times4)}O_{(1\times4)} = C_8H_8O_4$$

7.49 In exactly 100. g of nicotine, there are 74.0 g of C, 8.7 g of H, and 17.3 g of N.

$$74.0\ \cancel{\text{g C}} \times \frac{1\ \text{mol C}}{12.01\ \cancel{\text{g C}}} = 6.16\ \text{mol of C}$$

$$8.7\ \cancel{\text{g H}} \times \frac{1\ \text{mol H}}{1.008\ \cancel{\text{g H}}} = 8.6\ \text{mol of H}$$

$$17.3\ \cancel{\text{g N}} \times \frac{1\ \text{mol N}}{14.01\ \cancel{\text{g N}}} = 1.23\ \text{mol of N (smallest number of moles)}$$

$$\frac{6.16\ \text{mol C}}{1.23} = 5.00\ \text{mol of C} \qquad \frac{8.6\ \text{mol H}}{1.23} = 7.0\ \text{mol of H}$$

$$\frac{1.23\ \text{mol N}}{1.23} = 1.00\ \text{mol of N}$$

$\therefore$ Empirical formula $= C_{5.00}H_{7.0}N_{1.00} \rightarrow C_5H_7N$

Empirical formula mass of $C_5H_7N = 5(12.01) + 7(1.008) + 14.01 = 81.12$ g

$$\text{Small integer} = \frac{\text{molar mass of nicotine}}{\text{empirical formula mass of } C_5H_7N} = \frac{162\ \cancel{\text{g}}}{81.12\ \cancel{\text{g}}} = 2$$

$\therefore$ Molecular formula of nicotine $= C_{(5\times2)}H_{(7\times2)}N_{(1\times2)} = C_{10}H_{14}N_2$

7.51 **a.** $C_{10}H_8N_2O_2S_2 \rightarrow C_{(10\div2)}H_{(8\div2)}N_{(2\div2)}O_{(2\div2)}S_{(2\div2)} = C_5H_4NOS$ (empirical formula)

b.
$$10\ \cancel{\text{mol C}} \times \frac{12.01\ \text{g C}}{1\ \cancel{\text{mol C}}} = 120.1\ \text{g of C}$$

$$8\ \cancel{\text{mol H}} \times \frac{1.008\ \text{g H}}{1\ \cancel{\text{mol H}}} = 8.064\ \text{g of H}$$

$$2\ \cancel{\text{mol N}} \times \frac{14.01\ \text{g N}}{1\ \cancel{\text{mol N}}} = 28.02\ \text{g of N}$$

$$2\ \cancel{\text{mol O}} \times \frac{16.00\ \text{g O}}{1\ \cancel{\text{mol O}}} = 32.00\ \text{g of O}$$

$$2\ \cancel{\text{mol S}} \times \frac{32.07\ \text{g S}}{1\ \cancel{\text{mol S}}} = 64.14\ \text{g of S}$$

$\therefore$ Molar mass of $C_{10}H_8N_2O_2S_2 = 252.3$ g

c.
$$\text{Mass \% C} = \frac{120.1\ \text{g C}}{252.3\ \text{g } C_{10}H_8N_2O_2S_2} \times 100\% = 47.60\%\ \text{C (4 SFs)}$$

$$\text{Mass \% H} = \frac{8.064\ \text{g H}}{252.3\ \text{g } C_{10}H_8N_2O_2S_2} \times 100\% = 3.20\%\ \text{H (rounded off to hundredths)}$$

$$\text{Mass \% N} = \frac{28.02\ \text{g N}}{252.3\ \text{g } C_{10}H_8N_2O_2S_2} \times 100\% = 11.11\%\ \text{N (4 SFs)}$$

$$\text{Mass \% O} = \frac{32.00\ \text{g O}}{252.3\ \text{g } C_{10}H_8N_2O_2S_2} \times 100\% = 12.68\%\ \text{O (4 SFs)}$$

$$\text{Mass \% S} = \frac{64.14\ \text{g S}}{252.3\ \text{g } C_{10}H_8N_2O_2S_2} \times 100\% = 25.42\%\ \text{S (4 SFs)}$$

d. $25.0\ \cancel{g\ C_{10}H_8N_2O_2S_2} \times \frac{1\ \cancel{mol\ C_{10}H_8N_2O_2S_2}}{252.3\ \cancel{g\ C_{10}H_8N_2O_2S_2}} \times \frac{10\ \cancel{mol\ C}}{1\ \cancel{mol\ C_{10}H_8N_2O_2S_2}}$

$\times \frac{6.022 \times 10^{23}\ \text{atoms C}}{1\ \cancel{mol\ C}}$

$= 5.97 \times 10^{23}$ atoms of C (3 SFs)

e. $25.0\ \cancel{g\ C_{10}H_8N_2O_2S_2} \times \frac{1\ \text{mol}\ C_{10}H_8N_2O_2S_2}{252.3\ \cancel{g\ C_{10}H_8N_2O_2S_2}}$

$= 0.0991$ mol of $C_{10}H_8N_2O_2S_2$ (3 SFs)

7.53 (1) **a.** molecular formula = S_2Cl_2

b. $S_2Cl_2 \rightarrow S_{(2 \div 2)}Cl_{(2 \div 2)}$ = SCl (empirical formula)

c. $2\ \cancel{mol\ S} \times \frac{32.07\ g\ S}{1\ \cancel{mol\ S}} = 64.14$ g of S

$2\ \cancel{mol\ Cl} \times \frac{35.45\ g\ Cl}{1\ \cancel{mol\ Cl}} = 70.90$ g of Cl

∴ Molar mass of S_2Cl_2 = 135.04 g

d. Mass % S $= \frac{64.14\ g\ S}{135.04\ g\ S_2Cl_2} \times 100\% = 47.50\%$ S (4 SFs)

Mass % Cl $= \frac{70.90\ g\ Cl}{135.04\ g\ S_2Cl_2} \times 100\% = 52.50\%$ Cl (4 SFs)

(2) **a.** molecular formula = C_6H_6

b. $C_6H_6 \rightarrow C_{(6 \div 6)}H_{(6 \div 6)}$ = CH (empirical formula)

c. $6\ \cancel{mol\ C} \times \frac{12.01\ g\ C}{1\ \cancel{mol\ C}} = 72.06$ g of C

$6\ \cancel{mol\ H} \times \frac{1.008\ g\ H}{1\ \cancel{mol\ H}} = 6.048$ g of H

∴ Molar mass of C_6H_6 = 78.11 g

d. Mass % C $= \frac{72.06\ g\ C}{78.11\ g\ C_6H_6} \times 100\% = 92.25\%$ C (4 SFs)

Mass % H $= \frac{6.048\ g\ H}{78.11\ g\ C_6H_6} \times 100\% = 7.74\%$ H (rounded off to hundredths)

7.55 **a.** $1\ \cancel{mol\ Zn} \times \frac{65.41\ g\ Zn}{1\ \cancel{mol\ Zn}} = 65.41$ g of Zn

$1\ \cancel{mol\ S} \times \frac{32.07\ g\ S}{1\ \cancel{mol\ S}} = 32.07$ g of S

$4\ \cancel{mol\ O} \times \frac{16.00\ g\ O}{1\ \cancel{mol\ O}} = 64.00$ g of O

1 mol of Zn = 65.41 g of Zn
1 mol of S = 32.07 g of S
4 mol of O = 64.00 g of O

∴ Molar mass of $ZnSO_4$ = 161.48 g

b. $1\ \cancel{\text{mol Ca}} \times \dfrac{40.08\ \text{g Ca}}{1\ \cancel{\text{mol Ca}}} = 40.08\ \text{g of Ca}$

$2\ \cancel{\text{mol I}} \times \dfrac{126.9\ \text{g I}}{1\ \cancel{\text{mol I}}} = 253.8\ \text{g of I}$

$6\ \cancel{\text{mol O}} \times \dfrac{16.00\ \text{g O}}{1\ \cancel{\text{mol O}}} = 96.00\ \text{g of O}$

1 mol of Ca = 40.08 g of Ca
2 mol of I = 253.8 g of I
6 mol of O = 96.00 g of O

$\therefore$ Molar mass of $Ca(IO_3)_2$ = 389.9 g

c. $5\ \cancel{\text{mol C}} \times \dfrac{12.01\ \text{g C}}{1\ \cancel{\text{mol C}}} = 60.05\ \text{g of C}$

$8\ \cancel{\text{mol H}} \times \dfrac{1.008\ \text{g H}}{1\ \cancel{\text{mol H}}} = 8.064\ \text{g of H}$

$1\ \cancel{\text{mol N}} \times \dfrac{14.01\ \text{g N}}{1\ \cancel{\text{mol N}}} = 14.01\ \text{g of N}$

$1\ \cancel{\text{mol Na}} \times \dfrac{22.99\ \text{g Na}}{1\ \cancel{\text{mol Na}}} = 22.99\ \text{g of Na}$

$4\ \cancel{\text{mol O}} \times \dfrac{16.00\ \text{g O}}{1\ \cancel{\text{mol O}}} = 64.00\ \text{g of O}$

5 mol of C = 60.05 g of C
8 mol of H = 8.064 g of H
1 mol of N = 14.01 g of N
1 mol of Na = 22.99 g of Na
4 mol of O = 64.00 g of O

$\therefore$ Molar mass of $C_5H_8NNaO_4$ = 169.11 g

d. $6\ \cancel{\text{mol C}} \times \dfrac{12.01\ \text{g C}}{1\ \cancel{\text{mol C}}} = 72.06\ \text{g of C}$

$12\ \cancel{\text{mol H}} \times \dfrac{1.008\ \text{g H}}{1\ \cancel{\text{mol H}}} = 12.10\ \text{g of H}$

$2\ \cancel{\text{mol O}} \times \dfrac{16.00\ \text{g O}}{1\ \cancel{\text{mol O}}} = 32.00\ \text{g of O}$

6 mol of C = 72.06 g of C
12 mol of H = 12.10 g of H
2 mol of O = 32.00 g of O

$\therefore$ Molar mass of $C_6H_{12}O_2$ = 116.16 g

7.57 Mass percent of an element $= \dfrac{\text{mass of an element}}{\text{total mass of the compound}} \times 100\%$

a. Mass of compound = 3.85 g Ca + 3.65 g F = 7.50 g of compound

$\text{Mass \% Ca} = \dfrac{3.85\ \text{g Ca}}{7.50\ \text{g compound}} \times 100\% = 51.3\%\ \text{Ca (3 SFs)}$

$\text{Mass \% F} = \dfrac{3.65\ \text{g F}}{7.50\ \text{g compound}} \times 100\% = 48.7\%\ \text{F (3 SFs)}$

b. Mass of compound $= 0.389\text{ g Na} + 0.271\text{ g O} = 0.660\text{ g of compound}$

$$\text{Mass \% Na} = \frac{0.389\text{ g Na}}{0.660\text{ g compound}} \times 100\% = 58.9\%\text{ Na (3 SFs)}$$

$$\text{Mass \% O} = \frac{0.271\text{ g O}}{0.660\text{ g compound}} \times 100\% = 41.1\%\text{ O (3 SFs)}$$

c. Mass of compound $= 12.4\text{ g K} + 17.4\text{ g Mn} + 20.3\text{ g O} = 50.1\text{ g of compound}$

$$\text{Mass \% K} = \frac{12.4\text{ g K}}{50.1\text{ g compound}} \times 100\% = 24.8\%\text{ K (3 SFs)}$$

$$\text{Mass \% Mn} = \frac{17.4\text{ g Mn}}{50.1\text{ g compound}} \times 100\% = 34.7\%\text{ Mn (3 SFs)}$$

$$\text{Mass \% O} = \frac{20.3\text{ g O}}{50.1\text{ g compound}} \times 100\% = 40.5\%\text{ O (3 SFs)}$$

7.59 a. $2\text{ mol K} \times \frac{39.10\text{ g K}}{1\text{ mol K}} = 78.20\text{ g of K}$

$1\text{ mol Cr} \times \frac{52.00\text{ g Cr}}{1\text{ mol Cr}} = 52.00\text{ g of Cr}$

$4\text{ mol O} \times \frac{16.00\text{ g O}}{1\text{ mol O}} = 64.00\text{ g of O}$

$\therefore$ Molar mass of $K_2CrO_4 = 194.20\text{ g}$

$$\text{Mass \% K} = \frac{78.20\text{ g K}}{194.20\text{ g }K_2CrO_4} \times 100\% = 40.27\%\text{ K (4 SFs)}$$

$$\text{Mass \% Cr} = \frac{52.00\text{ g Cr}}{194.20\text{ g }K_2CrO_4} \times 100\% = 26.78\%\text{ Cr (4 SFs)}$$

$$\text{Mass \% O} = \frac{64.00\text{ g O}}{194.20\text{ g }K_2CrO_4} \times 100\% = 32.96\%\text{ O (4 SFs)}$$

b. $1\text{ mol Al} \times \frac{26.98\text{ g Al}}{1\text{ mol Al}} = 26.98\text{ g of Al}$

$3\text{ mol H} \times \frac{1.008\text{ g H}}{1\text{ mol H}} = 3.024\text{ g of H}$

$3\text{ mol C} \times \frac{12.01\text{ g C}}{1\text{ mol C}} = 36.03\text{ g of C}$

$9\text{ mol O} \times \frac{16.00\text{ g O}}{1\text{ mol O}} = 144.0\text{ g of O}$

$\therefore$ Molar mass of $Al(HCO_3)_3 = 210.0\text{ g}$

$$\text{Mass \% Al} = \frac{26.98\text{ g Al}}{210.0\text{ g }Al(HCO_3)_3} \times 100\% = 12.85\%\text{ Al (4 SFs)}$$

$$\text{Mass \% H} = \frac{3.024\text{ g H}}{210.0\text{ g }Al(HCO_3)_3} \times 100\% = 1.44\%\text{ H (rounded off to hundredths)}$$

$$\text{Mass \% C} = \frac{36.03\text{ g C}}{210.0\text{ g }Al(HCO_3)_3} \times 100\% = 17.16\%\text{ C (4 SFs)}$$

$$\text{Mass \% O} = \frac{144.0\text{ g O}}{210.0\text{ g }Al(HCO_3)_3} \times 100\% = 68.57\%\text{ O (4 SFs)}$$

c. $6 \text{ mol C} \times \frac{12.01 \text{ g C}}{1 \text{ mol C}} = 72.06 \text{ g of C}$

$12 \text{ mol H} \times \frac{1.008 \text{ g H}}{1 \text{ mol H}} = 12.10 \text{ g of H}$

$6 \text{ mol O} \times \frac{16.00 \text{ g O}}{1 \text{ mol O}} = 96.00 \text{ g of O}$

$\therefore$ Molar mass of $C_6H_{12}O_6 = 180.16 \text{ g}$

$$\text{Mass \% C} = \frac{72.06 \text{ g C}}{180.16 \text{ g } C_6H_{12}O_6} \times 100\% = 40.00\% \text{ C (4 SFs)}$$

$$\text{Mass \% H} = \frac{12.10 \text{ g H}}{180.16 \text{ g } C_6H_{12}O_6} \times 100\% = 6.72\% \text{ H (rounded off to hundredths)}$$

$$\text{Mass \% O} = \frac{96.00 \text{ g O}}{180.16 \text{ g } C_6H_{12}O_6} \times 100\% = 53.29\% \text{ O (4 SFs)}$$

7.61 **a.** $9 \text{ mol C} \times \frac{12.01 \text{ g C}}{1 \text{ mol C}} = 108.1 \text{ g of C}$

$8 \text{ mol H} \times \frac{1.008 \text{ g H}}{1 \text{ mol H}} = 8.064 \text{ g of H}$

$4 \text{ mol O} \times \frac{16.00 \text{ g O}}{1 \text{ mol O}} = 64.00 \text{ g of O}$

$\therefore$ Molar mass of $C_9H_8O_4 = 180.2 \text{ g}$

$$\text{Mass \% C} = \frac{108.1 \text{ g C}}{180.2 \text{ g } C_9H_8O_4} \times 100\% = 59.99\% \text{ C (4 SFs)}$$

$$\text{Mass \% H} = \frac{8.064 \text{ g H}}{180.2 \text{ g } C_9H_8O_4} \times 100\% = 4.48\% \text{ H (rounded off to hundredths)}$$

$$\text{Mass \% O} = \frac{64.00 \text{ g O}}{180.2 \text{ g } C_9H_8O_4} \times 100\% = 35.52\% \text{ O (4 SFs)}$$

b. $5.0 \times 10^{24} \text{ atoms C} \times \frac{1 \text{ mol C}}{6.022 \times 10^{23} \text{ atoms C}} \times \frac{1 \text{ mol } C_9H_8O_4}{9 \text{ mol C}}$

$= 0.92 \text{ mol of } C_9H_8O_4$ (aspirin) (2 SFs)

c. $7.50 \text{ g } C_9H_8O_4 \times \frac{1 \text{ mol } C_9H_8O_4}{180.2 \text{ g } C_9H_8O_4} \times \frac{4 \text{ mol O}}{1 \text{ mol } C_9H_8O_4}$

$\times \frac{6.022 \times 10^{23} \text{ atoms O}}{1 \text{ mol O}}$

$= 1.00 \times 10^{23}$ atoms of O (3 SFs)

d. $2.50 \text{ g H} \times \frac{1 \text{ mol H}}{1.008 \text{ g H}} \times \frac{1 \text{ mol } C_9H_8O_4}{8 \text{ mol H}} \times \frac{6.022 \times 10^{23} \text{ molecules } C_9H_8O_4}{1 \text{ mol } C_9H_8O_4}$

$= 1.87 \times 10^{23}$ molecules of $C_9H_8O_4$ (3 SFs)

7.63 **a.** $0.250\ \cancel{\text{mol Mn}_2\text{O}_3} \times \dfrac{3\ \text{mol O}}{1\ \cancel{\text{mol Mn}_2\text{O}_3}} = 0.750\ \text{mol of O}$

Molar mass of MnO_2

$= 1\ \text{mol of Mn} + 2\ \text{mol of O} = 54.94\ \text{g} + 2(16.00\ \text{g}) = 86.94\ \text{g}$

$20.0\ \cancel{\text{g MnO}_2} \times \dfrac{1\ \cancel{\text{mol MnO}_2}}{86.94\ \cancel{\text{g MnO}_2}} \times \dfrac{2\ \text{mol O}}{1\ \cancel{\text{mol MnO}_2}} = 0.460\ \text{mol of O}$

$\therefore$ Total moles of O $= 0.750\ \text{mol O} + 0.460\ \text{mol O} = 1.210\ \text{mol of O}$

$1.210\ \cancel{\text{mol O}} \times \dfrac{6.022 \times 10^{23}\ \text{atoms O}}{1\ \cancel{\text{mol O}}} = 7.29 \times 10^{23}\ \text{atoms of O (3 SFs)}$

b. $0.250\ \cancel{\text{mol Mn}_2\text{O}_3} \times \dfrac{2\ \cancel{\text{mol Mn}}}{1\ \cancel{\text{mol Mn}_2\text{O}_3}} \times \dfrac{54.94\ \text{g Mn}}{1\ \cancel{\text{mol Mn}}} = 27.5\ \text{g of Mn}$

$20.0\ \cancel{\text{g MnO}_2} \times \dfrac{1\ \cancel{\text{mol MnO}_2}}{86.94\ \cancel{\text{g MnO}_2}} \times \dfrac{1\ \cancel{\text{mol Mn}}}{1\ \cancel{\text{mol MnO}_2}} \times \dfrac{54.94\ \text{g Mn}}{1\ \cancel{\text{mol Mn}}} = 12.6\ \text{g of Mn}$

$\therefore$ Total grams of Mn $= 27.5\ \text{g Mn} + 12.6\ \text{g Mn} = 40.1\ \text{g of Mn (3 SFs)}$

7.65 **a.** $C_5H_5N_5 \rightarrow C_{(5\div5)}H_{(5\div5)}N_{(5\div5)} = CHN$ (empirical formula)

b. $FeC_2O_4 \rightarrow Fe_{(1\div1)}C_{(2\div1)}O_{(4\div1)} = FeC_2O_4$ (empirical formula)

c. $C_{16}H_{16}N_4 \rightarrow C_{(16\div4)}H_{(16\div4)}N_{(4\div4)} = C_4H_4N$ (empirical formula)

d. $C_6H_{14}N_2O_2 \rightarrow C_{(6\div2)}H_{(14\div2)}N_{(2\div2)}O_{(2\div2)} = C_3H_7NO$ (empirical formula)

7.67 **a.** $2.20\ \cancel{\text{g S}} \times \dfrac{1\ \text{mol S}}{32.07\ \cancel{\text{g S}}} = 0.0686\ \text{mol of S (smaller number of moles)}$

$7.81\ \cancel{\text{g F}} \times \dfrac{1\ \text{mol F}}{19.00\ \cancel{\text{g F}}} = 0.411\ \text{mol of F}$

$\dfrac{0.0686\ \text{mol S}}{0.0686} = 1.00\ \text{mol of S}$ $\qquad \dfrac{0.411\ \text{mol F}}{0.0686} = 5.99\ \text{mol of F}$

$\therefore$ Empirical formula $= S_{1.00}F_{5.99} \rightarrow SF_6$

b. $6.35\ \cancel{\text{g Ag}} \times \dfrac{1\ \text{mol Ag}}{107.9\ \cancel{\text{g Ag}}} = 0.0589\ \text{mol of Ag (smallest number of moles)}$

$0.825\ \cancel{\text{g N}} \times \dfrac{1\ \text{mol N}}{14.01\ \cancel{\text{g N}}} = 0.0589\ \text{mol of N}$

$2.83\ \cancel{\text{g O}} \times \dfrac{1\ \text{mol O}}{16.00\ \cancel{\text{g O}}} = 0.177\ \text{mol of O}$

$\dfrac{0.0589\ \text{mol Ag}}{0.0589} = 1.00\ \text{mol of Ag}$ $\qquad \dfrac{0.0589\ \text{mol N}}{0.0589} = 1.00\ \text{mol of N}$

$\dfrac{0.177\ \text{mol O}}{0.0589} = 3.01\ \text{mol of O}$

$\therefore$ Empirical formula $= Ag_{1.00}N_{1.00}O_{3.01} \rightarrow AgNO_3$

c. In exactly 100. g of compound, there are 43.6 g P and 56.4 g O.

$$43.6\ \cancel{g\,P} \times \frac{1\text{ mol P}}{30.97\ \cancel{g\,P}} = 1.41\text{ mol of P (smaller number of moles)}$$

$$56.4\ \cancel{g\,O} \times \frac{1\text{ mol O}}{16.00\ \cancel{g\,O}} = 3.53\text{ mol of O}$$

$$\frac{1.41\text{ mol P}}{1.41} = 1.00\text{ mol of P} \qquad \frac{3.53\text{ mol O}}{1.41} = 2.50\text{ mol of O}$$

$\therefore$ Empirical formula $= P_{1.00}O_{2.50} \rightarrow P_{(1.00\times2)}O_{(2.50\times2)} \rightarrow P_2O_5$

d. In exactly 100. g of compound, there are 22.1 g Al, 25.4 g P, and 52.5 g O.

$$22.1\ \cancel{g\,Al} \times \frac{1\text{ mol Al}}{26.98\ \cancel{g\,Al}} = 0.819\text{ mol of Al (smallest number of moles)}$$

$$25.4\ \cancel{g\,P} \times \frac{1\text{ mol P}}{30.97\ \cancel{g\,P}} = 0.820\text{ mol of P}$$

$$52.5\ \cancel{g\,O} \times \frac{1\text{ mol O}}{16.00\ \cancel{g\,O}} = 3.28\text{ mol of O}$$

$$\frac{0.819\text{ mol Al}}{0.819} = 1.00\text{ mol of Al} \qquad \frac{0.820\text{ mol P}}{0.819} = 1.00\text{ mol of P}$$

$$\frac{3.28\text{ mol O}}{0.819} = 4.00\text{ mol of O}$$

$\therefore$ Empirical formula $= Al_{1.00}P_{1.00}O_{4.00} \rightarrow AlPO_4$

7.69 a. In exactly 100. g of oleic acid, there are 76.54 g of C, 12.13 g of H, and 11.33 g of O.

$$76.54\ \cancel{g\,C} \times \frac{1\text{ mol C}}{12.01\ \cancel{g\,C}} = 6.373\text{ mol of C}$$

$$12.13\ \cancel{g\,H} \times \frac{1\text{ mol H}}{1.008\ \cancel{g\,H}} = 12.03\text{ mol of H}$$

$$11.33\ \cancel{g\,O} \times \frac{1\text{ mol O}}{16.00\ \cancel{g\,O}} = 0.7081\text{ mol of O (smallest number of moles)}$$

$$\frac{6.373\text{ mol C}}{0.7081} = 9.000\text{ mol of C} \qquad \frac{12.03\text{ mol H}}{0.7081} = 16.99\text{ mol of H}$$

$$\frac{0.7081\text{ mol O}}{0.7081} = 1.000\text{ mol of O}$$

$\therefore$ Empirical formula $= C_{9.000}H_{16.99}O_{1.000} \rightarrow C_9H_{17}O$

Empirical formula mass of $C_9H_{17}O = 9(12.01) + 17(1.008) + 16.00 = 141.2$ g

$$\text{Small integer} = \frac{\text{molar mass of oleic acid}}{\text{empirical formula mass of } C_9H_{17}O} = \frac{282\ \cancel{g}}{141.2\ \cancel{g}} = 2$$

$\therefore$ Molecular formula of oleic acid $= C_{(9\times2)}H_{(17\times2)}O_{(1\times2)} = C_{18}H_{34}O_2$

b. Molar mass of oleic acid ($C_{18}H_{34}O_2$)
= 18 mol of C + 34 mol of H + 2 mol of O
= 18(12.01 g) + 34(1.008 g) + 2(16.00 g) = 282.5 g

$$3.00\ \cancel{\text{mL } C_{18}H_{34}O_2} \times \frac{0.895\ \cancel{\text{g } C_{18}H_{34}O_2}}{1\ \cancel{\text{mL } C_{18}H_{34}O_2}} \times \frac{1\ \cancel{\text{mol } C_{18}H_{34}O_2}}{282.5\ \cancel{\text{g } C_{18}H_{34}O_2}} \times \frac{6.022 \times 10^{23}\text{ molecules } C_{18}H_{34}O_2}{1\ \cancel{\text{mol } C_{18}H_{34}O_2}}$$

$= 5.72 \times 10^{21}$ molecules of $C_{18}H_{34}O_2$ (3 SFs)

7.71 In exactly 100. g of succinic acid, there are 40.7 g of C, 5.12 g of H, and 54.2 g of O.

$$40.7\ \cancel{\text{g C}} \times \frac{1\text{ mol C}}{12.01\ \cancel{\text{g C}}} = 3.39\text{ mol of C}$$

$$5.12\ \cancel{\text{g H}} \times \frac{1\text{ mol H}}{1.008\ \cancel{\text{g H}}} = 5.08\text{ mol of H}$$

$$54.2\ \cancel{\text{g O}} \times \frac{1\text{ mol O}}{16.00\ \cancel{\text{g O}}} = 3.39\text{ mol of O (smallest number of moles)}$$

$$\frac{3.39\text{ mol C}}{3.39} = 1.00\text{ mol of C} \qquad \frac{5.08\text{ mol H}}{3.39} = 1.50\text{ mol of H}$$

$$\frac{3.39\text{ mol O}}{3.39} = 1.00\text{ mol of O}$$

$\therefore$ Empirical formula $= C_{1.00}H_{1.50}O_{1.00} \rightarrow C_{(1.00\times2)}H_{(1.50\times2)}O_{(1.00\times2)}$
$= C_{2.00}H_{3.00}O_{2.00} \rightarrow C_2H_3O_2$

Empirical formula mass of $C_2H_3O_2$ = 2(12.01) + 3(1.008) + 2(16.00) = 59.04 g

$$\text{Small integer} = \frac{\text{molar mass of succinic acid}}{\text{empirical formula mass of } C_2H_3O_2} = \frac{118\ \cancel{g}}{59.04\ \cancel{g}} = 2$$

$\therefore$ Molecular formula of succinic acid $= C_{(2\times2)}H_{(3\times2)}O_{(2\times2)} = C_4H_6O_4$

7.73 In the sample of compound, there are 1.65×10^{23} atoms of C, 0.552 g of H, and 4.39 g of O.

$$1.65 \times 10^{23}\ \cancel{\text{atoms C}} \times \frac{1\text{ mol C}}{6.022 \times 10^{23}\ \cancel{\text{atoms C}}} = 0.274\text{ mol of C (smallest number of moles)}$$

$$0.552\ \cancel{\text{g H}} \times \frac{1\text{ mol H}}{1.008\ \cancel{\text{g H}}} = 0.548\text{ mol of H}$$

$$4.39\ \cancel{\text{g O}} \times \frac{1\text{ mol O}}{16.00\ \cancel{\text{g O}}} = 0.274\text{ mol of O}$$

$$\frac{0.274\text{ mol C}}{0.274} = 1.00\text{ mol of C} \qquad \frac{0.548\text{ mol H}}{0.274} = 2.00\text{ mol of H}$$

$$\frac{0.274\text{ mol O}}{0.274} = 1.00\text{ mol of O}$$

$\therefore$ Empirical formula $= C_{1.00}H_{2.00}O_{1.00} \rightarrow CH_2O$

Since 1 mol of compound contains 4 mol of O,
$\therefore$ Molecular formula $= C_{(1\times4)}H_{(2\times4)}O_{(1\times4)} = C_4H_8O_4$

Molar mass of compound ($C_4H_8O_4$)

$= 4$ mol of C + 8 mol of H + 4 mol of O
$= 4(12.01\text{ g}) + 8(1.008\text{ g}) + 4(16.00\text{ g}) = 120.10\text{ g}$

7.75 **a.** Molar mass of NaF
$= 1$ mol of Na + 1 mol of F $= 22.99\text{ g} + 19.00\text{ g} = 41.99\text{ g}$

$$1\ \cancel{\text{tube}} \times \frac{119\ \cancel{\text{g toothpaste}}}{1\ \cancel{\text{tube}}} \times \frac{0.240\ \cancel{\text{g NaF}}}{100\ \cancel{\text{g toothpaste}}} \times \frac{1\text{ mol NaF}}{41.99\ \cancel{\text{g NaF}}}$$

$= 0.00680$ mol of NaF (3 SFs)

b. $$1\ \cancel{\text{tube}} \times \frac{119\ \cancel{\text{g toothpaste}}}{1\ \cancel{\text{tube}}} \times \frac{0.240\ \cancel{\text{g NaF}}}{100\ \cancel{\text{g toothpaste}}} \times \frac{1\ \cancel{\text{mol NaF}}}{41.99\ \cancel{\text{g NaF}}} \times \frac{1\ \cancel{\text{mol F}^-\text{ ions}}}{1\ \cancel{\text{mol NaF}}}$$

$$\times \frac{6.022 \times 10^{23}\ \text{F}^-\text{ ions}}{1\ \cancel{\text{mol F}^-\text{ ions}}} = 4.10 \times 10^{21}\ \text{F}^-\text{ ions (3 SFs)}$$

c. $$1.50\ \cancel{\text{g toothpaste}} \times \frac{0.240\ \cancel{\text{g NaF}}}{100\ \cancel{\text{g toothpaste}}} \times \frac{1\ \cancel{\text{mol NaF}}}{41.99\ \cancel{\text{g NaF}}} \times \frac{1\ \cancel{\text{mol Na}^+}}{1\ \cancel{\text{mol NaF}}} \times \frac{22.99\text{ g Na}^+}{1\ \cancel{\text{mol Na}^+}}$$

$= 0.00197$ g of Na^+ ions (3 SFs)

d. $$12\ \cancel{\text{mol C}} \times \frac{12.01\text{ g C}}{1\ \cancel{\text{mol C}}} = 144.1\text{ g of C}$$

$$7\ \cancel{\text{mol H}} \times \frac{1.008\text{ g H}}{1\ \cancel{\text{mol H}}} = 7.056\text{ g of H}$$

$$3\ \cancel{\text{mol Cl}} \times \frac{35.45\text{ g Cl}}{1\ \cancel{\text{mol Cl}}} = 106.4\text{ g of Cl}$$

$$2\ \cancel{\text{mol O}} \times \frac{16.00\text{ g O}}{1\ \cancel{\text{mol O}}} = 32.00\text{ g of O}$$

$\therefore$ Molar mass of $C_{12}H_7Cl_3O_2 = 289.6\text{ g}$

$$1\ \cancel{\text{tube}} \times \frac{119\ \cancel{\text{g toothpaste}}}{1\ \cancel{\text{tube}}} \times \frac{0.30\ \cancel{\text{g C}_{12}\text{H}_7\text{Cl}_3\text{O}_2}}{100\ \cancel{\text{g toothpaste}}} \times \frac{1\ \cancel{\text{mol C}_{12}\text{H}_7\text{Cl}_3\text{O}_2}}{289.6\ \cancel{\text{g C}_{12}\text{H}_7\text{Cl}_3\text{O}_2}}$$

$$\times \frac{6.022 \times 10^{23}\text{ molecules C}_{12}\text{H}_7\text{Cl}_3\text{O}_2}{1\ \cancel{\text{mol C}_{12}\text{H}_7\text{Cl}_3\text{O}_2}}$$

$= 7.4 \times 10^{20}$ molecules of $C_{12}H_7Cl_3O_2$ (triclosan) (2 SFs)

e. $$\text{Mass \% C} = \frac{144.1\text{ g C}}{289.6\text{ g C}_{12}\text{H}_7\text{Cl}_3\text{O}_2} \times 100\% = 49.76\%\text{ C (4 SFs)}$$

$$\text{Mass \% H} = \frac{7.056\text{ g H}}{289.6\text{ g C}_{12}\text{H}_7\text{Cl}_3\text{O}_2} \times 100\% = 2.44\%\text{ H (rounded off to hundredths)}$$

$$\text{Mass \% Cl} = \frac{106.4\text{ g Cl}}{289.6\text{ g C}_{12}\text{H}_7\text{Cl}_3\text{O}_2} \times 100\% = 36.74\%\text{ Cl (4 SFs)}$$

$$\text{Mass \% O} = \frac{32.00\text{ g O}}{289.6\text{ g C}_{12}\text{H}_7\text{Cl}_3\text{O}_2} \times 100\% = 11.05\%\text{ O (4 SFs)}$$

7.77 In exactly 100. g of iron(III) chromate, there are 24.3 g of Fe, 33.9 g of Cr, and 41.8 g of O.

$$24.3\ \text{g Fe} \times \frac{1\ \text{mol Fe}}{55.85\ \text{g Fe}} = 0.435\ \text{mol of Fe (smallest number of moles)}$$

$$33.9\ \text{g Cr} \times \frac{1\ \text{mol Cr}}{52.00\ \text{g Cr}} = 0.652\ \text{mol of Cr}$$

$$41.8\ \text{g O} \times \frac{1\ \text{mol O}}{16.00\ \text{g O}} = 2.61\ \text{mol of O}$$

$$\frac{0.435\ \text{mol Fe}}{0.435} = 1.00\ \text{mol of Fe} \qquad \frac{0.652\ \text{mol Cr}}{0.435} = 1.50\ \text{mol of Cr}$$

$$\frac{2.61\ \text{mol O}}{0.435} = 6.00\ \text{mol of O}$$

$$\therefore \text{Empirical formula} = Fe_{1.00}Cr_{1.50}O_{6.00} \rightarrow Fe_{(1.00\times2)}Cr_{(1.50\times2)}O_{(6.00\times2)}$$

$$= Fe_{2.00}Cr_{3.00}O_{12.0} \rightarrow Fe_2Cr_3O_{12}$$

Empirical formula mass of $Fe_2Cr_3O_{12} = 2(55.85) + 3(52.00) + 12(16.00) = 459.7\ \text{g}$

$$\text{Small integer} = \frac{\text{molar mass of compound}}{\text{empirical formula mass of } Fe_2Cr_3O_{12}} = \frac{460\ \text{g}}{459.7\ \text{g}} = 1$$

$$\therefore \text{Molecular formula} = Fe_{(2\times1)}Cr_{(3\times1)}O_{(12\times1)} = Fe_2Cr_3O_{12}$$

Answers to Combining Ideas from Chapters 4 to 7

CI.7 **a.** X is a metal; elements in Group 2A (2) are metals.
b. Y is a nonmetal; elements in Group 7A (17) are nonmetals.
c. X^{2+}, Y^-
d. $X = 1s^22s^22p^63s^2$ $Y = 1s^22s^22p^63s^23p^5$
e. $X^{2+} = 1s^22s^22p^6$ $Y^- = 1s^22s^22p^63s^23p^6$
f. $MgCl_2$, magnesium chloride

CI.9 **a.** In exactly 100. g of oxalic acid, there are 26.7 g of C, 2.24 g of H, and 71.1 g of O.

$$26.7\ \cancel{g\,C} \times \frac{1\text{ mol C}}{12.01\ \cancel{g\,C}} = 2.22\text{ mol of C}$$

$$2.24\ \cancel{g\,H} \times \frac{1\text{ mol H}}{1.008\ \cancel{g\,H}} = 2.22\text{ mol of H (smallest number of moles)}$$

$$71.1\ \cancel{g\,O} \times \frac{1\text{ mol O}}{16.00\ \cancel{g\,O}} = 4.44\text{ mol of O}$$

$$\frac{2.22\text{ mol C}}{2.22} = 1.00\text{ mol of C} \qquad \frac{2.22\text{ mol H}}{2.22} = 1.00\text{ mol of H}$$

$$\frac{4.44\text{ mol O}}{2.22} = 2.00\text{ mol of O}$$

$\therefore$ Empirical formula $= C_{1.00}H_{1.00}O_{2.00} \rightarrow CHO_2$

b. Empirical formula mass of $CHO_2 = 12.01 + 1.008 + 2(16.00) = 45.02$ g

$$\text{Small integer} = \frac{\text{molar mass of oxalic acid}}{\text{empirical formula mass of } CHO_2} = \frac{90.\ \cancel{g}}{45.02\ \cancel{g}} = 2$$

$\therefore$ Molecular formula $= C_{(1\times2)}H_{(1\times2)}O_{(2\times2)} = C_2H_2O_4$

c. $$160\ \cancel{lb} \times \frac{1\ \cancel{kg}}{2.205\ \cancel{lb}} \times \frac{375\ \cancel{\text{mg oxalic acid}}}{1\ \cancel{kg}} \times \frac{1\text{ g oxalic acid}}{1000\ \cancel{\text{mg oxalic acid}}}$$
$= 27$ g of oxalic acid (2 SFs)

d. $$160\ \cancel{lb} \times \frac{1\ \cancel{kg}}{2.205\ \cancel{lb}} \times \frac{375\ \cancel{\text{mg oxalic acid}}}{1\ \cancel{kg}} \times \frac{1\ \cancel{\text{g oxalic acid}}}{1000\ \cancel{\text{mg oxalic acid}}} \times \frac{100\ \cancel{\text{g rhubarb leaves}}}{0.5\ \cancel{\text{g oxalic acid}}} \times \frac{1\text{ kg}}{1000\ \cancel{g}}$$
$= 5$ kg of rhubarb leaves (1 SF)

CI.11 **a.** $C_{16}H_{28}N_2O_4 \rightarrow C_{(16\div2)}H_{(28\div2)}N_{(2\div2)}O_{(4\div2)} = C_8H_{14}NO_2$ (empirical formula)

b. $$16\ \cancel{\text{mol C}} \times \frac{12.01\text{ g C}}{1\ \cancel{\text{mol C}}} = 192.2\text{ g of C}$$

$$28\ \cancel{\text{mol H}} \times \frac{1.008\text{ g H}}{1\ \cancel{\text{mol H}}} = 28.22\text{ g of H}$$

$$2\ \cancel{\text{mol N}} \times \frac{14.01\text{ g N}}{1\ \cancel{\text{mol N}}} = 28.02\text{ g of N}$$

$$4\ \cancel{\text{mol O}} \times \frac{16.00\text{ g O}}{1\ \cancel{\text{mol O}}} = 64.00\text{ g of O}$$

$\therefore$ Molar mass of $C_{16}H_{28}N_2O_4 = 312.4$ g

$$\text{Mass \% C} = \frac{192.2 \text{ g C}}{312.4 \text{ g } C_{16}H_{28}N_2O_4} \times 100\% = 61.52\% \text{ C (4 SFs)}$$

$$\text{Mass \% H} = \frac{28.22 \text{ g H}}{312.4 \text{ g } C_{16}H_{28}N_2O_4} \times 100\% = 9.03\% \text{ H (rounded off to hundredths)}$$

$$\text{Mass \% N} = \frac{28.02 \text{ g N}}{312.4 \text{ g } C_{16}H_{28}N_2O_4} \times 100\% = 8.97\% \text{ N (rounded off to hundredths)}$$

$$\text{Mass \% O} = \frac{64.00 \text{ g O}}{312.4 \text{ g } C_{16}H_{28}N_2O_4} \times 100\% = 20.49\% \text{ O (4 SFs)}$$

c. $C_7H_{10}O_5$

d. Molar mass of shikimic acid ($C_7H_{10}O_5$)
= 7 mol of C + 10 mol of H + 5 mol of O
= 7(12.01 g) + 10(1.008 g) + 5(16.00 g) = 174.15 g

$$1.3 \cancel{\text{ g } C_7H_{10}O_5} \times \frac{1 \text{ mol } C_7H_{10}O_5}{174.15 \cancel{\text{ g } C_7H_{10}O_5}} = 0.0075 \text{ mol of } C_7H_{10}O_5 \text{ (shikimic acid) (2 SFs)}$$

e. $$155 \cancel{\text{ g star anise}} \times \frac{0.13 \cancel{\text{ g } C_7H_{10}O_5}}{2.6 \cancel{\text{ g star anise}}} \times \frac{75 \cancel{\text{ mg Tamiflu}}}{0.13 \cancel{\text{ g } C_7H_{10}O_5}} \times \frac{1 \text{ capsule}}{75 \cancel{\text{ mg Tamiflu}}}$$

= 59 capsules of Tamiflu (2 SFs)

f. $$1 \cancel{\text{ capsule}} \times \frac{75 \cancel{\text{ mg } C_{16}H_{28}N_2O_4}}{1 \cancel{\text{ capsule}}} \times \frac{1 \cancel{\text{ g } C_{16}H_{28}N_2O_4}}{1000 \cancel{\text{ mg } C_{16}H_{28}N_2O_4}} \times \frac{1 \cancel{\text{ mol } C_{16}H_{28}N_2O_4}}{312.4 \cancel{\text{ g } C_{16}H_{28}N_2O_4}}$$

$$\times \frac{16 \cancel{\text{ mol C}}}{1 \cancel{\text{ mol } C_{16}H_{28}N_2O_4}} \times \frac{12.01 \text{ g C}}{1 \cancel{\text{ mol C}}} = 0.046 \text{ g of C (2 SFs)}$$

g. $$500\,000 \cancel{\text{ people}} \times \frac{2 \cancel{\text{ capsules}}}{1 \cancel{\text{ day}} \text{ } 1 \cancel{\text{ person}}} \times 5 \cancel{\text{ days}}$$

$$\times \frac{75 \cancel{\text{ mg Tamiflu}}}{1 \cancel{\text{ capsule}}} \times \frac{1 \cancel{\text{ g Tamiflu}}}{1000 \cancel{\text{ mg Tamiflu}}} \times \frac{1 \text{ kg Tamiflu}}{1000 \cancel{\text{ g Tamiflu}}}$$

$= 4 \times 10^2$ kg of Tamiflu (1 SF)

8

Chemical Reactions

Learning Goals

- Identify a balanced chemical equation and determine the number of atoms in the reactants and products.
- Write a balanced chemical equation from the formulas of the reactants and products for a chemical reaction.
- Identify a reaction as a combination, decomposition, single replacement, double replacement, or combustion.
- Classify organic molecules according to their functional groups.
- Draw the condensed structural formulas for the products of reactions of organic compounds.
- Identify the type of a biochemical compound from its structure and name its functional groups.

Chapter Outline

Core Chemistry Skills

- Balancing a Chemical Equation (8.2)
- Classifying Types of Chemical Reactions (8.3)
- Balancing Combustion Reactions (8.3)
- Identifying Functional Groups (8.4)

Answers and Solutions to Text Problems

8.1 **a.** reactants/products: 7 O atoms
b. reactants/products: 4 O atoms
c. reactants/products: 10 O atoms
d. reactants/products: 4 O atoms

8.3 An equation is balanced when there are an equal number of atoms of each element on the reactant side and on the product side.

a. not balanced ($2\,O \neq 3\,O$) **b.** balanced
c. not balanced ($6\,O \neq 5\,O, 4\,H \neq 2\,H$) **d.** balanced

8.5 **a.** reactants/products: 2 Na atoms, 2 Cl atoms
b. reactants/products: 1 P atom, 3 Cl atoms, 6 H atoms
c. reactants/products: 4 P atoms, 16 O atoms, 12 H atoms

8.7 Place coefficients in front of formulas until you make the atoms of each element equal on each side of the equation.

a. $N_2(g) + O_2(g) \longrightarrow 2NO(g)$ **b.** $2HgO(s) \longrightarrow 2Hg(l) + O_2(g)$
c. $4Fe(s) + 3O_2(g) \longrightarrow 2Fe_2O_3(s)$ **d.** $2Na(s) + Cl_2(g) \longrightarrow 2NaCl(s)$
e. $2Cu_2O(s) + O_2(g) \longrightarrow 4CuO(s)$

8.9 **a.** $Mg(s) + 2AgNO_3(aq) \longrightarrow Mg(NO_3)_2(aq) + 2Ag(s)$
b. $CuCO_3(s) \longrightarrow CuO(s) + CO_2(g)$
c. $2Al(s) + 3CuSO_4(aq) \longrightarrow 3Cu(s) + Al_2(SO_4)_3(aq)$
d. $Pb(NO_3)_2(aq) + 2NaCl(aq) \longrightarrow PbCl_2(s) + 2NaNO_3(aq)$
e. $2Al(s) + 6HCl(aq) \longrightarrow 2AlCl_3(aq) + 3H_2(g)$

8.11 **a.** $Fe_2O_3(s) + 3CO(g) \longrightarrow 2Fe(s) + 3CO_2(g)$
b. $2Li_3N(s) \longrightarrow 6Li(s) + N_2(g)$
c. $2Al(s) + 6HBr(aq) \longrightarrow 2AlBr_3(aq) + 3H_2(g)$
d. $3Ba(OH)_2(aq) + 2Na_3PO_4(aq) \longrightarrow Ba_3(PO_4)_2(s) + 6NaOH(aq)$
e. $As_4S_6(s) + 9O_2(g) \longrightarrow As_4O_6(s) + 6SO_2(g)$

8.13 **a.** $2Li(s) + 2H_2O(l) \longrightarrow H_2(g) + 2LiOH(aq)$
b. $2P(s) + 5Cl_2(g) \longrightarrow 2PCl_5(s)$
c. $FeO(s) + CO(g) \longrightarrow Fe(s) + CO_2(g)$
d. $2C_5H_{10}(l) + 15O_2(g) \xrightarrow{\Delta} 10CO_2(g) + 10H_2O(g)$
e. $3H_2S(g) + 2FeCl_3(s) \longrightarrow Fe_2S_3(s) + 6HCl(g)$

8.15 **a.** This is a decomposition reaction because a single reactant splits into two simpler substances.
b. This is a single replacement reaction because one element in the reacting compound (I in BaI_2) is replaced by the other reactant (Br in Br_2).
c. This is a combustion reaction because a carbon-containing fuel burns in oxygen to produce carbon dioxide and water.
d. This is a double replacement reaction because the positive ions in the two reactants exchange places to form two products.
e. This is a combination reaction because atoms of two different elements bond to form one product.

8.17 **a.** combination **b.** single replacement
c. decomposition **d.** combustion
e. double replacement

8.19 **a.** Since this is a combination reaction, the product is a single compound.
$Mg(s) + Cl_2(g) \longrightarrow MgCl_2(s)$
b. Since this is a decomposition reaction, the single reactant breaks down into more than one simpler substance.
$2HBr(g) \longrightarrow H_2(g) + Br_2(g)$
c. Since this is a single replacement reaction, the uncombined metal element takes the place of the metal element in the compound.
$Mg(s) + Zn(NO_3)_2(aq) \longrightarrow Zn(s) + Mg(NO_3)_2(aq)$

d. Since this is a double replacement reaction, the positive ions in the reacting compounds switch places.
$K_2S(aq) + Pb(NO_3)_2(aq) \longrightarrow PbS(s) + 2KNO_3(aq)$

e. Since this is a combustion reaction, a fuel burns in oxygen to produce carbon dioxide and water.
$2C_5H_{10}(l) + 15O_2(g) \xrightarrow{\Delta} 10CO_2(g) + 10H_2O(g)$

8.21 **a.** Alcohols contain a hydroxyl ($-OH$) group attached to a carbon chain.
b. Alkenes have carbon–carbon double bonds.
c. Aldehydes contain a carbonyl ($C{=}O$) group attached to at least one H atom.

8.23 **a.** Alcohols have a hydroxyl ($-OH$) group.
b. Esters contain a $-COO-$ group attached to two carbon atoms.
c. Carboxylic acids have a carboxyl ($-COOH$) group.
d. Amines contain a nitrogen atom bonded to at least one carbon atom.

8.25 **a.** $C_3H_8(g) + 5O_2(g) \xrightarrow{\Delta} 3CO_2(g) + 4H_2O(g) + \text{energy}$
b. $2C_8H_{18}(l) + 25O_2(g) \xrightarrow{\Delta} 16CO_2(g) + 18H_2O(g) + \text{energy}$
c. $2C_3H_8O(l) + 9O_2(g) \xrightarrow{\Delta} 6CO_2(g) + 8H_2O(g) + \text{energy}$
d. $C_6H_{12}(l) + 9O_2(g) \xrightarrow{\Delta} 6CO_2(g) + 6H_2O(g) + \text{energy}$

8.27 **a.** $CH_3{-}CH_2{-}CH_2{-}CH_2{-}CH_3$ **b.** $CH_3{-}CH_2{-}CH_2{-}CH_3$

8.29 **a.** Erythrulose contains several hydroxyl ($-OH$) groups and a carbonyl group on carbon 2 (ketone).
b. Ribose contains several hydroxyl ($-OH$) groups and a carbonyl group on carbon 1 (aldehyde).

8.31 **a.** A polymer of amino acids is a **3.** protein.
b. A biochemical compound that is not soluble in water is a **2.** lipid.
c. A compound that contains many alcohol groups and an aldehyde or ketone group is a **1.** carbohydrate.
d. Fats are **2.** lipid.
e. A polysaccharide is a **1.** carbohydrate.

8.33 **a.** reactants NO and O_2; product NO_2 **b.** $2NO(g) + O_2(g) \longrightarrow 2NO_2(g)$
c. combination

8.35 **a.** reactant: NI_3; products N_2 and I_2 **b.** $2NI_3(s) \longrightarrow N_2(g) + 3I_2(g)$
c. decomposition

8.37 **a.** reactants Cl_2 and O_2; product OCl_2 **b.** $2Cl_2(g) + O_2(g) \longrightarrow 2OCl_2(g)$
c. combination

8.39 **a.** 1, 1, 2; combination **b.** 2, 2, 1; decomposition

8.41 **a.** ketone **b.** alkene
c. ester **d.** amine

8.43 **a.** An alcohol contains a hydroxyl group.
b. An alkene contains one or more carbon–carbon double bonds.
c. An aldehyde contains a carbonyl group bonded to a hydrogen.
d. Alkanes are hydrocarbons that contain only carbon–carbon single bonds.

8.45 **a.** aromatic, aldehyde **b.** aromatic, alkene, aldehyde
c. ketone

8.47 amine, carboxylic acid, amide, aromatic, ester

8.49 **a.** True
b. False; lipids are not soluble in water.
c. False; polysaccharides are a type of carbohydrate.
d. True

8.51 **a.** combination **b.** combustion
c. double replacement **d.** decomposition
e. single replacement

8.53
a. $NH_3(g) + HCl(g) \longrightarrow NH_4Cl(s)$ combination
b. $C_4H_8(g) + 6O_2(g) \xrightarrow{\Delta} 4CO_2(g) + 4H_2O(g)$ combustion
c. $2Sb(s) + 3Cl_2(g) \longrightarrow 2SbCl_3(s)$ combination
d. $2NI_3(s) \longrightarrow N_2(g) + 3I_2(g)$ decomposition
e. $2KBr(aq) + Cl_2(g) \longrightarrow 2KCl(aq) + Br_2(l)$ single replacement
f. $2Fe(s) + 3H_2SO_4(aq) \longrightarrow Fe_2(SO_4)_3(aq) + 3H_2(g)$ single replacement
g. $Al_2(SO_4)_3(aq) + 6NaOH(aq) \longrightarrow 3Na_2SO_4(aq) + 2Al(OH)_3(s)$ double replacement

8.55
a. $Zn(s) + 2HCl(aq) \longrightarrow ZnCl_2(aq) + H_2(g)$
b. $BaCO_3(s) \xrightarrow{\Delta} BaO(s) + CO_2(g)$
c. $NaOH(aq) + HCl(aq) \longrightarrow NaCl(aq) + H_2O(l)$
d. $2Al(s) + 3F_2(g) \longrightarrow 2AlF_3(s)$

8.57
a. $4Na(s) + O_2(g) \longrightarrow 2Na_2O(s)$ combination
b. $NaCl(aq) + AgNO_3(aq) \longrightarrow AgCl(s) + NaNO_3(aq)$ double replacement
c. $C_2H_6O(g) + 3O_2(g) \longrightarrow 2CO_2(g) + 3H_2O(g)$ combustion

8.59 alcohol, amide, carboxylic acid

8.61 amine, aromatic, ester

8.63
a. $3Pb(NO_3)_2(aq) + 2Na_3PO_4(aq) \longrightarrow Pb_3(PO_4)_2(s) + 6NaNO_3(aq)$ double replacement
b. $4Ga(s) + 3O_2(g) \xrightarrow{\Delta} 2Ga_2O_3(s)$ combination
c. $2NaNO_3(s) \xrightarrow{\Delta} 2NaNO_2(s) + O_2(g)$ decomposition

8.65 **a.** reactants: X and Y_2; products: XY_3 **b.** $2X + 3Y_2 \longrightarrow 2XY_3$
c. combination

8.67 aromatic, ketone, alkene, amine, carboxylic acid

8.69 **a.** alcohol **b.** amine **c.** ketone

8.71
a. $C_5H_{12} + 8O_2 \xrightarrow{\Delta} 5CO_2 + 6H_2O + \text{energy}$
b. $CH_3{-}CH_2{-}CH{=}CH{-}CH_2{-}CH_3 + H_2 \xrightarrow{Pt} CH_3{-}CH_2{-}CH_2{-}CH_2{-}CH_2{-}CH_3$

9

Chemical Quantities in Reactions

Learning Goals

- Given a quantity in moles of reactant or product, use a mole–mole factor from the balanced equation to calculate the number of moles of another substance in the reaction.
- Given the mass in grams of a substance in a reaction, calculate the mass in grams of another substance in the reaction.
- Identify a limiting reactant when given the quantities of two reactants; calculate the amount of product formed from the limiting reactant.
- Given the actual quantity of product, determine the percent yield for a reaction.
- Given the heat of reaction (enthalpy change), calculate the loss or gain of heat for an exothermic or endothermic reaction.

Chapter Outline

Chapter Opener: Environmental Scientist

9.1 Mole Relationships in Chemical Equations

9.2 Mass Calculations for Reactions

9.3 Limiting Reactants

9.4 Percent Yield

9.5 Energy in Chemical Reactions

Chemistry Link to Health: Cold Packs and Hot Packs

Core Chemistry Skills

- Using Mole–Mole Factors (9.1)
- Converting Grams to Grams (9.2)
- Calculating Quantity of Product from a Limiting Reactant (9.3)
- Calculating Percent Yield (9.4)
- Using the Heat of Reaction (9.5)

Answers and Solutions to Text Problems

9.1 **a.** Reactants: 2 mol of SO_2 and 1 mol of O_2 $= 2\ \cancel{\text{mol}}\ (64.07\ \text{g}/\cancel{\text{mol}}) + 1\ \cancel{\text{mol}}\ (32.00\ \text{g}/\cancel{\text{mol}})$
$= 128.14\ \text{g} + 32.00\ \text{g} = 160.14$ g of reactants
Products: 2 mol of SO_3 $= 2\ \cancel{\text{mol}}\ (80.07\ \text{g}/\cancel{\text{mol}}) = 160.14$ g of products

b. Reactants: 4 mol of P and 5 mol of O_2 $= 4\ \cancel{\text{mol}}\ (30.97\ \text{g}/\cancel{\text{mol}}) + 5\ \cancel{\text{mol}}\ (32.00\ \text{g}/\cancel{\text{mol}})$
$= 123.88\ \text{g} + 160.00\ \text{g} = 283.88$ g of reactants
Products: 2 mol of P_2O_5 $= 2\ \cancel{\text{mol}}\ (141.94\ \text{g}/\cancel{\text{mol}}) = 283.88$ g of products

9.3 **a.** $\frac{2\ \text{mol}\ SO_2}{1\ \text{mol}\ O_2}$ and $\frac{1\ \text{mol}\ O_2}{2\ \text{mol}\ SO_2}$; $\frac{2\ \text{mol}\ SO_2}{2\ \text{mol}\ SO_3}$ and $\frac{2\ \text{mol}\ SO_3}{2\ \text{mol}\ SO_2}$; $\frac{1\ \text{mol}\ O_2}{2\ \text{mol}\ SO_3}$ and $\frac{2\ \text{mol}\ SO_3}{1\ \text{mol}\ O_2}$

b. $\frac{4\ \text{mol P}}{5\ \text{mol}\ O_2}$ and $\frac{5\ \text{mol}\ O_2}{4\ \text{mol P}}$; $\frac{4\ \text{mol P}}{2\ \text{mol}\ P_2O_5}$ and $\frac{2\ \text{mol}\ P_2O_5}{4\ \text{mol P}}$; $\frac{5\ \text{mol}\ O_2}{2\ \text{mol}\ P_2O_5}$ and $\frac{2\ \text{mol}\ P_2O_5}{5\ \text{mol}\ O_2}$

9.5 **a.** $\cancel{\text{mol SO}_2} \times \frac{2 \text{ mol SO}_3}{2 \cancel{\text{ mol SO}_2}} = \text{mol of SO}_3$

b. $\cancel{\text{mol P}} \times \frac{5 \text{ mol O}_2}{4 \cancel{\text{ mol P}}} = \text{mol of O}_2$

9.7 **a.** $2.0 \cancel{\text{ mol H}_2} \times \frac{1 \text{ mol O}_2}{2 \cancel{\text{ mol H}_2}} = 1.0 \text{ mol of O}_2 \text{ (2 SFs)}$

b. $5.0 \cancel{\text{ mol O}_2} \times \frac{2 \text{ mol H}_2}{1 \cancel{\text{ mol O}_2}} = 10. \text{ mol of H}_2 \text{ (2 SFs)}$

c. $2.5 \cancel{\text{ mol O}_2} \times \frac{2 \text{ mol H}_2\text{O}}{1 \cancel{\text{ mol O}_2}} = 5.0 \text{ mol of H}_2\text{O (2 SFs)}$

9.9 **a.** $0.500 \cancel{\text{ mol SO}_2} \times \frac{5 \text{ mol C}}{2 \cancel{\text{ mol SO}_2}} = 1.25 \text{ mol of C (3 SFs)}$

b. $1.2 \cancel{\text{ mol C}} \times \frac{4 \text{ mol CO}}{5 \cancel{\text{ mol C}}} = 0.96 \text{ mol of CO (2 SFs)}$

c. $0.50 \cancel{\text{ mol CS}_2} \times \frac{2 \text{ mol SO}_2}{1 \cancel{\text{ mol CS}_2}} = 1.0 \text{ mol of SO}_2 \text{ (2 SFs)}$

d. $2.5 \cancel{\text{ mol C}} \times \frac{1 \text{ mol CS}_2}{5 \cancel{\text{ mol C}}} = 0.50 \text{ mol of CS}_2 \text{ (2 SFs)}$

9.11 **a.** $57.5 \cancel{\text{ g Na}} \times \frac{1 \cancel{\text{ mol Na}}}{22.99 \cancel{\text{ g Na}}} \times \frac{2 \cancel{\text{ mol Na}_2\text{O}}}{4 \cancel{\text{ mol Na}}} \times \frac{61.98 \text{ g Na}_2\text{O}}{1 \cancel{\text{ mol Na}_2\text{O}}} = 77.5 \text{ g of Na}_2\text{O (3 SFs)}$

b. $18.0 \cancel{\text{ g Na}} \times \frac{1 \cancel{\text{ mol Na}}}{22.99 \cancel{\text{ g Na}}} \times \frac{1 \cancel{\text{ mol O}_2}}{4 \cancel{\text{ mol Na}}} \times \frac{32.00 \text{ g O}_2}{1 \cancel{\text{ mol O}_2}} = 6.26 \text{ g of O}_2 \text{ (3 SFs)}$

c. $75.0 \cancel{\text{ g Na}_2\text{O}} \times \frac{1 \cancel{\text{ mol Na}_2\text{O}}}{61.98 \cancel{\text{ g Na}_2\text{O}}} \times \frac{1 \cancel{\text{ mol O}_2}}{2 \cancel{\text{ mol Na}_2\text{O}}} \times \frac{32.00 \text{ g O}_2}{1 \cancel{\text{ mol O}_2}} = 19.4 \text{ g of O}_2 \text{ (3 SFs)}$

9.13 **a.** $13.6 \cancel{\text{ g NH}_3} \times \frac{1 \cancel{\text{ mol NH}_3}}{17.03 \cancel{\text{ g NH}_3}} \times \frac{3 \cancel{\text{ mol O}_2}}{4 \cancel{\text{ mol NH}_3}} \times \frac{32.00 \text{ g O}_2}{1 \cancel{\text{ mol O}_2}} = 19.2 \text{ g of O}_2 \text{ (3 SFs)}$

b. $6.50 \cancel{\text{ g O}_2} \times \frac{1 \cancel{\text{ mol O}_2}}{32.00 \cancel{\text{ g O}_2}} \times \frac{2 \cancel{\text{ mol N}_2}}{3 \cancel{\text{ mol O}_2}} \times \frac{28.02 \text{ g N}_2}{1 \cancel{\text{ mol N}_2}} = 3.79 \text{ g of N}_2 \text{ (3 SFs)}$

c. $34.0 \cancel{\text{ g NH}_3} \times \frac{1 \cancel{\text{ mol NH}_3}}{17.03 \cancel{\text{ g NH}_3}} \times \frac{6 \cancel{\text{ mol H}_2\text{O}}}{4 \cancel{\text{ mol NH}_3}} \times \frac{18.02 \text{ g H}_2\text{O}}{1 \cancel{\text{ mol H}_2\text{O}}} = 54.0 \text{ g of H}_2\text{O (3 SFs)}$

9.15 **a.** $28.0 \cancel{\text{ g NO}_2} \times \frac{1 \cancel{\text{ mol NO}_2}}{46.01 \cancel{\text{ g NO}_2}} \times \frac{1 \cancel{\text{ mol H}_2\text{O}}}{3 \cancel{\text{ mol NO}_2}} \times \frac{18.02 \text{ g H}_2\text{O}}{1 \cancel{\text{ mol H}_2\text{O}}} = 3.66 \text{ g of H}_2\text{O (3 SFs)}$

b. $15.8 \cancel{\text{ g H}_2\text{O}} \times \frac{1 \cancel{\text{ mol H}_2\text{O}}}{18.02 \cancel{\text{ g H}_2\text{O}}} \times \frac{1 \cancel{\text{ mol NO}}}{1 \cancel{\text{ mol H}_2\text{O}}} \times \frac{30.01 \text{ g NO}}{1 \cancel{\text{ mol NO}}} = 26.3 \text{ g of NO (3 SFs)}$

c. $8.25 \cancel{\text{ g NO}_2} \times \frac{1 \cancel{\text{ mol NO}_2}}{46.01 \cancel{\text{ g NO}_2}} \times \frac{2 \cancel{\text{ mol HNO}_3}}{3 \cancel{\text{ mol NO}_2}} \times \frac{63.02 \text{ g HNO}_3}{1 \cancel{\text{ mol HNO}_3}} = 7.53 \text{ g of HNO}_3 \text{ (3 SFs)}$

9.17 **a.** $2\text{PbS}(s) + 3\text{O}_2(g) \rightarrow 2\text{PbO}(s) + 2\text{SO}_2(g)$

b. $29.9 \cancel{\text{ g PbS}} \times \frac{1 \cancel{\text{ mol PbS}}}{239.3 \cancel{\text{ g PbS}}} \times \frac{3 \cancel{\text{ mol O}_2}}{2 \cancel{\text{ mol PbS}}} \times \frac{32.00 \text{ g O}_2}{1 \cancel{\text{ mol O}_2}} = 6.00 \text{ g of O}_2 \text{ (3 SFs)}$

c. $65.0 \cancel{\text{ g PbS}} \times \frac{1 \cancel{\text{ mol PbS}}}{239.3 \cancel{\text{ g PbS}}} \times \frac{2 \cancel{\text{ mol SO}_2}}{2 \cancel{\text{ mol PbS}}} \times \frac{64.07 \text{ g SO}_2}{1 \cancel{\text{ mol SO}_2}} = 17.4 \text{ g of SO}_2 \text{ (3 SFs)}$

d. $$128\ \text{g}\ \cancel{\text{PbO}} \times \frac{1\ \cancel{\text{mol PbO}}}{223.2\ \text{g}\ \cancel{\text{PbO}}} \times \frac{2\ \cancel{\text{mol PbS}}}{2\ \cancel{\text{mol PbO}}} \times \frac{239.3\ \text{g PbS}}{1\ \cancel{\text{mol PbS}}} = 137\ \text{g of PbS (3 SFs)}$$

9.19 **a.** The limiting factor is the number of drivers: with only eight drivers available, only eight taxis can be used to pick up passengers.

b. The limiting factor is the number of taxis: only seven taxis are in working condition to be driven.

9.21 **a.** $$3.0\ \cancel{\text{mol N}_2} \times \frac{2\ \text{mol NH}_3}{1\ \cancel{\text{mol N}_2}} = 6.0\ \text{mol of NH}_3$$

$$5.0\ \cancel{\text{mol H}_2} \times \frac{2\ \text{mol NH}_3}{3\ \cancel{\text{mol H}_2}} = 3.3\ \text{mol of NH}_3\ \text{(smaller amount of product)}$$

The limiting reactant is 5.0 mol of H_2. (2 SFs)

b. $$8.0\ \cancel{\text{mol N}_2} \times \frac{2\ \text{mol NH}_3}{1\ \cancel{\text{mol N}_2}} = 16\ \text{mol of NH}_3$$

$$4.0\ \cancel{\text{mol H}_2} \times \frac{2\ \text{mol NH}_3}{3\ \cancel{\text{mol H}_2}} = 2.7\ \text{mol of NH}_3\ \text{(smaller amount of product)}$$

The limiting reactant is 4.0 mol of H_2. (2 SFs)

c. $$3.0\ \cancel{\text{mol N}_2} \times \frac{2\ \text{mol NH}_3}{1\ \cancel{\text{mol N}_2}} = 6.0\ \text{mol of NH}_3\ \text{(smaller amount of product)}$$

$$12.0\ \cancel{\text{mol H}_2} \times \frac{2\ \text{mol NH}_3}{3\ \cancel{\text{mol H}_2}} = 8.0\ \text{mol of NH}_3$$

The limiting reactant is 3.0 mol of N_2. (2 SFs)

9.23 **a.** $$20.0\ \text{g}\ \cancel{\text{Al}} \times \frac{1\ \cancel{\text{mol Al}}}{26.98\ \text{g}\ \cancel{\text{Al}}} \times \frac{2\ \cancel{\text{mol AlCl}_3}}{2\ \cancel{\text{mol Al}}} \times \frac{133.33\ \text{g AlCl}_3}{1\ \cancel{\text{mol AlCl}_3}} = 98.8\ \text{g of AlCl}_3$$

$$20.0\ \text{g}\ \cancel{\text{Cl}_2} \times \frac{1\ \cancel{\text{mol Cl}_2}}{70.90\ \text{g}\ \cancel{\text{Cl}_2}} \times \frac{2\ \cancel{\text{mol AlCl}_3}}{3\ \cancel{\text{mol Cl}_2}} \times \frac{133.33\ \text{g AlCl}_3}{1\ \cancel{\text{mol AlCl}_3}} = 25.1\ \text{g of AlCl}_3\ \text{(smaller amount of product)}$$

∴ Cl_2 is the limiting reactant, and 25.1 g of $AlCl_3$ (3 SFs) can be produced.

b. $$20.0\ \text{g}\ \cancel{\text{NH}_3} \times \frac{1\ \cancel{\text{mol NH}_3}}{17.03\ \text{g}\ \cancel{\text{NH}_3}} \times \frac{6\ \cancel{\text{mol H}_2\text{O}}}{4\ \cancel{\text{mol NH}_3}} \times \frac{18.02\ \text{g H}_2\text{O}}{1\ \cancel{\text{mol H}_2\text{O}}} = 31.7\ \text{g of H}_2\text{O}$$

$$20.0\ \text{g}\ \cancel{\text{O}_2} \times \frac{1\ \cancel{\text{mol O}_2}}{32.00\ \text{g}\ \cancel{\text{O}_2}} \times \frac{6\ \cancel{\text{mol H}_2\text{O}}}{5\ \cancel{\text{mol O}_2}} \times \frac{18.02\ \text{g H}_2\text{O}}{1\ \cancel{\text{mol H}_2\text{O}}} = 13.5\ \text{g of H}_2\text{O}\ \text{(smaller amount of product)}$$

∴ O_2 is the limiting reactant, and 13.5 g of H_2O (3 SFs) can be produced.

c. $$20.0\ \text{g}\ \cancel{\text{CS}_2} \times \frac{1\ \cancel{\text{mol CS}_2}}{76.15\ \text{g}\ \cancel{\text{CS}_2}} \times \frac{2\ \cancel{\text{mol SO}_2}}{1\ \cancel{\text{mol CS}_2}} \times \frac{64.07\ \text{g SO}_2}{1\ \cancel{\text{mol SO}_2}} = 33.7\ \text{g of SO}_2$$

$$20.0\ \text{g}\ \cancel{\text{O}_2} \times \frac{1\ \cancel{\text{mol O}_2}}{32.00\ \text{g}\ \cancel{\text{O}_2}} \times \frac{2\ \cancel{\text{mol SO}_2}}{3\ \cancel{\text{mol O}_2}} \times \frac{64.07\ \text{g SO}_2}{1\ \cancel{\text{mol SO}_2}} = 26.7\ \text{g of SO}_2\ \text{(smaller amount of product)}$$

∴ O_2 is the limiting reactant, and 26.7 g of SO_2 (3 SFs) can be produced.

9.25 **a.** $25.0\text{ g } SO_2 \times \frac{1\text{ mol } SO_2}{64.07\text{ g } SO_2} \times \frac{2\text{ mol } SO_3}{2\text{ mol } SO_2} \times \frac{80.07\text{ g } SO_3}{1\text{ mol } SO_3} = 31.2\text{ g of } SO_3$ (smaller amount of product)

$40.0\text{ g } O_2 \times \frac{1\text{ mol } O_2}{32.00\text{ g } O_2} \times \frac{2\text{ mol } SO_3}{1\text{ mol } O_2} \times \frac{80.07\text{ g } SO_3}{1\text{ mol } SO_3} = 200.\text{ g of } SO_3$

∴ SO_2 is the limiting reactant, and 31.2 g of SO_3 (3 SFs) can be produced.

b. $25.0\text{ g Fe} \times \frac{1\text{ mol Fe}}{55.85\text{ g Fe}} \times \frac{1\text{ mol } Fe_3O_4}{3\text{ mol Fe}} \times \frac{231.6\text{ g } Fe_3O_4}{1\text{ mol } Fe_3O_4} = 34.6\text{ g of } Fe_3O_4$ (smaller amount of product)

$40.0\text{ g } H_2O \times \frac{1\text{ mol } H_2O}{18.02\text{ g } H_2O} \times \frac{1\text{ mol } Fe_3O_4}{4\text{ mol } H_2O} \times \frac{231.6\text{ g } Fe_3O_4}{1\text{ mol } Fe_3O_4} = 129\text{ g of } Fe_3O_4$

∴ Fe is the limiting reactant, and 34.6 g of Fe_3O_4 (3 SFs) can be produced.

c. $25.0\text{ g } C_7H_{16} \times \frac{1\text{ mol } C_7H_{16}}{100.20\text{ g } C_7H_{16}} \times \frac{7\text{ mol } CO_2}{1\text{ mol } C_7H_{16}} \times \frac{44.01\text{ g } CO_2}{1\text{ mol } CO_2} = 76.9\text{ g of } CO_2$

$40.0\text{ g } O_2 \times \frac{1\text{ mol } O_2}{32.00\text{ g } O_2} \times \frac{7\text{ mol } CO_2}{11\text{ mol } O_2} \times \frac{44.01\text{ g } CO_2}{1\text{ mol } CO_2} = 35.0\text{ g of } CO_2$ (smaller amount of product)

∴ O_2 is the limiting reactant, and 35.0 g of CO_2 (3 SFs) can be produced.

9.27 **a.** Theoretical yield of CS_2:

$40.0\text{ g C} \times \frac{1\text{ mol C}}{12.01\text{ g C}} \times \frac{1\text{ mol } CS_2}{5\text{ mol C}} \times \frac{76.15\text{ g } CS_2}{1\text{ mol } CS_2} = 50.7\text{ g of } CS_2$

Percent yield: $\frac{36.0\text{ g } CS_2\text{ (actual)}}{50.7\text{ g } CS_2\text{ (theoretical)}} \times 100\% = 71.0\%$ (3 SFs)

b. Theoretical yield of CS_2:

$32.0\text{ g } SO_2 \times \frac{1\text{ mol } SO_2}{64.07\text{ g } SO_2} \times \frac{1\text{ mol } CS_2}{2\text{ mol } SO_2} \times \frac{76.15\text{ g } CS_2}{1\text{ mol } CS_2} = 19.0\text{ g of } CS_2$

Percent yield: $\frac{12.0\text{ g } CS_2\text{ (actual)}}{19.0\text{ g } CS_2\text{ (theoretical)}} \times 100\% = 63.2\%$ (3 SFs)

9.29 Theoretical yield of Al_2O_3:

$50.0\text{ g Al} \times \frac{1\text{ mol Al}}{26.98\text{ g Al}} \times \frac{2\text{ mol } Al_2O_3}{4\text{ mol Al}} \times \frac{101.96\text{ g } Al_2O_3}{1\text{ mol } Al_2O_3} = 94.5\text{ g of } Al_2O_3$

Use the percent yield to convert theoretical to actual:

$94.5\text{ g } Al_2O_3 \times \frac{75.0\text{ g } Al_2O_3}{100\text{ g } Al_2O_3} = 70.9\text{ g of } Al_2O_3$ (actual) (3 SFs)

9.31 Theoretical yield of CO_2:

$30.0\text{ g C} \times \frac{1\text{ mol C}}{12.01\text{ g C}} \times \frac{2\text{ mol CO}}{3\text{ mol C}} \times \frac{28.01\text{ g CO}}{1\text{ mol CO}} = 46.6\text{ g of CO}$

Percent yield: $\frac{28.2\text{ g CO (actual)}}{46.6\text{ g CO (theoretical)}} \times 100\% = 60.5\%$ (3 SFs)

9.33 In exothermic reactions, the energy of the products is lower than that of the reactants.

9.35 **a.** An exothermic reaction releases energy.
b. An endothermic reaction has a higher energy level for the products than the reactants.
c. Metabolism is an exothermic reaction, providing energy for the body.

9.37 **a.** Heat is released, which makes the reaction exothermic with $\Delta H = -802$ kJ.
b. Heat is absorbed, which makes the reaction endothermic with $\Delta H = +65.3$ kJ.
c. Heat is released, which makes the reaction exothermic with $\Delta H = -850$ kJ.

9.39 **a.** $125\text{ g }\cancel{Cl_2} \times \dfrac{1\text{ }\cancel{\text{mol }Cl_2}}{70.90\text{ }\cancel{\text{g }Cl_2}} \times \dfrac{-657\text{ kJ}}{2\text{ }\cancel{\text{mol }Cl_2}} = -579\text{ kJ (3 SFs)}$

$\therefore$ 579 kJ are released.

b. $278\text{ g }\cancel{PCl_5} \times \dfrac{1\text{ }\cancel{\text{mol }PCl_5}}{208.2\text{ }\cancel{\text{g }PCl_5}} \times \dfrac{+67\text{ kJ}}{1\text{ }\cancel{\text{mol }PCl_5}} = +89\text{ kJ (2 SFs)}$

$\therefore$ 89 kJ are absorbed.

9.41 **a.** $2NO + O_2 \rightarrow 2NO_2$
b. NO is the limiting reactant.

9.43 **a.** $N_2 + 3H_2 \rightarrow 2NH_3$
b. diagram A; N_2 is the excess reactant.

9.45 **a.** $2NI_3 \rightarrow N_2 + 3I_2$
b. Theoretical yield: from the $6NI_3$ we could obtain

$6\cancel{NI_3} \times \dfrac{1N_2}{2\cancel{NI_3}} = 3N_2 \qquad 6\cancel{NI_3} \times \dfrac{3I_2}{2\cancel{NI_3}} = 9I_2$

Actual yield: in the actual products, we obtain $2N_2$ and $6I_2$.

Percent yield: $\dfrac{2N_2 \text{ and } 6I_2 \text{ (actual)}}{3N_2 \text{ and } 9I_2 \text{ (theoretical)}} \times 100\%$

$= \dfrac{2\ \cancel{(N_2 \text{ and } 3I_2)}\text{ (actual)}}{3\ \cancel{(N_2 \text{ and } 3I_2)}\text{ (theoretical)}} \times 100\% = 67\%$

9.47 **a.** $4.00\text{ }\cancel{\text{mol HF}} \times \dfrac{2\text{ mol }NH_3}{6\text{ }\cancel{\text{mol HF}}} = 1.33\text{ mol of }NH_3\text{ (3 SFs)}$

$4.00\text{ }\cancel{\text{mol HF}} \times \dfrac{5\text{ mol }F_2}{6\text{ }\cancel{\text{mol HF}}} = 3.33\text{ mol of }F_2\text{ (3 SFs)}$

b. $25.5\text{ g }\cancel{NH_3} \times \dfrac{1\text{ }\cancel{\text{mol }NH_3}}{17.03\text{ }\cancel{\text{g }NH_3}} \times \dfrac{5\text{ }\cancel{\text{mol }F_2}}{2\text{ }\cancel{\text{mol }NH_3}} \times \dfrac{38.00\text{ g }F_2}{1\text{ }\cancel{\text{mol }F_2}} = 142\text{ g of }F_2\text{ (3 SFs)}$

c. $3.40\text{ g }\cancel{NH_3} \times \dfrac{1\text{ }\cancel{\text{mol }NH_3}}{17.03\text{ }\cancel{\text{g }NH_3}} \times \dfrac{1\text{ }\cancel{\text{mol }N_2F_4}}{2\text{ }\cancel{\text{mol }NH_3}} \times \dfrac{104.02\text{ g }N_2F_4}{1\text{ }\cancel{\text{mol }N_2F_4}} = 10.4\text{ g of }N_2F_4\text{ (3 SFs)}$

9.49 **a.** $3.00\text{ }\cancel{\text{mol }H_2O} \times \dfrac{2\text{ mol }H_2O_2}{2\text{ }\cancel{\text{mol }H_2O}} = 3.00\text{ mol of }H_2O_2\text{ (3 SFs)}$

b. $36.5\text{ g }\cancel{O_2} \times \dfrac{1\text{ }\cancel{\text{mol }O_2}}{32.00\text{ }\cancel{\text{g }O_2}} \times \dfrac{2\text{ }\cancel{\text{mol }H_2O_2}}{1\text{ }\cancel{\text{mol }O_2}} \times \dfrac{34.02\text{ g }H_2O_2}{1\text{ }\cancel{\text{mol }H_2O_2}} = 77.6\text{ g of }H_2O_2\text{ (3 SFs)}$

c. $12.2\text{ g }\cancel{H_2O_2} \times \dfrac{1\text{ }\cancel{\text{mol }H_2O_2}}{34.02\text{ }\cancel{\text{g }H_2O_2}} \times \dfrac{2\text{ }\cancel{\text{mol }H_2O}}{2\text{ }\cancel{\text{mol }H_2O_2}} \times \dfrac{18.02\text{ g }H_2O}{1\text{ }\cancel{\text{mol }H_2O}} = 6.46\text{ g of }H_2O\text{ (3 SFs)}$

9.51 $12.8\ \text{g Na} \times \frac{1\ \text{mol Na}}{22.99\ \text{g Na}} \times \frac{2\ \text{mol NaCl}}{2\ \text{mol Na}} \times \frac{58.44\ \text{g NaCl}}{1\ \text{mol NaCl}} = 32.5\ \text{g of NaCl}$

$10.2\ \text{g}\ Cl_2 \times \frac{1\ \text{mol}\ Cl_2}{70.90\ \text{g}\ Cl_2} \times \frac{2\ \text{mol NaCl}}{1\ \text{mol}\ Cl_2} \times \frac{58.44\ \text{g NaCl}}{1\ \text{mol NaCl}} = 16.8\ \text{g of NaCl}$ (smaller amount of product)

∴ Cl_2 is the limiting reactant, and 16.8 g of NaCl (3 SFs) can be produced.

9.53 **a.** $4.00\ \text{mol}\ H_2O \times \frac{1\ \text{mol}\ C_5H_{12}}{6\ \text{mol}\ H_2O} = 0.667\ \text{mol of}\ C_5H_{12}$ (3 SFs)

b. $32.0\ \text{g}\ O_2 \times \frac{1\ \text{mol}\ O_2}{32.00\ \text{g}\ O_2} \times \frac{5\ \text{mol}\ CO_2}{8\ \text{mol}\ O_2} \times \frac{44.01\ \text{g}\ CO_2}{1\ \text{mol}\ CO_2} = 27.5\ \text{g of}\ CO_2$ (3 SFs)

c. $44.5\ \text{g}\ C_5H_{12} \times \frac{1\ \text{mol}\ C_5H_{12}}{72.15\ \text{g}\ C_5H_{12}} \times \frac{5\ \text{mol}\ CO_2}{1\ \text{mol}\ C_5H_{12}} \times \frac{44.01\ \text{g}\ CO_2}{1\ \text{mol}\ CO_2} = 136\ \text{g of}\ CO_2$

$108\ \text{g}\ O_2 \times \frac{1\ \text{mol}\ O_2}{32.00\ \text{g}\ O_2} \times \frac{5\ \text{mol}\ CO_2}{8\ \text{mol}\ O_2} \times \frac{44.01\ \text{g}\ CO_2}{1\ \text{mol}\ CO_2} = 92.8\ \text{g of}\ CO_2$ (smaller amount of product)

∴ O_2 is the limiting reactant, and 92.8 g of CO_2 (3 SFs) can be produced.

9.55 **a.** Theoretical yield of CO_2:

$22.0\ \text{g}\ C_2H_2 \times \frac{1\ \text{mol}\ C_2H_2}{26.04\ \text{g}\ C_2H_2} \times \frac{4\ \text{mol}\ CO_2}{2\ \text{mol}\ C_2H_2} \times \frac{44.01\ \text{g}\ CO_2}{1\ \text{mol}\ CO_2} = 74.4\ \text{g of}\ CO_2$ (3 SFs)

b. Percent yield: $\frac{64.0\ \text{g}\ CO_2\ (\text{actual})}{74.4\ \text{g}\ CO_2\ (\text{theoretical})} \times 100\% = 86.0\%$ (3 SFs)

9.57 Theoretical yield of C_2H_6:

$28.0\ \text{g}\ C_2H_2 \times \frac{1\ \text{mol}\ C_2H_2}{26.04\ \text{g}\ C_2H_2} \times \frac{1\ \text{mol}\ C_2H_6}{1\ \text{mol}\ C_2H_2} \times \frac{30.07\ \text{g}\ C_2H_6}{1\ \text{mol}\ C_2H_6} = 32.3\ \text{g of}\ C_2H_6$

Percent yield: $\frac{24.5\ \text{g}\ C_2H_6\ (\text{actual})}{32.3\ \text{g}\ C_2H_6\ (\text{theoretical})} \times 100\% = 75.9\%$ (3 SFs)

9.59 **a.** Use the limiting reactant to calculate the theoretical yield of NH_3:

$50.0\ \text{g}\ N_2 \times \frac{1\ \text{mol}\ N_2}{28.02\ \text{g}\ N_2} \times \frac{2\ \text{mol}\ NH_3}{1\ \text{mol}\ N_2} \times \frac{17.03\ \text{g}\ NH_3}{1\ \text{mol}\ NH_3} = 60.8\ \text{g of}\ NH_3$ (smaller amount of product)

$20.0\ \text{g}\ H_2 \times \frac{1\ \text{mol}\ H_2}{2.016\ \text{g}\ H_2} \times \frac{2\ \text{mol}\ NH_3}{3\ \text{mol}\ H_2} \times \frac{17.03\ \text{g}\ NH_3}{1\ \text{mol}\ NH_3} = 113\ \text{g of}\ NH_3$

∴ N_2 is the limiting reactant, and 60.8 g of NH_3 (3 SFs) can be produced (theoretical yield).

b. Use the percent yield to calculate the actual yield of NH_3:

$60.8\ \text{g}\ NH_3\ (\text{theoretical}) \times \frac{62.0\ \text{g}\ NH_3\ (\text{actual})}{100\ \text{g}\ NH_3\ (\text{theoretical})} = 37.7\ \text{g of}\ NH_3$ (3 SFs)

9.61 **a.** $3.00\ \text{g NO} \times \frac{1\ \text{mol NO}}{30.01\ \text{g NO}} \times \frac{+90.2\ \text{kJ}}{2\ \text{mol NO}} = +4.51\ \text{kJ}$ (3 SFs)

∴ 4.51 kJ are required.

b. The equation for the decomposition of NO will be the reverse of the equation for the formation of NO.

$2NO(g) \rightarrow N_2(g) + O_2(g) + 90.2\ \text{kJ}$

c. $5.00\text{ g NO} \times \dfrac{1\text{ mol NO}}{30.01\text{ g NO}} \times \dfrac{-90.2\text{ kJ}}{2\text{ mol NO}} = -7.51\text{ kJ (3 SFs)}$

∴ 7.51 kJ are released.

9.63 **a.** Heat is released, which makes the reaction exothermic.

b. Heat is absorbed, which makes the reaction endothermic.

9.65 **a.** $4.50\text{ mol Cr} \times \dfrac{3\text{ mol } O_2}{4\text{ mol Cr}} = 3.38\text{ mol of } O_2\text{ (3 SFs)}$

b. $24.8\text{ g Cr} \times \dfrac{1\text{ mol Cr}}{52.00\text{ g Cr}} \times \dfrac{2\text{ mol } Cr_2O_3}{4\text{ mol Cr}} \times \dfrac{152.0\text{ g } Cr_2O_3}{1\text{ mol } Cr_2O_3} = 36.2\text{ g of } Cr_2O_3\text{ (3 SFs)}$

c. $26.0\text{ g Cr} \times \dfrac{1\text{ mol Cr}}{52.00\text{ g Cr}} \times \dfrac{2\text{ mol } Cr_2O_3}{4\text{ mol Cr}} \times \dfrac{152.0\text{ g } Cr_2O_3}{1\text{ mol } Cr_2O_3} = 38.0\text{ g of } Cr_2O_3$

$8.00\text{ g } O_2 \times \dfrac{1\text{ mol } O_2}{32.00\text{ g } O_2} \times \dfrac{2\text{ mol } Cr_2O_3}{3\text{ mol } O_2} \times \dfrac{152.0\text{ g } Cr_2O_3}{1\text{ mol } Cr_2O_3} = 25.3\text{ g of } Cr_2O_3$ (smaller amount of product)

∴ O_2 is the limiting reactant, and 25.3 g of Cr_2O_3 (3 SFs) can be produced.

d. $74.0\text{ g Cr} \times \dfrac{1\text{ mol Cr}}{52.00\text{ g Cr}} \times \dfrac{2\text{ mol } Cr_2O_3}{4\text{ mol Cr}} \times \dfrac{152.0\text{ g } Cr_2O_3}{1\text{ mol } Cr_2O_3} = 108\text{ g of } Cr_2O_3$ (smaller amount of product)

$62.0\text{ g } O_2 \times \dfrac{1\text{ mol } O_2}{32.00\text{ g } O_2} \times \dfrac{2\text{ mol } Cr_2O_3}{3\text{ mol } O_2} \times \dfrac{152.0\text{ g } Cr_2O_3}{1\text{ mol } Cr_2O_3} = 196\text{ g of } Cr_2O_3$

∴ Cr is the limiting reactant, and 108 g of Cr_2O_3 (3 SFs) can be produced (theoretical yield).

Percent yield: $\dfrac{87.3\text{ g } Cr_2O_3\text{ (actual)}}{108\text{ g } Cr_2O_3\text{ (theoretical)}} \times 100\% = 80.8\%\text{ (3 SFs)}$

9.67 **a.** $0.225\text{ mol } C_3H_4 \times \dfrac{4\text{ mol } O_2}{1\text{ mol } C_3H_4} = 0.900\text{ mol of } O_2\text{ (3 SFs)}$

b. $64.0\text{ g } O_2 \times \dfrac{1\text{ mol } O_2}{32.00\text{ g } O_2} \times \dfrac{2\text{ mol } H_2O}{4\text{ mol } O_2} \times \dfrac{18.02\text{ g } H_2O}{1\text{ mol } H_2O} = 18.0\text{ g of } H_2O\text{ (3 SFs)}$

c. $78.0\text{ g } C_3H_4 \times \dfrac{1\text{ mol } C_3H_4}{40.06\text{ g } C_3H_4} \times \dfrac{3\text{ mol } CO_2}{1\text{ mol } C_3H_4} \times \dfrac{44.01\text{ g } CO_2}{1\text{ mol } CO_2} = 257\text{ g of } CO_2\text{ (3 SFs)}$

d. Percent yield: $\dfrac{186\text{ g } CO_2\text{ (actual)}}{257\text{ g } CO_2\text{ (theoretical)}} \times 100\% = 72.4\%\text{ (3 SFs)}$

9.69 **a.** The heat of reaction (ΔH) is positive, so the reaction is endothermic.

b. $1.5\text{ mol } SO_3 \times \dfrac{+790\text{ kJ}}{2\text{ mol } SO_3} = +590\text{ kJ (2 SFs)}$

∴ 590 kJ are required.

c. $150\text{ g } O_2 \times \dfrac{1\text{ mol } O_2}{32.00\text{ g } O_2} \times \dfrac{+790\text{ kJ}}{3\text{ mol } O_2} = +1200\text{ kJ (2 SFs)}$

∴ 1200 kJ are required.

10

Properties of Solids and Liquids

Learning Goals

- Draw the electron-dot formulas for molecular compounds or polyatomic ions with multiple bonds and show resonance structures.
- Predict the three-dimensional structure of a molecule or polyatomic ion and classify it as polar or nonpolar.
- Use electronegativity to determine the polarity of a bond or a molecule.
- Describe the attractive forces between ions, polar covalent molecules, and nonpolar covalent molecules.
- Describe the changes of state between solids, liquids, and gases; calculate the energy involved.

Chapter Outline

Chapter Opener: Histologist

10.1 Electron-Dot Formulas

10.2 Shapes of Molecules and Ions (VSEPR Theory)

10.3 Electronegativity and Polarity

10.4 Attractive Forces in Compounds

Chemistry Link to Health: Attractive Forces in Biological Compounds

10.5 Changes of State

Chemistry Link to Health: Steam Burns

Core Chemistry Skills

- Drawing Electron-Dot Formulas (10.1)
- Drawing Resonance Structures (10.1)
- Predicting Shape (10.2)
- Using Electronegativity (10.3)
- Identifying Polarity of Molecules (10.3)
- Identifying Attractive Forces (10.4)
- Calculating Heat for Change of State (10.5)

Answers and Solutions to Text Problems

10.1 **a.** $2\,H(1\,e^-) + 1\,S(6\,e^-) = 2 + 6 = 8$ valence electrons
b. $2\,I(7\,e^-) = 14$ valence electrons
c. $1\,C(4\,e^-) + 4\,Cl(7\,e^-) = 4 + 28 = 32$ valence electrons
d. $1\,O(6\,e^-) + 1\,H(1\,e^-) + 1\,e^-$(negative charge) $= 6 + 1 + 1 = 8$ valence electrons

10.3 **a.** $1\,H(1\,e^-) + 1\,F(7\,e^-) = 1 + 7 = 8$ valence electrons

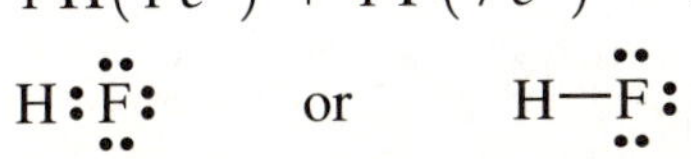

b. $1\,S(6\,e^-) + 2\,F(7\,e^-) = 6 + 14 = 20$ valence electrons

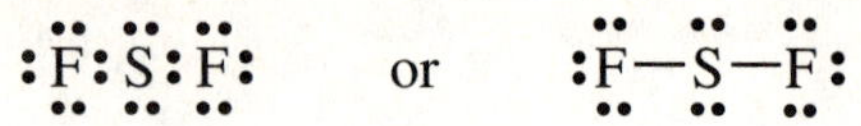

c. $1\,N(5\,e^-) + 3\,Br(7\,e^-) = 5 + 21 = 26$ valence electrons

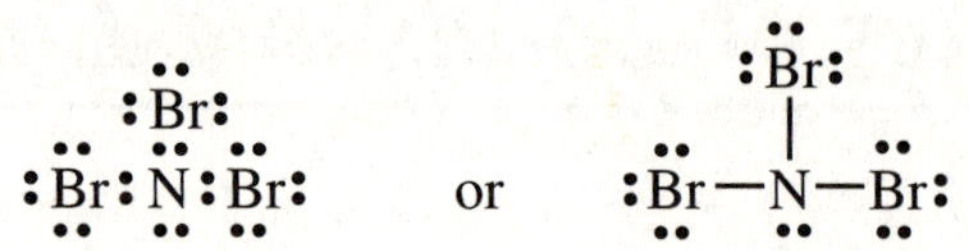

d. $1\,B(3\,e^-) + 4\,H(1\,e^-) + 1\,e^-$(negative charge) $= 3 + 4 + 1 = 8$ valence electrons

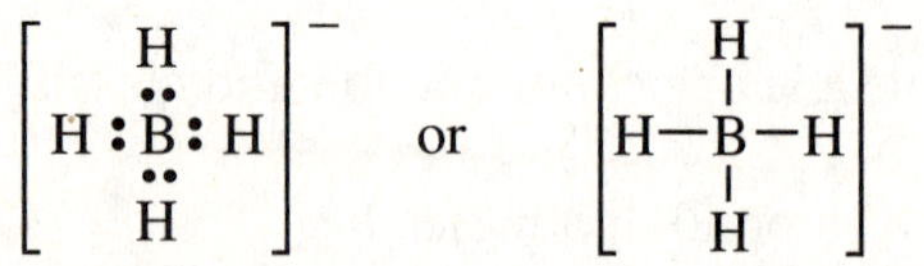

e. $1\,C(4\,e^-) + 4\,H(1\,e^-) + 1\,O(6\,e^-) = 4 + 4 + 6 = 14$ valence electrons

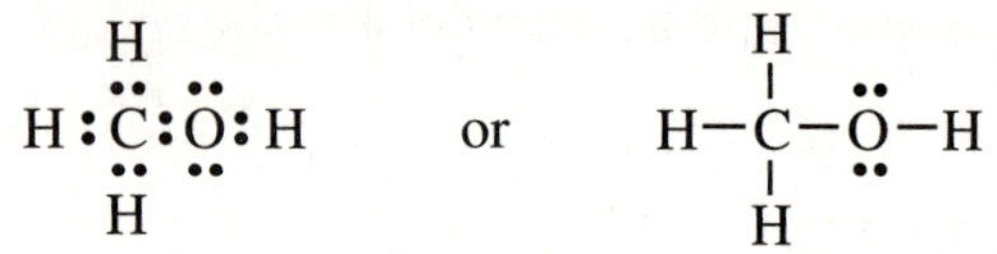

f. $2\,N(5\,e^-) + 4\,H(1\,e^-) = 10 + 4 = 14$ valence electrons

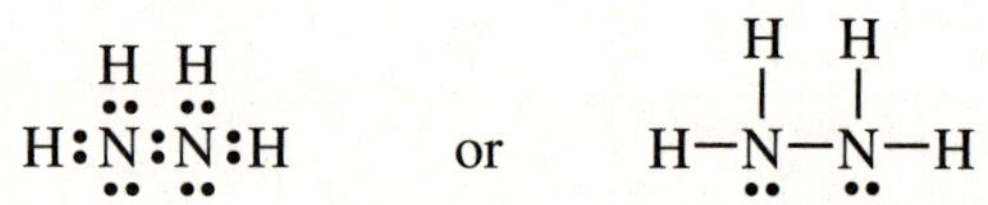

10.5 If complete octets cannot be formed by using all the valence electrons, it is necessary to draw multiple bonds.

10.7 Resonance occurs when we can draw two or more electron-dot formulas for the same molecule or ion.

10.9 **a.** $1\,C(4\,e^-) + 1\,O(6\,e^-) = 4 + 6 = 10$ valence electrons

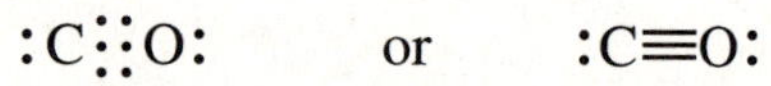

b. $1\,H(1\,e^-) + 1\,C(4\,e^-) + 1\,N(5\,e^-) = 1 + 4 + 5 = 10$ valence electrons

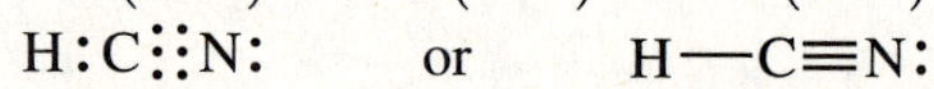

c. $2\,H(1\,e^-) + 1\,C(4\,e^-) + 1\,O(6\,e^-) = 2 + 4 + 6 = 12$ valence electrons

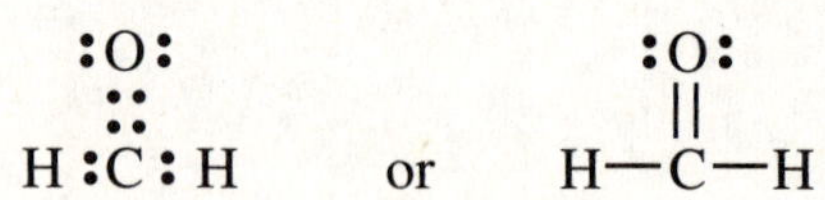

10.11 **a.** $1\,Cl(7\,e^-) + 1\,N(5\,e^-) + 2\,O(6\,e^-) = 7 + 5 + 12 = 24$ valence electrons

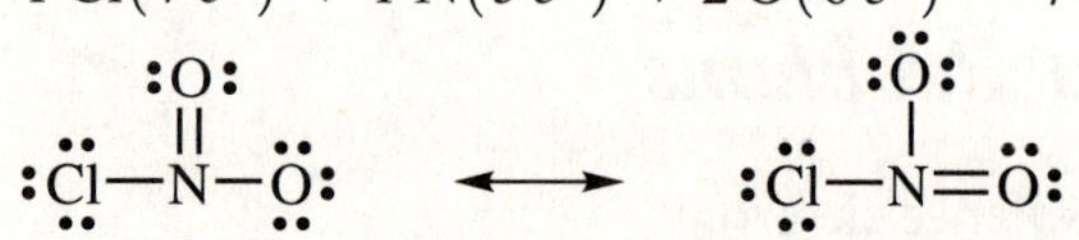

b. $1\,O(6\,e^-) + 1\,C(4\,e^-) + 1\,N(5\,e^-) + 1\,e^-$(negative charge)
$= 6 + 4 + 5 + 1 = 16$ valence electrons

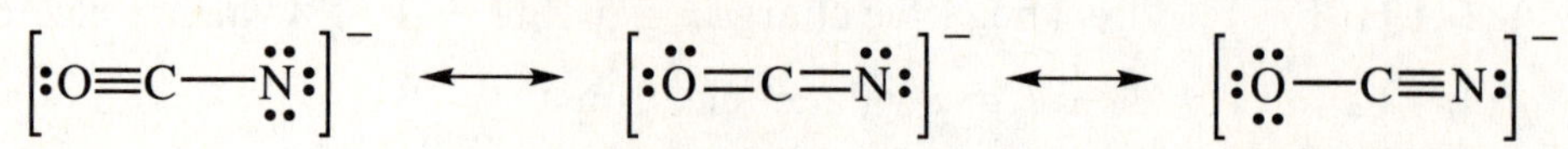

10.13 **a.** 6; Four electron groups around a central atom have a tetrahedral electron arrangement. With four bonded atoms, the shape of the molecule is tetrahedral.

b. 5; Four electron groups around a central atom have a tetrahedral electron arrangement. With three bonded atoms (and one lone pair of electrons), the shape of the molecule is trigonal pyramidal.

c. 3; Three electron groups around a central atom have a trigonal planar electron arrangement. With three bonded atoms, the shape of the molecule is trigonal planar.

10.15 SeO_3 $1\,Se(6\,e^-) + 3\,O(6\,e^-) = 6 + 18 = 24$ valence electrons

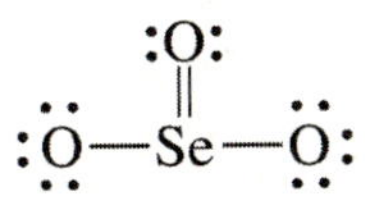

a. There are <u>three</u> electron groups around the central Se atom.

b. The electron-group geometry is <u>trigonal planar</u>.

c. The number of atoms attached to the central Se atom is <u>three</u>.

d. The shape of the molecule is <u>trigonal planar</u>.

10.17 In CF_4, the central atom C has four bonded atoms and no lone pairs of electrons, which gives CF_4 a tetrahedral shape. In NF_3, the central atom N has three bonded atoms and one lone pair of electrons, which gives NF_3 a trigonal pyramidal shape.

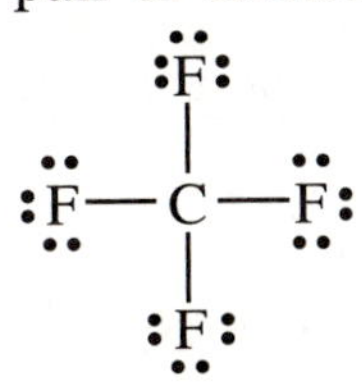

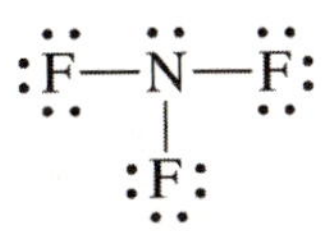

10.19 **a.** The central atom Ga has three electron groups bonded to three hydrogen atoms; GaH_3 has a trigonal planar shape.

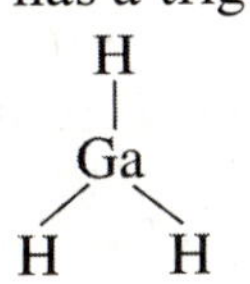

b. The central O atom has four electron groups with two bonded atoms and two lone pairs of electrons, which gives OF_2 a bent shape (109°).

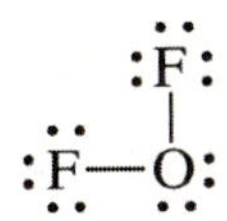

c. The central atom C has two electron groups bonded to two atoms; HCN is linear.

H—C≡N:

d. The central atom C has four electron groups bonded to four chlorine atoms; CCl_4 has a tetrahedral shape.

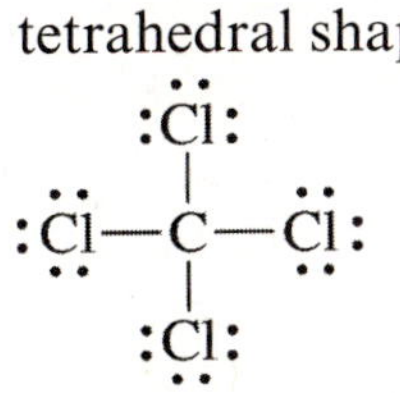

10.21 To find the total valence electrons for an ion, add the total valence electrons for each atom and add the number of electrons indicated by a negative charge.

a. 1 C(4 e^-) + 3 O(6 e^-) + 2 e^-(negative charge) = 4 + 18 + 2 = 24 valence electrons

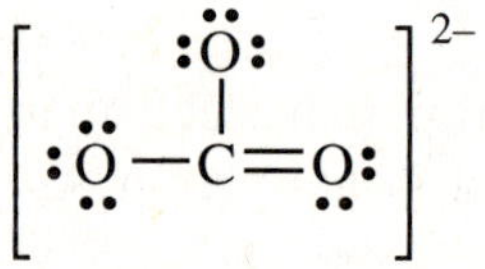

three electron groups around C bonded to three atoms; trigonal planar shape

b. 1 S(6 e^-) + 4 O(6 e^-) + 2 e^-(negative charge) = 6 + 24 + 2 = 32 valence electrons

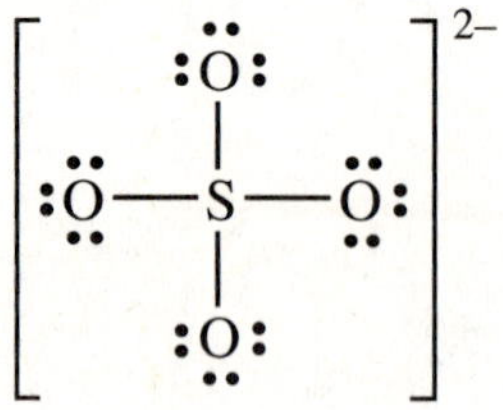

four electron groups around S bonded to four atoms; tetrahedral shape

c. 1 N(5 e^-) + 4 H(1 e^-) − 1 e^-(positive charge) = 5 + 4 − 1 = 8 valence electrons

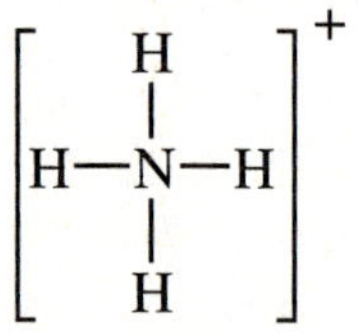

four electron groups around N bonded to four atoms; tetrahedral shape

d. 1 N(5 e^-) + 2 O(6 e^-) − 1 e^-(positive charge) = 5 + 12 − 1 = 16 valence electrons

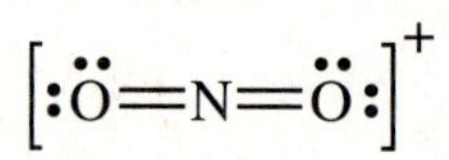

two electron groups around N bonded to two atoms; linear shape

10.23 **a.** The electronegativity values increase going from left to right across Period 2 from B to F.
b. The electronegativity values decrease going down Group 2A (2) from Mg to Ba.
c. The electronegativity values decrease going down Group 7A (17) from F to I.

10.25 A nonpolar covalent bond would have an electronegativity difference between 0.0 and 0.4. (**a.**)

10.27 **a.** Electronegativity increases going up a group: K, Na, Li.
b. Electronegativity increases going left to right across a period: Na, P, Cl.
c. Electronegativity increases going across a period and at the top of a group: Ca, Se, O.

10.29 A dipole arrow points from the atom with the lower electronegativity value (more positive) to the atom in the bond that has the higher electronegativity value (more negative).

a. $\overset{\delta^+}{\text{N}}$—$\overset{\delta^-}{\text{F}}$

b. $\overset{\delta^+}{\text{Si}}$—$\overset{\delta^-}{\text{Br}}$

c. $\overset{\delta^+}{\text{C}}$—$\overset{\delta^-}{\text{O}}$

d. $\overset{\delta^+}{\text{P}}$—$\overset{\delta^-}{\text{Br}}$

e. $\overset{\delta^-}{\text{N}}$—$\overset{\delta^+}{\text{P}}$

10.31 a. Si—Br electronegativity difference: 2.8 − 1.8 = 1.0, polar covalent
b. Li—F electronegativity difference: 4.0 − 1.0 = 3.0, ionic
c. Br—F electronegativity difference: 4.0 − 2.8 = 1.2, polar covalent
d. I—I electronegativity difference: 2.5 − 2.5 = 0.0, nonpolar covalent
e. N—P electronegativity difference: 3.0 − 2.1 = 0.9, polar covalent
f. C—P electronegativity difference: 2.5 − 2.1 = 0.4, nonpolar covalent

10.33 F_2 is a nonpolar molecule because there is a nonpolar covalent bond between F atoms, which have identical electronegativity values. In HF, the bond is a polar bond because there is a large electronegativity difference, which makes HF a polar molecule.

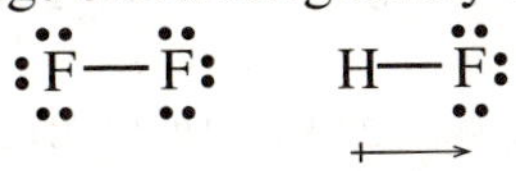

nonpolar polar

10.35 a. The molecule CS_2 contains two polar C—S bonds whose dipoles point in opposite directions. As a result, the dipoles cancel out, which makes CS_2 a nonpolar molecule.

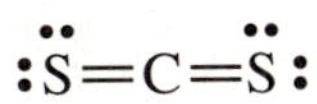

nonpolar

b. The molecule NF_3 contains three polar covalent N—F bonds and a lone pair of electrons on the central N atom. This asymmetric trigonal pyramidal shape makes NF_3 a polar molecule.

polar

c. In the molecule CHF_3, there are three polar covalent C—F bonds and one nonpolar covalent C—H bond, which makes CHF_3 a polar molecule.

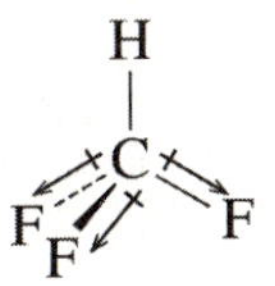

polar

d. In the molecule SO_3, there are three polar covalent S—O bonds arranged in a trigonal planar shape. This symmetric shape makes SO_3 a nonpolar molecule.

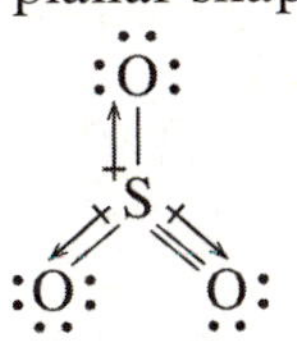

nonpolar

10.37 In the molecule CO_2, the two C—O dipoles cancel, resulting in a nonpolar molecule; in CO, there is only one dipole, making it a polar molecule.

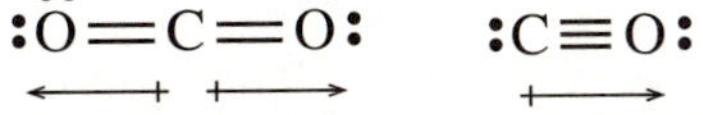

nonpolar polar

10.39 a. BrF is a polar molecule. An attraction between the positive end of one polar molecule and the negative end of another polar molecule is called a dipole–dipole attraction.
b. An ionic bond is an attraction between a positive and negative ion, as in KCl.
c. NF_3 is a polar molecule. An attraction between the positive end of one polar molecule and the negative end of another polar molecule is called a dipole–dipole attraction.
d. Cl_2 is a nonpolar molecule. The weak attractions that occur between temporary dipoles in nonpolar molecules are called dispersion forces.

10.41 a. Hydrogen bonds are strong dipole–dipole attractions that occur between a partially positive hydrogen atom of one molecule and one of the strongly electronegative atoms F, O, or N in another, as is seen with CH_3OH molecules.

b. CO is a polar molecule. Dipole–dipole attractions occur between dipoles in polar molecules.

c. Dispersion forces occur between temporary dipoles in nonpolar CF_4 molecules.

d. Dispersion forces occur between temporary dipoles in nonpolar $CH_3—CH_3$ molecules.

10.43 a. HF would have a higher boiling point than HBr; the hydrogen bonds in HF are stronger than the dipole–dipole attractions in HBr.

b. NaF would have a higher boiling point than HF; the ionic bonds in NaF are stronger than the hydrogen bonds in HF.

c. $MgBr_2$ would have a higher boiling point than PBr_3; the ionic bonds in $MgBr_2$ are stronger than the dipole–dipole attractions in PBr_3.

d. CH_3OH would have a higher boiling point than CH_4; the hydrogen bonds in CH_3OH are stronger than the dispersion forces in CH_4.

10.45 a. $65.0\ \cancel{\text{g ice}} \times \frac{334\ \text{J}}{1\ \cancel{\text{g ice}}} = 21\ 700\ \text{J (3 SFs) is absorbed}$

b. $17.0\ \cancel{\text{g ethanol}} \times \frac{109\ \text{J}}{1\ \cancel{\text{g ethanol}}} = 1850\ \text{J (3 SFs) is absorbed}$

c. $225\ \cancel{\text{g acetone}} \times \frac{99.3\ \cancel{\text{J}}}{1\ \cancel{\text{g acetone}}} \times \frac{1\ \text{kJ}}{1000\ \cancel{\text{J}}} = 22.3\ \text{kJ (3 SFs) is released}$

d. $0.0500\ \cancel{\text{kg water}} \times \frac{1000\ \cancel{\text{g water}}}{1\ \cancel{\text{kg water}}} \times \frac{334\ \cancel{\text{J}}}{1\ \cancel{\text{g water}}} \times \frac{1\ \text{kJ}}{1000\ \cancel{\text{J}}} = 16.7\ \text{kJ (3 SFs) is released}$

10.47 a. $10.0\ \cancel{\text{g water}} \times \frac{2260\ \text{J}}{1\ \cancel{\text{g water}}} = 22\ 600\ \text{J (3 SFs) is absorbed}$

b. $50.0\ \cancel{\text{g ethanol}} \times \frac{841\ \cancel{\text{J}}}{1\ \cancel{\text{g ethanol}}} \times \frac{1\ \text{kJ}}{1000\ \cancel{\text{J}}} = 42.1\ \text{kJ (3 SFs) is absorbed}$

c. $8.00\ \cancel{\text{g acetic acid}} \times \frac{390\ \text{J}}{1\ \cancel{\text{g acetic acid}}} = 3100\ \text{J (2 SFs) is released}$

d. $0.175\ \cancel{\text{kg ammonia}} \times \frac{1000\ \cancel{\text{g ammonia}}}{1\ \cancel{\text{kg ammonia}}} \times \frac{1380\ \cancel{\text{J}}}{1\ \cancel{\text{g ammonia}}} \times \frac{1\ \text{kJ}}{1000\ \cancel{\text{J}}}$

$= 242\ \text{kJ (3 SFs) is released}$

10.49 a. water 15 °C → 72 °C: $\Delta T = 72\ °\text{C} - 15\ °\text{C} = 57\ °\text{C}$;

$20.0\ \cancel{\text{g}} \times 57\ \cancel{°\text{C}} \times \frac{4.184\ \text{J}}{\cancel{\text{g}}\ \cancel{°\text{C}}}$

$= 4800\ \text{J (2 SFs)}$

b. Two calculations are needed:

(1) ice 0 °C → water 0 °C: $50.0\ \cancel{\text{g ice}} \times \frac{334\ \text{J}}{1\ \cancel{\text{g ice}}} = 16\ 700\ \text{J (3 SFs)}$

(2) water 0 °C → 65.0 °C: $\Delta T = 65.0\ °\text{C} - 0\ °\text{C} = 65.0\ °\text{C}$;

$50.0\ \cancel{\text{g}} \times 65.0\ \cancel{°\text{C}} \times \frac{4.184\ \text{J}}{\cancel{\text{g}}\ \cancel{°\text{C}}} = 13\ 600\ \text{J (3 SFs)}$

∴ Total heat needed = 16 700 J + 13 600 J = 30 300 J (3 SFs)

c. Two calculations are needed:

(1) steam 100 °C → water 100 °C: $15.0\ \text{g steam} \times \dfrac{2260\ \text{J}}{1\ \text{g steam}} \times \dfrac{1\ \text{kJ}}{1000\ \text{J}} = 33.9\ \text{kJ}$ (3 SFs)

(2) water 100 °C → 0 °C: $15.0\ \text{g water} \times 100.\ ^\circ\text{C} \times \dfrac{4.184\ \text{J}}{\text{g}\ ^\circ\text{C}} \times \dfrac{1\ \text{kJ}}{1000\ \text{J}} = 6.28\ \text{kJ}$ (3 SFs)

∴ Total heat released = 33.9 kJ + 6.28 kJ = 40.2 kJ (3 SFs)

d. Three calculations are needed:

(1) ice 0 °C → water 0 °C: $24.0\ \text{g ice} \times \dfrac{334\ \text{J}}{1\ \text{g ice}} \times \dfrac{1\ \text{kJ}}{1000\ \text{J}} = 8.02\ \text{kJ}$ (3 SFs)

(2) water 0 °C → 100 °C: $24.0\ \text{g water} \times 100.\ ^\circ\text{C} \times \dfrac{4.184\ \text{J}}{\text{g}\ ^\circ\text{C}} \times \dfrac{1\ \text{kJ}}{1000\ \text{J}} = 10.0\ \text{kJ}$ (3 SFs)

(3) water 100 °C → steam 100 °C: $24.0\ \text{g water} \times \dfrac{2260\ \text{J}}{1\ \text{g water}} \times \dfrac{1\ \text{kJ}}{1000\ \text{J}} = 54.2\ \text{kJ}$ (3 SFs)

∴ Total heat needed = 8.02 kJ + 10.0 kJ + 54.2 kJ = 72.2 kJ (3 SFs)

10.51 Two calculations are needed:

(1) ice 0 °C → water 0 °C: $275\ \text{g ice} \times \dfrac{334\ \text{J}}{1\ \text{g ice}} \times \dfrac{1\ \text{kJ}}{1000\ \text{J}} = 91.9\ \text{kJ}$ (3 SFs)

(2) water 0 °C → 24.0 °C: $\Delta T = 24.0\ ^\circ\text{C} - 0\ ^\circ\text{C} = 24.0\ ^\circ\text{C}$;

$275\ \text{g water} \times 24.0\ ^\circ\text{C} \times \dfrac{4.184\ \text{J}}{\text{g}\ ^\circ\text{C}} \times \dfrac{1\ \text{kJ}}{1000\ \text{J}} = 27.6\ \text{kJ}$ (3 SFs)

∴ Total heat absorbed = 91.9 kJ + 27.6 kJ = 119.5 kJ (4 SFs)

10.53 **a.** 2; trigonal pyramidal shape, polar molecule
b. 1; bent shape (109°), polar molecule
c. 3; tetrahedral shape, nonpolar molecule

10.55 **a.** C—O (EN 3.5 − 2.5 = 1.0) and N—O (EN 3.5 − 3.0 = 0.5) are polar covalent bonds.
b. O—O (EN 3.5 − 3.5 = 0.0) is a nonpolar covalent bond.
c. Ca—O (EN 3.5 − 1.0 = 2.5) and K—O (EN 3.5 − 0.8 = 2.7) are ionic bonds.
d. C—O, N—O, O—O

10.57 **a.** PH_3 is a nonpolar molecule. Dispersion forces occur between temporary dipoles in nonpolar molecules.
b. NO_2 is a polar molecule. Dipole–dipole attractions occur between dipoles in polar molecules.
c. Hydrogen bonds are strong dipole–dipole attractions that occur between a partially positive hydrogen atom of one molecule and one of the strongly electronegative atoms F, O, or N in another.
d. Dispersion forces occur between temporary dipoles in Ar atoms.

10.59 **a.** The heat from the skin is used to evaporate the water (perspiration). Therefore, the skin is cooled.
b. On a hot day, there are more liquid water molecules in the damp towels that have sufficient energy to become water vapor. Thus, water evaporates from the towels more readily on a hot day.
c. In a closed plastic bag, some water molecules evaporate, but they cannot escape and will condense back to liquid; the clothes will not dry.

10.61

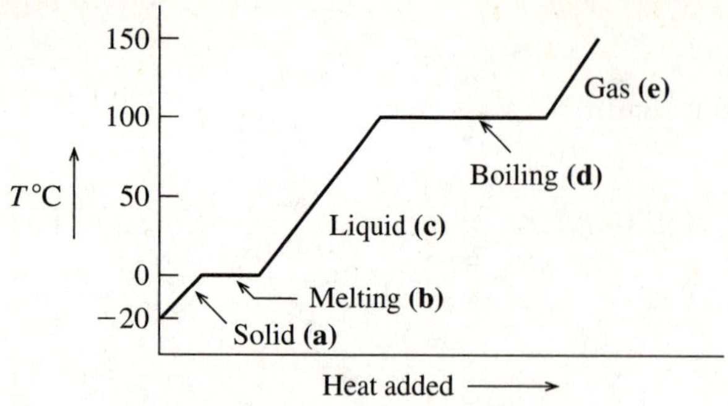

10.63 **a.** The melting point of chloroform is about −60 °C.

b. The boiling point of chloroform is about 60 °C.

c. The diagonal line **A** represents the solid state as temperature increases. The horizontal line **B** represents the change from solid to liquid, or melting of the substance. The diagonal line **C** represents the liquid state as temperature increases. The horizontal line **D** represents the change from liquid to gas, or boiling of the liquid. The diagonal line **E** represents the gas state as temperature increases.

d. At −80 °C, it is solid; at −40 °C, it is liquid; at 25 °C, it is liquid; at 80 °C, it is gas.

10.65 **a.** $1\,H(1\,e^-) + 1\,N(5\,e^-) + 2\,O(6\,e^-) = 1 + 5 + 12 = 18$ valence electrons

b. $2\,C(4\,e^-) + 4\,H(1\,e^-) + 1\,O(6\,e^-) = 8 + 4 + 6 = 18$ valence electrons

c. $1\,P(5\,e^-) + 4\,H(1\,e^-) - 1\,e^-$(positive charge) $= 5 + 4 - 1 = 8$ valence electrons

d. $1\,S(6\,e^-) + 3\,O(6\,e^-) + 2\,e^-$(negative charge) $= 6 + 18 + 2 = 26$ valence electrons

10.67 **a.** $1\,B(3\,e^-) + 4\,F(7\,e^-) + 1\,e^-$(negative charge) $= 3 + 28 + 1 = 32$ valence electrons

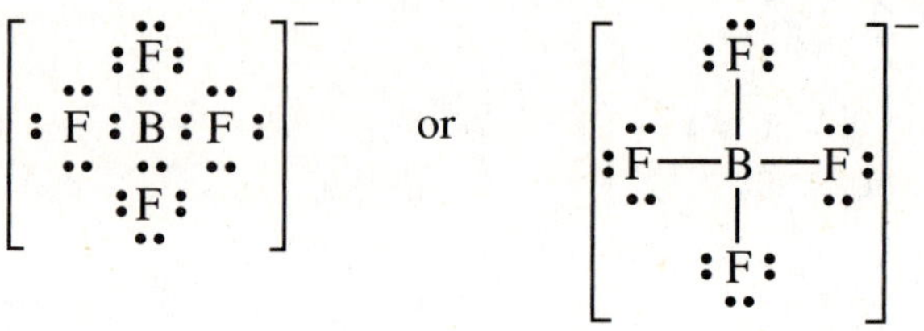

b. $2\,Cl(7\,e^-) + 1\,O(6\,e^-) = 14 + 6 = 20$ valence electrons

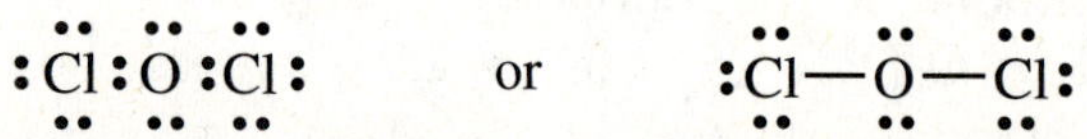

c. $1\,N(5\,e^-) + 3\,H(1\,e^-) + 1\,O(6\,e^-) = 5 + 3 + 6 = 14$ valence electrons

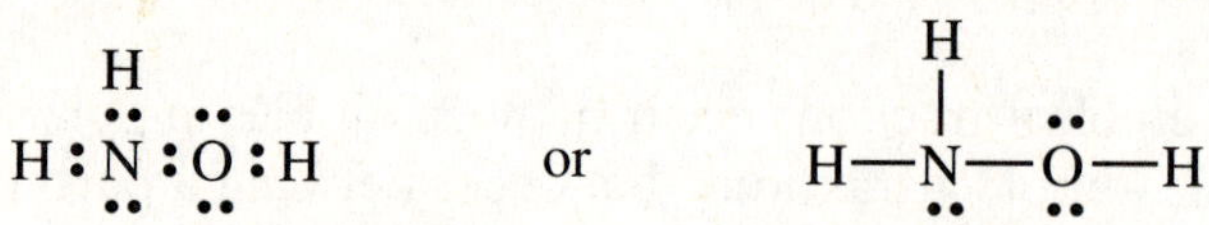

d. $2\,C(4\,e^-) + 2\,H(1\,e^-) + 2\,Cl(7\,e^-) = 8 + 2 + 14 = 24$ valence electrons

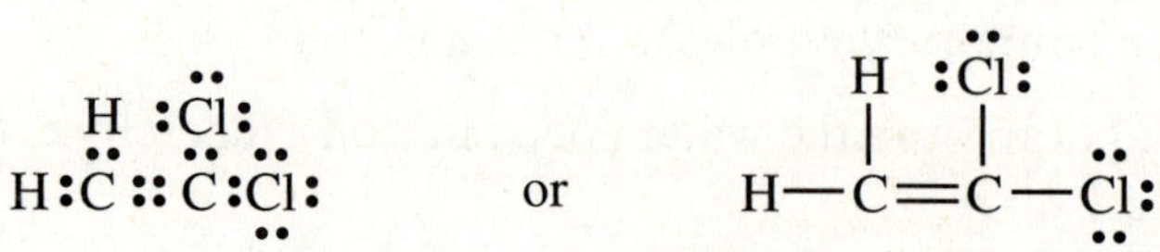

10.69 **a.** $3\,N(5\,e^-) + 1\,e^-$(negative charge) $= 15 + 1 = 16$ valence electrons

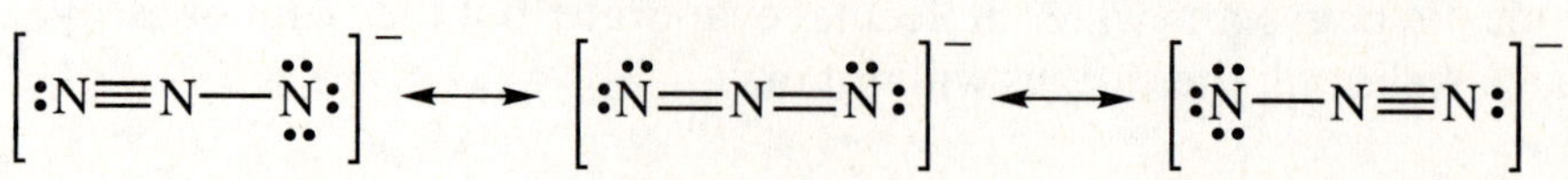

b. 1 N($5\,e^-$) + 2 O($6\,e^-$) − 1 e^-(positive charge) = 5 + 12 − 1 = 16 valence electrons

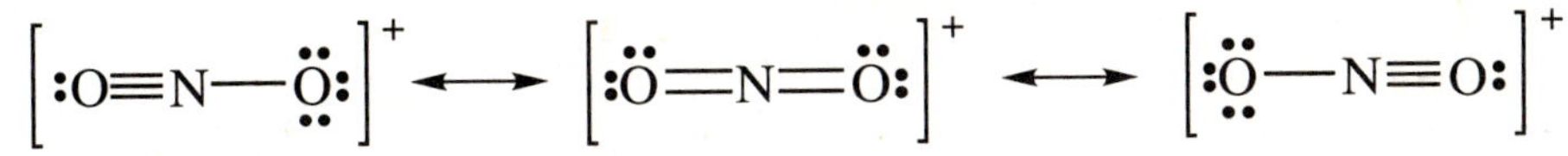

c. 1 C($4\,e^-$) + 1 N($5\,e^-$) + 1 S($6\,e^-$) + 1 e^-(negative charge)
= 4 + 5 + 6 + 1 = 16 valence electrons

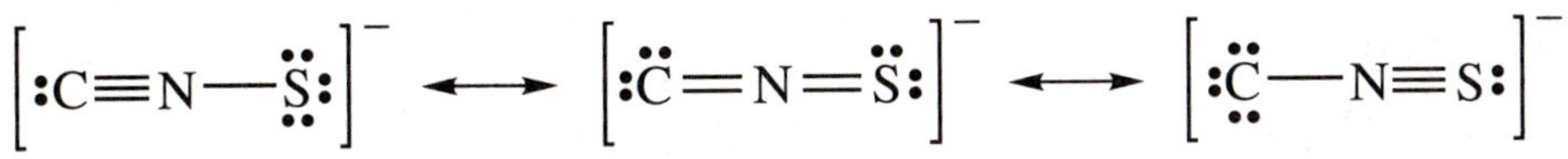

10.71 a. Electronegativity increases going up a group: I, Cl, F
b. Electronegativity increases going left to right across a period and at the top of a group: K, Li, S, Cl
c. Electronegativity increases going up a group: Ba, Sr, Mg, Be

10.73 a. C—O(3.5 − 2.5 = 1.0) is more polar than C—N(3.5 − 3.0 = 0.5).
b. N—F (4.0 − 3.0 = 1.0) is more polar than N—Br (3.0 − 2.8 = 0.2).
c. S—Cl (3.0 − 2.5 = 0.5) is more polar than Br—Cl (3.0 − 2.8 = 0.2).
d. Br—I (2.8 − 2.5 = 0.3) is more polar than Br—Cl (3.0 − 2.8 = 0.2).
e. N—F (4.0 − 3.0 = 1.0) is more polar than N—O (3.5 − 3.0 = 0.5).

10.75 A dipole arrow points from the atom with the lower electronegativity value (more positive) to the atom in the bond that has the higher electronegativity value (more negative).

a. 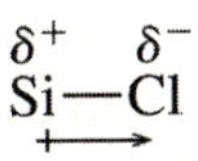

b. 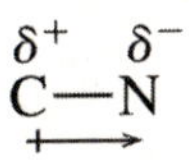

c. 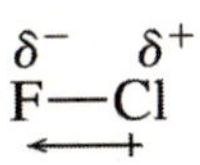

d. 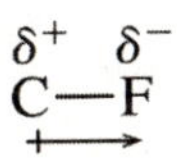

e. 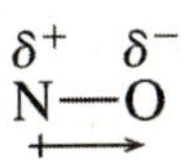

10.77 a. Si—Cl electronegativity difference: 3.0 − 1.8 = 1.2, polar covalent
b. C—C electronegativity difference: 2.5 − 2.5 = 0.0, nonpolar covalent
c. Na—Cl electronegativity difference: 3.0 − 0.9 = 2.1, ionic
d. C—H electronegativity difference: 2.5 − 2.1 = 0.4, nonpolar covalent
e. F—F electronegativity difference: 4.0 − 4.0 = 0.0, nonpolar covalent

10.79 a. 1 N($5\,e^-$) + 3 F($7\,e^-$) = 5 + 21 = 26 valence electrons

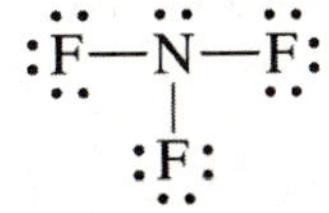

The central atom N has four electron groups with three bonded atoms and one lone pair of electrons, which gives NF_3 a trigonal pyramidal shape.

b. 1 Si($4\,e^-$) + 4 Br($7\,e^-$) = 4 + 28 = 32 valence electrons

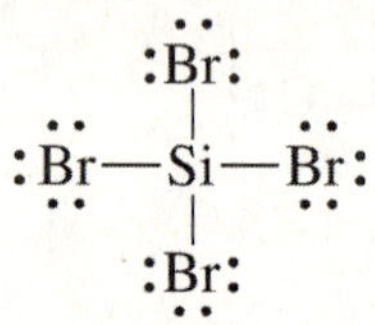

The central atom Si has four electron groups bonded to four bromine atoms; $SiBr_4$ has a tetrahedral shape.

c. 1 C($4\,e^-$) + 2 Se($6\,e^-$) = 4 + 12 = 16 valence electrons

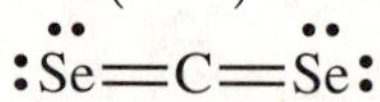

The central atom C has two electron groups bonded to two selenium atoms; CSe_2 has a linear shape.

d. 1 S($6\,e^-$) + 2 O($6\,e^-$) = 6 + 12 = 18 valence electrons

The central atom S has three electron groups with two bonded atoms and one lone pair of electrons, which gives SO_2 a bent shape (120°).

10.81 a. 1 Br($7\,e^-$) + 2 O($6\,e^-$) + 1 e^-(negative charge) = 7 + 12 + 1 = 20 valence electrons

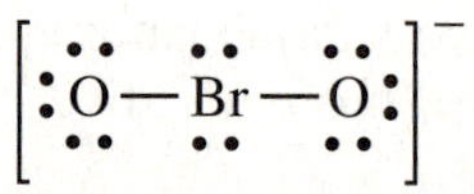

The central atom Br has four electron groups with two bonded atoms and two lone pairs of electrons, which gives BrO_2^- a bent shape (109°).

b. 1 O($6\,e^-$) + 2 H($1\,e^-$) = 6 + 2 = 8 valence electrons

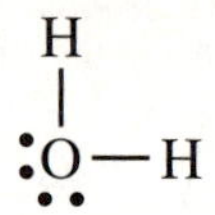

The central atom O has four electron groups with two bonded atoms and two lone pairs of electrons, which gives H_2O a bent shape (109°).

c. 1 C($4\,e^-$) + 4 Br($7\,e^-$) = 4 + 28 = 32 valence electrons

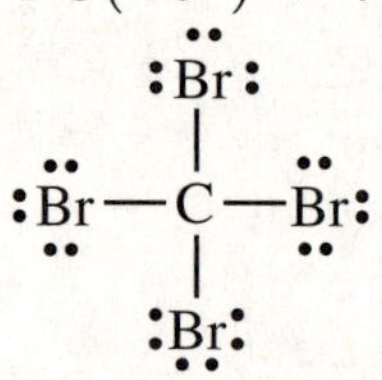

The central atom C has four electron groups bonded to four bromine atoms, which gives CBr_4 a tetrahedral shape.

d. 1 P($5\,e^-$) + 3 O($6\,e^-$) + 3 e^-(negative charge) = 5 + 18 + 3 = 26 valence electrons

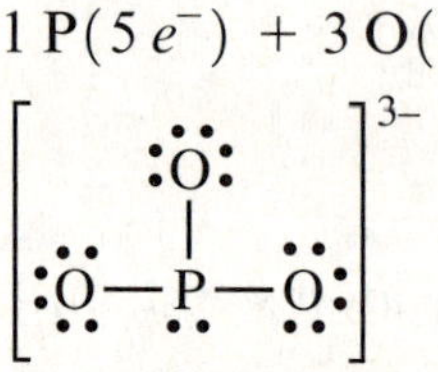

The central atom P has four electron groups with three bonded atoms and one lone pair of electrons, which gives PO_3^{3-} a trigonal pyramidal shape.

10.83 a. The molecule HBr contains only a single polar covalent H—Br bond (EN 2.8 − 2.1 = 0.7); it is a polar molecule.

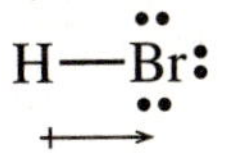

b. The molecule SiO_2 contains two polar covalent Si—O bonds (EN 3.5 − 1.8 = 1.7) and has a linear shape. The two equal dipoles directed away from each other at 180° will cancel, resulting in a nonpolar molecule.

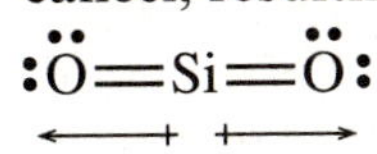

c. The molecule NCl_3 contains three nonpolar covalent N—Cl bonds (EN 3.0 − 3.0 = 0.0) and a lone pair of electrons on the central N atom, which gives the molecule a trigonal pyramidal shape. Since the molecule contains only nonpolar bonds, NCl_3 is a nonpolar molecule.

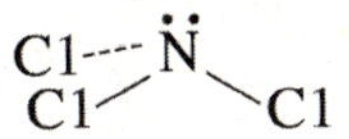

d. The molecule CH_3Cl consists of a central atom, C, with three nonpolar covalent C—H bonds (EN 2.5 − 2.1 = 0.4) and one polar covalent C—Cl bond (EN 3.0 − 2.5 = 0.5). The molecule has a tetrahedral shape, but the single dipole makes CH_3Cl a polar molecule.

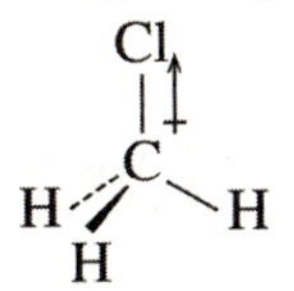

e. The molecule NI_3 contains three polar covalent N—I bonds (EN 3.0 − 2.5 = 0.5) and a lone pair of electrons on the central N atom, which gives the molecule a trigonal pyramidal shape. This asymmetric shape does not allow the dipoles to cancel, which makes NI_3 a polar molecule.

f. The molecule H_2O contains two polar covalent O—H bonds (EN 3.5 − 2.1 = 1.4) and two lone pairs of electrons on the central O atom. The molecule has a bent shape (109°) and the asymmetric arrangement of the O—H dipoles makes H_2O a polar molecule.

10.85 a. NF_3 is a polar molecule. Dipole–dipole attractions (2) occur between dipoles in polar molecules.

b. ClF is a polar molecule (EN 4.0 − 3.0 = 1.0). Dipole–dipole attractions (2) occur between dipoles in polar molecules.

c. Dispersion forces (4) occur between temporary dipoles in nonpolar Br_2 molecules.

d. Ionic bonds (1) are strong attractions between positive and negative ions, as in Cs_2O.

e. Dispersion forces (4) occur between temporary dipoles in nonpolar C_4H_{10} molecules.

f. CH_3OH is a polar molecule and contains the polar O—H bond. Hydrogen bonding (3) involves strong dipole–dipole attractions that occur between a partially positive hydrogen atom of one polar molecule and one of the strongly electronegative atoms F, O, or N in another, as is seen with CH_3OH molecules.

10.87 When water vapor condenses or liquid water freezes, heat is released, which warms the air.

10.89 Given 1540 J absorbed; heat of fusion for water = 334 J/g H_2O

Need grams of ice that will melt at 0 °C

Plan J $\rightarrow$ g of H_2O $\quad \frac{1 \text{ g } H_2O}{334 \text{ J}}$

Set Up $1540 \cancel{\text{J}} \times \frac{1 \text{ g } H_2O}{334 \cancel{\text{J}}} = 4.61 \text{ g of } H_2O$ (3 SFs)

10.91 Given 5.25 kJ removed; heat of fusion for acetic acid = 192 J/g acetic acid

Need grams of acetic acid that will freeze at 17 °C

Plan kJ $\rightarrow$ J $\rightarrow$ g of acetic acid $\quad \frac{1000 \text{ J}}{1 \text{ kJ}} \quad \frac{1 \text{ g acetic acid}}{192 \text{ J}}$

Set Up $5.25 \cancel{\text{kJ}} \times \frac{1000 \cancel{\text{J}}}{1 \cancel{\text{kJ}}} \times \frac{1 \text{ g acetic acid}}{192 \cancel{\text{J}}} = 27.3 \text{ g of acetic acid}$ (3 SFs)

10.93 a. $3\,H(1\,e^-) + 1\,N(5\,e^-) + 1\,C(4\,e^-) + 1\,O(6\,e^-) = 3 + 5 + 4 + 6 = 18$ valence electrons

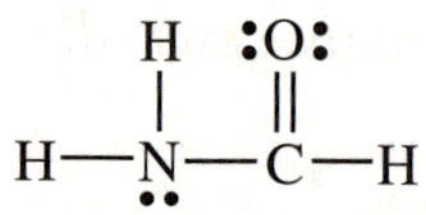

b. $1\,Cl(7\,e^-) + 2\,H(1\,e^-) + 2\,C(4\,e^-) + 1\,N(5\,e^-) = 7 + 2 + 8 + 5 = 22$ valence electrons

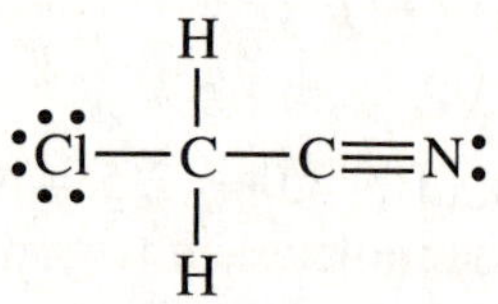

c. $2\,H(1\,e^-) + 2\,N(5\,e^-) = 2 + 10 = 12$ valence electrons

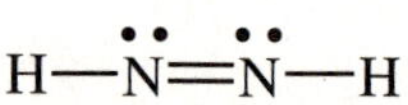

d. $1\,Cl(7\,e^-) + 2\,C(4\,e^-) + 2\,O(6\,e^-) + 3\,H(1\,e^-) = 7 + 8 + 12 + 3 = 30$ valence electrons

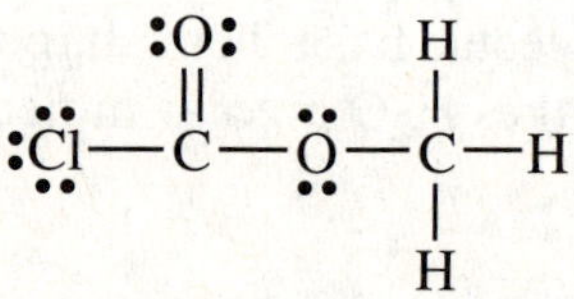

10.95 a. The central atom N has four electron groups with three bonded atoms and one lone pair of electrons, which gives NH_2Cl a trigonal pyramidal shape.

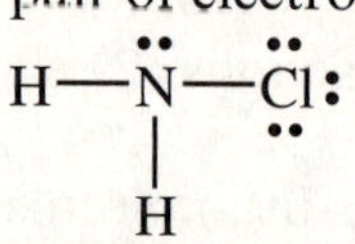

b. The central atom P has four electron pairs bonded to four H atoms; PH_4^+ has a tetrahedral shape.

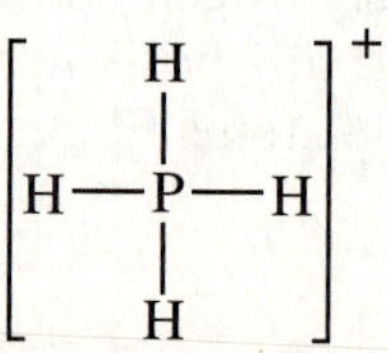

c. The central atom C has two electron groups bonded to two atoms and no lone pairs, which gives SCN^- a linear shape.

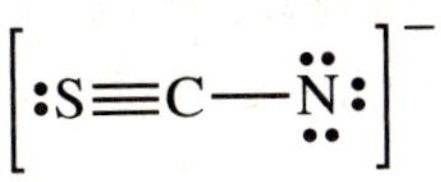

10.97

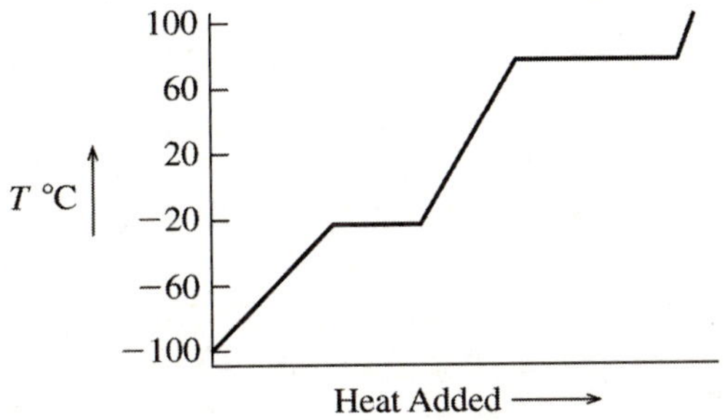

a. Carbon tetrachloride is a solid at −50 °C.
b. At −23 °C, solid carbon tetrachloride melts (solid → liquid).
c. Carbon tetrachloride is a liquid at 20 °C.
d. At 90 °C, carbon tetrachloride is a gas.
e. Both solid and liquid carbon tetrachloride will be present at the melting temperature of −23 °C.

10.99 Two calculations are required:

(1) ice 0 °C → water 0 °C:

$$\text{Heat} = \text{mass} \times \text{heat of fusion} = 45.0\ \cancel{g} \times \frac{334\ \text{J}}{\cancel{g}} = 1.50 \times 10^4\ \text{J}\ (3\ \text{SFs})$$

(2) water 8.0 °C → 0.0 °C:

$$\text{Heat} = \text{mass} \times \Delta T \times SH$$

$$\therefore \text{Mass}_{\text{water}} = \frac{\text{heat}}{\Delta T \times SH} = \frac{1.50 \times 10^4\ \cancel{J}}{(8.0\ \cancel{°C})\left(\dfrac{4.184\ \cancel{J}}{1\ \text{g}\ \cancel{°C}}\right)} = 450\ \text{g of water}\ (2\ \text{SFs})$$

10.101 Two calculations are required:

$$(1)\ \text{water } 25\ °\text{C} \rightarrow 0\ °\text{C}:\quad 325\ \cancel{g} \times 25\ \cancel{°C} \times \frac{4.184\ \cancel{J}}{\cancel{g}\ \cancel{°C}} \times \frac{1\ \text{kJ}}{1000\ \cancel{J}} = 34\ \text{kJ}$$

$$(2)\ \text{water } 0\ °\text{C} \rightarrow \text{ice } 0\ °\text{C}:\quad 325\ \cancel{g} \times \frac{334\ \cancel{J}}{1\ \cancel{g}} \times \frac{1\ \text{kJ}}{1000\ \cancel{J}} = 109\ \text{kJ}$$

$$\therefore \text{Total heat removed} = 34\ \text{kJ} + 109\ \text{kJ} = 143\ \text{kJ removed}\ (3\ \text{SFs})$$

Answers to Combining Ideas from Chapters 8 to 10

CI.13 a. $8.56\ \cancel{\text{g Cu}} \times \dfrac{1\ \cancel{\text{mol Cu}}}{63.55\ \cancel{\text{g Cu}}} \times \dfrac{6.022 \times 10^{23}\ \text{Cu atoms}}{1\ \cancel{\text{mol Cu}}} = 8.11 \times 10^{22}$ atoms of Cu (3 SFs)

b. $2\text{Cu}(s) + \text{O}_2(g) \longrightarrow 2\text{CuO}(s)$

c. The reaction of the elements copper and oxygen to form copper(II) oxide is a combination reaction.

d. $8.56\ \cancel{\text{g Cu}} \times \dfrac{1\ \cancel{\text{mol Cu}}}{63.55\ \cancel{\text{g Cu}}} \times \dfrac{1\ \cancel{\text{mol O}_2}}{2\ \cancel{\text{mol Cu}}} \times \dfrac{32.00\ \text{g O}_2}{1\ \cancel{\text{mol O}_2}} = 2.16$ g of O_2 (3 SFs)

e. From part **d**, we know that 8.56 g of Cu reacts with 2.16 g of O_2
$\therefore$ 3.72 g of O_2 is excess O_2; Cu is the limiting reactant.
Molar mass of CuO = 1 mol of Cu + 1 mol of O = 63.55 g + 16.00 g = 79.55 g

$8.56\ \cancel{\text{g Cu}} \times \dfrac{1\ \cancel{\text{mol Cu}}}{63.55\ \cancel{\text{g Cu}}} \times \dfrac{2\ \cancel{\text{mol CuO}}}{2\ \cancel{\text{mol Cu}}} \times \dfrac{79.55\ \text{g CuO}}{1\ \cancel{\text{mol CuO}}} = 10.7$ g of CuO (3 SFs)

f. Theoretical yield of CuO (from part **e**) = 10.7 g of CuO
Use the percent yield to convert theoretical to actual:

$10.7\ \cancel{\text{g CuO}} \times \dfrac{85.0\ \text{g CuO}}{100\ \cancel{\text{g CuO}}} = 9.10$ g of CuO (actual) (3 SFs)

CI.15 a. The formula of sodium hypochlorite is NaClO.
Molar mass of NaClO
= 1 mol of Na + 1 mol of Cl + 1 mol of O = 22.99 g + 35.45 g + 16.00 g = 74.44 g
1 Cl($7\,e^-$) + 1 O($6\,e^-$) + 1 e^-(negative charge) = 7 + 6 + 1 = 14 valence electrons

b. $\left[:\underset{\cdot\cdot}{\overset{\cdot\cdot}{\text{Cl}}}-\underset{\cdot\cdot}{\overset{\cdot\cdot}{\text{O}}}:\right]^-$

c. Given gallons of bleach **Need** number of ClO^- ions
Plan gal → qt → mL → g of solution → g of NaClO → mol of NaClO → mol of ClO^- → number of ClO^- ions

Set Up $1.00\ \cancel{\text{gal}} \times \dfrac{4\ \cancel{\text{qt}}}{1\ \cancel{\text{gal}}} \times \dfrac{946.3\ \cancel{\text{mL}}}{1\ \cancel{\text{qt}}} \times \dfrac{1.08\ \cancel{\text{g solution}}}{1\ \cancel{\text{mL solution}}} \times \dfrac{5.25\ \cancel{\text{g NaClO}}}{100\ \cancel{\text{g solution}}}$

$\times \dfrac{1\ \cancel{\text{mol NaClO}}}{74.44\ \cancel{\text{g NaClO}}} \times \dfrac{1\ \cancel{\text{mol ClO}^-}}{1\ \cancel{\text{mol NaClO}}} \times \dfrac{6.022 \times 10^{23}\ \text{ClO}^-\ \text{ions}}{1\ \cancel{\text{mol ClO}^-}}$

$= 1.74 \times 10^{24}\ \text{ClO}^-$ ions (3 SFs)

d. $2\text{NaOH}(aq) + \text{Cl}_2(g) \longrightarrow \text{NaClO}(aq) + \text{NaCl}(aq) + \text{H}_2\text{O}(l)$

e. Given gallons of bleach **Need** grams of NaOH
Plan gal → qt → mL → g of solution → g of NaClO → mol of NaClO → mol of NaOH → g of NaOH

Set Up $1.00\ \cancel{\text{gal}} \times \dfrac{4\ \cancel{\text{qt}}}{1\ \cancel{\text{gal}}} \times \dfrac{946.3\ \cancel{\text{mL}}}{1\ \cancel{\text{qt}}} \times \dfrac{1.08\ \cancel{\text{g solution}}}{1\ \cancel{\text{mL solution}}} \times \dfrac{5.25\ \cancel{\text{g NaClO}}}{100\ \cancel{\text{g solution}}}$

$\times \dfrac{1\ \cancel{\text{mol NaClO}}}{74.44\ \cancel{\text{g NaClO}}} \times \dfrac{2\ \cancel{\text{mol NaOH}}}{1\ \cancel{\text{mol NaClO}}} \times \dfrac{40.00\ \text{g NaOH}}{1\ \cancel{\text{mol NaOH}}}$

= 231 g of NaOH (3 SFs)

f. $165\ \text{g Cl}_2 \times \frac{1\ \text{mol Cl}_2}{70.90\ \text{g Cl}_2} \times \frac{1\ \text{mol NaClO}}{1\ \text{mol Cl}_2} \times \frac{74.44\ \text{g NaClO}}{1\ \text{mol NaClO}} = 173\ \text{g of NaClO}$ (smaller amount of product)

$275\ \text{g NaOH} \times \frac{1\ \text{mol NaOH}}{40.00\ \text{g NaOH}} \times \frac{1\ \text{mol NaClO}}{2\ \text{mol NaOH}} \times \frac{74.44\ \text{g NaClO}}{1\ \text{mol NaClO}} = 256\ \text{g of NaClO}$

∴ Cl_2 is the limiting reactant and 173 g of NaClO can be produced (theoretical).

∴ Percent yield: $\frac{162\ \text{g NaClO (actual)}}{173\ \text{g NaClO (theoretical)}} \times 100\% = 93.6\%$ (3 SFs)

CI.17 a. $3\,\text{Cl}(7\,e^-) + 2\,\text{C}(4\,e^-) + 2\,\text{O}(6\,e^-) + 3\,\text{H}(1\,e^-) = 21 + 8 + 12 + 3$
$= 44$ valence electrons

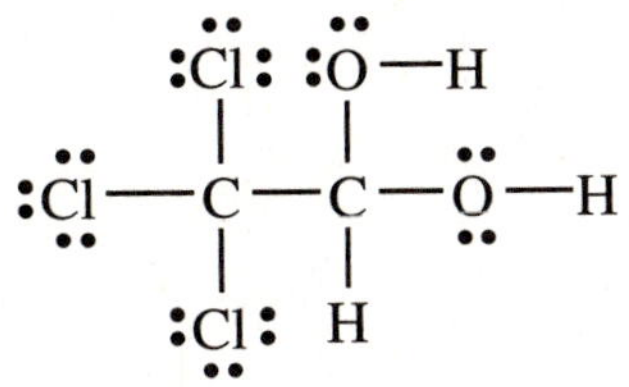

Chloral hydrate

$3\,\text{Cl}(7\,e^-) + 2\,\text{C}(4\,e^-) + 1\,\text{O}(6\,e^-) + 1\,\text{H}(1\,e^-) = 21 + 8 + 6 + 1$
$= 36$ valence electrons

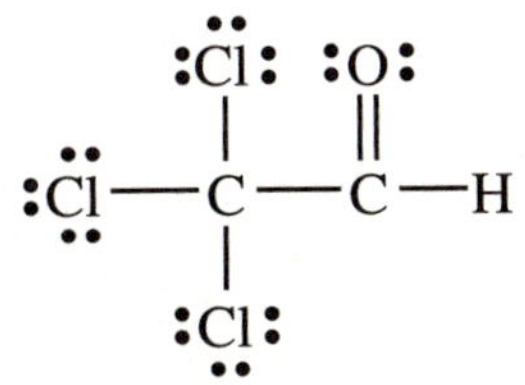

Chloral

b. Chloral hydrate has two hydroxyl groups, and chloral has an aldehyde (carbonyl) group.

c. Chloral hydrate, $C_2H_3O_2Cl_3 \rightarrow C_{(2\div1)}H_{(3\div1)}O_{(2\div1)}Cl_{(3\div1)} = C_2H_3O_2Cl_3$ (empirical formula) Chloral, $C_2HOCl_3 \rightarrow C_{(2\div1)}H_{(1\div1)}O_{(1\div1)}Cl_{(3\div1)} = C_2HOCl_3$ (empirical formula)

d. $2\ \text{mol C} \times \frac{12.01\ \text{g C}}{1\ \text{mol C}} = 24.02\ \text{g of C}$

$3\ \text{mol H} \times \frac{1.008\ \text{g H}}{1\ \text{mol H}} = 3.024\ \text{g of H}$

$2\ \text{mol O} \times \frac{16.00\ \text{g O}}{1\ \text{mol O}} = 32.00\ \text{g of O}$

$3\ \text{mol Cl} \times \frac{35.45\ \text{g Cl}}{1\ \text{mol Cl}} = 106.4\ \text{g of Cl}$

∴ Molar mass of $C_2H_3O_2Cl_3 = 165.4\ \text{g}$

$\text{Mass \% Cl} = \frac{106.4\ \text{g Cl}}{165.4\ \text{g } C_2H_3O_2Cl_3} \times 100\% = 64.33\%\ \text{Cl}$ (4 SFs)

CI.19 a.

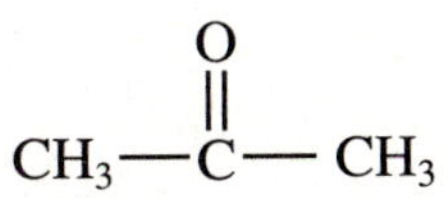

b. The molecular formula of acetone is C_3H_6O.
Molar mass of C_3H_6O
$= 3\ \text{mol of C} + 6\ \text{mol of H} + 1\ \text{mol of O} = 3(12.01\ \text{g}) + 6(1.008\ \text{g}) + 16.00\ \text{g} = 58.08\ \text{g}$

c. $C_3H_6O(l) + 4O_2(g) \xrightarrow{\Delta} 3CO_2(g) + 3H_2O(g) + 1790\ \text{kJ}$

d. Combustion is an exothermic reaction.

e. $2.58\ \text{g}\ \cancel{C_3H_6O} \times \dfrac{1\ \cancel{\text{mol}\ C_3H_6O}}{58.08\ \cancel{\text{g}\ C_3H_6O}} \times \dfrac{1790\ \text{kJ}}{1\ \cancel{\text{mol}\ C_3H_6O}} = 79.5\ \text{kJ}\ (3\ \text{SFs})$

f. $15.0\ \cancel{\text{mL}\ C_3H_6O} \times \dfrac{0.786\ \cancel{\text{g}\ C_3H_6O}}{1\ \cancel{\text{mL}\ C_3H_6O}} \times \dfrac{1\ \cancel{\text{mol}\ C_3H_6O}}{58.08\ \cancel{\text{g}\ C_3H_6O}} \times \dfrac{4\ \cancel{\text{mol}\ O_2}}{1\ \cancel{\text{mol}\ C_3H_6O}} \times \dfrac{32.00\ \text{g}\ O_2}{1\ \cancel{\text{mol}\ O_2}}$
$= 26.0\ \text{g of}\ O_2\ (3\ \text{SFs})$

11

Gases

Learning Goals

- Describe the kinetic molecular theory of gases and the properties of gases.
- Describe the units of measurement used for pressure, and change from one unit to another.
- Use the pressure–volume relationship (Boyle's law) to determine the final pressure or volume when the temperature and amount of gas are constant.
- Use the temperature–volume relationship (Charles's law) to determine the final temperature or volume when the pressure and amount of gas are constant.
- Use the temperature–pressure relationship (Gay-Lussac's law) to determine the final temperature or pressure when the volume and amount of gas are constant.
- Use the combined gas law to find the final pressure, volume, or temperature of a gas when changes in two of these properties are given and the amount of gas is constant.
- Use Avogadro's law to determine the amount or volume of a gas when the pressure and temperature are constant.
- Use the ideal gas law equation to solve for P, V, T, or n of a gas when given three of the four values in the ideal gas law equation.
- Calculate density, molar mass, or volume of a gas in a chemical reaction.
- Determine the mass or volume of a gas that reacts or forms in a chemical reaction.
- Use Dalton's law of partial pressures to calculate the total pressure of a mixture of gases.

Chapter Outline

Core Chemistry Skills

- Using the Gas Laws (11.3)
- Using the Ideal Gas Law (11.8)
- Calculating Mass or Volume of a Gas in a Chemical Reaction (11.9)
- Calculating Partial Pressure (11.10)

Answers and Solutions to Text Problems

11.1 **a.** At a higher temperature, gas particles have greater kinetic energy, which makes them move faster.

b. Because there are great distances between the particles of a gas, they can be pushed closer together and still remain a gas.

c. Gas particles are very far apart, meaning that the mass of a gas in a certain volume is very small, resulting in a low density.

11.3 **a.** The temperature of a gas can be expressed in kelvins.

b. The volume of a gas can be expressed in milliliters.

c. The amount of a gas can be expressed in grams.

d. Pressure can be expressed in millimeters of mercury (mmHg).

11.5 Statements **a**, **d**, and **e** describe the pressure of a gas.

11.7 **a.** $2.00\ \cancel{\text{atm}} \times \dfrac{760\ \text{torr}}{1\ \cancel{\text{atm}}} = 1520\ \text{torr}\ (3\ \text{SFs})$

b. $2.00\ \cancel{\text{atm}} \times \dfrac{14.7\ \text{lb/in.}^2}{1\ \cancel{\text{atm}}} = 29.4\ \text{lb/in.}^2\ (3\ \text{SFs})$

c. $2.00\ \cancel{\text{atm}} \times \dfrac{760\ \text{mmHg}}{1\ \cancel{\text{atm}}} = 1520\ \text{mmHg}\ (3\ \text{SFs})$

d. $2.00\ \cancel{\text{atm}} \times \dfrac{101.325\ \text{kPa}}{1\ \cancel{\text{atm}}} = 203\ \text{kPa}\ (3\ \text{SFs})$

11.9 As the scuba diver ascends to the surface, external pressure decreases. If the air in the lungs, which is at a higher pressure, were not exhaled, its volume would expand and severely damage the lungs. The pressure of the gas in the lungs must adjust to changes in the external pressure.

11.11 **a.** The pressure is greater in cylinder A. According to Boyle's law, a decrease in volume pushes the gas particles closer together, which will cause an increase in the pressure.

b.

Property	Conditions 1	Conditions 2	Know	Predict
Pressure (P)	$P_1 = 650$ mmHg	$P_2 = 1.2$ atm	P increases	
Volume (V)	$V_1 = 220$ mL	$V_2 = ?$ mL		V decreases

According to Boyle's law, $P_1V_1 = P_2V_2$, then

$$V_2 = V_1 \times \frac{P_1}{P_2} = 220\ \text{mL} \times \frac{650\ \cancel{\text{mmHg}}}{1.2\ \cancel{\text{atm}}} \times \frac{1\ \cancel{\text{atm}}}{760\ \cancel{\text{mmHg}}} = 160\ \text{mL}\ (2\ \text{SFs})$$

11.13 **a.** The pressure of the gas *increases* to two times the original pressure when the volume is halved.

b. The pressure *decreases* to one-third the original pressure when the volume expands to three times its initial value.

c. The pressure *increases* to ten times the original pressure when the volume decreases to one-tenth of the initial volume.

11.15 From Boyle's law, we know that pressure is inversely related to volume (for example, pressure increases when volume decreases).

a. Volume increases; pressure must decrease.

$$P_2 = P_1 \times \frac{V_1}{V_2} = 655\ \text{mmHg} \times \frac{10.0\ \cancel{\text{L}}}{20.0\ \cancel{\text{L}}} = 328\ \text{mmHg}\ (3\ \text{SFs})$$

b. Volume decreases; pressure must increase.

$$P_2 = P_1 \times \frac{V_1}{V_2} = 655\ \text{mmHg} \times \frac{10.0\ \cancel{\text{L}}}{2.50\ \cancel{\text{L}}} = 2620\ \text{mmHg}\ (3\ \text{SFs})$$

c. The mL units must be converted to L for unit cancellation in the calculation, and because the volume increases, pressure must decrease.

$$P_2 = P_1 \times \frac{V_1}{V_2} = 655\ \text{mmHg} \times \frac{10.0\ \cancel{\text{L}}}{13\ 800\ \cancel{\text{mL}}} \times \frac{1000\ \cancel{\text{mL}}}{1\ \cancel{\text{L}}} = 475\ \text{mmHg}\ (3\ \text{SFs})$$

d. The mL units must be converted to L for unit cancellation in the calculation, and because the volume decreases, pressure must increase.

$$P_2 = P_1 \times \frac{V_1}{V_2} = 655\ \text{mmHg} \times \frac{10.0\ \cancel{\text{L}}}{1250\ \cancel{\text{mL}}} \times \frac{1000\ \cancel{\text{mL}}}{1\ \cancel{\text{L}}} = 5240\ \text{mmHg}\ (3\ \text{SFs})$$

11.17 From Boyle's law, we know that pressure is inversely related to volume.

a. Pressure decreases; volume must increase.

$$V_2 = V_1 \times \frac{P_1}{P_2} = 50.0\ \text{L} \times \frac{760.\ \cancel{\text{mmHg}}}{725\ \cancel{\text{mmHg}}} = 52.4\ \text{L}\ (3\ \text{SFs})$$

b. The mmHg units must be converted to atm for unit cancellation in the calculation, and because the pressure increases, volume must decrease.

$$P_1 = 760.\ \cancel{\text{mmHg}} \times \frac{1\ \text{atm}}{760\ \cancel{\text{mmHg}}} = 1.00\ \text{atm}$$

$$V_2 = V_1 \times \frac{P_1}{P_2} = 50.0\ \text{L} \times \frac{1.00\ \cancel{\text{atm}}}{2.0\ \cancel{\text{atm}}} = 25\ \text{L}\ (2\ \text{SFs})$$

c. The mmHg units must be converted to atm for unit cancellation in the calculation, and because the pressure decreases, volume must increase.

$$P_1 = 760.\ \cancel{\text{mmHg}} \times \frac{1\ \text{atm}}{760\ \cancel{\text{mmHg}}} = 1.00\ \text{atm}$$

$$V_2 = V_1 \times \frac{P_1}{P_2} = 50.0\ \text{L} \times \frac{1.00\ \cancel{\text{atm}}}{0.500\ \cancel{\text{atm}}} = 100.\ \text{L}\ (3\ \text{SFs})$$

d. The mmHg units must be converted to torr for unit cancellation in the calculation, and because the pressure increases, volume must decrease.

$$P_1 = 760.\ \cancel{\text{mmHg}} \times \frac{760\ \text{torr}}{760\ \cancel{\text{mmHg}}} = 760.\ \text{torr}$$

$$V_2 = V_1 \times \frac{P_1}{P_2} = 50.0\ \text{L} \times \frac{760.\ \cancel{\text{torr}}}{850\ \cancel{\text{torr}}} = 45\ \text{L}\ (2\ \text{SFs})$$

11.19 Pressure decreases; volume must increase.

$$V_2 = V_1 \times \frac{P_1}{P_2} = 5.0\ \text{L} \times \frac{5.0\ \cancel{\text{atm}}}{1.0\ \cancel{\text{atm}}} = 25\ \text{L}\ (2\ \text{SFs})$$

11.21 a. *Inspiration* begins when the diaphragm flattens, causing the lungs to expand. The increased volume of the thoracic cavity reduces the pressure in the lungs such that air flows into the lungs.

b. *Expiration* occurs as the diaphragm relaxes, causing a decrease in the volume of the lungs. The pressure of the air in the lungs increases, and air flows out of the lungs.

c. *Inspiration* occurs when the pressure within the lungs is less than that of the atmosphere.

11.23 According to Charles's law, there is a direct relationship between Kelvin temperature and volume (for example, volume increases when temperature increases, if the pressure and amount of gas remain constant).

a. Diagram C shows an increased volume corresponding to an increase in temperature.

b. Diagram A shows a decreased volume corresponding to a decrease in temperature.

c. Diagram B shows no change in volume, which corresponds to no net change in temperature.

11.25 According to Charles's law, the volume of a gas is directly related to the Kelvin temperature. In all gas law computations, temperatures must be in kelvins. (Temperatures in °C are converted to K by the addition of 273.) The initial temperature for all cases here is $T_1 = 15\,°\text{C} + 273 = 288\text{ K}$.

a. Volume increases; temperature must have increased.

$$T_2 = T_1 \times \frac{V_2}{V_1} = 288\text{ K} \times \frac{5.00\cancel{\text{L}}}{2.50\cancel{\text{L}}} = 576\text{ K} \quad 576\text{ K} - 273 = 303\,°\text{C (3 SFs)}$$

b. Volume decreases; temperature must have decreased.

$$T_2 = T_1 \times \frac{V_2}{V_1} = 288\text{ K} \times \frac{1250\cancel{\text{mL}}}{2.50\cancel{\text{L}}} \times \frac{1\cancel{\text{L}}}{1000\cancel{\text{mL}}} = 144\text{ K} \quad 144\text{ K} - 273 = -129\,°\text{C (3 SFs)}$$

c. Volume increases; temperature must have increased.

$$T_2 = T_1 \times \frac{V_2}{V_1} = 288\text{ K} \times \frac{7.50\cancel{\text{L}}}{2.50\cancel{\text{L}}} = 864\text{ K} \quad 864\text{ K} - 273 = 591\,°\text{C (3 SFs)}$$

d. Volume increases; temperature must have increased.

$$T_2 = T_1 \times \frac{V_2}{V_1} = 288\text{ K} \times \frac{3550\cancel{\text{mL}}}{2.50\cancel{\text{L}}} \times \frac{1\cancel{\text{L}}}{1000\cancel{\text{mL}}} = 409\text{ K} \quad 409\text{ K} - 273 = 136\,°\text{C (3 SFs)}$$

11.27 According to Charles's law, the volume of a gas is directly related to the Kelvin temperature. In all gas law computations, temperatures must be in kelvins. (Temperatures in °C are converted to K by the addition of 273.) The initial temperature for all cases here is $T_1 = 75\,°\text{C} + 273 = 348\text{ K}$.

a. When temperature decreases, volume must also decrease.

$T_2 = 55\,°\text{C} + 273 = 328\text{ K}$

$$V_2 = V_1 \times \frac{T_2}{T_1} = 2500\text{ mL} \times \frac{328\cancel{\text{K}}}{348\cancel{\text{K}}} = 2400\text{ mL (2 SFs)}$$

b. When temperature increases, volume must also increase.

$$V_2 = V_1 \times \frac{T_2}{T_1} = 2500\text{ mL} \times \frac{680.\cancel{\text{K}}}{348\cancel{\text{K}}} = 4900\text{ mL (2 SFs)}$$

c. When temperature decreases, volume must also decrease.

$T_2 = -25\,°\text{C} + 273 = 248\text{ K}$

$$V_2 = V_1 \times \frac{T_2}{T_1} = 2500\text{ mL} \times \frac{248\cancel{\text{K}}}{348\cancel{\text{K}}} = 1800\text{ mL (2 SFs)}$$

d. When temperature decreases, volume must also decrease.

$$V_2 = V_1 \times \frac{T_2}{T_1} = 2500\text{ mL} \times \frac{240.\ \cancel{\text{K}}}{348\ \cancel{\text{K}}} = 1700\text{ mL (2 SFs)}$$

11.29 According to Gay-Lussac's law, temperature is directly related to pressure. In all gas law computations, temperatures must be in kelvins. (Temperatures in °C are converted to K by the addition of 273.)

a. When temperature decreases, pressure must also decrease.

$T_1 = 155\ °\text{C} + 273 = 428\text{ K} \qquad T_2 = 0\ °\text{C} + 273 = 273\text{ K}$

$$P_2 = P_1 \times \frac{T_2}{T_1} = 1200\ \cancel{\text{torr}} \times \frac{1\text{ mmHg}}{1\ \cancel{\text{torr}}} \times \frac{273\ \cancel{\text{K}}}{428\ \cancel{\text{K}}} = 770\text{ mmHg (2 SFs)}$$

b. When temperature increases, pressure must also increase.

$T_1 = 12\ °\text{C} + 273 = 285\text{ K} \qquad T_2 = 35\ °\text{C} + 273 = 308\text{ K}$

$$P_2 = P_1 \times \frac{T_2}{T_1} = 1.40\ \cancel{\text{atm}} \times \frac{760\text{ mmHg}}{1\ \cancel{\text{atm}}} \times \frac{308\ \cancel{\text{K}}}{285\ \cancel{\text{K}}} = 1150\text{ mmHg (3 SFs)}$$

11.31 According to Gay-Lussac's law, temperature is directly related to pressure. For example, temperature increases when the pressure increases. In all gas law computations, temperatures must be in kelvins. (Temperatures in °C are converted to K by the addition of 273.)

a. Pressure decreases; temperature must have decreased.

$T_1 = 25\ °\text{C} + 273 = 298\text{ K}$

$$T_2 = T_1 \times \frac{P_2}{P_1} = 298\text{ K} \times \frac{620.\ \cancel{\text{mmHg}}}{740.\ \cancel{\text{mmHg}}} = 250.\text{ K} \quad 250.\text{ K} - 273 = -23\ °\text{C (2 SFs)}$$

b. Pressure increases; temperature must have increased.

$T_1 = -18\ °\text{C} + 273 = 255\text{ K}$

$$T_2 = T_1 \times \frac{P_2}{P_1} = 255\text{ K} \times \frac{1250\ \cancel{\text{torr}}}{0.950\ \cancel{\text{atm}}} \times \frac{1\ \cancel{\text{atm}}}{760\ \cancel{\text{torr}}} = 441\text{ K} \quad 441\text{ K} - 273 = 168\ °\text{C (3 SFs)}$$

11.33 a. The *boiling point* is the temperature at which bubbles of vapor appear within the liquid.

b. *Vapor pressure* is the pressure exerted by a gas above the surface of its liquid.

11.35 a. On the top of a mountain, water boils below 100 °C because the atmospheric (external) pressure is less than 1 atm. The boiling point is the temperature at which the vapor pressure of a liquid becomes equal to the external (in this case, atmospheric) pressure.

b. Because the pressure inside a pressure cooker is greater than 1 atm, water boils above 100 °C. The higher temperature of the boiling water allows food to cook more quickly.

11.37 $\frac{P_1V_1}{T_1} = \frac{P_2V_2}{T_2} \qquad \frac{\cancel{P_1}\cancel{V_1}}{\cancel{T_1}} \times \frac{T_2\cancel{T_1}}{\cancel{P_1}\cancel{V_1}} = \frac{P_2V_2}{\cancel{T_2}} \times \frac{\cancel{T_2}T_1}{P_1V_1} \qquad \therefore T_2 = T_1 \times \frac{P_2}{P_1} \times \frac{V_2}{V_1}$

11.39 $T_1 = 25\ °\text{C} + 273 = 298\text{ K}; \quad V_1 = 6.50\text{ L}; \quad P_1 = 845\text{ mmHg}$

a. $T_2 = 325\text{ K}; \quad V_2 = 1850\text{ mL} = 1.85\text{ L}; \quad P_2 = ?\text{ atm}$

$$P_2 = P_1 \times \frac{V_1}{V_2} \times \frac{T_2}{T_1} = 845\ \cancel{\text{mmHg}} \times \frac{6.50\ \cancel{\text{L}}}{1.85\ \cancel{\text{L}}} \times \frac{325\ \cancel{\text{K}}}{298\ \cancel{\text{K}}} \times \frac{1\text{ atm}}{760\ \cancel{\text{mmHg}}} = 4.26\text{ atm (3 SFs)}$$

b. $T_2 = 12\ °\text{C} + 273 = 285\text{ K}; \quad V_2 = 2.25\text{ L}; \quad P_2 = ?\text{ atm}$

$$P_2 = P_1 \times \frac{V_1}{V_2} \times \frac{T_2}{T_1} = 845\ \cancel{\text{mmHg}} \times \frac{6.50\ \cancel{\text{L}}}{2.25\ \cancel{\text{L}}} \times \frac{285\ \cancel{\text{K}}}{298\ \cancel{\text{K}}} \times \frac{1\text{ atm}}{760\ \cancel{\text{mmHg}}} = 3.07\text{ atm (3 SFs)}$$

c. $T_2 = 47\,°C + 273 = 320.\ K$; $V_2 = 12.8\ L$; $P_2 = ?\ atm$

$$P_2 = P_1 \times \frac{V_1}{V_2} \times \frac{T_2}{T_1} = 845\ \cancel{mmHg} \times \frac{6.50\ \cancel{L}}{12.8\ \cancel{L}} \times \frac{320.\ \cancel{K}}{298\ \cancel{K}} \times \frac{1\ atm}{760\ \cancel{mmHg}} = 0.606\ atm\ (3\ SFs)$$

11.41 $T_1 = 212\,°C + 273 = 485\ K$; $V_1 = 124\ mL$; $P_1 = 1.80\ atm$

$T_2 = ?\,°C$; $V_2 = 138\ mL$; $P_2 = 0.800\ atm$

$$T_2 = T_1 \times \frac{P_2}{P_1} \times \frac{V_2}{V_1} = 485\ K \times \frac{0.800\ \cancel{atm}}{1.80\ \cancel{atm}} \times \frac{138\ \cancel{mL}}{124\ \cancel{mL}} = 240.\ K \quad 240.\ K - 273 = -33\,°C\ (2\ SFs)$$

11.43 The volume increases because the number of gas particles in the tire or basketball is increased.

11.45 According to Avogadro's law, the volume of a gas is directly related to the number of moles of gas.

$n_1 = 1.50$ mol of Ne; $V_1 = 8.00$ L

a. $$V_2 = V_1 \times \frac{n_2}{n_1} = 8.00\ L \times \frac{\frac{1}{2}(1.50)\ \cancel{mol\ Ne}}{1.50\ \cancel{mol\ Ne}} = 4.00\ L\ (3\ SFs)$$

b. $n_2 = 1.50$ mol Ne + 3.50 mol Ne = 5.00 mol of Ne

$$V_2 = V_1 \times \frac{n_2}{n_1} = 8.00\ L \times \frac{5.00\ \cancel{mol\ Ne}}{1.50\ \cancel{mol\ Ne}} = 26.7\ L\ (3\ SFs)$$

c. $$25.0\ \cancel{g\ Ne} \times \frac{1\ mol\ Ne}{20.18\ \cancel{g\ Ne}} = 1.24\ mol\ of\ Ne\ added$$

$n_2 = 1.50$ mol Ne + 1.24 mol Ne = 2.74 mol of Ne

$$V_2 = V_1 \times \frac{n_2}{n_1} = 8.00\ L \times \frac{2.74\ \cancel{mol\ Ne}}{1.50\ \cancel{mol\ Ne}} = 14.6\ L\ (3\ SFs)$$

11.47 At STP, 1 mol of any gas occupies a volume of 22.4 L.

a. $$44.8\ \cancel{L\ O_2\ (STP)} \times \frac{1\ mol\ O_2}{22.4\ \cancel{L\ O_2\ (STP)}} = 2.00\ mol\ of\ O_2\ (3\ SFs)$$

b. $$4.00\ \cancel{L\ CO_2\ (STP)} \times \frac{1\ mol\ CO_2}{22.4\ \cancel{L\ CO_2\ (STP)}} = 0.179\ mol\ of\ CO_2\ (3\ SFs)$$

c. $$6.40\ \cancel{g\ O_2} \times \frac{1\ \cancel{mol\ O_2}}{32.00\ \cancel{g\ O_2}} \times \frac{22.4\ L\ O_2\ (STP)}{1\ \cancel{mol\ O_2}} = 4.48\ L\ at\ STP\ (3\ SFs)$$

d. $$50.0\ \cancel{g\ Ne} \times \frac{1\ \cancel{mol\ Ne}}{20.18\ \cancel{g\ Ne}} \times \frac{22.4\ \cancel{L\ Ne\ (STP)}}{1\ \cancel{mol\ Ne}} \times \frac{1000\ mL}{1\ \cancel{L}} = 55\,500\ mL\ at\ STP\ (3\ SFs)$$

11.49 At STP, 1 mol of any gas occupies a volume of 22.4 L.

a. For F_2, molar mass = 2(19.00 g) = 38.00 g

$$\text{Density of } F_2 = \frac{\text{molar mass}}{\text{molar volume}} = \frac{\dfrac{38.00\ g\ F_2}{1\ \cancel{mol\ F_2}}}{\dfrac{22.4\ L\ F_2}{1\ \cancel{mol\ F_2}}} = 1.70\ g/L\ (3\ SFs)$$

b. For CH_4, molar mass = 12.01 g + 4(1.008 g) = 16.04 g

$$\text{Density of } CH_4 = \frac{\text{molar mass}}{\text{molar volume}} = \frac{\dfrac{16.04 \text{ g } CH_4}{1 \cancel{\text{mol } CH_4}}}{\dfrac{22.4 \text{ L } CH_4}{1 \cancel{\text{mol } CH_4}}} = 0.716 \text{ g/L (3 SFs)}$$

c. For Ne, molar mass = 20.18 g

$$\text{Density of Ne} = \frac{\text{molar mass}}{\text{molar volume}} = \frac{\dfrac{20.18 \text{ g Ne}}{1 \cancel{\text{mol Ne}}}}{\dfrac{22.4 \text{ L Ne}}{1 \cancel{\text{mol Ne}}}} = 0.901 \text{ g/L (3 SFs)}$$

d. For SO_2, molar mass = 32.07 g + 2(16.00 g) = 64.07 g

$$\text{Density of } SO_2 = \frac{\text{molar mass}}{\text{molar volume}} = \frac{\dfrac{64.07 \text{ g } SO_2}{1 \cancel{\text{mol } SO_2}}}{\dfrac{22.4 \text{ L } SO_2}{1 \cancel{\text{mol } SO_2}}} = 2.86 \text{ g/L (3 SFs)}$$

11.51 Analyze the Problem: Using the Ideal Gas Law to Calculate Pressure

Property	*P*	*V*	*n*	*R*	*T*
Given		10.0 L	2.00 mol	$\frac{0.0821 \text{ L} \cdot \text{atm}}{\text{mol} \cdot \text{K}}$	27 °C 27 °C + 273 = 300. K
Need	? atm				

$$P = \frac{nRT}{V} = \frac{(2.00 \cancel{\text{mol}})\left(\dfrac{0.0821 \cancel{\text{L}} \cdot \text{atm}}{\cancel{\text{mol}} \cdot \cancel{\text{K}}}\right)(300. \cancel{\text{K}})}{(10.0 \cancel{\text{L}})} = 4.93 \text{ atm (3 SFs)}$$

11.53 Analyze the Problem: Using the Ideal Gas Law to Calculate Moles

Property	*P*	*V*	*n*	*R*	*T*
Given	845 mmHg	20.0 L		$\frac{62.4 \text{ L} \cdot \text{mmHg}}{\text{mol} \cdot \text{K}}$	22 °C 22 °C + 273 = 295 K
Need			? mol (? g)		

$$n = \frac{PV}{RT} = \frac{(845 \cancel{\text{mmHg}})(20.0 \cancel{\text{L}})}{\left(\dfrac{62.4 \cancel{\text{L}} \cdot \cancel{\text{mmHg}}}{\text{mol} \cdot \cancel{\text{K}}}\right)(295 \cancel{\text{K}})} = 0.918 \text{ mol of } O_2$$

$$0.918 \cancel{\text{mol } O_2} \times \frac{32.00 \text{ g } O_2}{1 \cancel{\text{mol } O_2}} = 29.4 \text{ g of } O_2 \text{ (3 SFs)}$$

11.55 $n = 25.0\ \cancel{g\ N_2} \times \dfrac{1\ \text{mol}\ N_2}{28.02\ \cancel{g\ N_2}} = 0.892\ \text{mol of}\ N_2$

Analyze the Problem: Using the Ideal Gas Law to Calculate Temperature

Property	*P*	*V*	*n*	*R*	*T*
Given	630. mmHg	50.0 L	25.0 g (0.892 mol)	$\dfrac{62.4\ \text{L} \cdot \text{mmHg}}{\text{mol} \cdot \text{K}}$	
Need					? K (? °C)

$$T = \frac{PV}{nR} = \frac{(630.\ \cancel{\text{mmHg}})(50.0\ \cancel{\text{L}})}{(0.892\ \cancel{\text{mol}})\left(\dfrac{62.4\ \cancel{\text{L}} \cdot \cancel{\text{mmHg}}}{\cancel{\text{mol}} \cdot \text{K}}\right)} = 566\ \text{K} - 273 = 293\ °\text{C}\ (3\ \text{SFs})$$

11.57 a. $n = 450\ \cancel{\text{mL (STP)}} \times \dfrac{1\ \cancel{\text{L}}}{1000\ \cancel{\text{mL}}} \times \dfrac{1\ \text{mol}}{22.4\ \cancel{\text{L (STP)}}} = 0.020\ \text{mol}$

$\text{Molar mass} = \dfrac{\text{mass}}{\text{moles}} = \dfrac{0.84\ \text{g}}{0.020\ \text{mol}} = 42\ \text{g/mol}\ (2\ \text{SFs})$

b. $n = 1.00\ \cancel{\text{L (STP)}} \times \dfrac{1\ \text{mol}}{22.4\ \cancel{\text{L (STP)}}} = 0.0446\ \text{mol}$

$\text{Molar mass} = \dfrac{\text{mass}}{\text{moles}} = \dfrac{1.28\ \text{g}}{0.0446\ \text{mol}} = 28.7\ \text{g/mol}\ (3\ \text{SFs})$

c. Analyze the Problem: Using the Ideal Gas Law to Calculate Moles

Property	*P*	*V*	*n*	*R*	*T*
Given	685 mmHg	1.00 L		$\dfrac{62.4\ \text{L} \cdot \text{mmHg}}{\text{mol} \cdot \text{K}}$	22 °C 22 °C + 273 = 295 K
Need			? mol (? molar mass)		

$$n = \frac{PV}{RT} = \frac{(685\ \cancel{\text{mmHg}})(1.00\ \cancel{\text{L}})}{\left(\dfrac{62.4\ \cancel{\text{L}} \cdot \cancel{\text{mmHg}}}{\text{mol} \cdot \cancel{\text{K}}}\right)(295\ \cancel{\text{K}})} = 0.0372\ \text{mol}$$

$\text{Molar mass} = \dfrac{\text{mass}}{\text{moles}} = \dfrac{1.48\ \text{g}}{0.0372\ \text{mol}} = 39.8\ \text{g/mol}\ (3\ \text{SFs})$

d. Analyze the Problem: Using the Ideal Gas Law to Calculate Moles

Property	*P*	*V*	*n*	*R*	*T*
Given	0.95 atm	2.30 L		$\dfrac{0.0821\ \text{L} \cdot \text{atm}}{\text{mol} \cdot \text{K}}$	24 °C 24 °C + 273 = 297 K
Need			? mol (? molar mass)		

$$n = \frac{PV}{RT} = \frac{(0.95\ \cancel{\text{atm}})(2.30\ \cancel{\text{L}})}{\left(\dfrac{0.0821\ \cancel{\text{L}} \cdot \cancel{\text{atm}}}{\text{mol} \cdot \cancel{\text{K}}}\right)(297\ \cancel{\text{K}})} = 0.090\ \text{mol}$$

$\text{Molar mass} = \dfrac{\text{mass}}{\text{moles}} = \dfrac{2.96\ \text{g}}{0.090\ \text{mol}} = 33\ \text{g/mol}\ (2\ \text{SFs})$

11.59 a. $8.25\ \text{g Mg} \times \frac{1\ \text{mol Mg}}{24.31\ \text{g Mg}} \times \frac{1\ \text{mol } H_2}{1\ \text{mol Mg}} \times \frac{22.4\ \text{L (STP)}}{1\ \text{mol } H_2} = 7.60\ \text{L of } H_2$ released at STP (3 SFs)

b. Analyze the Problem: Using the Ideal Gas Law to Calculate Moles

Property	*P*	*V*	*n*	*R*	*T*
Given	735 mmHg	5.00 L		$\frac{62.4\ \text{L} \cdot \text{mmHg}}{\text{mol} \cdot \text{K}}$	18 °C 18 °C + 273 = 291 K
Need			? mol H_2		

$$n = \frac{PV}{RT} = \frac{(735\ \text{mmHg})(5.00\ \text{L})}{\left(\frac{62.4\ \text{L} \cdot \text{mmHg}}{\text{mol} \cdot \text{K}}\right)(291\ \text{K})} = 0.202\ \text{mol of } H_2$$

Analyze the Problem: Reactant and Product

	Reactant	Product
Mass	? g of Mg	
Molar Mass	1 mol of Mg = 24.31 g of Mg	
Moles		0.202 mol of H_2
Equation	$Mg(s) + 2HCl(aq) \longrightarrow H_2(g) + MgCl_2(aq)$	

$$0.202\ \text{mol } H_2 \times \frac{1\ \text{mol Mg}}{1\ \text{mol } H_2} \times \frac{24.31\ \text{g Mg}}{1\ \text{mol Mg}} = 4.91\ \text{g of Mg (3 SFs)}$$

11.61 Analyze the Problem: Reactant and Reactant

	Reactant 1	Reactant 2
Mass	55.2 g of C_4H_{10}	
Molar Mass	1 mol of C_4H_{10} = 58.12 g of C_4H_{10}	
Moles		? mol O_2
Equation	$2C_4H_{10}(g) + 13O_2(g) \xrightarrow{\Delta} 8CO_2(g) + 10H_2O(g)$	

$$55.2\ \text{g } C_4H_{10} \times \frac{1\ \text{mol } C_4H_{10}}{58.12\ \text{g } C_4H_{10}} \times \frac{13\ \text{mol } O_2}{2\ \text{mol } C_4H_{10}} = 6.17\ \text{mol of } O_2$$

Analyze the Problem: Using the Ideal Gas Law to Calculate Volume

Property	*P*	*V*	*n*	*R*	*T*
Given	0.850 atm		6.17 mol O_2	$\frac{0.0821\ \text{L} \cdot \text{atm}}{\text{mol} \cdot \text{K}}$	25 °C 25 °C + 273 = 298 K
Need		? L			

$$V = \frac{nRT}{P} = \frac{(6.17\ \text{mol})\left(\frac{0.0821\ \text{L} \cdot \text{atm}}{\text{mol} \cdot \text{K}}\right)(298\ \text{K})}{(0.850\ \text{atm})} = 178\ \text{L of } O_2\ \text{(3 SFs)}$$

11.63 Analyze the Problem: Reactant and Reactant

	Reactant 1	Reactant 2
Mass	5.4 g of Al	
Molar Mass	1 mol of Al = 26.98 g of Al	
Moles		? mol O_2 (? L O_2)
Molar Volume		1 mol of O_2 = 22.4 L of O_2 (STP)
Equation	$4Al(s) + 3O_2(g) \xrightarrow{\Delta} 2Al_2O_3(s)$	

$$5.4\ \cancel{g\ Al} \times \frac{1\ \cancel{mol\ Al}}{26.98\ \cancel{g\ Al}} \times \frac{3\ \cancel{mol\ O_2}}{4\ \cancel{mol\ Al}} \times \frac{22.4\ L\ O_2\ (STP)}{1\ \cancel{mol\ O_2}} = 3.4\ L\ of\ O_2\ at\ STP\ (2\ SFs)$$

11.65 To obtain the total pressure of a gas mixture, add up all of the partial pressures using the same pressure unit.

$P_{total} = P_{nitrogen} + P_{oxygen} + P_{helium} = 425\ torr + 115\ torr + 225\ torr = 765\ torr\ (3\ SFs)$

11.67 To obtain the total pressure of a gas mixture, add up all of the partial pressures using the same pressure unit.

$P_{total} = P_{oxygen} + P_{nitrogen} + P_{carbon\ dioxide} + P_{water\ vapor}$

$= 93\ mmHg + 565\ mmHg + 38\ mmHg + 47\ mmHg$

$= 743\ mmHg$

$$743\ \cancel{mmHg} \times \frac{1\ atm}{760\ \cancel{mmHg}} = 0.978\ atm\ (3\ SFs)$$

11.69 Because the total pressure of a gas mixture is the sum of the partial pressures using the same pressure unit, addition and subtraction are used to obtain the "missing" partial pressure.

$P_{nitrogen} = P_{total} - (P_{oxygen} + P_{helium}) = 925\ torr - (425\ torr + 75\ torr) = 425\ torr\ (3\ SFs)$

11.71 a. $P_{O_2} = P_{total} - P_{H_2O} = 765\ mmHg - 22\ mmHg = 743\ mmHg\ (3\ SFs)$

b. $V = 256\ \cancel{mL} \times \frac{1\ L}{1000\ \cancel{mL}} = 0.256\ L; \quad T = 24\ °C + 273 = 297\ K$

$$n = \frac{PV}{RT} = \frac{(743\ \cancel{mmHg})(0.256\ \cancel{L})}{\left(\frac{62.4\ \cancel{L} \cdot \cancel{mmHg}}{mol \cdot \cancel{K}}\right)(297\ \cancel{K})} = 0.0103\ mol\ of\ O_2\ (3\ SFs)$$

11.73 a. True. The flask containing helium has more moles of helium and thus more atoms of helium to collide with the walls of the container to give a higher pressure.

b. True. The mass and volume of each are the same, meaning that the mass/volume ratio or density is the same in both flasks.

11.75 a. 2; The fewest number of gas particles will exert the lowest pressure.

b. 1; The greatest number of gas particles will exert the highest pressure.

11.77 a. A; Volume decreases when temperature decreases (at constant *P* and *n*).

b. C; Volume increases when pressure decreases (at constant *n* and *T*).

c. A; Volume decreases when the number of moles of gas decreases (at constant *T* and *P*).

d. B; Doubling the Kelvin temperature would double the volume, but when half of the gas escapes, the volume would decrease by half. These two opposing effects cancel each other, and there is no overall change in the volume (at constant *P*).

e. C; Increasing the moles of gas causes an increase in the volume to keep *T* and *P* constant.

11.79 a. $P_{O_2} = 650.\ \cancel{\text{mmHg total}} \times \dfrac{21\ \text{mmHg O}_2}{100\ \cancel{\text{mmHg total}}} = 140\ \text{mmHg (2 SFs)}$

b. $P_{\text{total}} = 100.\ \cancel{\text{mmHg O}_2} \times \dfrac{100\ \text{mmHg total}}{21\ \cancel{\text{mmHg O}_2}} = 480\ \text{mmHg (2 SFs)}$

11.81 $T_1 = -8\ ^\circ\text{C} + 273 = 265\ \text{K}; \quad P_1 = 658\ \text{mmHg}; \quad V_1 = 31\ 000\ \text{L}$

$T_2 = 0\ ^\circ\text{C} + 273 = 273\ \text{K}; \quad P_2 = 760\ \text{mmHg}; \quad V_2 = ?\ \text{L}$

$V_2 = V_1 \times \dfrac{P_1}{P_2} \times \dfrac{T_2}{T_1} = 31\ 000\ \text{L} \times \dfrac{658\ \cancel{\text{mmHg}}}{760\ \cancel{\text{mmHg}}} \times \dfrac{273\ \cancel{\text{K}}}{265\ \cancel{\text{K}}} = 28\ 000\ \text{L at STP}$

$28\ 000\ \cancel{\text{L H}_2\ \text{(STP)}} \times \dfrac{1\ \cancel{\text{mol H}_2}}{22.4\ \cancel{\text{L H}_2\ \text{(STP)}}} \times \dfrac{2.02\ \cancel{\text{g H}_2}}{1\ \cancel{\text{mol H}_2}} \times \dfrac{1\ \text{kg H}_2}{1000\ \cancel{\text{g H}_2}} = 2.5\ \text{kg of H}_2\ \text{(2 SFs)}$

11.83 $T_1 = 15\ ^\circ\text{C} + 273 = 288\ \text{K}; \quad P_1 = 745\ \text{mmHg}; \quad V_1 = 4250\ \text{mL}$

$T_2 = ?\ ^\circ\text{C}; \quad P_2 = 1.20\ \cancel{\text{atm}} \times \dfrac{760\ \text{mmHg}}{1\ \cancel{\text{atm}}} = 912\ \text{mmHg};$

$V_2 = 2.50\ \cancel{\text{L}} \times \dfrac{1000\ \text{mL}}{1\ \cancel{\text{L}}} = 2.50 \times 10^3\ \text{mL}$

$T_2 = T_1 \times \dfrac{V_2}{V_1} \times \dfrac{P_2}{P_1} = 288\ \text{K} \times \dfrac{2.50 \times 10^3\ \cancel{\text{mL}}}{4250\ \cancel{\text{mL}}} \times \dfrac{912\ \cancel{\text{mmHg}}}{745\ \cancel{\text{mmHg}}} = 207\ \text{K} - 273$

$= -66\ ^\circ\text{C (2 SFs)}$

11.85 Analyze the Problem: Using the Ideal Gas Law to Calculate Moles

Property	*P*	*V*	*n*	*R*	*T*
Given	2500. mmHg	2.00 L		$\dfrac{62.4\ \text{L}\cdot\text{mmHg}}{\text{mol}\cdot\text{K}}$	18 °C 18 °C + 273 = 291 K
Need			? mol		

$n = \dfrac{PV}{RT} = \dfrac{(2500.\ \cancel{\text{mmHg}})(2.00\ \cancel{\text{L}})}{\left(\dfrac{62.4\ \cancel{\text{L}}\cdot\cancel{\text{mmHg}}}{\text{mol}\cdot\cancel{\text{K}}}\right)(291\ \cancel{\text{K}})} = 0.275\ \text{mol of CH}_4$

$0.275\ \cancel{\text{mol CH}_4} \times \dfrac{16.04\ \text{g CH}_4}{1\ \cancel{\text{mol CH}_4}} = 4.41\ \text{g of CH}_4\ \text{(3 SFs)}$

11.87 $V = 941\ \cancel{\text{mL}} \times \dfrac{1\ \text{L}}{1000\ \cancel{\text{mL}}} = 0.941\ \text{L}; \quad P = 748\ \cancel{\text{torr}} \times \dfrac{1\ \text{mmHg}}{1\ \cancel{\text{torr}}} = 748\ \text{mmHg}$

Analyze the Problem: Using the Ideal Gas Law to Calculate Moles

Property	*P*	*V*	*n*	*R*	*T*
Given	748 torr (748 mmHg)	941 mL (0.941 L)		$\dfrac{62.4\ \text{L}\cdot\text{mmHg}}{\text{mol}\cdot\text{K}}$	20.0 °C 20.0 °C + 273 = 293 K
Need			? mol (? molar mass)		

$n = \dfrac{PV}{RT} = \dfrac{(748\ \cancel{\text{mmHg}})(0.941\ \cancel{\text{L}})}{\left(\dfrac{62.4\ \cancel{\text{L}}\cdot\cancel{\text{mmHg}}}{\text{mol}\cdot\cancel{\text{K}}}\right)(293\ \cancel{\text{K}})} = 0.0385\ \text{mol}$

$\text{Molar mass} = \dfrac{\text{mass}}{\text{moles}} = \dfrac{1.62\ \text{g}}{0.0385\ \text{mol}} = 42.1\ \text{g/mol (3 SFs)}$

11.89 Analyze the Problem: Using the Ideal Gas Law to Calculate Moles

Property	*P*	*V*	*n*	*R*	*T*
Given	1.20 atm	35.0 L		$\frac{0.0821\ \text{L}\cdot\text{atm}}{\text{mol}\cdot\text{K}}$	5 °C 5 °C + 273 = 278 K
Need			? mol		

$$n = \frac{PV}{RT} = \frac{(1.20\ \text{atm})(35.0\ \text{L})}{\left(\frac{0.0821\ \text{L}\cdot\text{atm}}{\text{mol}\cdot\text{K}}\right)(278\ \text{K})} = 1.84\ \text{mol of } CO_2$$

$$1.84\ \text{mol } CO_2 \times \frac{44.01\ \text{g } CO_2}{1\ \text{mol } CO_2} = 81.0\ \text{g of } CO_2\ (3\ \text{SFs})$$

11.91 $25.0\ \text{g Zn} \times \frac{1\ \text{mol Zn}}{65.41\ \text{g Zn}} \times \frac{1\ \text{mol } H_2}{1\ \text{mol Zn}} \times \frac{22.4\ \text{L (STP)}}{1\ \text{mol } H_2} = 8.56\ \text{L of } H_2$ released at STP (3 SFs)

11.93 Analyze the Problem: Using the Ideal Gas Law to Calculate Moles

Property	*P*	*V*	*n*	*R*	*T*
Given	725 mmHg	5.00 L		$\frac{62.4\ \text{L}\cdot\text{mmHg}}{\text{mol}\cdot\text{K}}$	375 °C 375 °C + 273 = 648 K
Need			? mol H_2O		

$$n = \frac{PV}{RT} = \frac{(725\ \text{mmHg})(5.00\ \text{L})}{\left(\frac{62.4\ \text{L}\cdot\text{mmHg}}{\text{mol}\cdot\text{K}}\right)(648\ \text{K})} = 0.0896\ \text{mol of } H_2O$$

Analyze the Problem: Reactant and Product

	Reactant	Product
Mass		? g NH_3
Molar Mass		1 mol of NH_3 = 17.03 g of NH_3
Moles	0.0896 mol of H_2O	
Equation	$4NO_2(g) + 6H_2O(g) \longrightarrow$	$7O_2(g) + 4NH_3(g)$

$$0.0896\ \text{mol } H_2O \times \frac{4\ \text{mol } NH_3}{6\ \text{mol } H_2O} \times \frac{17.03\ \text{g } NH_3}{1\ \text{mol } NH_3} = 1.02\ \text{g of } NH_3\ (3\ \text{SFs})$$

11.95 The partial pressure of each gas is proportional to the number of particles of each type of gas that is present. Thus, a ratio of partial pressure to total pressure is equal to the ratio of moles of that gas to the total number of moles of gases that are present:

$$\frac{P_{helium}}{P_{total}} = \frac{n_{helium}}{n_{total}}$$

Solving the equation for the partial pressure of helium yields:

$$P_{helium} = P_{total} \times \frac{n_{helium}}{n_{total}} = 2400\ \text{torr} \times \frac{2.0\ \text{mol}}{8.0\ \text{mol}} = 600\ \text{torr}\ (6.0 \times 10^2\ \text{torr})\ (2\ \text{SFs})$$

$$P_{oxygen} = P_{total} \times \frac{n_{oxygen}}{n_{total}} = 2400\ \text{torr} \times \frac{6.0\ \text{mol}}{8.0\ \text{mol}} = 1800\ \text{torr}\ (1.8 \times 10^3\ \text{torr})\ (2\ \text{SFs})$$

11.97 Because the partial pressure of nitrogen is to be reported in torr, the atm and mmHg units (for oxygen and argon, respectively) must be converted to torr, as follows:

$$P_{\text{oxygen}} = 0.60\ \cancel{\text{atm}} \times \frac{760\ \text{torr}}{1\ \cancel{\text{atm}}} = 460\ \text{torr} \quad \text{and} \quad P_{\text{argon}} = 425\ \cancel{\text{mmHg}} \times \frac{1\ \text{torr}}{1\ \cancel{\text{mmHg}}} = 425\ \text{torr}$$

$$\therefore P_{\text{nitrogen}} = P_{\text{total}} - (P_{\text{oxygen}} + P_{\text{argon}})$$
$$= 1250\ \text{torr} - (460\ \text{torr} + 425\ \text{torr}) = 370\ \text{torr}\ (2\ \text{SFs})$$

11.99 a. $P_{\text{hydrogen}} = P_{\text{total}} - P_{\text{water}} = 755\ \text{mmHg} - 21\ \text{mmHg} = 734\ \text{mmHg}\ (2\ \text{SFs})$

b. $V = 415\ \cancel{\text{mL}} \times \dfrac{1\ \text{L}}{1000\ \cancel{\text{mL}}} = 0.415\ \text{L}$

Analyze the Problem: Using the Ideal Gas Law to Calculate Moles

Property	***P***	***V***	***n***	***R***	***T***
Given	734 mmHg	415 mL (0.415 L)		$\dfrac{62.4\ \text{L}\cdot\text{mmHg}}{\text{mol}\cdot\text{K}}$	23 °C 23 °C + 273 = 296 K
Need			? mol		

$$n = \frac{PV}{RT} = \frac{(734\ \cancel{\text{mmHg}})(0.415\ \cancel{\text{L}})}{\left(\dfrac{62.4\ \cancel{\text{L}}\cdot\cancel{\text{mmHg}}}{\text{mol}\cdot\cancel{\text{K}}}\right)(296\ \cancel{\text{K}})} = 0.0165\ \text{mol of}\ H_2\ (3\ \text{SFs})$$

c. Analyze the Problem: Reactant and Product

	Reactant	**Product**
Mass	? g Al	
Molar Mass	1 mol of Al = 26.98 g of Al	
Moles		0.0165 mol of H_2
Equation	$2Al(s) + 3H_2SO_4(aq) \longrightarrow$	$3H_2(g) + Al_2(SO_4)_3(aq)$

$$0.0165\ \cancel{\text{mol}\ H_2} \times \frac{2\ \cancel{\text{mol Al}}}{3\ \cancel{\text{mol}\ H_2}} \times \frac{26.98\ \text{g Al}}{1\ \cancel{\text{mol Al}}} = 0.297\ \text{g of Al}\ (3\ \text{SFs})$$

11.101 a. Analyze the Problem: Using the Ideal Gas Law to Calculate Moles

Property	***P***	***V***	***n***	***R***	***T***
Given	760. mmHg	762 mL (0.762 L)		$\dfrac{62.4\ \text{L}\cdot\text{mmHg}}{\text{mol}\cdot\text{K}}$	0.0 °C 0.0 °C + 273 = 273 K
Need			? mol		

$$n = \frac{PV}{RT} = \frac{(760.\ \cancel{\text{mmHg}})(0.762\ \cancel{\text{L}})}{\left(\dfrac{62.4\ \cancel{\text{L}}\cdot\cancel{\text{mmHg}}}{\text{mol}\cdot\cancel{\text{K}}}\right)(273\ \cancel{\text{K}})} = 0.0340\ \text{mol}$$

$$\text{Molar mass} = \frac{\text{mass}}{\text{moles}} = \frac{1.020\ \text{g}}{0.0340\ \text{mol}} = 30.0\ \text{g/mol}$$

b. $0.815\ \cancel{\text{g C}} \times \dfrac{1\ \text{mol C}}{12.01\ \cancel{\text{g C}}} = 0.0679\ \text{mol of C}$ (smaller number of moles)

$$0.205\ \cancel{\text{g H}} \times \frac{1\ \text{mol H}}{1.008\ \cancel{\text{g H}}} = 0.203\ \text{mol of H}$$

$$\frac{0.0679\ \text{mol C}}{0.0679} = 1.00\ \text{mol of C} \qquad \frac{0.203\ \text{mol H}}{0.0679} = 2.99\ \text{mol of H}$$

$$\therefore \text{Empirical formula} = C_{1.00}H_{2.99} \rightarrow CH_3$$

$$\text{Empirical formula mass of } CH_3 = 12.01 + 3(1.008) = 15.03 \text{ g}$$

$$\text{Small integer} = \frac{\text{molar mass}}{\text{empirical formula mass of } CH_3} = \frac{30.0 \cancel{g}}{15.03 \cancel{g}} = 2$$

$$\therefore \text{Molecular formula} = C_{(1\times2)}H_{(3\times2)} = C_2H_6$$

11.103 Analyze the Problem: Reactant and Product

	Reactant	Product
Mass	881 g of C_3H_8	
Molar Mass	1 mol of C_3H_8 = 44.09 g of C_3H_8	
Moles		? mol CO_2 (? L CO_2)
Molar Volume		1 mol of CO_2 = 22.4 L of CO_2 (STP)
Equation	$C_3H_8(g) + 5O_2(g) \xrightarrow{\Delta} 3CO_2(g) + 4H_2O(g)$	

$$881 \text{ g } \cancel{C_3H_8} \times \frac{1 \cancel{\text{mol } C_3H_8}}{44.09 \cancel{\text{g } C_3H_8}} \times \frac{3 \cancel{\text{mol } CO_2}}{1 \cancel{\text{mol } C_3H_8}} \times \frac{22.4 \text{ L } CO_2 \text{ (STP)}}{1 \cancel{\text{mol } CO_2}}$$

$= 1340$ L CO_2 at STP (3 SFs)

11.105 Analyze the Problem: Using the Ideal Gas Law to Calculate Moles

Property	*P*	*V*	*n*	*R*	*T*
Given	1.00 atm	7.50 L		$\frac{0.0821 \text{ L} \cdot \text{atm}}{\text{mol} \cdot \text{K}}$	37 °C 37 °C + 273 = 310 K
Need			? mol O_2		

$$n = \frac{PV}{RT} = \frac{(1.00 \cancel{\text{atm}})(7.50 \cancel{\text{L}})}{\left(\frac{0.0821 \cancel{\text{L}} \cdot \cancel{\text{atm}}}{\text{mol} \cdot \cancel{\text{K}}}\right)(310 \cancel{\text{K}})} = 0.295 \text{ mol of } O_2$$

$$0.295 \cancel{\text{mol } O_2} \times \frac{6 \cancel{\text{mol } H_2O}}{6 \cancel{\text{mol } O_2}} \times \frac{18.02 \text{ g } H_2O}{1 \cancel{\text{mol } H_2O}} = 5.32 \text{ g of } H_2O \text{ (smaller amount of product)}$$

$$18.0 \cancel{\text{g } C_6H_{12}O_6} \times \frac{1 \cancel{\text{mol } C_6H_{12}O_6}}{180.16 \cancel{\text{g } C_6H_{12}O_6}} \times \frac{6 \cancel{\text{mol } H_2O}}{1 \cancel{\text{mol } C_6H_{12}O_6}} \times \frac{18.02 \text{ g } H_2O}{1 \cancel{\text{mol } H_2O}} = 10.8 \text{ g of } H_2O$$

Thus, O_2 is the limiting reactant, and 5.32 g of H_2O (3 SFs) can be produced.

12
Solutions

Learning Goals

- Identify the solute and solvent in a solution; describe the formation of a solution.
- Identify solutes as electrolytes or nonelectrolytes.
- Define solubility; distinguish between an unsaturated and a saturated solution. Identify a salt as soluble or insoluble.
- Calculate the concentration of a solute in a solution; use concentration units to calculate the amount of solute or solution.
- Calculate the new concentration or volume of a diluted solution. Given the volume and concentration of a solution, calculate the amount of another reactant or product in a reaction.
- Identify a mixture as a solution, a colloid, or a suspension. Describe how the number of particles in a solution affects the freezing point, boiling point, and osmotic pressure.

Chapter Outline

Core Chemistry Skills

- Using Solubility Rules (12.3)
- Calculating Concentration (12.4)
- Using Concentration as a Conversion Factor (12.4)
- Calculating the Quantity of a Reactant or Product for a Chemical Reaction in Solution (12.5)
- Calculating Temperature Change (12.6)

Answers and Solutions to Text Problems

12.1 The component present in the smaller amount is the solute; the component present in the larger amount is the solvent.

a. NaCl, solute; water, solvent
b. water, solute; ethanol, solvent
c. oxygen, solute; nitrogen, solvent

12.3 **a.** $NaNO_3$ (an ionic solute) would be soluble in water (a polar solvent).
b. I_2 (a nonpolar solute) would be soluble in CCl_4 (a nonpolar solvent).
c. Sucrose (a polar solute) would be soluble in water (a polar solvent).
d. Octane (a nonpolar solute) would be soluble in CCl_4 (a nonpolar solvent).

12.5 The K^+ and I^- ions at the surface of the solid are pulled into solution by the polar water molecules, where the hydration process surrounds separate ions with water molecules.

12.7 The strong electrolyte KF completely dissociates into K^+ and F^- ions when it dissolves in water. When the weak electrolyte HF dissolves in water, there are a few ions of H^+ and F^- present, but mostly dissolved HF molecules.

12.9 Strong electrolytes dissociate completely into ions.
a. $KCl(s) \xrightarrow{H_2O} K^+(aq) + Cl^-(aq)$
b. $CaCl_2(s) \xrightarrow{H_2O} Ca^{2+}(aq) + 2Cl^-(aq)$
c. $K_3PO_4(s) \xrightarrow{H_2O} 3K^+(aq) + PO_4^{3-}(aq)$
d. $Fe(NO_3)_3(s) \xrightarrow{H_2O} Fe^{3+}(aq) + 3NO_3^-(aq)$

12.11 **a.** $CH_3{-}COOH(l) \overset{H_2O}{\rightleftarrows} H^+(aq) + CH_3{-}COO^-(aq)$ mostly molecules, a few ions
An aqueous solution of a weak electrolyte like acetic acid will contain mostly $CH_3{-}COOH$ molecules, with a few H^+ ions and a few $CH_3{-}COO^-$ ions.

b. $NaBr(s) \xrightarrow{H_2O} Na^+(aq) + Br^-(aq)$ ions only
An aqueous solution of a strong electrolyte like NaBr will contain only the ions Na^+ and Br^-.

c. $C_6H_{12}O_6(s) \xrightarrow{H_2O} C_6H_{12}O_6(aq)$ molecules only
An aqueous solution of a nonelectrolyte like fructose will contain only $C_6H_{12}O_6$ molecules.

12.13 **a.** K_2SO_4 is a strong electrolyte because only ions are present in the K_2SO_4 solution.
b. NH_4OH is a weak electrolyte because only a few NH_4^+ and OH^- ions are present in the solution.
c. $C_6H_{12}O_6$ is a nonelectrolyte because only $C_6H_{12}O_6$ molecules are present in the solution.

12.15 **a.** The solution must be saturated because no additional solute dissolves.
b. The solution was unsaturated because the sugar cube dissolves completely.

12.17 **a.** At 20 °C, KCl has a solubility of 34 g of KCl in 100 g of H_2O. Because 25 g of KCl is less than the maximum amount that can dissolve in 100 g of H_2O at 20 °C, the KCl solution is unsaturated.

b. At 20 °C, $NaNO_3$ has a solubility of 88 g of $NaNO_3$ in 100 g of H_2O. Using the solubility as a conversion factor, we can calculate the maximum amount of $NaNO_3$ that can dissolve in 25 g of H_2O:

$$25\ \text{g}\ \cancel{H_2O} \times \frac{88\ \text{g}\ NaNO_3}{100\ \text{g}\ \cancel{H_2O}} = 22\ \text{g of}\ NaNO_3\ (2\ \text{SFs})$$

Because 11 g of $NaNO_3$ is less than the maximum amount that can dissolve in 25 g of H_2O at 20 °C, the $NaNO_3$ solution is unsaturated.

c. At 20 °C, sugar has a solubility of 204 g of $C_{12}H_{22}O_{11}$ in 100 g of H_2O. Using the solubility as a conversion factor, we can calculate the maximum amount of sugar that can dissolve in 125 g of H_2O:

$$125\ \text{g}\ \cancel{H_2O} \times \frac{204\ \text{g sugar}}{100\ \text{g}\ \cancel{H_2O}} = 255\ \text{g of sugar (3 SFs)}$$

Because 400. g of $C_{12}H_{22}O_{11}$ exceeds the maximum amount that can dissolve in 125 g of H_2O at 20 °C, the sugar solution is saturated, and excess undissolved sugar will be present on the bottom of the container.

12.19 a. At 20 °C, KCl has a solubility of 34 g of KCl in 100 g of H_2O.

$\therefore$ 200. g of H_2O will dissolve:

$$200.\ \text{g}\ \cancel{H_2O} \times \frac{34\ \text{g KCl}}{100\ \text{g}\ \cancel{H_2O}} = 68\ \text{g of KCl (2 SFs)}$$

At 20 °C, 68 g of KCl will remain in solution.

b. Since 80. g of KCl dissolves at 50 °C and 68 g remains in solution at 20 °C, the mass of solid KCl that crystallizes after cooling is (80. g KCl − 68 g KCl) = 12 g of KCl. (2 SFs)

12.21 a. In general, the solubility of solid solutes (like sugar) increases as temperature is increased.

b. The solubility of a gaseous solute (CO_2) is less at a higher temperature.

c. The solubility of a gaseous solute is less at a higher temperature, and the CO_2 pressure in the can is increased. When the can of warm soda is opened, more CO_2 is released, producing more spray.

12.23 a. Salts containing Li^+ ions are soluble.

b. The Cl^- salt containing Ag^+ is insoluble.

c. Salts containing CO_3^{2-} ions are usually insoluble.

d. Salts containing K^+ ions are soluble.

e. Salts containing NO_3^- ions are soluble.

12.25 a. No solid forms; salts containing K^+ and Na^+ are soluble.

b. Solid silver sulfide (Ag_2S) forms:

$$2AgNO_3(aq) + K_2S(aq) \longrightarrow Ag_2S(s) + 2KNO_3(aq)$$

$$2Ag^+(aq) + \cancel{2NO_3^-(aq)} + \cancel{2K^+(aq)} + S^{2-}(aq) \longrightarrow Ag_2S(s) + \cancel{2K^+(aq)} + \cancel{2NO_3^-(aq)}$$

$$2Ag^+(aq) + S^{2-}(aq) \longrightarrow Ag_2S(s) \quad \text{Net ionic equation}$$

c. Solid calcium sulfate ($CaSO_4$) forms:

$$CaCl_2(aq) + Na_2SO_4(aq) \longrightarrow CaSO_4(s) + 2NaCl(aq)$$

$$Ca^{2+}(aq) + \cancel{2Cl^-(aq)} + \cancel{2Na^+(aq)} + SO_4^{2-}(aq) \longrightarrow CaSO_4(s) + \cancel{2Na^+(aq)} + \cancel{2Cl^-(aq)}$$

$$Ca^{2+}(aq) + SO_4^{2-}(aq) \longrightarrow CaSO_4(s) \quad \text{Net ionic equation}$$

d. Solid copper phosphate ($Cu_3(PO_4)_2$) forms:

$$3CuCl_2(aq) + 2Li_3PO_4(aq) \longrightarrow Cu_3(PO_4)_2(s) + 6LiCl(aq)$$

$$3Cu^{2+}(aq) + \cancel{6Cl^-(aq)} + \cancel{6Li^+(aq)} + 2PO_4^{3-}(aq) \longrightarrow Cu_3(PO_4)_2(s) + \cancel{6Li^+(aq)} + \cancel{6Cl^-(aq)}$$

$$3Cu^{2+}(aq) + 2PO_4^{3-}(aq) \longrightarrow Cu_3(PO_4)_2(s) \quad \text{Net ionic equation}$$

12.27 A 5.00% (m/m) glucose solution can be made by adding 5.00 g of glucose to 95.00 g of water, while a 5.00% (m/v) glucose solution can be made by adding 5.00 g of glucose to enough water to make 100.0 mL of solution.

12.29 $\text{mass percent (m/m)} = \dfrac{\text{mass of solute (g)}}{\text{mass of solution (g)}} \times 100\%$

a. mass of solution = 25 g of KCl + 125 g of H_2O = 150. g of solution

$$\frac{25\text{ g KCl}}{150.\text{ g solution}} \times 100\% = 17\%\text{ (m/m) KCl solution (2 SFs)}$$

b. $\dfrac{8.0\text{ g } CaCl_2}{80.0\text{ g solution}} \times 100\% = 10.\%\text{ (m/m) } CaCl_2\text{ solution (2 SFs)}$

c. $\dfrac{12\text{ g sucrose}}{225\text{ g solution}} \times 100\% = 5.3\%\text{ (m/m) sucrose solution (2 SFs)}$

12.31 $\text{mass/volume percent (m/v)} = \dfrac{\text{mass of solute (g)}}{\text{volume of solution (mL)}} \times 100\%$

a. $\dfrac{75\text{ g } Na_2SO_4}{250\text{ mL solution}} \times 100\% = 30.\%\text{ (m/v) } Na_2SO_4\text{ solution (2 SFs)}$

b. $\dfrac{39\text{ g sucrose}}{355\text{ mL solution}} \times 100\% = 11\%\text{ (m/v) sucrose solution (2 SFs)}$

12.33 $\text{molarity (M)} = \dfrac{\text{moles of solute}}{\text{liters of solution}}$

a. $\dfrac{2.00\text{ mol glucose}}{4.00\text{ L solution}} = 0.500\text{ M glucose solution (3 SFs)}$

b. $\dfrac{5.85\text{ g NaCl}}{40.0\text{ mL solution}} \times \dfrac{1\text{ mol NaCl}}{58.44\text{ g NaCl}} \times \dfrac{1000\text{ mL solution}}{1\text{ L solution}} = 2.50\text{ M NaCl solution (3 SFs)}$

c. $\dfrac{4.00\text{ g KOH}}{2.00\text{ L solution}} \times \dfrac{1\text{ mol KOH}}{56.11\text{ g KOH}} = 0.0356\text{ M KOH solution (3 SFs)}$

12.35 a. $50.0\text{ g solution} \times \dfrac{5.0\text{ g KCl}}{100\text{ g solution}} = 2.5\text{ g of KCl (2 SFs)}$

b. $1250\text{ g solution} \times \dfrac{4.0\text{ g } NH_4Cl}{100\text{ g solution}} = 50.\text{ g of } NH_4Cl\text{ (2 SFs)}$

c. $250\text{ mL solution} \times \dfrac{10.0\text{ mL acetic acid}}{100\text{ mL solution}} = 25\text{ mL of acetic acid (2 SFs)}$

12.37 $355\text{ mL solution} \times \dfrac{22.5\text{ mL alcohol}}{100\text{ mL solution}} = 79.9\text{ mL of alcohol (3 SFs)}$

12.39 a. $5.0\text{ g } LiNO_3 \times \dfrac{100\text{ g solution}}{25\text{ g } LiNO_3} = 20.\text{ g of } LiNO_3\text{ solution (2 SFs)}$

b. $40.0\text{ g KOH} \times \dfrac{100\text{ mL solution}}{10.0\text{ g KOH}} = 400.\text{ mL of KOH solution (3 SFs)}$

c. $2.0\text{ mL formic acid} \times \dfrac{100\text{ mL solution}}{10.0\text{ mL formic acid}} = 20.\text{ mL of formic acid solution (2 SFs)}$

12.41 a. $2.00\ \cancel{\text{L solution}} \times \dfrac{1.50\ \cancel{\text{mol NaOH}}}{1\ \cancel{\text{L solution}}} \times \dfrac{40.00\ \text{g NaOH}}{1\ \cancel{\text{mol NaOH}}} = 120.\ \text{g of NaOH}\ (3\ \text{SFs})$

b. $125\ \cancel{\text{mL solution}} \times \dfrac{1\ \cancel{\text{L solution}}}{1000\ \cancel{\text{mL solution}}} \times \dfrac{0.200\ \cancel{\text{mol KCl}}}{1\ \cancel{\text{L solution}}} \times \dfrac{74.55\ \text{g KCl}}{1\ \cancel{\text{mol KCl}}}$
$= 1.86\ \text{g of KCl}\ (3\ \text{SFs})$

c. $25.0\ \cancel{\text{mL solution}} \times \dfrac{1\ \cancel{\text{L solution}}}{1000\ \cancel{\text{mL solution}}} \times \dfrac{3.50\ \cancel{\text{mol HCl}}}{1\ \cancel{\text{L solution}}} \times \dfrac{36.46\ \text{g HCl}}{1\ \cancel{\text{mol HCl}}}$
$= 3.19\ \text{g of HCl}\ (3\ \text{SFs})$

12.43 a. $3.00\ \cancel{\text{mol NaOH}} \times \dfrac{1\ \text{L solution}}{2.00\ \cancel{\text{mol NaOH}}} = 1.50\ \text{L of solution}\ (3\ \text{SFs})$

b. $15.0\ \cancel{\text{mol NaCl}} \times \dfrac{1\ \text{L solution}}{1.50\ \cancel{\text{mol NaCl}}} = 10.0\ \text{L of solution}\ (3\ \text{SFs})$

c. $0.0500\ \cancel{\text{mol Ca(NO}_3)_2} \times \dfrac{1\ \cancel{\text{L solution}}}{0.800\ \cancel{\text{mol Ca(NO}_3)_2}} \times \dfrac{1000\ \text{mL solution}}{1\ \cancel{\text{L solution}}}$
$= 62.5\ \text{mL of solution}\ (3\ \text{SFs})$

12.45 a. $12.5\ \cancel{\text{g Na}_2\text{CO}_3} \times \dfrac{1\ \cancel{\text{mol Na}_2\text{CO}_3}}{105.99\ \cancel{\text{g Na}_2\text{CO}_3}} \times \dfrac{1\ \cancel{\text{L solution}}}{0.120\ \cancel{\text{mol Na}_2\text{CO}_3}} \times \dfrac{1000\ \text{mL solution}}{1\ \cancel{\text{L solution}}}$
$= 983\ \text{mL of Na}_2\text{CO}_3\ \text{solution}\ (3\ \text{SFs})$

b. $0.850\ \cancel{\text{mol NaNO}_3} \times \dfrac{1\ \cancel{\text{L solution}}}{0.500\ \cancel{\text{mol NaNO}_3}} \times \dfrac{1000\ \text{mL solution}}{1\ \cancel{\text{L solution}}}$
$= 1700\ \text{mL}\ (1.70 \times 10^3\ \text{mL})\ \text{of NaNO}_3\ \text{solution}\ (3\ \text{SFs})$

c. $30.0\ \cancel{\text{g LiOH}} \times \dfrac{1\ \cancel{\text{mol LiOH}}}{23.95\ \cancel{\text{g LiOH}}} \times \dfrac{1\ \cancel{\text{L solution}}}{2.70\ \cancel{\text{mol LiOH}}} \times \dfrac{1000\ \text{mL solution}}{1\ \cancel{\text{L solution}}}$
$= 464\ \text{mL of LiOH solution}\ (3\ \text{SFs})$

12.47 $C_1V_1 = C_2V_2$

a. $C_2 = C_1 \times \dfrac{V_1}{V_2} = 6.00\ \text{M} \times \dfrac{0.150\ \cancel{\text{L}}}{0.500\ \cancel{\text{L}}} = 1.80\ \text{M HCl solution}\ (3\ \text{SFs})$

b. $C_2 = C_1 \times \dfrac{V_1}{V_2} = 12\ \text{M} \times \dfrac{0.50\ \cancel{\text{L}}}{3.0\ \cancel{\text{L}}} = 2.0\ \text{M NaOH solution}\ (2\ \text{SFs})$

c. $C_2 = C_1 \times \dfrac{V_1}{V_2} = 25\% \times \dfrac{10.0\ \cancel{\text{mL}}}{100.0\ \cancel{\text{mL}}} = 2.5\%\ (\text{m/v})\ \text{KOH solution}\ (2\ \text{SFs})$

d. $C_2 = C_1 \times \dfrac{V_1}{V_2} = 15\% \times \dfrac{50.0\ \cancel{\text{mL}}}{250\ \cancel{\text{mL}}} = 3.0\%\ (\text{m/v})\ \text{H}_2\text{SO}_4\ \text{solution}\ (2\ \text{SFs})$

12.49 $C_1V_1 = C_2V_2$

a. $V_2 = V_1 \times \dfrac{C_1}{C_2} = 20.0\ \text{mL} \times \dfrac{6.00\ \cancel{\text{M}}}{1.50\ \cancel{\text{M}}} = 80.0\ \text{mL of diluted HCl solution}\ (2\ \text{SFs})$

b. $V_2 = V_1 \times \dfrac{C_1}{C_2} = 50.0\ \text{mL} \times \dfrac{10.0\cancel{\%}}{2.0\cancel{\%}} = 250\ \text{mL of diluted LiCl solution}\ (2\ \text{SFs})$

c. $V_2 = V_1 \times \dfrac{C_1}{C_2} = 50.0\text{ mL} \times \dfrac{6.00\text{ M}}{0.500\text{ M}} = 600.\text{ mL of diluted } H_3PO_4 \text{ solution (3 SFs)}$

d. $V_2 = V_1 \times \dfrac{C_1}{C_2} = 75\text{ mL} \times \dfrac{12\%}{5.0\%} = 180\text{ mL of diluted glucose solution (2 SFs)}$

12.51 $C_1V_1 = C_2V_2$

a. $V_1 = V_2 \times \dfrac{C_2}{C_1} = 255\text{ mL} \times \dfrac{0.200\text{ M}}{4.00\text{ M}} = 12.8\text{ mL of the } HNO_3 \text{ solution (3 SFs)}$

b. $V_1 = V_2 \times \dfrac{C_2}{C_1} = 715\text{ mL} \times \dfrac{0.100\text{ M}}{6.00\text{ M}} = 11.9\text{ mL of the } MgCl_2 \text{ solution (3 SFs)}$

c. $V_2 = 0.100\text{ L} \times \dfrac{1000\text{ mL}}{1\text{ L}} = 100.\text{ mL}$

$V_1 = V_2 \times \dfrac{C_2}{C_1} = 100.\text{ mL} \times \dfrac{0.150\text{ M}}{8.00\text{ M}} = 1.88\text{ mL of the KCl solution (3 SFs)}$

12.53 a. $50.0\text{ mL solution} \times \dfrac{1\text{ L solution}}{1000\text{ mL solution}} \times \dfrac{1.50\text{ mol KCl}}{1\text{ L solution}} = 0.0750\text{ mol of KCl}$

$0.0750\text{ mol KCl} \times \dfrac{1\text{ mol } PbCl_2}{2\text{ mol KCl}} \times \dfrac{278.1\text{ g } PbCl_2}{1\text{ mol } PbCl_2} = 10.4\text{ g of } PbCl_2 \text{ (3 SFs)}$

b. $50.0\text{ mL solution} \times \dfrac{1\text{ L solution}}{1000\text{ mL solution}} \times \dfrac{1.50\text{ mol KCl}}{1\text{ L solution}} = 0.0750\text{ mol of KCl}$

$0.0750\text{ mol KCl} \times \dfrac{1\text{ mol } Pb(NO_3)_2}{2\text{ mol KCl}} \times \dfrac{1\text{ L solution}}{2.00\text{ mol } Pb(NO_3)_2} \times \dfrac{1000\text{ mL solution}}{1\text{ L solution}}$

$= 18.8\text{ mL of } Pb(NO_3)_2 \text{ solution (3 SFs)}$

c. $30.0\text{ mL solution} \times \dfrac{1\text{ L solution}}{1000\text{ mL solution}} \times \dfrac{0.400\text{ mol } Pb(NO_3)_2}{1\text{ L solution}} \times \dfrac{2\text{ mol KCl}}{1\text{ mol } Pb(NO_3)_2}$

$= 0.0240\text{ mol of KCl}$

$20.0\text{ mL solution} \times \dfrac{1\text{ L solution}}{1000\text{ mL solution}} = 0.0200\text{ L of solution}$

$\text{molarity (M)} = \dfrac{\text{moles of solute}}{\text{liters of solution}} = \dfrac{0.0240\text{ mol KCl}}{0.0200\text{ L solution}} = 1.20\text{ M KCl solution (3 SFs)}$

12.55 a. $15.0\text{ g Mg} \times \dfrac{1\text{ mol Mg}}{24.31\text{ g Mg}} \times \dfrac{2\text{ mol HCl}}{1\text{ mol Mg}} \times \dfrac{1\text{ L solution}}{6.00\text{ mol HCl}} \times \dfrac{1000\text{ mL solution}}{1\text{ L solution}}$

$= 206\text{ mL of HCl solution (3 SFs)}$

b. $0.500\text{ L solution} \times \dfrac{2.00\text{ mol HCl}}{1\text{ L solution}} \times \dfrac{1\text{ mol } H_2}{2\text{ mol HCl}} \times \dfrac{22.4\text{ L } H_2\text{ (STP)}}{1\text{ mol } H_2}$

$= 11.2\text{ L of } H_2 \text{ at STP (3 SFs)}$

c. $n = \dfrac{PV}{RT} = \dfrac{(735\text{ mmHg})(5.20\text{ L})}{\left(\dfrac{62.4\text{ L}\cdot\text{mmHg}}{\text{mol}\cdot\text{K}}\right)(298\text{ K})} = 0.206\text{ mol of } H_2 \text{ gas}$

$0.206\text{ mol } H_2 \times \dfrac{2\text{ mol HCl}}{1\text{ mol } H_2} = 0.412\text{ mol of HCl}$

$$45.2\ \cancel{\text{mL solution}} \times \frac{1\ \text{L solution}}{1000\ \cancel{\text{mL solution}}} = 0.0452\ \text{L of solution}$$

$$\text{molarity (M)} = \frac{\text{moles of solute}}{\text{liters of solution}} = \frac{0.412\ \text{mol HCl}}{0.0452\ \text{L solution}} = 9.12\ \text{M HCl solution (3 SFs)}$$

12.57 a. A solution cannot be separated by a semipermeable membrane.

b. A suspension settles out upon standing.

12.59 a. When 1.0 mol of glycerol (a nonelectrolyte) dissolves in water, it does not dissociate into ions and so will only produce 1.0 mol of particles. Similarly, 2.0 mol of ethylene glycol (also a nonelectrolyte) dissolves as molcules to produce only 2.0 mol of particles in water. The ethylene glycol solution has more particles in 1.0 kg of water and will thus have a lower freezing point.

b. When 0.50 mol of the strong electrolyte KCl dissolves in water, it will produce 1.0 mol of particles because each formula unit of KCl dissociates to give two particles, K^+ and Cl^-. When 0.50 mol of the strong electrolyte $MgCl_2$ dissolves in water, it will produce 1.5 mol of particles because each formula unit of $MgCl_2$ dissociates to give three particles, Mg^{2+} and $2Cl^-$. Thus, a solution of 0.50 mol of $MgCl_2$ in 2.0 kg of water will have the lower freezing point.

12.61 a. $$325\ \cancel{\text{g CH}_3\text{—OH}} \times \frac{1\ \text{mol CH}_3\text{—OH}}{32.04\ \cancel{\text{g CH}_3\text{—OH}}} = 10.1\ \text{mol of CH}_3\text{—OH}$$

$$455\ \cancel{\text{g H}_2\text{O}} \times \frac{1\ \text{kg H}_2\text{O}}{1000\ \cancel{\text{g H}_2\text{O}}} = 0.455\ \text{kg of H}_2\text{O}$$

$$\text{molality}\ (m) = \frac{\text{moles of solute}}{\text{kilograms of water}} = \frac{10.1\ \text{mol CH}_3\text{—OH}}{0.455\ \text{kg H}_2\text{O}} = 22.2\ m\ \text{CH}_3\text{—OH solution (3 SFs)}$$

b. $$640.\ \cancel{\text{g C}_3\text{H}_8\text{O}_2} \times \frac{1\ \text{mol C}_3\text{H}_8\text{O}_2}{76.09\ \cancel{\text{g C}_3\text{H}_8\text{O}_2}} = 8.41\ \text{mol of C}_3\text{H}_8\text{O}_2$$

$$\text{molality}\ (m) = \frac{\text{moles of solute}}{\text{kilograms of water}} = \frac{8.41\ \text{mol C}_3\text{H}_8\text{O}_2}{1.22\ \text{kg H}_2\text{O}} = 6.89\ m\ \text{C}_3\text{H}_8\text{O}_2\ \text{solution (3 SFs)}$$

12.63 a. 22.2 m CH_3—OH solution; freezing-point lowering: $\Delta T_f = m\,K_f$; boiling-point elevation: $\Delta T_b = m\,K_b$

$$\Delta T_f = m\,K_f = 22.2\ \cancel{m} \times \frac{1.86\ ^\circ\text{C}}{1\ \cancel{m}} = 41.3\ ^\circ\text{C (3 SFs)}$$

freezing point = 0.0 °C − 41.3 °C = −41.3 °C

$$\Delta T_b = m\,K_b = 22.2\ \cancel{m} \times \frac{0.52\ ^\circ\text{C}}{1\ \cancel{m}} = 11.5\ ^\circ\text{C (3 SFs)}$$

boiling point = 100.0 °C + 11.5 °C = 111.5 °C

b. 6.89 m $C_3H_8O_2$ solution; freezing-point lowering: $\Delta T_f = m\,K_f$; boiling-point elevation: $\Delta T_b = m\,K_b$

$$\Delta T_f = m\,K_f = 6.89\ \cancel{m} \times \frac{1.86\ ^\circ\text{C}}{1\ \cancel{m}} = 12.8\ ^\circ\text{C (3 SFs)}$$

freezing point = 0.0 °C − 12.8 °C = −12.8 °C

$$\Delta T_b = m\,K_b = 6.89\ \cancel{m} \times \frac{0.52\ ^\circ\text{C}}{1\ \cancel{m}} = 3.58\ ^\circ\text{C (3 SFs)}$$

boiling point = 100.0 °C + 3.58 °C = 103.6 °C

12.65 Water will flow from a region of higher solvent concentration)which corresponds to a lower solute concentration) to a region of lower solvent concentration (which corresponds to a higher solute concentration).

a. B; the volume will rise as water flows into compartment B, which contains the 10% (m/v) starch solution.

b. B; the volume will rise as water flows into compartment B, which contains the 8% (m/v) albumin solution.

c. A; the volume will rise as water flows into compartment A, which contains the 10% (m/v) sucrose solution.

12.67 a. The 10% (m/v) starch solution has the higher solute concentration, more solute particles, and therefore the higher osmotic pressure.

b. Initially, water will flow out of the 1% (m/v) starch solution into the more concentrated 10% (m/v) starch solution.

c. The volume of the 10% (m/v) starch solution will increase due to inflow of water.

12.69 A red blood cell has the same osmotic pressure as a 5% (m/v) glucose solution or a 0.9% (m/v) NaCl solution. In a hypotonic solution (lower osmotic pressure), solvent flows from the hypotonic solution into the red blood cell. When a red blood cell is placed into a hypertonic solution (higher osmotic pressure), solvent (water) flows from the red blood cell to the hypertonic solution. Isotonic solutions have the same osmotic pressure, and a red blood cell in an isotonic solution will not change volume because the flow of solvent into and out of the cell is equal.

a. Distilled water is a hypotonic solution when compared with a red blood cell's contents.

b. A 1% (m/v) glucose solution is a hypotonic solution.

c. A 0.9% (m/v) NaCl solution is isotonic with a red blood cell's contents.

d. A 15% (m/v) glucose solution is a hypertonic solution.

12.71 a. 3; A nonelectrolyte will show no dissociation.

b. 1; A weak electrolyte will show some dissociation, producing a few ions, but mostly remaining as molecules.

c. 2; A strong electrolyte will be completely dissociated into ions.

12.73 a. 2; To halve the mass/volume percent (m/v), the volume would double.

b. 3; To go to one-fourth the mass/volume percent (m/v), the volume would be four times the initial volume.

12.75 a. Beaker 3; Solid silver chloride (AgCl) will precipitate when the two solutions are mixed.

b. $NaCl(aq) + AgNO_3(aq) \longrightarrow AgCl(s) + NaNO_3(aq)$

$\cancel{Na^+(aq)} + Cl^-(aq) + Ag^+(aq) + \cancel{NO_3^-(aq)} \longrightarrow AgCl(s) + \cancel{Na^+(aq)} + \cancel{NO_3^-(aq)}$

c. $Ag^+(aq) + Cl^-(aq) \longrightarrow AgCl(s)$ Net ionic equation

12.77 A "brine" saltwater solution has a high concentration of Na^+ and Cl^- ions and is hypertonic to the cucumber. The skin of the cucumber acts like a semipermeable membrane; therefore, water flows from the more dilute solution inside the cucumber into the more concentrated brine solution that surrounds it. The loss of water causes the cucumber to become a wrinkled pickle.

12.79 a. 2; Water will flow into the B (8% starch solution) compartment.

b. 1; Water will continue to flow equally in both directions; no change in volumes.

c. 3; Water will flow into the A (5% sucrose solution) compartment.

d. 2; Water will flow into the B (1% sucrose solution) compartment.

12.81 Because iodine is a nonpolar molecule, it will dissolve in hexane, a nonpolar solvent. Iodine does not dissolve in water because water is a polar solvent.

12.83 $80.0\ \cancel{\text{g NaCl}} \times \dfrac{100\ \text{g water}}{36.0\ \cancel{\text{g NaCl}}} = 222$ g of water needed (3 SFs)

12.85 At 20 °C, KNO_3 has a solubility of 32 g of KNO_3 in 100 g of H_2O.

a. 200. g of H_2O will dissolve:

$$200.\ \cancel{\text{g H}_2\text{O}} \times \frac{32\ \text{g KNO}_3}{100\ \cancel{\text{g H}_2\text{O}}} = 64\ \text{g of KNO}_3\ \ (2\ \text{SFs})$$

Because 32 g of KNO_3 is less than the maximum amount that can dissolve in 200. g of H_2O at 20 °C, the KNO_3 solution is unsaturated.

b. 50. g of H_2O will dissolve:

$$50.\ \cancel{\text{g H}_2\text{O}} \times \frac{32\ \text{g KNO}_3}{100\ \cancel{\text{g H}_2\text{O}}} = 16\ \text{g of KNO}_3\ \ (2\ \text{SFs})$$

Because 19 g of KNO_3 exceeds the maximum amount that can dissolve in 50. g of H_2O at 20 °C, the KNO_3 solution is saturated, and excess undissolved KNO_3 will be present on the bottom of the container.

c. 150. g of H_2O will dissolve:

$$150.\ \cancel{\text{g H}_2\text{O}} \times \frac{32\ \text{g KNO}_3}{100\ \cancel{\text{g H}_2\text{O}}} = 48\ \text{g of KNO}_3\ \ (2\ \text{SFs})$$

Because 68 g of KNO_3 exceeds the maximum amount that can dissolve in 150. g of H_2O at 20 °C, the KNO_3 solution is saturated, and excess undissolved KNO_3 will be present on the bottom of the container.

12.87 a. Most CO_3^{2-} salts are insoluble.
b. Na^+ salts are soluble.
c. Most PO_4^{3-} salts are insoluble.
d. Salts containing NH_4^+ ions are soluble.
e. Most O^{2-} salts are insoluble.
f. Most OH^- salts are insoluble.

12.89 a. Solid silver chloride (AgCl) forms:

$$AgNO_3(aq) + LiCl(aq) \longrightarrow AgCl(s) + LiNO_3(aq)$$
$$Ag^+(aq) + \cancel{NO_3^-(aq)} + \cancel{Li^+(aq)} + Cl^-(aq) \longrightarrow AgCl(s) + \cancel{Li^+(aq)} + \cancel{NO_3^-(aq)}$$
$$Ag^+(aq) + Cl^-(aq) \longrightarrow AgCl(s) \quad \text{Net ionic equation}$$

b. none; no solid forms; salts containing K^+ and Na^+ are soluble.

c. Solid barium sulfate ($BaSO_4$) forms:

$$Na_2SO_4(aq) + BaCl_2(aq) \longrightarrow BaSO_4(s) + 2NaCl(aq)$$
$$\cancel{2Na^+(aq)} + SO_4^{2-}(aq) + Ba^{2+}(aq) + \cancel{2Cl^-(aq)} \longrightarrow BaSO_4(s) + \cancel{2Na^+(aq)} + \cancel{2Cl^-(aq)}$$
$$Ba^{2+}(aq) + SO_4^{2-}(aq) \longrightarrow BaSO_4(s) \quad \text{Net ionic equation}$$

12.91 mass of solution = 15.5 g of Na_2SO_4 + 75.5 g of H_2O = 91.0 g of solution

$$\frac{15.5\ \text{g Na}_2\text{SO}_4}{91.0\ \text{g solution}} \times 100\% = 17.0\%\ (\text{m/m})\ \text{Na}_2\text{SO}_4\ \text{solution (3 SFs)}$$

12.93 $\dfrac{8.0\ \cancel{\text{g NaOH}}}{400.\ \cancel{\text{mL solution}}} \times \dfrac{1\ \text{mol NaOH}}{40.00\ \cancel{\text{g NaOH}}} \times \dfrac{1000\ \cancel{\text{mL solution}}}{1\ \text{L solution}} = 0.50$ M NaOH solution (2 SFs)

12.95 $4.5\ \cancel{\text{mL propyl alcohol}} \times \dfrac{100\ \text{mL solution}}{12\ \cancel{\text{mL propyl alcohol}}} = 38\ \text{mL of propyl alcohol solution (2 SFs)}$

12.97 **a.** $2.52\ \cancel{\text{L solution}} \times \dfrac{3.00\ \cancel{\text{mol KNO}_3}}{1\ \cancel{\text{L solution}}} \times \dfrac{101.11\ \text{g KNO}_3}{1\ \cancel{\text{mol KNO}_3}}$

$= 764\ \text{g of KNO}_3\ \text{(3 SFs)}$

b. $75.0\ \cancel{\text{mL solution}} \times \dfrac{1\ \cancel{\text{L solution}}}{1000\ \cancel{\text{mL solution}}} \times \dfrac{0.506\ \cancel{\text{mol Na}_2\text{SO}_4}}{1\ \cancel{\text{L solution}}} \times \dfrac{142.05\ \text{g Na}_2\text{SO}_4}{1\ \cancel{\text{mol Na}_2\text{SO}_4}}$

$= 5.39\ \text{g of Na}_2\text{SO}_4\ \text{(3 SFs)}$

c. $45.2\ \cancel{\text{mL solution}} \times \dfrac{1\ \cancel{\text{L solution}}}{1000\ \cancel{\text{mL solution}}} \times \dfrac{1.80\ \cancel{\text{mol HCl}}}{1\ \cancel{\text{L solution}}} \times \dfrac{36.46\ \text{g HCl}}{1\ \cancel{\text{mol HCl}}}$

$= 2.97\ \text{g of HCl (3 SFs)}$

12.99 $15.2\ \cancel{\text{g LiCl}} \times \dfrac{1\ \cancel{\text{mol LiCl}}}{42.39\ \cancel{\text{g LiCl}}} \times \dfrac{1\ \cancel{\text{L solution}}}{1.75\ \cancel{\text{mol LiCl}}} \times \dfrac{1000\ \text{mL solution}}{1\ \cancel{\text{L solution}}}$

$= 205\ \text{mL of LiCl solution (3 SFs)}$

12.101 $60.0\ \cancel{\text{g KNO}_3} \times \dfrac{1\ \cancel{\text{mol KNO}_3}}{101.11\ \cancel{\text{g KNO}_3}} \times \dfrac{1\ \text{L solution}}{2.50\ \cancel{\text{mol KNO}_3}} = 0.237\ \text{L of KNO}_3\ \text{solution (3 SFs)}$

12.103 $60.0\ \cancel{\text{mL solution}} \times \dfrac{1\ \cancel{\text{L solution}}}{1000\ \cancel{\text{mL solution}}} \times \dfrac{2.00\ \cancel{\text{mol Al(OH)}_3}}{1\ \cancel{\text{L solution}}} \times \dfrac{3\ \cancel{\text{mol HCl}}}{1\ \cancel{\text{mol Al(OH)}_3}}$

$\times \dfrac{1\ \cancel{\text{L solution}}}{6.00\ \cancel{\text{mol HCl}}} \times \dfrac{1000\ \text{mL HCl solution}}{1\ \cancel{\text{L solution}}}$

$= 60.0\ \text{mL of HCl solution (3 SFs)}$

12.105 Since this is a limiting reactant question, we must determine whether $CaCO_3$ or HCl is limiting (i.e., produces the smaller amount of product):

$500.0\ \cancel{\text{mg CaCO}_3} \times \dfrac{1\ \cancel{\text{g CaCO}_3}}{1000\ \cancel{\text{mg CaCO}_3}} \times \dfrac{1\ \cancel{\text{mol CaCO}_3}}{100.09\ \cancel{\text{g CaCO}_3}} \times \dfrac{1\ \cancel{\text{mol CO}_2}}{1\ \cancel{\text{mol CaCO}_3}}$

$\times \dfrac{22.4\ \text{L CO}_2\ \text{(STP)}}{1\ \cancel{\text{mol CO}_2}} = 0.112\ \text{L of CO}_2\ \text{at STP}$

$20.0\ \cancel{\text{mL HCl solution}} \times \dfrac{1\ \cancel{\text{L HCl solution}}}{1000\ \cancel{\text{mL HCl solution}}} \times \dfrac{0.100\ \cancel{\text{mol HCl}}}{1\ \cancel{\text{L HCl solution}}}$

$\times \dfrac{1\ \cancel{\text{mol CO}_2}}{2\ \cancel{\text{mol HCl}}} \times \dfrac{22.4\ \text{L CO}_2\ \text{(STP)}}{1\ \cancel{\text{mol CO}_2}}$

$= 0.0224\ \text{L of CO}_2\ \text{at STP (smaller amount of product)}$

Thus, HCl is the limiting reactant, and 0.0224 L of CO_2 at STP (3 SFs) can be produced.

12.107 $C_1V_1 = C_2V_2$

a. $C_2 = C_1 \times \dfrac{V_1}{V_2} = 0.200\ \text{M} \times \dfrac{25.0\ \cancel{\text{mL}}}{50.0\ \cancel{\text{mL}}} = 0.100\ \text{M NaBr solution (3 SFs)}$

b. $C_2 = C_1 \times \frac{V_1}{V_2} = 12.0\% \times \frac{15.0\ \cancel{mL}}{40.0\ \cancel{mL}} = 4.50\%\ (m/v)\ K_2SO_4$ solution (3 SFs)

c. $C_2 = C_1 \times \frac{V_1}{V_2} = 6.00\ M \times \frac{75.0\ \cancel{mL}}{255\ \cancel{mL}} = 1.76\ M\ NaOH$ solution (3 SFs)

12.109 $C_1V_1 = C_2V_2$

a. $V_2 = V_1 \times \frac{C_1}{C_2} = 25.0\ mL \times \frac{10.0\ \cancel{\%}}{2.50\ \cancel{\%}} = 100.\ mL$ of diluted HCl solution (3 SFs)

b. $V_2 = V_1 \times \frac{C_1}{C_2} = 25.0\ mL \times \frac{5.00\ \cancel{M}}{1.00\ \cancel{M}} = 125\ mL$ of diluted HCl solution (3 SFs)

c. $V_2 = V_1 \times \frac{C_1}{C_2} = 25.0\ mL \times \frac{6.00\ \cancel{M}}{0.500\ \cancel{M}} = 300.\ mL$ of diluted HCl solution (3 SFs)

12.111 **a.** Solid silver sulfate (Ag_2SO_4) forms:

$2AgNO_3(aq) + Na_2SO_4(aq) \longrightarrow Ag_2SO_4(s) + 2NaNO_3(aq)$

$2Ag^+(aq) + \cancel{2NO_3^-(aq)} + \cancel{2Na^+(aq)} + SO_4^{2-}(aq) \longrightarrow Ag_2SO_4(s) + \cancel{2Na^+(aq)} + \cancel{2NO_3^-(aq)}$

$2Ag^+(aq) + SO_4^{2-}(aq) \longrightarrow Ag_2SO_4(s)$ Net ionic equation

b. Solid lead(II) chloride ($PbCl_2$) forms:

$2KCl(aq) + Pb(NO_3)_2(aq) \longrightarrow PbCl_2(s) + 2KNO_3(aq)$

$\cancel{2K^+(aq)} + 2Cl^-(aq) + Pb^{2+}(aq) + \cancel{2NO_3^-(aq)} \longrightarrow PbCl_2(s) + \cancel{2K^+(aq)} + \cancel{2NO_3^-(aq)}$

$Pb^{2+}(aq) + 2Cl^-(aq) \longrightarrow PbCl_2(s)$ Net ionic equation

c. Solid calcium phosphate ($Ca_3(PO_4)_2$) forms:

$3CaCl_2(aq) + 2(NH_4)_3PO_4(aq) \longrightarrow Ca_3(PO_4)_2(s) + 6NH_4Cl(aq)$

$3Ca^{2+}(aq) + \cancel{6Cl^-(aq)} + \cancel{6NH_4^+(aq)} + 2PO_4^{3-}(aq) \longrightarrow Ca_3(PO_4)_2(s) + \cancel{6NH_4^+(aq)} + \cancel{6Cl^-(aq)}$

$3Ca^{2+}(aq) + 2PO_4^{3-}(aq) \longrightarrow Ca_3(PO_4)_2(s)$ Net ionic equation

d. Solid barium sulfate ($BaSO_4$) forms:

$K_2SO_4(aq) + BaCl_2(aq) \longrightarrow BaSO_4(s) + 2KCl(aq)$

$\cancel{2K^+(aq)} + SO_4^{2-}(aq) + Ba^{2+}(aq) + \cancel{2Cl^-(aq)} \longrightarrow BaSO_4(s) + \cancel{2K^+(aq)} + \cancel{2Cl^-(aq)}$

$Ba^{2+}(aq) + SO_4^{2-}(aq) \longrightarrow BaSO_4(s)$ Net ionic equation

12.113 mass of solution $= 70.0$ g of $HNO_3 + 130.0$ g of $H_2O = 200.0$ g of solution

a. $\frac{70.0\ g\ HNO_3}{200.0\ g\ solution} \times 100\% = 35.0\%\ (m/m)\ HNO_3$ solution (3 SFs)

b. $200.0\ \cancel{g\ solution} \times \frac{1\ mL\ solution}{1.21\ \cancel{g\ solution}} = 165\ mL$ of solution (3 SFs)

c. $\frac{70.0\ g\ HNO_3}{165\ mL\ solution} \times 100\% = 42.4\%\ (m/v)\ HNO_3$ solution (3 SFs)

d. $\frac{70.0\ \cancel{g\ HNO_3}}{165\ \cancel{mL\ solution}} \times \frac{1\ mol\ HNO_3}{63.02\ \cancel{g\ HNO_3}} \times \frac{1000\ \cancel{mL\ solution}}{1\ L\ solution} = 6.73\ M\ HNO_3$ solution (3 SFs)

12.115 **a.** mass of NaCl = 25.50 g − 24.10 g = 1.40 g of NaCl
mass of solution = 36.15 g − 24.10 g = 12.05 g of solution

$$\text{mass percent (m/m)} = \frac{1.40 \text{ g NaCl}}{12.05 \text{ g solution}} \times 100\% = 11.6\% \text{ (m/m) NaCl solution (3 SFs)}$$

b. $$\text{molarity (M)} = \frac{1.40 \text{ g NaCl}}{10.0 \text{ mL solution}} \times \frac{1 \text{ mol NaCl}}{58.44 \text{ g NaCl}} \times \frac{1000 \text{ mL solution}}{1 \text{ L solution}}$$

= 2.40 M NaCl solution (3 SFs)

c. $C_1V_1 = C_2V_2 \quad C_2 = C_1 \times \frac{V_1}{V_2} = 2.40 \text{ M} \times \frac{10.0 \text{ mL}}{60.0 \text{ mL}} = 0.400 \text{ M NaCl solution (3 SFs)}$

12.117 **Analyze the Problem: Using the Ideal Gas Law to Calculate Moles**

Property	*P*	*V*	*n*	*R*	*T*
Given	745 mmHg	4.20 L		$\frac{62.4 \text{ L} \cdot \text{mmHg}}{\text{mol} \cdot \text{K}}$	35 °C 35 °C + 273 = 308 K
Need			? mol		

$$n = \frac{PV}{RT} = \frac{(745 \text{ mmHg})(4.20 \text{ L})}{\left(\frac{62.4 \text{ L} \cdot \text{mmHg}}{\text{mol} \cdot \text{K}}\right)(308 \text{ K})} = 0.163 \text{ mol of } H_2 \text{ gas}$$

$$0.163 \text{ mol } H_2 \times \frac{2 \text{ mol HCl}}{1 \text{ mol } H_2} = 0.326 \text{ mol of HCl}$$

$$355 \text{ mL solution} \times \frac{1 \text{ L solution}}{1000 \text{ mL solution}} = 0.355 \text{ L of solution}$$

$$\text{molarity (M)} = \frac{\text{moles of solute}}{\text{liters of solution}} = \frac{0.326 \text{ mol HCl}}{0.355 \text{ L solution}} = 0.918 \text{ M HCl solution (3 SFs)}$$

12.119 **a.** When the strong electrolyte NaCl dissolves in water, each formula unit of NaCl dissociates to give two particles, Na^+ and Cl^-. Since freezing-point lowering depends on the number of particles in solution, only 0.60 mol of NaCl would be needed to have the same effect in 1 kg of water as 1.2 mol of the nonelectrolyte ethylene glycol.

b. When the strong electrolyte K_3PO_4 dissolves in water, each formula unit of K_3PO_4 dissociates to give four particles, three K^+ and one PO_4^{3-}. Since freezing-point lowering depends on the number of particles in solution, only 0.30 mol of K_3PO_4 would be needed to have the same effect in 1 kg of water as 1.2 mol of the nonelectrolyte ethylene glycol.

12.121 **a.** The boiling point of the NaCl solution is given as 101.04 °C. Since the boiling point for pure water is 100.0 °C, the ΔT_b is 101.04 °C − 100.0 °C = 1.04 °C.
Rearranging $\Delta T_b = m\,K_b$ to solve for m:

$$m = \frac{\Delta T_b}{K_b} = \frac{1.04 \text{ °C}}{0.52 \text{ °C}/m} = 2.0\,m \text{ NaCl solution (2 SFs)}$$

b. $\Delta T_f = m\,K_f = 2.0\,m \times \frac{1.86 \text{ °C}}{1\,m} = 3.7 \text{ °C (2 SFs)}$

freezing point = 0.0 °C − 3.7 °C = −3.7 °C

13

Reaction Rates and Chemical Equilibrium

Learning Goals

- Describe how temperature, concentration, and catalysts affect the rate of a reaction.
- Use the concept of reversible reactions to explain chemical equilibrium.
- Write the equilibrium constant expression for a reaction.
- Calculate the equilibrium constant for a reversible reaction given the concentrations of reactants and products at equilibrium.
- Use an equilibrium constant to predict the extent of reaction and to calculate equilibrium concentrations.
- Use Le Châtelier's principle to describe the changes made in equilibrium concentrations when reaction conditions change.
- Write the solubility product expression for a slightly soluble salt and calculate the K_{sp}; use the K_{sp} to determine the solubility.

Chapter Outline

Core Chemistry Skills

- Writing the Equilibrium Constant Expression (13.3)
- Calculating an Equilibrium Constant (13.3)
- Calculating Equilibrium Concentrations (13.4)
- Using Le Châtelier's Principle (13.5)
- Writing the Solubility Product Expression (13.6)
- Calculating the Solubility Product Constant (13.6)
- Calculating the Molar Solubility (13.6)

Answers and Solutions to Text Problems

13.1 **a.** The rate of a reaction indicates how fast the products form or how fast the reactants are used up.

b. At room temperature, more of the reactants will have the energy necessary to proceed to products (the activation energy) than at the lower temperature of the refrigerator, so the rate of formation of bread mold will be faster.

13.3 Adding $Br_2(g)$ molecules increases the concentration of reactants, which increases the number of collisions that take place between the reactants.

13.5 **a.** Adding more reactant increases the number of collisions that take place between the reactants, which increases the reaction rate.
b. Raising the temperature increases the kinetic energy of the reactant molecules, which increases the number of collisions and makes more collisions effective. The rate of reaction will be increased.
c. Adding a catalyst lowers the energy of activation, which increases the reaction rate.
d. Removing a reactant decreases the number of collisions that take place between the reactants, which decreases the reaction rate.

13.7 A reversible reaction is one in which a forward reaction converts reactants to products, while a reverse reaction converts products to reactants.

13.9 **a.** When the rate of the forward reaction is faster than the rate of the reverse reaction, the process is not at equilibrium.
b. When the concentrations of the reactants and the products do not change, the process is at equilibrium.
c. When the rate of either the forward or reverse reaction does not change, the process is at equilibrium.

13.11 In the expression for K_c, the products are divided by the reactants, with each concentration raised to a power equal to its coefficient in the balanced chemical equation:

a. $K_c = \dfrac{[CS_2][H_2]^4}{[CH_4][H_2S]^2}$ **b.** $K_c = \dfrac{[N_2][O_2]}{[NO]^2}$ **c.** $K_c = \dfrac{[CS_2][O_2]^4}{[SO_3]^2[CO_2]}$

13.13 **a.** only one state (gas) is present; homogeneous equilibrium
b. solid and gaseous states present; heterogeneous equilibrium
c. only one state (gas) is present; homogeneous equilibrium
d. gaseous, liquid, and solid states present; heterogeneous equilibrium

13.15 **a.** $K_c = \dfrac{[O_2]^3}{[O_3]^2}$ **b.** $K_c = [CO_2][H_2O]$

c. $K_c = \dfrac{[H_2]^3[CO]}{[CH_4][H_2O]}$ **d.** $K_c = \dfrac{[H_2]^2}{[HCl]^4}$

13.17 $K_c = \dfrac{[NO_2]^2}{[N_2O_4]} = \dfrac{[0.21]^2}{[0.030]} = 1.5$ (2 SFs)

13.19 $K_c = \dfrac{[CH_4][H_2O]}{[CO][H_2]^3} = \dfrac{[1.8][2.0]}{[0.51][0.30]^3} = 260$ (2 SFs)

13.21 $K_c = \dfrac{[CO_2]}{[CO]} = \dfrac{[0.052]}{[0.20]} = 0.26$ (2 SFs)

13.23 **a.** A large K_c value indicates that the equilibrium mixture contains mostly products.
b. A large K_c value indicates that the equilibrium mixture contains mostly products.
c. A small K_c value indicates that the equilibrium mixture contains mostly reactants.

13.25 $K_c = \dfrac{[HI]^2}{[H_2][I_2]} = 54$

Rearrange the K_c expression to solve for $[H_2]$ and substitute in known values.

$$[H_2] = \frac{[HI]^2}{K_c[I_2]} = \frac{[0.030]^2}{54[0.015]} = 1.1 \times 10^{-3}\ \text{M (2 SFs)}$$

13.27 $K_c = \dfrac{[NO]^2[Br_2]}{[NOBr]^2} = 2.0$

Rearrange the K_c expression to solve for [NOBr] and substitute in known values.

$$[NOBr]^2 = \frac{[NO]^2[Br_2]}{K_c} = \frac{[2.0]^2[1.0]}{2.0} = 2.0$$

Take the square root of both sides of the equation.

$$[NOBr] = \sqrt{2.0} = 1.4\ \text{M (2 SFs)}$$

13.29 **a.** Adding more reactant shifts equilibrium in the direction of the product.
b. Adding more product shifts equilibrium in the direction of the reactants.
c. Raising the temperature of an endothermic reaction shifts equilibrium in the direction of the product.
d. Increasing the volume of the container shifts the equilibrium in the direction of the reactants, which has more moles of gas.
e. No shift in equilibrium occurs when a catalyst is added.

13.31 **a.** Adding more reactant shifts equilibrium in the direction of the product.
b. Raising the temperature of an endothermic reaction shifts equilibrium in the direction of the product to remove heat.
c. Removing product shifts equilibrium in the direction of the product.
d. No shift in equilibrium occurs when a catalyst is added.
e. Removing reactant shifts equilibrium in the direction of the reactants.

13.33 **a.** $BaCO_3(s) \rightleftharpoons Ba^{2+}(aq) + CO_3^{2-}(aq);\ K_{sp} = [Ba^{2+}][CO_3^{2-}]$
b. $CaF_2(s) \rightleftharpoons Ca^{2+}(aq) + 2F^-(aq);\ K_{sp} = [Ca^{2+}][F^-]^2$
c. $Ag_3PO_4(s) \rightleftharpoons 3Ag^+(aq) + PO_4^{3-}(aq);\ K_{sp} = [Ag^+]^3[PO_4^{3-}]$

13.35 $BaSO_4(s) \rightleftharpoons Ba^{2+}(aq) + SO_4^{2-}(aq);$
$K_{sp} = [Ba^{2+}][SO_4^{2-}] = [1 \times 10^{-5}][1 \times 10^{-5}] = 1 \times 10^{-10}$ (1 SF)

13.37 $Ag_2CO_3(s) \rightleftharpoons 2Ag^+(aq) + CO_3^{2-}(aq);$
$K_{sp} = [Ag^+]^2[CO_3^{2-}] = [2.6 \times 10^{-4}]^2[1.3 \times 10^{-4}] = 8.8 \times 10^{-12}$ (2 SFs)

13.39 $CuI(s) \rightleftharpoons Cu^+(aq) + I^-(aq);\ K_{sp} = [Cu^+][I^-]$

Substitute S for the molarity of each ion into the K_{sp} expression.

$K_{sp} = S \times S = S^2 = 1 \times 10^{-12}$

Calculate the molar solubility (S) by taking the square root of both sides of the equation.

$S = \sqrt{1 \times 10^{-12}} = 1 \times 10^{-6}$ M

$\therefore [Cu^+] = [I^-] = 1 \times 10^{-6}$ M (1 SF)

13.41 $AgCl(s) \rightleftarrows Ag^+(aq) + Cl^-(aq)$; $K_{sp} = [Ag^+][Cl^-]$

Rearrange the K_{sp} expression to solve for $[Cl^-]$ and substitute in known values.

$$[Cl^-] = \frac{K_{sp}}{[Ag^+]} = \frac{1.8 \times 10^{-10}}{[2.0 \times 10^{-3}]} = 9.0 \times 10^{-8}\text{ M (2 SFs)}$$

13.43 **a.** $K_c = \dfrac{[CO_2][H_2O]^2}{[CH_4][O_2]^2}$

b. $K_c = \dfrac{[N_2]^2[H_2O]^6}{[NH_3]^4[O_2]^3}$

c. $K_c = \dfrac{[CH_4]}{[H_2]^2}$

13.45 At equilibrium, the diagram shows mostly products and a few reactants, so the equilibrium constant K_c for the reaction would have a large value.

13.47 T_2 is lower than T_1. This would cause the exothermic reaction shown to shift in the direction of the products to add heat; more product is seen in the T_2 diagram.

13.49 **a.** Raising the temperature of an exothermic reaction shifts equilibrium in the direction of the reactants.

b. Decreasing volume favors the side of the reaction with fewer moles of gas, so there is a shift in the direction of the products.

c. Adding a catalyst does not shift equilibrium.

d. Adding more reactant shifts equilibrium in the direction of the products.

13.51 **a.** A large K_c value indicates that the equilibrium mixture contains mostly products.

b. A K_c value close to 1 indicates that the equilibrium mixture contains both reactants and products.

c. A small K_c value indicates that the equilibrium mixture contains mostly reactants.

d. A K_c value close to 1 indicates that the equilibrium mixture contains both reactants and products.

13.53 **a.** $K_c = \dfrac{[N_2][H_2]^3}{[NH_3]^2}$

b. $K_c = \dfrac{[3.0][0.50]^3}{[0.20]^2} = 9.4\text{ (2 SFs)}$

13.55 $K_c = \dfrac{[N_2O_4]}{[NO_2]^2} = 5.0$

Rearrange the K_c expression to solve for $[N_2O_4]$ and substitute in known values.

$$[N_2O_4] = K_c[NO_2]^2 = 5.0[0.50]^2 = 1.3\text{ M (2 SFs)}$$

13.57 **a.** When the reactant $[O_2]$ increases, the rate of the forward reaction increases to shift the equilibrium in the direction of the products.

b. When the product $[O_2]$ increases, the rate of the reverse reaction increases to shift the equilibrium in the direction of the reactants.

c. When the reactant $[O_2]$ increases, the rate of the forward reaction increases to shift the equilibrium in the direction of the products.

d. When the product $[O_2]$ increases, the rate of the reverse reaction increases to shift the equilibrium in the direction of the reactants.

13.59 Decreasing the volume of an equilibrium mixture shifts the equilibrium in the direction of the side of the reaction that has the fewer number of moles of gas. No shift occurs when there are an equal number of moles of gas on both sides of the equation.

a. With 3 mol of gas on the reactant side and 2 mol of gas on the product side, decreasing the volume will shift equilibrium in the direction of products.

b. With 2 mol of gas on the reactant side and 3 mol of gas on the product side, decreasing the volume will shift equilibrium in the direction of reactants.

c. With 6 mol of gas on reactant side and 0 mol of gas on the product side, decreasing the volume will shift equilibrium in the direction of products.

d. With 4 mol of gas on the reactant side and 5 mol of gas on the product side, decreasing the volume will shift equilibrium in the direction of reactants.

13.61 a. $CuCO_3(s) \rightleftharpoons Cu^{2+}(aq) + CO_3^{2-}(aq); K_{sp} = [Cu^{2+}][CO_3^{2-}]$

b. $PbF_2(s) \rightleftharpoons Pb^{2+}(aq) + 2F^-(aq); K_{sp} = [Pb^{2+}][F^-]^2$

c. $Fe(OH)_3(s) \rightleftharpoons Fe^{3+}(aq) + 3OH^-(aq); K_{sp} = [Fe^{3+}][OH^-]^3$

13.63 $FeS(s) \rightleftharpoons Fe^{2+}(aq) + S^{2-}(aq); K_{sp} = [Fe^{2+}][S^{2-}]$

$K_{sp} = [7.7 \times 10^{-10}][7.7 \times 10^{-10}] = 5.9 \times 10^{-19}$ (2 SFs)

13.65 $Mn(OH)_2(s) \rightleftharpoons Mn^{2+}(aq) + 2OH^-(aq); K_{sp} = [Mn^{2+}][OH^-]^2$

$K_{sp} = [3.7 \times 10^{-5}][7.4 \times 10^{-5}]^2 = 2.0 \times 10^{-13}$ (2 SFs)

13.67 $CdS(s) \rightleftharpoons Cd^{2+}(aq) + S^{2-}(aq); K_{sp} = [Cd^{2+}][S^{2-}]$

Substitute S for the molarity of each ion into the K_{sp} expression.

$K_{sp} = S \times S = 1.0 \times 10^{-24}$

Calculate the molar solubility (S) by taking the square root of both sides of the equation.

$S = \sqrt{1.0 \times 10^{-24}} = 1.0 \times 10^{-12}$ M

$\therefore [Cd^{2+}] = [S^{2-}] = 1.0 \times 10^{-12}$ M (2 SFs)

13.69 $BaSO_4(s) \rightleftharpoons Ba^{2+}(aq) + SO_4^{2-}(aq); K_{sp} = [Ba^{2+}][SO_4^{2-}]$

Rearrange the K_{sp} expression to solve for $[SO_4^{2-}]$ and substitute in known values.

$$S = [SO_4^{2-}] = \frac{K_{sp}}{[Ba^{2+}]} = \frac{1.1 \times 10^{-10}}{[1.0 \times 10^{-3}]} = 1.1 \times 10^{-7} \text{ M (2 SFs)}$$

13.71 a. A small K_c value indicates that the equilibrium mixture contains mostly reactants.

b. A large K_c value indicates that the equilibrium mixture contains mostly products.

13.73 a. $K_c = \dfrac{[NO]^2[Br_2]}{[NOBr]^2}$

b. When the concentrations are substituted into the expression, the result is 1.0, which is not equal to K_c (2.0). Therefore, the system is not at equilibrium.

$$K_c = \frac{[NO]^2[Br_2]}{[NOBr]^2} = \frac{[1.0]^2[1.0]}{[1.0]^2} = 1.0 \text{ (2 SFs)}$$

c. Since the calculated value in part **b** is less than K_c, the rate of the forward reaction will initially increase.

d. When the system has reestablished equilibrium, the $[Br_2]$ and $[NO]$ will have increased, and the $[NOBr]$ will have decreased.

13.75 **a.** When more product molecules are added to an equilibrium mixture, the system shifts in the direction of the reactants. This will cause a decrease in the equilibrium concentration of the product CO.

b. When the temperature is increased for an endothermic reaction, the system shifts in the direction of the products to remove heat. This will cause an increase in the equilibrium concentration of the product CO.

c. Increasing the volume of the reaction container favors the side of the reaction with the greater number of moles of gas, so this system shifts in the direction of the products. This will cause an increase in the equilibrium concentration of the product CO.

d. Decreasing the volume of the reaction container favors the side of the reaction with the fewer moles of gas, so this system shifts in the direction of the reactants. This will cause a decrease in the equilibrium concentration of the product CO.

e. When a catalyst is added, the rates of both forward and reverse reactions increase; the equilibrium position does not change. No change will be observed in the equilibrium concentration of the product CO.

13.77 Increasing the volume of an equilibrium mixture shifts the system toward the side of the reaction that has the greater number of moles of gas; decreasing the volume shifts the equilibrium toward the side of the reaction that has the fewer moles of gas.

a. Since the product side has the greater number of moles of gas, increasing the volume of the container will increase the yield of products.

b. Since the product side has the greater number of moles of gas, increasing the volume of the container will increase the yield of products.

c. Since the product side has the fewer moles of gas, decreasing the volume of the container will increase the yield of products.

13.79 $Mg(OH)_2(s) \rightleftharpoons Mg^{2+}(aq) + 2OH^-(aq)$; $K_{sp} = [Mg^{2+}][OH^-]^2$

$$\frac{9.7 \times 10^{-3}\ \cancel{\text{g Mg(OH)}_2}}{1\ \text{L solution}} \times \frac{1\ \text{mol Mg(OH)}_2}{58.33\ \cancel{\text{g Mg(OH)}_2}} = 1.7 \times 10^{-4}\ \text{M Mg(OH)}_2\ \text{solution}$$

$$\frac{1.7 \times 10^{-4}\ \cancel{\text{mol Mg(OH)}_2}}{1\ \text{L solution}} \times \frac{1\ \text{mol Mg}^{2+}}{1\ \cancel{\text{mol Mg(OH)}_2}} = 1.7 \times 10^{-4}\ \text{M Mg}^{2+} = [\text{Mg}^{2+}]$$

$$\frac{1.7 \times 10^{-4}\ \cancel{\text{mol Mg(OH)}_2}}{1\ \text{L solution}} \times \frac{2\ \text{mol OH}^-}{1\ \cancel{\text{mol Mg(OH)}_2}} = 3.4 \times 10^{-4}\ \text{M OH}^- = [\text{OH}^-]$$

$$K_{sp} = [\text{Mg}^{2+}][\text{OH}^-]^2 = [1.7 \times 10^{-4}][3.4 \times 10^{-4}]^2 = 2.0 \times 10^{-11}\ (2\ \text{SFs})$$

14

Acids and Bases

Learning Goals

- Describe and name Arrhenius acids and bases, and organic acids.
- Identify conjugate acid–base pairs for Brønsted–Lowry acids and bases.
- Write equations for the dissociation of strong and weak acids; identify the direction of reaction.
- Write the expression for the dissociation constant of a weak acid or weak base.
- Use the ion product of water to calculate the $[H_3O^+]$ and $[OH^-]$ in an aqueous solution.
- Calculate pH from $[H_3O^+]$; given the pH, calculate $[H_3O^+]$ and $[OH^-]$ of a solution.
- Write balanced equations for reactions of acids with metals, carbonates, and bases.
- Calculate the molarity or volume of an acid or base from titration information.
- Describe the role of buffers in maintaining the pH of a solution; calculate the pH of a buffer.

Chapter Outline

Key Math Skills

- Calculating pH from $[H_3O^+]$ (14.6)
- Calculating $[H_3O^+]$ from pH (14.6)

Core Chemistry Skills

- Identifying Conjugate Acid–Base Pairs (14.2)
- Calculating $[H_3O^+]$ and $[OH^-]$ in Solutions (14.5)
- Writing Equations for Reactions of Acids and Bases (14.7)
- Calculating Molarity or Volume of an Acid or Base in a Titration (14.8)
- Calculating the pH of a Buffer (14.9)

Answers and Solutions to Text Problems

14.1 **a.** Acids taste sour.
b. Acids neutralize bases.
c. Acids produce H^+ ions in water.
d. Barium hydroxide is the name of a base.
e. Both acids and bases are electrolytes.

14.3 Acids containing a simple nonmetal anion use the prefix *hydro*, followed by the name of the anion with its *ide* ending changed to *ic acid*. When the anion is an oxygen-containing polyatomic ion, the *ate* ending of the polyatomic anion is replaced with *ic acid*. Acids with one oxygen less than the common *ic acid* name are named as *ous acids*. Carboxylic acids are named using IUPAC rules where the alkane name of the longest carbon chain containing the carboxyl functional group has the *e* ending replaced with *oic acid*; many carboxylic acids also have common names derived from their natural sources. Bases are named as ionic compounds containing hydroxide anions.
a. hydrochloric acid **b.** calcium hydroxide **c.** carbonic acid
d. nitric acid **e.** sulfurous acid **f.** bromous acid
g. ethanoic acid (IUPAC); acetic acid (common)

14.5 **a.** $RbOH$ **b.** HF **c.** HCH_2O
d. $LiOH$ **e.** NH_4OH **f.** HIO_4

14.7 A Brønsted–Lowry acid donates a hydrogen ion (H^+), whereas a Brønsted–Lowry base accepts a hydrogen ion.
a. HI is the acid (H^+ donor); H_2O is the base (H^+ acceptor).
b. H_2O is the acid (H^+ donor); F^- is the base (H^+ acceptor).
c. H_2S is the acid (H^+ donor); $CH_3—CH_2—NH_2$ is the base (H^+ acceptor).

14.9 To form the conjugate base, remove a hydrogen ion (H^+) from the acid.
a. F^- **b.** OH^- **c.** HPO_3^{2-}
d. SO_4^{2-} **e.** ClO_2^-

14.11 To form the conjugate acid, add a hydrogen ion (H^+) to the base.
a. HCO_3^- **b.** H_3O^+ **c.** H_3PO_4
d. HBr **e.** $HClO_4$

14.13 The conjugate acid is an H^+ donor, and the conjugate base is an H^+ acceptor.
a. In the reaction, the acid H_2CO_3 donates an H^+ to the base H_2O. The conjugate acid–base pairs are H_2CO_3/HCO_3^- and H_3O^+/H_2O.
b. In the reaction, the acid NH_4^+ donates an H^+ to the base H_2O. The conjugate acid–base pairs are NH_4^+/NH_3 and H_3O^+/H_2O.
c. In the reaction, the acid HCN donates an H^+ to the base NO_2^-. The conjugate acid–base pairs are HCN/CN^- and HNO_2/NO_2^-.
d. In the reaction, the acid HF donates an H^+ to the base CHO_2^-. The conjugate acid–base pairs are HF/F^- and $HCHO_2/CHO_2^-$.

14.15 $NH_4^+(aq) + H_2O(l) \rightleftharpoons NH_3(aq) + H_3O^+(aq)$

14.17 A strong acid is a good H^+ donor, whereas its conjugate base is a poor H^+ acceptor.

14.19 Use Table 14.4 to answer (the stronger acid will be closer to the top of the table).
a. HBr is the stronger acid.
b. HSO_4^- is the stronger acid.
c. H_2CO_3 is the stronger acid.

14.21 Use Table 14.4 to answer (the weaker acid will be closer to the bottom of the table).
a. HSO_4^- is the weaker acid.
b. HNO_2 is the weaker acid.
c. HCO_3^- is the weaker acid.

14.23 a. From Table 14.4, we see that H_2CO_3 is a weaker acid than H_3O^+ and that H_2O is a weaker base than HCO_3^-. Thus, the solution will contain mostly reactants at equilibrium.
b. From Table 14.4, we see that NH_4^+ is a weaker acid than H_3O^+ and that H_2O is a weaker base than NH_3. Thus, the solution will contain mostly reactants at equilibrium.
c. From Table 14.4, we see that NH_4^+ is a weaker acid than HNO_2 and that NO_2^- is a weaker base than NH_3. Thus, the solution will contain mostly products at equilibrium.

14.25 $NH_4^+(aq) + SO_4^{2-}(aq) \rightleftarrows NH_3(aq) + HSO_4^-(aq)$
This equilibrium contains mostly reactants because NH_4^+ is a weaker acid than HSO_4^- and because SO_4^{2-} is a weaker base than NH_3.

14.27 a. True
b. False; a strong acid has a large value of K_a.
c. False; a strong acid has a weak conjugate base.
d. True
e. False; a strong acid is completely ionized in aqueous solution.

14.29 The smaller the K_a value, the weaker the acid. The weaker acid has the stronger conjugate base.
a. H_2SO_3, which has a larger K_a value than HS^-, is the stronger acid.
b. The conjugate base HSO_3^- is formed by removing an H^+ from the acid H_2SO_3.
c. The stronger acid, H_2SO_3, has the weaker conjugate base, HSO_3^-.
d. The weaker acid, HS^-, has the stronger conjugate base, S^{2-}.
e. The stronger acid, H_2SO_3, dissociates more and produces more ions.

14.31 $H_3PO_4(aq) + H_2O(l) \rightleftarrows H_3O^+(aq) + H_2PO_4^-(aq)$
The K_a expression is the ratio of the [products] divided by the [reactants] with $[H_2O]$ considered constant and part of the K_a:

$$K_a = \frac{[H_3O^+][H_2PO_4^-]}{[H_3PO_4]} = 7.5 \times 10^{-3}$$

14.33 In pure water, a small fraction of the water molecules break apart to form H^+ and OH^-. The H^+ combines with H_2O to form H_3O^+. Every time an H^+ is formed, an OH^- is also formed. Therefore, the concentration of the two must be equal in pure water.

14.35 In an acidic solution, the $[H_3O^+]$ is greater than the $[OH^-]$, which means that at 25 °C, the $[H_3O^+]$ is greater than 1.0×10^{-7} M and the $[OH^-]$ is less than 1.0×10^{-7} M.

14.37 The value of $K_w = [H_3O^+][OH^-] = 1.0 \times 10^{-14}$ at 25 °C.

If $[H_3O^+]$ needs to be calculated from $[OH^-]$, then rearranging the K_w for $[H_3O^+]$ gives

$$[H_3O^+] = \frac{1.0 \times 10^{-14}}{[OH^-]}.$$

If $[OH^-]$ needs to be calculated from $[H_3O^+]$, then rearranging the K_w for $[OH^-]$ gives

$$[OH^-] = \frac{1.0 \times 10^{-14}}{[H_3O^+]}.$$

A neutral solution has $[OH^-] = [H_3O^+]$. If the $[OH^-] > [H_3O^+]$, the solution is basic; if the $[H_3O^+] > [OH^-]$, the solution is acidic.

a. $[OH^-] = \dfrac{1.0 \times 10^{-14}}{[H_3O^+]} = \dfrac{1.0 \times 10^{-14}}{[2.0 \times 10^{-5}]} = 5.0 \times 10^{-10}$ M;

since $[H_3O^+] > [OH^-]$, the solution is acidic.

b. $[OH^-] = \dfrac{1.0 \times 10^{-14}}{[H_3O^+]} = \dfrac{1.0 \times 10^{-14}}{[1.4 \times 10^{-9}]} = 7.1 \times 10^{-6}$ M;

since $[OH^-] > [H_3O^+]$, the solution is basic.

c. $[H_3O^+] = \dfrac{1.0 \times 10^{-14}}{[OH^-]} = \dfrac{1.0 \times 10^{-14}}{[8.0 \times 10^{-3}]} = 1.3 \times 10^{-12}$ M;

since $[OH^-] > [H_3O^+]$, the solution is basic.

d. $[H_3O^+] = \dfrac{1.0 \times 10^{-14}}{[OH^-]} = \dfrac{1.0 \times 10^{-14}}{[3.5 \times 10^{-10}]} = 2.9 \times 10^{-5}$ M;

since $[H_3O^+] > [OH^-]$, the solution is acidic.

14.39 The value of $K_w = [H_3O^+][OH^-] = 1.0 \times 10^{-14}$ at 25 °C.

When $[OH^-]$ is known, the $[H_3O^+]$ can be calculated by rearranging the K_w for $[H_3O^+]$:

$$[H_3O^+] = \frac{1.0 \times 10^{-14}}{[OH^-]}$$

a. $[H_3O^+] = \dfrac{1.0 \times 10^{-14}}{[OH^-]} = \dfrac{1.0 \times 10^{-14}}{[1.0 \times 10^{-9}]} = 1.0 \times 10^{-5}$ M (2 SFs)

b. $[H_3O^+] = \dfrac{1.0 \times 10^{-14}}{[OH^-]} = \dfrac{1.0 \times 10^{-14}}{[1.0 \times 10^{-6}]} = 1.0 \times 10^{-8}$ M (2 SFs)

c. $[H_3O^+] = \dfrac{1.0 \times 10^{-14}}{[OH^-]} = \dfrac{1.0 \times 10^{-14}}{[2.0 \times 10^{-5}]} = 5.0 \times 10^{-10}$ M (2 SFs)

d. $[H_3O^+] = \dfrac{1.0 \times 10^{-14}}{[OH^-]} = \dfrac{1.0 \times 10^{-14}}{[4.0 \times 10^{-13}]} = 2.5 \times 10^{-2}$ M (2 SFs)

14.41 The value of $K_w = [H_3O^+][OH^-] = 1.0 \times 10^{-14}$ at 25 °C.

When $[H_3O^+]$ is known, the $[OH^-]$ can be calculated by rearranging the K_w for $[OH^-]$:

$$[OH^-] = \frac{1.0 \times 10^{-14}}{[H_3O^+]}$$

a. $[OH^-] = \dfrac{1.0 \times 10^{-14}}{[H_3O^+]} = \dfrac{1.0 \times 10^{-14}}{[1.0 \times 10^{-3}]} = 1.0 \times 10^{-11}$ M (2 SFs)

b. $[OH^-] = \dfrac{1.0 \times 10^{-14}}{[H_3O^+]} = \dfrac{1.0 \times 10^{-14}}{[5.0 \times 10^{-6}]} = 2.0 \times 10^{-9}$ M (2 SFs)

c. $[OH^-] = \dfrac{1.0 \times 10^{-14}}{[H_3O^+]} = \dfrac{1.0 \times 10^{-14}}{[1.8 \times 10^{-12}]} = 5.6 \times 10^{-3}$ M (2 SFs)

d. $[OH^-] = \dfrac{1.0 \times 10^{-14}}{[H_3O^+]} = \dfrac{1.0 \times 10^{-14}}{[4.0 \times 10^{-13}]} = 2.5 \times 10^{-2}$ M (2 SFs)

14.43 In a neutral solution, the $[H_3O^+] = 1.0 \times 10^{-7}$ M at 25 °C.
pH $= -\log[H_3O^+] = -\log[1.0 \times 10^{-7}] = 7.00$. The pH value contains two *decimal places*, which represent the two significant figures in the coefficient 1.0.

14.45 An acidic solution has a pH less than 7.0 (or a pOH greater than 7.0). A basic solution has a pH greater than 7.0 (or a pOH less than 7.0). A neutral solution has a pH (or pOH) equal to 7.0.

a. basic (pH 7.38 $>$ 7.0) **b.** acidic (pH 2.8 $<$ 7.0) **c.** basic (pOH 2.8 $<$ 7.0)
d. acidic (pH 5.52 $<$ 7.0) **e.** acidic (pH 4.2 $<$ 7.0) **f.** basic (pH 7.6 $>$ 7.0)

14.47 Since pH is a logarithmic scale, an increase or decrease of 1 pH unit changes the $[H_3O^+]$ by a factor of 10. Thus, a pH of 3 ($[H_3O^+] = 10^{-3}$ M, or 0.001 M) is ten times more acidic than a pH of 4 ($[H_3O^+] = 10^{-4}$ M, or 0.0001 M).

14.49 $pH = -\log[H_3O^+]$

Since the value of $K_w = [H_3O^+][OH^-] = 1.0 \times 10^{-14}$ at 25 °C, if $[H_3O^+]$ needs to be calculated from $[OH^-]$, rearranging the K_w for $[H_3O^+]$ gives $[H_3O^+] = \dfrac{1.0 \times 10^{-14}}{[OH^-]}$.

a. $pH = -\log[H_3O^+] = -\log[1.0 \times 10^{-4}] = 4.00$ (2 SFs on the right of the decimal point)

b. $pH = -\log[H_3O^+] = -\log[3.0 \times 10^{-9}] = 8.52$ (2 SFs on the right of the decimal point)

c. $[H_3O^+] = \dfrac{1.0 \times 10^{-14}}{[1.0 \times 10^{-5}]} = 1.0 \times 10^{-9}$ M

$pH = -\log[1.0 \times 10^{-9}] = 9.00$ (2 SFs on the right of the decimal point)

d. $[H_3O^+] = \dfrac{1.0 \times 10^{-14}}{[2.5 \times 10^{-11}]} = 4.0 \times 10^{-4}$ M

$pH = -\log[4.0 \times 10^{-4}] = 3.40$ (2 SFs on the right of the decimal point)

e. $pH = -\log[H_3O^+] = -\log[6.7 \times 10^{-8}] = 7.17$ (2 SFs on the right of the decimal point)

f. $[H_3O^+] = \dfrac{1.0 \times 10^{-14}}{[8.2 \times 10^{-4}]} = 1.2 \times 10^{-11}$ M

$pH = -\log[1.2 \times 10^{-11}] = 10.92$ (2 SFs on the right of the decimal point)

14.51 On a calculator, pH is calculated by entering $-log$, followed by the coefficient *EE* (*EXP*) key and the power of 10 followed by the change sign (+/−) key. On some calculators, the concentration is entered first (coefficient *EXP* –power) followed by *log* and +/− key.

$[H_3O^+] = \dfrac{1.0 \times 10^{-14}}{[OH^-]}$; $[OH^-] = \dfrac{1.0 \times 10^{-14}}{[H_3O^+]}$; $pH = -\log[H_3O^+]$; $pOH = -\log[OH^-]$

$[H_3O^+]$	$[OH^-]$	pH	pOH	Acidic, Basic, or Neutral?
1.0×10^{-8} M	1.0×10^{-6} M	8.00	6.00	Basic
3.2×10^{-4} M	3.1×10^{-11} M	3.49	10.51	Acidic
2.8×10^{-5} M	3.6×10^{-10} M	4.55	9.45	Acidic
1.0×10^{-12} M	1.0×10^{-2} M	12.00	2.00	Basic

$[H_3O^+]$ and $[OH^-]$ all have 2 SFs here, so all pH/pOH values have 2 decimal places on the right of the decimal point.

14.53 Acids react with active metals to form $H_2(g)$ and a salt of the metal. The reaction of acids with carbonates yields CO_2, H_2O, and a salt. In a neutralization reaction, an acid and a base react to form H_2O and a salt.

a. $ZnCO_3(s) + 2HBr(aq) \longrightarrow CO_2(g) + H_2O(l) + ZnBr_2(aq)$
b. $Zn(s) + 2HCl(aq) \longrightarrow H_2(g) + ZnCl_2(aq)$
c. $HCl(aq) + NaHCO_3(s) \longrightarrow CO_2(g) + H_2O(l) + NaCl(aq)$
d. $H_2SO_4(aq) + Mg(OH)_2(s) \longrightarrow 2H_2O(l) + MgSO_4(aq)$

14.55 In balancing a neutralization equation, the number of H^+ and OH^- must be equalized by placing coefficients in front of the formulas for the acid and base.

a. $2HCl(aq) + Mg(OH)_2(s) \longrightarrow 2H_2O(l) + MgCl_2(aq)$
b. $H_3PO_4(aq) + 3LiOH(aq) \longrightarrow 3H_2O(l) + Li_3PO_4(aq)$

14.57 The products of a neutralization are water and a salt. In balancing a neutralization equation, the number of H^+ and OH^- must be equalized by placing coefficients in front of the formulas for the acid and base.

a. $H_2SO_4(aq) + 2NaOH(aq) \longrightarrow 2H_2O(l) + Na_2SO_4(aq)$
b. $3HCl(aq) + Fe(OH)_3(s) \longrightarrow 3H_2O(l) + FeCl_3(aq)$
c. $H_2CO_3(aq) + Mg(OH)_2(s) \longrightarrow 2H_2O(l) + MgCO_3(s)$

14.59 To a measured volume of the formic acid solution, add a few drops of indicator. Place a solution of NaOH of known molarity in a buret. Add NaOH to the acid solution until one drop changes the color of the solution. Use the volume and molarity of the NaOH solution to calculate the moles of NaOH added to reach the endpoint. This equals the moles of formic acid in the sample. Then calculate the molarity of the formic acid solution from the calculated moles of formic acid and the known volume you started with.

14.61 In the titration equation, 1 mol of HCl reacts with 1 mol of NaOH.

$$28.6\ \cancel{\text{mL NaOH solution}} \times \frac{1\ \cancel{\text{L solution}}}{1000\ \cancel{\text{mL solution}}} \times \frac{0.145\ \cancel{\text{mol NaOH}}}{1\ \cancel{\text{L solution}}} \times \frac{1\ \text{mol HCl}}{1\ \cancel{\text{mol NaOH}}}$$
$$= 0.004\ 15\ \text{mol of HCl}$$

$$5.00\ \cancel{\text{mL HCl solution}} \times \frac{1\ \text{L solution}}{1000\ \cancel{\text{mL solution}}} = 0.005\ 00\ \text{L of HCl solution}$$

$$\text{molarity (M) of HCl} = \frac{\text{moles of solute}}{\text{liters of solution}} = \frac{0.004\ 15\ \text{mol HCl}}{0.005\ 00\ \text{L solution}}$$
$$= 0.830\ \text{M HCl solution (3 SFs)}$$

14.63 In the titration equation, 1 mol of H_2SO_4 reacts with 2 mol of KOH.

$$38.2\ \cancel{\text{mL KOH solution}} \times \frac{1\ \cancel{\text{L solution}}}{1000\ \cancel{\text{mL solution}}} \times \frac{0.163\ \cancel{\text{mol KOH}}}{1\ \cancel{\text{L solution}}} \times \frac{1\ \text{mol H}_2\text{SO}_4}{2\ \cancel{\text{mol KOH}}}$$
$$= 0.003\ 11\ \text{mol of H}_2\text{SO}_4$$

$$25.0\ \cancel{\text{mL H}_2\text{SO}_4\ \text{solution}} \times \frac{1\ \text{L solution}}{1000\ \cancel{\text{mL solution}}} = 0.0250\ \text{L of H}_2\text{SO}_4\ \text{solution}$$

$$\text{molarity (M) of H}_2\text{SO}_4 = \frac{\text{moles of solute}}{\text{liters of solution}} = \frac{0.003\ 11\ \text{mol H}_2\text{SO}_4}{0.0250\ \text{L solution}}$$
$$= 0.124\ \text{M H}_2\text{SO}_4\ \text{solution (3 SFs)}$$

14.65 In the titration equation, 1 mol of H_3PO_4 reacts with 3 mol of NaOH.

$$50.0\ \text{mL } \cancel{H_3PO_4 \text{ solution}} \times \frac{1\ \cancel{\text{L } H_3PO_4 \text{ solution}}}{1000\ \cancel{\text{mL } H_3PO_4 \text{ solution}}} \times \frac{0.0224\ \cancel{\text{mol } H_3PO_4}}{1\ \cancel{\text{L } H_3PO_4 \text{ solution}}} \times \frac{3\ \cancel{\text{mol NaOH}}}{1\ \cancel{\text{mol } H_3PO_4}}$$

$$\times \frac{1\ \cancel{\text{L NaOH solution}}}{0.204\ \cancel{\text{mol NaOH}}} \times \frac{1000\ \text{mL NaOH solution}}{1\ \cancel{\text{L NaOH solution}}} = 16.5\ \text{mL of NaOH solution (3 SFs)}$$

14.67 A buffer system contains a weak acid and a salt containing its conjugate base (or a weak base and a salt containing its conjugate acid).

a. This is not a buffer system because it only contains the strong base NaOH and the neutral salt NaCl.

b. This is a buffer system; it contains the weak acid H_2CO_3 and a salt containing its conjugate base HCO_3^-.

c. This is a buffer system; it contains the weak acid HF and a salt containing its conjugate base F^-.

d. This is not a buffer system because it only contains the neutral salts KCl and NaCl.

14.69 a. The purpose of this buffer system is to (3) maintain pH.

b. The salt of the weak acid is needed to (1) provide the conjugate base and (2) neutralize added H_3O^+.

c. The addition of OH^- is neutralized by (3) H_3O^+.

d. When H_3O^+ is added, the equilibrium shifts in the direction of the (1) reactants.

14.71 $HNO_2(aq) + H_2O(l) \rightleftarrows NO_2^-(aq) + H_3O^+(aq)$

Rearrange the K_a for $[H_3O^+]$ and use it to calculate the pH.

$$[H_3O^+] = K_a \times \frac{[HNO_2]}{[NO_2^-]} = 4.5 \times 10^{-4} \times \frac{[0.10\ \cancel{M}]}{[0.10\ \cancel{M}]} = 4.5 \times 10^{-4}\ M$$

$$\text{pH} = -\log[H_3O^+] = -\log[4.5 \times 10^{-4}] = 3.35 \text{ (2 SFs on the right of the decimal point)}$$

14.73 $HF(aq) + H_2O(l) \rightleftarrows F^-(aq) + H_3O^+(aq)$

Rearrange the K_a for $[H_3O^+]$ and use it to calculate the pH.

$$[H_3O^+] = K_a \times \frac{[HF]}{[F^-]} = 7.2 \times 10^{-4} \times \frac{[0.10\ \cancel{M}]}{[0.10\ \cancel{M}]} = 7.2 \times 10^{-4}\ M$$

$$\text{pH} = -\log[7.2 \times 10^{-4}] = 3.14 \text{ (2 SFs on the right of the decimal point)}$$

$$[H_3O^+] = K_a \times \frac{[HF]}{[F^-]} = 7.2 \times 10^{-4} \times \frac{[0.060\ \cancel{M}]}{[0.120\ \cancel{M}]} = 3.6 \times 10^{-4}\ M$$

$$\text{pH} = -\log[3.6 \times 10^{-4}] = 3.44 \text{ (2 SFs on the right of the decimal point)}$$

∴ The solution with 0.10 M HF/0.10 M NaF is more acidic.

14.75 a. This diagram represents a weak acid; only a few HX molecules dissociate into H_3O^+ and X^- ions.

b. This diagram represents a strong acid; all of the HX molecules dissociate into H_3O^+ and X^- ions.

c. This diagram represents a weak acid; only a few HX molecules dissociate into H_3O^+ and X^- ions.

14.77 a. Hyperventilation will lower the CO_2 concentration in the blood, which lowers the $[H_2CO_3]$, which decreases the $[H_3O^+]$ and increases the blood pH.

b. Breathing into a paper bag will increase the CO_2 concentration in the blood, increase the $[H_2CO_3]$, increase $[H_3O^+]$, and lower the blood pH back toward the normal range.

14.79 **a.** acid; bromous acid
b. base; rubidium hydroxide
c. salt; magnesium nitrate
d. acid; butanoic acid (IUPAC), butyric acid (common)
e. acid; perchloric acid

14.81

Acid	Conjugate Base
HI	I^-
HCl	Cl^-
NH_4^+	NH_3
H_2S	HS^-

14.83 Use Table 14.4 to answer (the stronger acid will be closer to the top of the table).
a. HF is the stronger acid.
b. H_3O^+ is the stronger acid.
c. HNO_2 is the stronger acid.
d. HCO_3^- is the stronger acid.

14.85 An acidic solution has a pH less than 7.0. A neutral solution has a pH equal to 7.0. A basic solution has a pH greater than 7.0.
a. acidic (pH $5.2 < 7.0$)
b. basic (pH $7.5 > 7.0$)
c. acidic (pH $3.8 < 7.0$)
d. acidic (pH $2.5 < 7.0$)
e. basic (pH $12.0 > 7.0$)

14.87 $[H_3O^+] = \dfrac{1.0 \times 10^{-14}}{[OH^-]}$; $[OH^-] = \dfrac{1.0 \times 10^{-14}}{[H_3O^+]}$;
$pH = -\log[H_3O^+]$; $pOH = -\log[OH^-]$; $pH + pOH = 14.00$

a. $pH = -\log[H_3O^+] = -\log[2.0 \times 10^{-8}] = 7.70$ (2 SFs on the right of the decimal point)
since $pH + pOH = 14.00$, $pOH = 14.00 - pH = 14.00 - 7.70 = 6.30$

b. $pH = -\log[5.0 \times 10^{-2}] = 1.30$ (2 SFs on the right of the decimal point)
since $pH + pOH = 14.00$, $pOH = 14.00 - pH = 14.00 - 1.30 = 12.70$

c. $[H_3O^+] = \dfrac{1.0 \times 10^{-14}}{[3.5 \times 10^{-4}]} = 2.9 \times 10^{-11}\ M$

$pH = -\log[2.9 \times 10^{-11}] = 10.54$ (2 SFs on the right of the decimal point)
$pOH = -\log[OH^-] = -\log[3.5 \times 10^{-4}] = 3.46$

d. $[H_3O^+] = \dfrac{1.0 \times 10^{-14}}{[0.0054]} = 1.9 \times 10^{-12}\ M$

$pH = -\log[1.9 \times 10^{-12}] = 11.73$ (2 SFs on the right of the decimal point)
$pOH = -\log[OH^-] = -\log[0.0054] = 2.27$

14.89 **a.** basic ($pH > 7.0$)
b. acidic ($pH < 7.0$)
c. basic ($pH > 7.0$)
d. basic ($pH > 7.0$)

14.91 If the pH is given, the $[H_3O^+]$ can be found by using the relationship $[H_3O^+] = 10^{-pH}$. The $[OH^-]$ can be found by rearranging $K_w = [H_3O^+][OH^-] = 1 \times 10^{-14}$.

a. $pH = 3.00;\ [H_3O^+] = 10^{-pH} = 10^{-3.00} = 1.0 \times 10^{-3}\ M$ (2 SFs)

$$[OH^-] = \frac{1.0 \times 10^{-14}}{[H_3O^+]} = \frac{1.0 \times 10^{-14}}{[1.0 \times 10^{-3}]} = 1.0 \times 10^{-11}\ M \text{ (2 SFs)}$$

b. $pH = 6.48;\ [H_3O^+] = 10^{-pH} = 10^{-6.48} = 3.3 \times 10^{-7}\ M$ (2 SFs)

$$[OH^-] = \frac{1.0 \times 10^{-14}}{[H_3O^+]} = \frac{1.0 \times 10^{-14}}{[3.3 \times 10^{-7}]} = 3.0 \times 10^{-8}\ M \text{ (2 SFs)}$$

c. $pH = 8.85;\ [H_3O^+] = 10^{-pH} = 10^{-8.85} = 1.4 \times 10^{-9}\ M$ (2 SFs)

$$[OH^-] = \frac{1.0 \times 10^{-14}}{[H_3O^+]} = \frac{1.0 \times 10^{-14}}{[1.4 \times 10^{-9}]} = 7.1 \times 10^{-6}\ M \text{ (2 SFs)}$$

d. $pH = 11.00;\ [H_3O^+] = 10^{-pH} = 10^{-11.00} = 1.0 \times 10^{-11}\ M$ (2 SFs)

$$[OH^-] = \frac{1.0 \times 10^{-14}}{[H_3O^+]} = \frac{1.0 \times 10^{-14}}{[1.0 \times 10^{-11}]} = 1.0 \times 10^{-3}\ M \text{ (2 SFs)}$$

14.93 a. Solution A, with a pH of 4.5, is more acidic.

b. In solution A, the $[H_3O^+] = 10^{-pH} = 10^{-4.5} = 3 \times 10^{-5}\ M$ (1 SF)

In solution B, the $[H_3O^+] = 10^{-pH} = 10^{-6.7} = 2 \times 10^{-7}\ M$ (1 SF)

c. In solution A, the $[OH^-] = \dfrac{1.0 \times 10^{-14}}{[H_3O^+]} = \dfrac{1.0 \times 10^{-14}}{[3 \times 10^{-5}]} = 3 \times 10^{-10}\ M$ (1 SF)

In solution B, the $[OH^-] = \dfrac{1.0 \times 10^{-14}}{[H_3O^+]} = \dfrac{1.0 \times 10^{-14}}{[2 \times 10^{-7}]} = 5 \times 10^{-8}\ M$ (1 SF)

14.95 The $[OH^-]$ can be calculated from the moles of NaOH (each NaOH produces 1 OH^-) and the volume of the solution (in L).

$$0.225\ \text{g NaOH} \times \frac{1\ \text{mol NaOH}}{40.00\ \text{g NaOH}} \times \frac{1\ \text{mol OH}^-}{1\ \text{mol NaOH}} = 0.005\ 63\ \text{mol of OH}^-$$

$$[OH^-] = \frac{0.005\ 63\ \text{mol OH}^-}{0.250\ \text{L solution}} = 0.0225\ M \text{ (3 SFs)}$$

14.97 $2.5\ \text{g HCl} \times \dfrac{1\ \text{mol HCl}}{36.46\ \text{g HCl}} \times \dfrac{1\ \text{mol H}_3\text{O}^+}{1\ \text{mol HCl}} = 0.069\ \text{mol of H}_3\text{O}^+$ (2 SFs)

$$[H_3O^+] = \frac{0.069\ \text{mol H}_3\text{O}^+}{0.425\ \text{L solution}} = 0.16\ M \text{ (2 SFs)}$$

$pH = -\log[H_3O^+] = -\log[0.16] = 0.80$ (2 SFs on the right of the decimal point)

$pOH = 14.00 - pH = 14.00 - 0.80 = 13.20$ (2 SFs on the right of the decimal point)

14.99 a. To form the conjugate base, remove a hydrogen ion (H^+) from the acid.

1. HS^- **2.** $H_2PO_4^-$

b. 1. $\dfrac{[H_3O^+][HS^-]}{[H_2S]}$ **2.** $\dfrac{[H_3O^+][H_2PO_4^-]}{[H_3PO_4]}$

c. H_2S (see Table 14.4; the weaker acid will be closer to the bottom of the table)

14.101 **a.** NH_4^+/NH_3 and HNO_3/NO_3^-
From Table 14.4, we see that NH_4^+ is a weaker acid than HNO_3 and that NO_3^- is a weaker base than NH_3. Thus, the equilibrium mixture will contain mostly products.

b. H_3O^+/H_2O and HBr/Br^-
From Table 14.4, we see that H_3O^+ is a weaker acid than HBr and that Br^- is a weaker base than H_2O. Thus, the equilibrium mixture will contain mostly products.

14.103 **a.** $ZnCO_3(s) + H_2SO_4(aq) \longrightarrow CO_2(g) + H_2O(l) + ZnSO_4(aq)$
b. $2Al(s) + 6HNO_3(aq) \longrightarrow 3H_2(g) + 2Al(NO_3)_3(aq)$

14.105 KOH (strong base) $\longrightarrow K^+(aq) + OH^-(aq)$ (100% dissociation)
$[OH^-] = 0.050\ M = 5.0 \times 10^{-2}\ M$

a. $$[H_3O^+] = \frac{1.0 \times 10^{-14}}{[OH^-]} = \frac{1.0 \times 10^{-14}}{[5.0 \times 10^{-2}]} = 2.0 \times 10^{-13}\ M \text{ (2 SFs)}$$

b. $pH = -\log[H_3O^+] = -\log[2.0 \times 10^{-13}] = 12.70$ (2 SFs on the right of the decimal point)

c. $pOH = -\log[OH^-] = -\log[5.0 \times 10^{-2}] = 1.30$,
or $pOH = 14.00 - 12.70 = 1.30$ (2 SFs on the right of the decimal point)

d. $H_2SO_4(aq) + 2KOH(aq) \longrightarrow 2H_2O(l) + K_2SO_4(aq)$

e. In the titration equation, 1 mol of H_2SO_4 reacts with 2 mol of KOH.

$$40.0\ \cancel{\text{mL } H_2SO_4 \text{ solution}} \times \frac{1\ \cancel{\text{L } H_2SO_4 \text{ solution}}}{1000\ \cancel{\text{mL } H_2SO_4 \text{ solution}}} \times \frac{0.035\ \cancel{\text{mol } H_2SO_4}}{1\ \cancel{\text{L } H_2SO_4 \text{ solution}}} \times \frac{2\ \cancel{\text{mol KOH}}}{1\ \cancel{\text{mol } H_2SO_4}}$$
$$\times \frac{1\ \cancel{\text{L KOH solution}}}{0.050\ \cancel{\text{mol KOH}}} \times \frac{1000 \text{ mL KOH solution}}{1\ \cancel{\text{L KOH solution}}} = 56 \text{ mL of KOH solution (2 SFs)}$$

14.107 **a.** The $[H_3O^+]$ can be found by using the relationship $[H_3O^+] = 10^{-pH}$.
$[H_3O^+] = 10^{-pH} = 10^{-4.2} = 6 \times 10^{-5}\ M$ (1 SF)

$$[OH^-] = \frac{1.0 \times 10^{-14}}{[H_3O^+]} = \frac{1.0 \times 10^{-14}}{[6 \times 10^{-5}]} = 2 \times 10^{-10}\ M \text{ (1 SF)}$$

b. $[H_3O^+] = 10^{-pH} = 10^{-6.5} = 3 \times 10^{-7}\ M$ (1 SF)

$$[OH^-] = \frac{1.0 \times 10^{-14}}{[H_3O^+]} = \frac{1.0 \times 10^{-14}}{[3 \times 10^{-7}]} = 3 \times 10^{-8}\ M \text{ (1 SF)}$$

c. In the titration equation, 1 mol of $CaCO_3$ reacts with 1 mol of H_2SO_4.

$$1.0\ \cancel{\text{kL solution}} \times \frac{1000\ \cancel{\text{L solution}}}{1\ \cancel{\text{kL solution}}} \times \frac{6 \times 10^{-5}\ \cancel{\text{mol } H_3O^+}}{1\ \cancel{\text{L solution}}}$$
$$\times \frac{1\ \cancel{\text{mol } H_2SO_4}}{2\ \cancel{\text{mol } H_3O^+}} \times \frac{1\ \cancel{\text{mol } CaCO_3}}{1\ \cancel{\text{mol } H_2SO_4}} \times \frac{100.09 \text{ g } CaCO_3}{1\ \cancel{\text{mol } CaCO_3}} = 3 \text{ g of } CaCO_3 \text{ (1 SF)}$$

14.109 **a.** In the titration equation, 1 mol of HCl reacts with 1 mol of NaOH.
$HCl(aq) + NaOH(aq) \longrightarrow H_2O(l) + NaCl(aq)$

$$25.0\ \cancel{\text{mL HCl solution}} \times \frac{1\ \cancel{\text{L HCl solution}}}{1000\ \cancel{\text{mL HCl solution}}} \times \frac{0.288\ \cancel{\text{mol HCl}}}{1\ \cancel{\text{L HCl solution}}} \times \frac{1\ \cancel{\text{mol NaOH}}}{1\ \cancel{\text{mol HCl}}}$$
$$\times \frac{1\ \cancel{\text{L NaOH solution}}}{0.150\ \cancel{\text{mol NaOH}}} \times \frac{1000 \text{ mL NaOH solution}}{1\ \cancel{\text{L NaOH solution}}} = 48.0 \text{ mL of NaOH solution (3 SFs)}$$

b. In the titration equation, 1 mol of H_2SO_4 reacts with 2 mol of NaOH.

$$H_2SO_4(aq) + 2NaOH(aq) \longrightarrow 2H_2O(l) + Na_2SO_4(aq)$$

$$10.0\ \text{mL}\ H_2SO_4\ \text{solution} \times \frac{1\ \text{L}\ H_2SO_4\ \text{solution}}{1000\ \text{mL}\ H_2SO_4\ \text{solution}} \times \frac{0.560\ \text{mol}\ H_2SO_4}{1\ \text{L}\ H_2SO_4\ \text{solution}} \times \frac{2\ \text{mol NaOH}}{1\ \text{mol}\ H_2SO_4}$$

$$\times \frac{1\ \text{L NaOH solution}}{0.150\ \text{mol NaOH}} \times \frac{1000\ \text{mL NaOH solution}}{1\ \text{L NaOH solution}} = 74.7\ \text{mL of NaOH solution (3 SFs)}$$

14.111 In the titration equation, 1 mol of H_2SO_4 reacts with 2 mol of NaOH.

$$45.6\ \text{mL NaOH solution} \times \frac{1\ \text{L solution}}{1000\ \text{mL solution}} \times \frac{0.205\ \text{mol NaOH}}{1\ \text{L solution}} \times \frac{1\ \text{mol}\ H_2SO_4}{2\ \text{mol NaOH}}$$

$$= 0.004\,67\ \text{mol of}\ H_2SO_4$$

$$20.0\ \text{mL}\ H_2SO_4\ \text{solution} \times \frac{1\ \text{L solution}}{1000\ \text{mL solution}} = 0.0200\ \text{L of}\ H_2SO_4\ \text{solution}$$

$$\text{molarity (M) of}\ H_2SO_4 = \frac{\text{moles of solute}}{\text{liters of solution}} = \frac{0.004\,67\ \text{mol}\ H_2SO_4}{0.0200\ \text{L solution}}$$

$$= 0.234\ \text{M}\ H_2SO_4\ \text{solution (3 SFs)}$$

14.113 This buffer solution is made from the weak acid H_3PO_4 and a salt containing its conjugate base $H_2PO_4^-$.

a. Add acid: $H_2PO_4^-(aq) + H_3O^+(aq) \longrightarrow H_3PO_4(aq) + H_2O(l)$

b. Add base: $H_3PO_4(aq) + OH^-(aq) \longrightarrow H_2PO_4^-(aq) + H_2O(l)$

c. $[H_3O^+] = K_a \times \dfrac{[H_3PO_4]}{[H_2PO_4^-]} = 7.5 \times 10^{-3} \times \dfrac{[0.50\ \text{M}]}{[0.20\ \text{M}]} = 1.9 \times 10^{-2}\ \text{M}$

$\text{pH} = -\log[1.9 \times 10^{-2}] = 1.72$ (2 SFs on the right of the decimal point)

Answers to Combining Ideas from Chapters 11 to 14

CI.21 a. CH_4 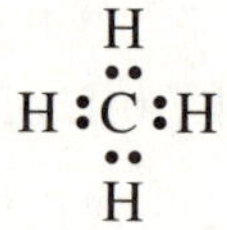 or 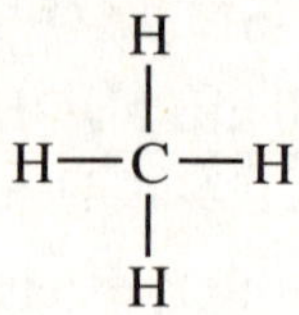

b. $7.0 \times 10^6\ \cancel{\text{gal}} \times \frac{4\ \cancel{\text{qt}}}{1\ \cancel{\text{gal}}} \times \frac{946.3\ \cancel{\text{mL}}}{1\ \cancel{\text{qt}}} \times \frac{0.45\ \cancel{\text{g}}}{1\ \cancel{\text{mL}}} \times \frac{1\ \text{kg}}{1000\ \cancel{\text{g}}} = 1.2 \times 10^7$ kg of LNG (2 SFs)

c. $7.0 \times 10^6\ \cancel{\text{gal}} \times \frac{4\ \cancel{\text{qt}}}{1\ \cancel{\text{gal}}} \times \frac{946.3\ \cancel{\text{mL}}}{1\ \cancel{\text{qt}}} \times \frac{0.45\ \cancel{\text{g}}}{1\ \cancel{\text{mL}}} \times \frac{1\ \cancel{\text{mol CH}_4}}{16.04\ \cancel{\text{g}}} \times \frac{22.4\ \text{L CH}_4\ \text{(STP)}}{1\ \cancel{\text{mol CH}_4}}$

$= 1.7 \times 10^{10}$ L of LNG (STP) (2 SFs)

d. $CH_4(g) + 2O_2(g) \xrightarrow{\Delta} CO_2(g) + 2H_2O(g) + 883\ \text{kJ}$

e. $7.0 \times 10^6\ \cancel{\text{gal}} \times \frac{4\ \cancel{\text{qt}}}{1\ \cancel{\text{gal}}} \times \frac{946.3\ \cancel{\text{mL}}}{1\ \cancel{\text{qt}}} \times \frac{0.45\ \cancel{\text{g}}}{1\ \cancel{\text{mL}}} \times \frac{1\ \cancel{\text{mol CH}_4}}{16.04\ \cancel{\text{g}}}$

$\times \frac{2\ \cancel{\text{mol O}_2}}{1\ \cancel{\text{mol CH}_4}} \times \frac{32.00\ \cancel{\text{g O}_2}}{1\ \cancel{\text{mol O}_2}} \times \frac{1\ \text{kg O}_2}{1000\ \cancel{\text{g O}_2}}$

$= 4.8 \times 10^7$ kg of O_2 (2 SFs)

f. $7.0 \times 10^6\ \cancel{\text{gal}} \times \frac{4\ \cancel{\text{qt}}}{1\ \cancel{\text{gal}}} \times \frac{946.3\ \cancel{\text{mL}}}{1\ \cancel{\text{qt}}} \times \frac{0.45\ \cancel{\text{g}}}{1\ \cancel{\text{mL}}} \times \frac{1\ \cancel{\text{mol CH}_4}}{16.04\ \cancel{\text{g}}} \times \frac{883\ \text{kJ}}{1\ \cancel{\text{mol CH}_4}}$

$= 6.6 \times 10^{11}$ kJ (2 SFs)

CI.23 a. $CS_2(g) + 3O_2(g) \xrightarrow{\Delta} CO_2(g) + 2SO_2(g)$

b. $25.0\ \cancel{\text{g CS}_2} \times \frac{1\ \cancel{\text{mol CS}_2}}{76.15\ \cancel{\text{g CS}_2}} \times \frac{1\ \cancel{\text{mol CO}_2}}{1\ \cancel{\text{mol CS}_2}} \times \frac{44.01\ \text{g CO}_2}{1\ \cancel{\text{mol CO}_2}} = 14.4$ g of CO_2

$30.0\ \cancel{\text{g O}_2} \times \frac{1\ \cancel{\text{mol O}_2}}{32.00\ \cancel{\text{g O}_2}} \times \frac{1\ \cancel{\text{mol CO}_2}}{3\ \cancel{\text{mol O}_2}} \times \frac{44.01\ \text{g CO}_2}{1\ \cancel{\text{mol CO}_2}}$

$= 13.8$ g of CO_2 (smaller amount of product)

∴ O_2 is the limiting reactant, and 13.8 g of CO_2 (3 SFs) can be produced.

c. $25.0\ \cancel{\text{g CS}_2} \times \frac{1\ \text{mol CS}_2}{76.15\ \cancel{\text{g CS}_2}} = 0.328$ mol of CS_2 initially

$30.0\ \cancel{\text{g O}_2} \times \frac{1\ \cancel{\text{mol O}_2}}{32.00\ \cancel{\text{g O}_2}} \times \frac{1\ \text{mol CS}_2}{3\ \cancel{\text{mol O}_2}} = 0.313$ mol of CS_2 consumed

∴ 0.328 mol − 0.313 mol = 0.015 mol of CS_2 remaining $= n$

and $T = 125\ °\text{C} + 273 = 398$ K; $V = 10.0$ L

$$P_{CS_2} = \frac{nRT}{V} = \frac{(0.015\ \cancel{\text{mol}})\left(\frac{62.4\ \cancel{\text{L}} \cdot \text{mmHg}}{\cancel{\text{mol}} \cdot \cancel{\text{K}}}\right)(398\ \cancel{\text{K}})}{10.0\ \cancel{\text{L}}} = 37\ \text{mmHg (2 SFs)}$$

d. $30.0\ \cancel{g\ O_2} \times \frac{1\ \cancel{mol\ O_2}}{32.00\ \cancel{g\ O_2}} \times \frac{1\ mol\ CO_2}{3\ \cancel{mol\ O_2}} = 0.313\ mol\ of\ CO_2$

$30.0\ \cancel{g\ O_2} \times \frac{1\ \cancel{mol\ O_2}}{32.00\ \cancel{g\ O_2}} \times \frac{2\ mol\ SO_2}{3\ \cancel{mol\ O_2}} = 0.625\ mol\ of\ SO_2$

and 0.015 mol of CS_2 remaining (excess)

$\therefore$ Total moles of gas $= 0.313\ mol\ CO_2 + 0.625\ mol\ SO_2 + 0.015\ mol\ CS_2 = 0.953\ mol$

$$P_{total} = \frac{nRT}{V} = \frac{(0.953\ \cancel{mol})\left(\frac{62.4\ \cancel{L}\cdot mmHg}{\cancel{mol}\cdot \cancel{K}}\right)(398\ \cancel{K})}{10.0\ \cancel{L}} = 2370\ mmHg\ (3\ SFs)$$

CI.25 a. $[H_2] = \frac{2.02\ \cancel{g\ H_2}}{10.0\ L} \times \frac{1\ mol\ H_2}{2.016\ \cancel{g\ H_2}} = 0.100\ M$

$[S_2] = \frac{10.3\ \cancel{g\ S_2}}{10.0\ L} \times \frac{1\ mol\ S_2}{64.14\ \cancel{g\ S_2}} = 0.0161\ M$

$[H_2S] = \frac{68.2\ \cancel{g\ H_2S}}{10.0\ L} \times \frac{1\ mol\ H_2S}{34.09\ \cancel{g\ H_2S}} = 0.200\ M$

$K_c = \frac{[H_2S]^2}{[H_2]^2[S_2]} = \frac{[0.200]^2}{[0.100]^2[0.0161]} = 248\ (3\ SFs)$

b. If more H_2 (a reactant) is added, the equilibrium will shift in the direction of the products.

c. If the volume decreases from 10.0 L to 5.00 L (at constant temperature), the equilibrium will shift in the direction of the products (fewer moles of gas).

d. $[H_2] = \frac{0.300\ mol\ H_2}{5.00\ L} = 0.0600\ M$

$[H_2S] = \frac{2.50\ mol\ H_2S}{5.00\ L} = 0.500\ M$

$K_c = \frac{[H_2S]^2}{[H_2]^2[S_2]} = 248$

Rearrange the expression to solve for $[S_2]$ and substitute in known values.

$[S_2] = \frac{[H_2S]^2}{[H_2]^2 K_c} = \frac{[0.500]^2}{[0.0600]^2(248)} = 0.280\ M\ (3\ SFs)$

CI.27 a. $2M(s) + 6HCl(aq) \longrightarrow 3H_2(g) + 2MCl_3(aq)$

b. $34.8\ \cancel{mL\ solution} \times \frac{1\ \cancel{L\ solution}}{1000\ \cancel{mL\ solution}} \times \frac{0.520\ \cancel{mol\ HCl}}{1\ \cancel{L\ solution}} \times \frac{3\ mol\ H_2}{6\ \cancel{mol\ HCl}}$

$= 0.00905$ mol of H_2 (3 SFs)

$T = 24\ °C + 273 = 297\ K$

Rearrange the ideal gas law $PV = nRT$ to solve for V,

$$V = \frac{nRT}{P} = \frac{(0.00905\ \cancel{mol})\left(\frac{62.4\ \cancel{L}\cdot \cancel{mmHg}}{\cancel{mol}\cdot \cancel{K}}\right)(297\ \cancel{K})}{(720.\ \cancel{mmHg})} \times \frac{1000\ mL}{1\ \cancel{L}} = 233\ mL\ of\ H_2\ (3\ SFs)$$

c. $34.8\ \cancel{mL\ solution} \times \frac{1\ \cancel{L\ solution}}{1000\ \cancel{mL\ solution}} \times \frac{0.520\ \cancel{mol\ HCl}}{1\ \cancel{L\ solution}} \times \frac{2\ mol\ M}{6\ \cancel{mol\ HCl}}$

$= 6.03 \times 10^{-3}$ mol of M (3 SFs)

d. $\frac{0.420\ g\ M}{6.03 \times 10^{-3}\ mol\ M} = 69.7$ g/mol of M (3 SFs); $\therefore$ metal M is gallium

e. $2Ga(s) + 6HCl(aq) \longrightarrow 3H_2(g) + 2GaCl_3(aq)$

15

Oxidation and Reduction

Learning Goals

- Identify what is oxidized and what is reduced in an oxidation–reduction reaction.
- Assign and use oxidation numbers to identify elements that are oxidized or reduced.
- Balance oxidation–reduction equations using the half-reaction method.
- Use the activity series to determine if an oxidation–reduction reaction is spontaneous.
- Write the half-reactions that occur in a voltaic cell and the cell notation.
- Describe the half-cell reactions and the overall reactions that occur in electrolysis.
- Classify alcohols as primary, secondary, or tertiary; write equations for the oxidation of alcohols.

Chapter Outline

Core Chemistry Skills

- Identifying Oxidized and Reduced Substances (15.1)
- Assigning Oxidation Numbers (15.2)
- Using Oxidation Numbers (15.2)
- Identifying Oxidizing and Reducing Agents (15.2)
- Using Half-Reactions to Balance Redox Equations (15.3)
- Identifying Spontaneous Reactions (15.4)

Answers and Solutions to Text Problems

15.1 Oxidation is the loss of electrons; reduction is the gain of electrons.

a. Na^+ gains an electron to form Na; this is a reduction.

b. Ni loses electrons to form Ni^{2+}; this is an oxidation.

c. Cr^{3+} gains electrons to form Cr; this is a reduction.

d. $2H^+$ gain electrons to form H_2; this is a reduction.

15.3 An oxidized substance has lost electrons; a reduced substance has gained electrons.
a. Zn loses electrons and is oxidized. Cl_2 gains electrons and is reduced.
b. Br^- (in NaBr) loses electrons and is oxidized. Cl_2 gains electrons and is reduced.
c. Pb loses electrons and is oxidized. O_2 gains electrons and is reduced.
d. Sn^{2+} loses electrons and is oxidized. Fe^{3+} gains electrons and is reduced.

15.5 **a.** An element has an oxidation number of zero (Rule 2); Cu = 0.
b. An element has an oxidation number of zero (Rule 2); F in F_2 = 0.
c. The oxidation number of a monatomic ion is equal to its charge (Rule 3); $Fe^{2+} = +2$.
d. The oxidation number of a monatomic ion is equal to its charge (Rule 3); $Cl^- = -1$.

15.7 **a.** Cl has an oxidation number of −1 (Rule 5). Because KCl is neutral, the oxidation number of K is calculated as +1. Oxidation numbers: K = +1, Cl = −1
$K + Cl = 0$
$K + (-1) = 0$
$\therefore K = +1$
b. In MnO_2, the oxidation number of O is −2 (Rule 6). Because MnO_2 is neutral, the oxidation number of Mn is calculated as +4. Oxidation numbers: Mn = +4, O = −2
$1Mn + 2O = 0$
$Mn + 2(-2) = 0$
$\therefore Mn = +4$
c. In NO, the oxidation number of O is −2 (Rule 6). Because NO is neutral, the oxidation number of N is calculated as +2. Oxidation numbers: N = +2, O = −2
$N + O = 0$
$N + (-2) = 0$
$\therefore N = +2$
d. In Mn_2O_3, the oxidation number of O is −2 (Rule 6). Because Mn_2O_3 is neutral, the oxidation number of Mn is calculated as +3. Oxidation numbers: Mn = +3, O = −2
$2Mn + 3O = 0$
$2Mn + 3(-2) = 0$
$2Mn = +6$
$\therefore Mn = +3$

15.9 **a.** Li_3PO_4 is an ionic compound composed of the Li^+ and PO_4^{3-} ions. The oxidation number of the monatomic ion Li^+ is +1 (Rule 3). In the polyatomic ion PO_4^{3-}, the oxidation number of O is −2 (Rule 6). Because the sum of the oxidation numbers in PO_4^{3-} must equal −3, the oxidation number of P is calculated as +5. Oxidation numbers: Li = +1, P = +5, O = −2
$P + 4O = -3$
$P + 4(-2) = -3$
$\therefore P = +5$
b. In SO_3^{2-}, the oxidation number of O is −2 (Rule 6). Because the sum of the oxidation numbers in the polyatomic ion SO_3^{2-} must equal −2 (Rule 1), the oxidation number of S is calculated as +4. Oxidation numbers: S = +4, O = −2
$S + 3O = -2$
$S + 3(-2) = -2$
$\therefore S = +4$
c. In Cr_2S_3, the oxidation number of S is −2 (Rule 6). Because Cr_2S_3 is neutral, the oxidation number of Cr is calculated as +3. Oxidation numbers: Cr = +3, S = −2
$2Cr + 3S = 0$
$2Cr + 3(-2) = 0$
$2Cr = +6$
$\therefore Cr = +3$

d. In NO_3^-, the oxidation number of O is −2 (Rule 6). Because the sum of the oxidation numbers in the polyatomic ion NO_3^- must equal −1 (Rule 1), the oxidation number of N is calculated as +5. Oxidation numbers: N = +5, O = −2
$N + 3O = -1$
$N + 3(-2) = -1$
$\therefore N = +5$

15.11 a. In HSO_4^-, the oxidation number of H is +1 (Rule 7) and the oxidation number of O is −2 (Rule 6). Because the sum of the oxidation numbers in the polyatomic ion HSO_4^- must equal −1 (Rule 1), the oxidation number of S is calculated as +6.
Oxidation numbers: H = +1, S = +6, O = −2
$H + S + 4O = -1$
$+1 + S + 4(-2) = -1$
$\therefore S = +6$

b. In H_3PO_3, the oxidation number of H is +1 (Rule 7) and the oxidation number of O is −2 (Rule 6). Because H_3PO_3 is neutral, the oxidation number of P is calculated as +3.
Oxidation numbers: H = +1, P = +3, O = −2
$3H + P + 3O = 0$
$3(+1) + P + 3(-2) = 0$
$\therefore P = +3$

c. In $Cr_2O_7^{2-}$, the oxidation number of O is −2 (Rule 6). Because the sum of the oxidation numbers in the polyatomic ion $Cr_2O_7^{2-}$ must equal −2 (Rule 1), the oxidation number of Cr is calculated as +6. Oxidation numbers: Cr = +6, O = −2
$2Cr + 7O = -2$
$2Cr + 7(-2) = -2$
$2Cr = +12$
$\therefore Cr = +6$

d. Na_2CO_3 is an ionic compound composed of the Na^+ and CO_3^{2-} ions. The oxidation number of the monatomic ion Na^+ is +1 (Rule 3). In the polyatomic ion CO_3^{2-}, the oxidation number of O is −2 (Rule 6). Because the sum of the oxidation numbers in CO_3^{2-} must equal −2, the oxidation number of C is calculated as +4.
Oxidation numbers: Na = +1, C = +4, O = −2
$2Na + C + 3O = 0$
$2(+1) + C + 3(-2) = 0$
$\therefore C = +4$

15.13 a. HNO_3 H = +1, O = −2
$HNO_3 \rightarrow (+1) + N + 3(-2) = 0 \quad \therefore N = +5$

b. C_3H_6 H = +1
$C_3H_6 \rightarrow 3C + 6(+1) = 0 \quad \therefore 3C = -6$ and $C = -2$

c. K_3PO_4 K = +1, O = −2
$K_3PO_4 \rightarrow 3(+1) + P + 4(-2) = 0 \quad \therefore P = +5$

d. CrO_4^{2-} O = −2
$CrO_4^{2-} \rightarrow Cr + 4(-2) = -2 \quad \therefore Cr = +6$

15.15 a. The substance that is oxidized is the reducing agent; it provides electrons for reduction.
b. The substance that gains electrons is the oxidizing agent; it accepts the electrons lost in an oxidation, and so is reduced.

15.17 a. Li is oxidized; Li is the reducing agent. Cl (in Cl_2) is reduced; Cl_2 is the oxidizing agent.
b. Br^- (in NaBr) is oxidized; NaBr is the reducing agent. Cl (in Cl_2) is reduced; Cl_2 is the oxidizing agent.
c. Pb is oxidized; Pb is the reducing agent. O (in O_2) is reduced; O_2 is the oxidizing agent.
d. Al is oxidized; Al is the reducing agent. Ag^+ is reduced; Ag^+ is the oxidizing agent.

15.19 Assign oxidation numbers and determine which one increases and which one decreases. The substance with an increase in oxidation number is oxidized and is also the reducing agent. The substance with a decrease in oxidation number is reduced and is also the oxidizing agent.

a. $2NiS + 3O_2 \rightarrow 2NiO + 2SO_2$

+2−2 0 +2−2 +4−2

Ni: $+2 \rightarrow +2$ No change
S: $-2 \rightarrow +4$ Oxidation number increases (oxidation)
O: $0 \rightarrow -2$ Oxidation number decreases (reduction)

∴ S^{2-} (in NiS) is oxidized; NiS is the reducing agent. O (in O_2) is reduced; O_2 is the oxidizing agent.

b. $Sn^{2+} + 2Fe^{3+} \rightarrow Sn^{4+} + 2Fe^{2+}$

+2 +3 +4 +2

Sn: $+2 \rightarrow +4$ Oxidation number increases (oxidation)
Fe: $+3 \rightarrow +2$ Oxidation number decreases (reduction)

∴ Sn^{2+} is oxidized; Sn^{2+} is the reducing agent. Fe^{3+} is reduced; Fe^{3+} is the oxidizing agent.

c. $CH_4 + 2O_2 \rightarrow CO_2 + 2H_2O$

−4+1 0 +4−2 +1−2

C: $-4 \rightarrow +4$ Oxidation number increases (oxidation)
H: $+1 \rightarrow +1$ No change
O: $0 \rightarrow -2$ Oxidation number decreases (reduction)

∴ C (in CH_4) is oxidized; CH_4 is the reducing agent. O (in O_2) is reduced; O_2 is the oxidizing agent.

d. $2Cr_2O_3 + 3Si \rightarrow 4Cr + 3SiO_2$

+3−2 0 0 +4−2

Cr: $+3 \rightarrow 0$ Oxidation number decreases (reduction)
O: $-2 \rightarrow -2$ No change
Si: $0 \rightarrow +4$ Oxidation number increases (oxidation)

∴ Si is oxidized; Si is the reducing agent. Cr^{3+} (in Cr_2O_3) is reduced; Cr_2O_3 is the oxidizing agent.

15.21 a. $Sn^{2+}(aq) \rightarrow Sn^{4+}(aq)$

Balance charge with e^-: $Sn^{2+}(aq) \rightarrow Sn^{4+}(aq) + 2\,e^-$

b. $Mn^{2+}(aq) \rightarrow MnO_4^-(aq)$

Balance O with H_2O: $Mn^{2+}(aq) + 4H_2O(l) \rightarrow MnO_4^-(aq)$
Balance H with H^+: $Mn^{2+}(aq) + 4H_2O(l) \rightarrow MnO_4^-(aq) + 8H^+(aq)$
Balance charge with e^-: $Mn^{2+}(aq) + 4H_2O(l) \rightarrow MnO_4^-(aq) + 8H^+(aq) + 5\,e^-$

c. $NO_2^-(aq) \rightarrow NO_3^-(aq)$

Balance O with H_2O: $NO_2^-(aq) + H_2O(l) \rightarrow NO_3^-(aq)$
Balance H with H^+: $NO_2^-(aq) + H_2O(l) \rightarrow NO_3^-(aq) + 2H^+(aq)$
Balance charge with e^-: $NO_2^-(aq) + H_2O(l) \rightarrow NO_3^-(aq) + 2H^+(aq) + 2\,e^-$

d. $ClO_3^-(aq) \rightarrow ClO_2(aq)$

Balance O with H_2O: $ClO_3^-(aq) \rightarrow ClO_2(aq) + H_2O(l)$
Balance H with H^+: $2H^+(aq) + ClO_3^-(aq) \rightarrow ClO_2(aq) + H_2O(l)$
Balance charge with e^-: $e^- + 2H^+(aq) + ClO_3^-(aq) \rightarrow ClO_2(aq) + H_2O(l)$

15.23 Write half-reactions, multiply by small numbers to equalize electrons lost and gained, add together, and combine common species.

a. $Ag(s) \rightarrow Ag^+(aq) + e^-$
$e^- + 2H^+(aq) + NO_3^-(aq) \rightarrow NO_2(g) + H_2O(l)$

Overall: $2H^+(aq) + Ag(s) + NO_3^-(aq) \rightarrow Ag^+(aq) + NO_2(g) + H_2O(l)$

b. $[3\,e^- + 4H^+(aq) + NO_3^-(aq) \rightarrow NO(g) + 2H_2O(l)] \times 4$
$[2H_2O(l) + S(s) \rightarrow SO_2(g) + 4H^+(aq) + 4\,e^-] \times 3$

Overall: $4H^+(aq) + 4NO_3^-(aq) + 3S(s) \rightarrow 4NO(g) + 3SO_2(g) + 2H_2O(l)$

c. $2S_2O_3^{2-}(aq) \rightarrow S_4O_6^{2-}(aq) + 2e^-$
$2e^- + Cu^{2+}(aq) \rightarrow Cu(s)$

Overall: $2S_2O_3^{2-}(aq) + Cu^{2+}(aq) \rightarrow S_4O_6^{2-}(aq) + Cu(s)$

15.25 Write half-reactions, multiply by small numbers to equalize electrons lost and gained, add together, and combine common species.

a. $2Fe(s) + 3H_2O(l) \rightarrow Fe_2O_3(s) + 6H^+(aq) + 6e^-$
$6e^- + 10H^+(aq) + 2CrO_4^{2-}(aq) \rightarrow Cr_2O_3(s) + 5H_2O(l)$

Overall (in acid): $4H^+(aq) + 2Fe(s) + 2CrO_4^{2-}(aq) \rightarrow Fe_2O_3(s) + Cr_2O_3(s) + 2H_2O(l)$

$\therefore$ in base: $\underbrace{4H^+(aq) + 4OH^-(aq)}_{4H_2O(l)} + 2Fe(s) + 2CrO_4^{2-}(aq) \rightarrow Fe_2O_3(s) + Cr_2O_3(s) + 2H_2O(l) + 4OH^-(aq)$

Overall (in base): $2Fe(s) + 2CrO_4^{2-}(aq) + 2H_2O(l) \rightarrow Fe_2O_3(s) + Cr_2O_3(s) + 4OH^-(aq)$

b. $[CN^-(aq) + H_2O(l) \rightarrow CNO^-(aq) + 2H^+(aq) + 2e^-] \times 3$
$[3e^- + 4H^+(aq) + MnO_4^-(aq) \rightarrow MnO_2(s) + 2H_2O(l)] \times 2$

Overall (in acid): $2H^+(aq) + 3CN^-(aq) + 2MnO_4^-(aq) \rightarrow 3CNO^-(aq) + 2MnO_2(s) + H_2O(l)$

$\therefore$ in base: $\underbrace{2H^+(aq) + 2OH^-(aq)}_{2H_2O(l)} + 3CN^-(aq) + 2MnO_4^-(aq) \rightarrow 3CNO^-(aq) + 2MnO_2(s) + H_2O(l) + 2OH^-(aq)$

Overall (in base): $3CN^-(aq) + 2MnO_4^-(aq) + H_2O(l) \rightarrow 3CNO^-(aq) + 2MnO_2(s) + 2OH^-(aq)$

15.27 In the activity series, a metal oxidizes spontaneously when combined with the ions in solutions of any metal below it.

a. Since Au is below H_2 in the activity series, the reaction will not be spontaneous.
b. Since Fe is above Ni in the activity series, the reaction will be spontaneous.
c. Since Ag is below Cu in the activity series, the reaction will not be spontaneous.

15.29 Oxidation occurs at the anode; reduction occurs at the cathode. In the shorthand cell notation, the oxidation half-cell is written on the left side of the double vertical line, and the reduction half-cell is written on the right side.

a. Anode reaction: $Pb(s) \mid Pb^{2+}(aq) = Pb(s) \rightarrow Pb^{2+}(aq) + 2e^-$
Cathode reaction: $Cu^{2+}(aq) \mid Cu(s) = Cu^{2+}(aq) + 2e^- \rightarrow Cu(s)$
Overall cell reaction: $Pb(s) + Cu^{2+}(aq) \rightarrow Pb^{2+}(aq) + Cu(s)$

b. Anode reaction: $Cr(s) \mid Cr^{2+}(aq) = [Cr(s) \rightarrow Cr^{2+}(aq) + 2e^-] \times 1$
Cathode reaction: $Ag^+(aq) \mid Ag(s) = [Ag^+(aq) + e^- \rightarrow Ag(s)] \times 2$
Overall cell reaction: $Cr(s) + 2Ag^+(aq) \rightarrow Cr^{2+}(aq) + 2Ag(s)$

15.31 a. The anode is a Cd metal electrode in a Cd^{2+} solution. The anode reaction is
$Cd(s) \rightarrow Cd^{2+}(aq) + 2e^-$
The cathode is a Sn metal electrode in a Sn^{2+} solution. The cathode reaction is
$Sn^{2+}(aq) + 2e^- \rightarrow Sn(s)$
The shorthand notation for this cell is
$Cd(s) \mid Cd^{2+}(aq) \parallel Sn^{2+}(aq) \mid Sn(s)$

b. The anode is a Zn metal electrode in a Zn^{2+} solution. The anode reaction is
$Zn(s) \rightarrow Zn^{2+}(aq) + 2e^-$
The cathode is a C(graphite) electrode, where Cl_2 gas is reduced to Cl^-.
The cathode reaction is
$Cl_2(g) + 2e^- \rightarrow 2Cl^-(aq)$
The shorthand notation for this cell is
$Zn(s) \mid Zn^{2+}(aq) \parallel Cl_2(g), Cl^-(aq) \mid C(\text{graphite})$

15.33 **a.** The $Cd(s)$ has lost electrons, which makes the half-reaction an oxidation.
b. Cd metal is oxidized.
c. Because this is oxidation, it takes place at the anode.

15.35 **a.** The $Zn(s)$ has lost electrons, which makes the half-reaction an oxidation.
b. Zn metal is oxidized.
c. Because this is oxidation, it takes place at the anode.

15.37 **a.** The half-reaction to plate tin is $Sn^{2+}(aq) + 2\,e^- \rightarrow Sn(s)$.
b. The reduction of Sn^{2+} to Sn occurs at the cathode, which is the iron can.
c. The oxidation of Sn to Sn^{2+} occurs at the anode, which is the tin bar.

15.39 Since Fe is above Sn in the activity series, if the Fe is exposed to air and water, Fe will be oxidized and rust will form. To protect iron, Sn would have to be *more* active than Fe and it is not.

15.41 **a.** One carbon group attached to the carbon atom bonded to the —OH group makes this a primary alcohol.
b. One carbon group attached to the carbon atom bonded to the —OH group makes this a primary alcohol.
c. Three carbon groups attached to the carbon atom bonded to the —OH group makes this a tertiary alcohol.

15.43 **a.** 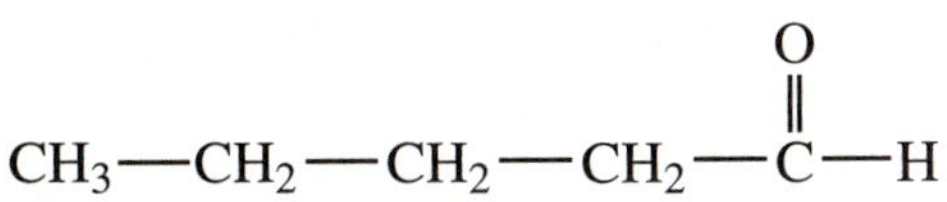

b. none

c. 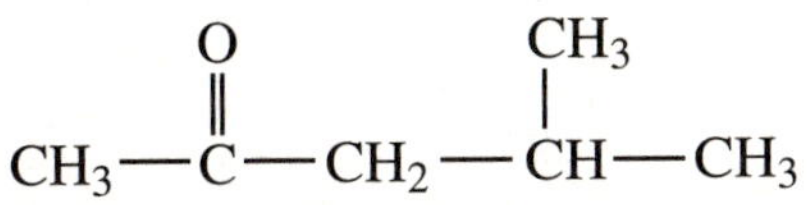

15.45 **a.** Electrons are lost in an oxidation.
b. An oxidizing agent undergoes reduction.
c. O_2 gains electrons to form OH^-; this is a reduction.
d. Br_2 gains electrons to form $2Br^-$; this is a reduction.
e. Sn^{2+} loses electrons to form Sn^{4+}; this is an oxidation.

15.47

a. VO_2	$O = -2$	Calculate: $V + 2(-2) = 0$	$\therefore V = +4$
b. Ag_2CrO_4	$Ag = +1, O = -2$	Calculate: $2(+1) + Cr + 4(-2) = 0$	$\therefore Cr = +6$
c. $S_2O_8^{2-}$	$O = -2$	Calculate: $2S + 8(-2) = -2$	$\therefore 2S = +14$ and $S = +7$
d. $FeSO_4$	$Fe = +2, O = -2$	Calculate: $(+2) + S + 4(-2) = 0$	$\therefore S = +6$

15.49 Reactions **a** and **c** both involve loss and gain of electrons; **a** and **c** are oxidation–reduction reactions.
a. Yes. Ca: $0 \rightarrow +2$ $\therefore$ oxidation H: $+1 \rightarrow 0$ $\therefore$ reduction
b. No. No change in oxidation numbers: $Ca = +2, C = +4, O = -2$
c. Yes. Al: $0 \rightarrow +3$ $\therefore$ oxidation O: $0 \rightarrow -2$ $\therefore$ reduction

15.51 $$\underset{+3-2}{2Cr_2O_3(s)} + \underset{0}{3Si(s)} \rightarrow \underset{0}{4Cr(s)} + \underset{+4-2}{3SiO_2(s)}$$
Cr: $+3 \rightarrow 0$ Oxidation number decreases (reduction)
Si: $0 \rightarrow +4$ Oxidation number increases (oxidation)
a. Cr^{3+} (in Cr_2O_3) is reduced.
b. Si is oxidized.
c. Cr_2O_3 is the oxidizing agent.
d. Si is the reducing agent.

15.53 **a.** Balance charge with e^-: $Zn(s) \rightarrow Zn^{2+}(aq) + 2\,e^-$

b. Balance O with H_2O: $SnO_2^{2-}(aq) + H_2O(l) \rightarrow SnO_3^{2-}(aq)$

Balance H with H^+: $SnO_2^{2-}(aq) + H_2O(l) \rightarrow SnO_3^{2-}(aq) + 2H^+(aq)$

Balance charge with e^-: $SnO_2^{2-}(aq) + H_2O(l) \rightarrow SnO_3^{2-}(aq) + 2H^+(aq) + 2\,e^-$

c. Balance O with H_2O: $SO_3^{2-}(aq) + H_2O(l) \rightarrow SO_4^{2-}(aq)$

Balance H with H^+: $SO_3^{2-}(aq) + H_2O(l) \rightarrow SO_4^{2-}(aq) + 2H^+(aq)$

Balance charge with e^-: $SO_3^{2-}(aq) + H_2O(l) \rightarrow SO_4^{2-}(aq) + 2H^+(aq) + 2\,e^-$

d. Balance O with H_2O: $NO_3^-(aq) \rightarrow NO(g) + 2H_2O(l)$

Balance H with H^+: $4H^+(aq) + NO_3^-(aq) \rightarrow NO(g) + 2H_2O(l)$

Balance charge with e^-: $3\,e^- + 4H^+(aq) + NO_3^-(aq) \rightarrow NO(g) + 2H_2O(l)$

15.55 **a.** $Fe(s) \rightarrow Fe^{2+}(aq) + 2\,e^-$

b. $Ni^{2+}(aq) + 2\,e^- \rightarrow Ni(s)$

c. Fe is the anode.

d. Ni is the cathode.

e. The electrons flow from Fe to Ni.

f. $Fe(s) + Ni^{2+}(aq) \rightarrow Fe^{2+}(aq) + Ni(s)$

g. $Fe(s)\,|\,Fe^{2+}(aq)\,||\,Ni^{2+}(aq)\,|\,Ni(s)$

15.57 **a.** One carbon group attached to the carbon atom bonded to the —OH group makes this a primary alcohol.

b. Two carbon groups attached to the carbon atom bonded to the —OH group makes this a secondary alcohol.

c. One carbon group attached to the carbon atom bonded to the —OH group makes this a primary alcohol.

15.59 **a.** 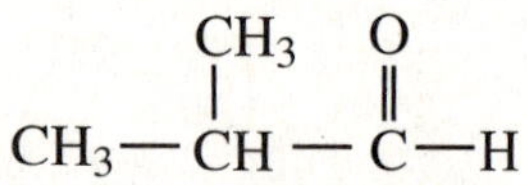

b. 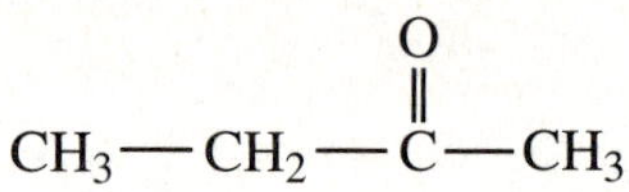

15.61 Reactions **b**, **c**, and **d** all involve loss and gain of electrons; **b**, **c**, and **d** are oxidation–reduction reactions.

a. No. No change in oxidation numbers: Ag = +1, N = +5, Na = +1, Cl = −1, O = −2

b. Yes. Li: $0 \rightarrow +1$ ∴ oxidation   N: $0 \rightarrow -3$ ∴ reduction

c. Yes. Ni: $0 \rightarrow +2$ ∴ oxidation   Pb: $+2 \rightarrow 0$ ∴ reduction

d. Yes. K: $0 \rightarrow +1$ ∴ oxidation   H: $+1 \rightarrow 0$ ∴ reduction

15.63 **a.** Fe^{3+} gains an electron to form Fe^{2+}; this is a reduction.

b. Fe^{2+} is loses an electron to form Fe^{3+}; this is an oxidation.

15.65 **a.** Co_2O_3 O = −2   Calculate: $2Co + 3(-2) = 0$ ∴ $2Co = +6$ and $Co = +3$

b. $KMnO_4$ K = +1, O = −2   Calculate: $+1 + Mn + 4(-2) = 0$ ∴ $Mn = +7$

c. $SbCl_5$ Cl = −1   Calculate: $Sb + 5(-1) = 0$ ∴ $Sb = +5$

d. ClO_3^- O = −2   Calculate: $Cl + 3(-2) = -1$ ∴ $Cl = +5$

e. PO_4^{3-} O = −2   Calculate: $P + 4(-2) = -3$ ∴ $P = +5$

15.67 **a.** $\underset{+2-1}{2FeCl_2(aq)} + \underset{0}{Cl_2(g)} \rightarrow \underset{+3-1}{2FeCl_3(aq)}$

Fe: $+2 \rightarrow +3$ Oxidation number increases (oxidation)

Cl: $0 \rightarrow -1$ Oxidation number decreases (reduction)

Fe (in $FeCl_2$) is oxidized; $FeCl_2$ is the reducing agent.

Cl (in Cl_2) is reduced; Cl_2 is the oxidizing agent.

b. $2H_2S(g) + 3O_2(g) \rightarrow 2H_2O(l) + 2SO_2(g)$

$\quad +1{-}2 \qquad\quad 0 \qquad\quad +1{-}2 \qquad +4{-}2$

H: $+1 \rightarrow +1$ No change
S: $-2 \rightarrow +4$ Oxidation number increases (oxidation)
O: $0 \rightarrow -2$ Oxidation number decreases (reduction)
S (in H_2S) is oxidized; H_2S is the reducing agent.
O (in O_2) is reduced; O_2 is the oxidizing agent.

c. $P_2O_5(s) + 5C(s) \rightarrow 2P(s) + 5CO(g)$

$\quad +5{-}2 \qquad\quad 0 \qquad\quad 0 \qquad\quad +2{-}2$

P: $+5 \rightarrow 0$ Oxidation number decreases (reduction)
O: $-2 \rightarrow -2$ No change
C: $0 \rightarrow +2$ Oxidation number increases (oxidation)
C is oxidized; C is the reducing agent.
P (in P_2O_5) is reduced; P_2O_5 is the oxidizing agent.

15.69 Write half-reactions, multiply by small numbers to equalize electrons lost and gained, add together, and combine common species.

a. $Zn(s) \rightarrow Zn^{2+}(aq) + 2\,e^-$
$[e^- + 2H^+(aq) + NO_3^-(aq) \rightarrow NO_2(g) + H_2O(l)] \times 2$

Overall: $4H^+(aq) + Zn(s) + 2NO_3^-(aq) \rightarrow Zn^{2+}(aq) + 2NO_2(g) + 2H_2O(l)$

b. $[5\,e^- + 8H^+(aq) + MnO_4^-(aq) \rightarrow Mn^{2+}(aq) + 4H_2O(l)] \times 2$
$[SO_3^{2-}(aq) + H_2O(l) \rightarrow SO_4^{2-}(aq) + 2H^+(aq) + 2\,e^-] \times 5$

Overall: $6H^+(aq) + 2MnO_4^-(aq) + 5SO_3^{2-}(aq) \rightarrow 2Mn^{2+}(aq) + 5SO_4^{2-}(aq) + 3H_2O(l)$

c. $[2I^-(aq) \rightarrow I_2(s) + 2\,e^-] \times 3$
$6\,e^- + 6H^+(aq) + ClO_3^-(aq) \rightarrow Cl^-(aq) + 3H_2O(l)$

Overall: $6H^+(aq) + ClO_3^-(aq) + 6I^-(aq) \rightarrow Cl^-(aq) + 3I_2(s) + 3H_2O(l)$

d. $[C_2O_4^{2-}(aq) \rightarrow 2CO_2(g) + 2\,e^-] \times 3$
$6\,e^- + 14H^+(aq) + Cr_2O_7^{2-}(aq) \rightarrow 2Cr^{3+}(aq) + 7H_2O(l)$

Overall: $14H^+(aq) + Cr_2O_7^{2-}(aq) + 3C_2O_4^{2-}(aq) \rightarrow 2Cr^{3+}(aq) + 6CO_2(g) + 7H_2O(l)$

15.71 a. Since Cr is above Ni in the activity series, the reaction will be spontaneous.
b. Since Cu is below Zn in the activity series, the reaction will not be spontaneous.
c. Since Zn is above Pb in the activity series, the reaction will be spontaneous.

15.73 a. The anode is Mg.
b. The cathode is Ni.
c. The half-reaction at the anode is $Mg(s) \rightarrow Mg^{2+}(aq) + 2\,e^-$.
d. The half-reaction at the cathode is $Ni^{2+}(aq) + 2\,e^- \rightarrow Ni(s)$.
e. The overall reaction is $Mg(s) + Ni^{2+}(aq) \rightarrow Mg^{2+}(aq) + Ni(s)$.
f. The shorthand cell notation is $Mg(s)\,|\,Mg^{2+}(aq)\,||\,Ni^{2+}(aq)\,|\,Ni(s)$.

15.75 In the activity series, Fe is above Ni, Pb, and Ag, which means it is oxidized. However, it is below Ca and Al, which means it cannot be oxidized and therefore Fe will not reduce Ca^{2+} or Al^{3+}.
a. $Ca^{2+}(aq)$ will not be reduced by an iron strip.
b. $Ag^+(aq)$ will be reduced by an iron strip.
c. $Ni^{2+}(aq)$ will be reduced by an iron strip.
d. $Al^{3+}(aq)$ will not be reduced by an iron strip.
e. $Pb^{2+}(aq)$ will be reduced by an iron strip.

15.77 a. $Pb(s) + SO_4^{2-}(aq) \rightarrow PbSO_4(s) + 2\,e^-$
b. Pb(*s*) loses electrons; it is oxidized.
c. The oxidation half-reaction takes place at the anode.

15.79

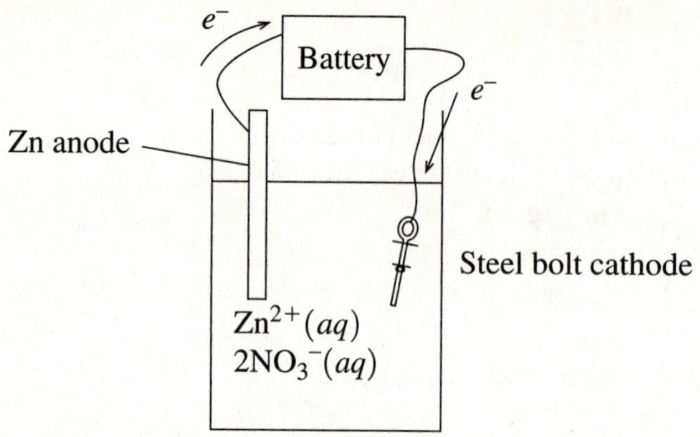

a. The anode is a bar of zinc.
b. The cathode is the steel bolt.
c. The half reaction at the anode is $Zn(s) \rightarrow Zn^{2+}(aq) + 2\,e^-$
d. The half reaction at the cathode is $Zn^{2+}(aq) + 2\,e^- \rightarrow Zn(s)$
e. The purpose of the zinc coating is to prevent rusting of the bolt by H_2O and O_2.

15.81 Primary alcohols oxidize to aldehydes. Secondary alcohols oxidize to ketones.

a. $CH_3-CH_2-\overset{\displaystyle O}{\overset{\|}{C}}-H$

b. $CH_3-\overset{\displaystyle O}{\overset{\|}{C}}-CH_2-CH_2-CH_3$

c. $CH_3-CH_2-CH_2-\overset{\displaystyle O}{\overset{\|}{C}}-H$

15.83
a. Br_2 $\therefore Br = 0$
b. $HBrO_2$ $H = +1, O = -2$ Calculate: $+1 + Br + 2(-2) = 0$ $\therefore Br = +3$
c. BrO_3^- $O = -2$ Calculate: $Br + 3(-2) = -1$ $\therefore Br = +5$
d. $NaBrO_4$ $Na = +1, O = -2$ Calculate: $+1 + Br + 4(-2) = 0$ $\therefore Br = +7$

15.85
$$6\,e^- + 14H^+(aq) + Cr_2O_7^{2-}(aq) \rightarrow 2Cr^{3+}(aq) + 7H_2O(l)$$
$$[NO_2^-(aq) + H_2O(l) \rightarrow NO_3^-(aq) + 2H^+(aq) + 2\,e^-] \times 3$$
Overall: $8H^+(aq) + Cr_2O_7^{2-}(aq) + 3NO_2^-(aq) \rightarrow 2Cr^{3+}(aq) + 3NO_3^-(aq) + 4H_2O(l)$

15.87
a.
$$[Ag(s) \rightarrow Ag^+(aq) + e^-] \times 3$$
$$3\,e^- + 4H^+(aq) + NO_3^-(aq) \rightarrow NO(g) + 2H_2O(l)$$
Overall: $4H^+(aq) + 3Ag(s) + NO_3^-(aq) \rightarrow 3Ag^+(aq) + NO(g) + 2H_2O(l)$

b. $15.0\ \text{g Ag} \times \dfrac{1\ \text{mol Ag}}{107.9\ \text{g Ag}} \times \dfrac{1\ \text{mol NO}}{3\ \text{mol Ag}} \times \dfrac{22.4\ \text{L NO (STP)}}{1\ \text{mol NO}} = 1.04$ L of $NO(g)$ at STP (3 SFs)

15.89
a.
$$[CuS(s) + 4H_2O(l) \rightarrow CuSO_4(aq) + 8H^+(aq) + 8\,e^-] \times 3$$
$$[3\,e^- + 3H^+(aq) + HNO_3(aq) \rightarrow NO(g) + 2H_2O(l)] \times 8$$
Overall: $3CuS(s) + 8HNO_3(aq) \rightarrow 3CuSO_4(aq) + 8NO(g) + 4H_2O(l)$

b. $24.8\ \text{g CuS} \times \dfrac{1\ \text{mol CuS}}{95.62\ \text{g CuS}} \times \dfrac{8\ \text{mol HNO}_3}{3\ \text{mol CuS}} \times \dfrac{1000\ \text{mL solution}}{16.0\ \text{mol HNO}_3}$
= 43.2 mL of HNO_3 solution (3 SFs)

15.91 $Br^-(aq) + 3H_2O(l) \rightarrow BrO_3^-(aq) + 6H^+(aq) + 6e^-$

$[3e^- + 4H^+(aq) + MnO_4^-(aq) \rightarrow MnO_2(s) + 2H_2O(l)] \times 2$

Overall (in acid): $2H^+(aq) + Br^-(aq) + 2MnO_4^-(aq) \rightarrow BrO_3^-(aq) + 2MnO_2(s) + H_2O(l)$

∴ in base: $\underbrace{2H^+(aq) + 2OH^-(aq)}_{2H_2O(l)} + Br^-(aq) + 2MnO_4^-(aq) \rightarrow BrO_3^-(aq) + 2MnO_2(s) + H_2O(l) + 2OH^-(aq)$

Overall (in base): $H_2O(l) + Br^-(aq) + 2MnO_4^-(aq) \rightarrow BrO_3^-(aq) + 2MnO_2(s) + 2OH^-(aq)$

15.93 **a.** Since Zn is below Ca on the activity series, the reaction will not be spontaneous.
b. Since Al is above Sn on the activity series, the reaction will be spontaneous.
c. Since Mg is above H_2 on the activity series, the reaction will be spontaneous.

15.95

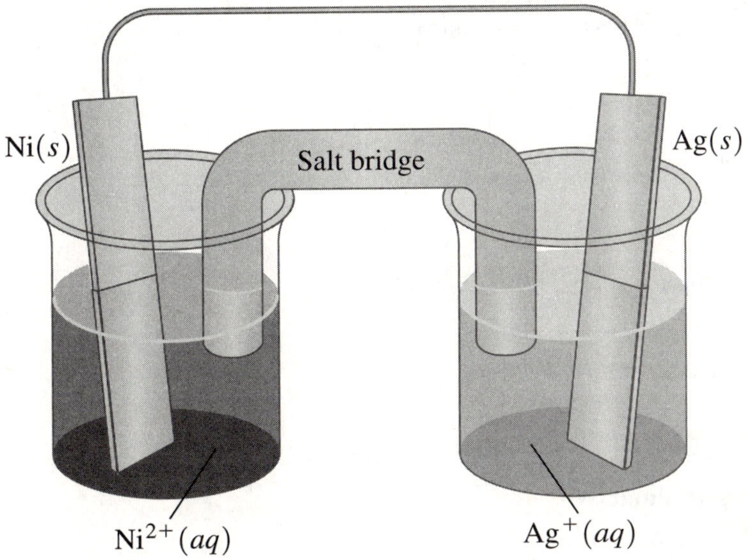

a. Ni(s) is the anode.
b. Ag(s) is the cathode.
c. The half-reaction at the anode is $Ni(s) \rightarrow Ni^{2+}(aq) + 2e^-$.
d. The half-reaction at the cathode is $Ag^+(aq) + e^- \rightarrow Ag(s)$.
e. The overall cell reaction is $Ni(s) + 2Ag^+(aq) \rightarrow Ni^{2+}(aq) + 2Ag(s)$.

16

Nuclear Chemistry

Learning Goals

- Describe alpha, beta, positron, and gamma radiation.
- Write a balanced nuclear equation showing mass numbers and atomic numbers for radioactive decay.
- Describe the detection and measurement of radiation.
- Given the half-life of a radioisotope, calculate the amount of radioisotope remaining after one or more half-lives.
- Describe the use of radioisotopes in medicine.
- Describe the processes of nuclear fission and fusion.

Chapter Outline

Core Chemistry Skills

- Writing Nuclear Equations (16.2)
- Using Half-Lives (16.4)

Answers and Solutions to Text Problems

16.1 **a.** ${}^{4}_{2}He$ is the symbol for an alpha particle.
b. ${}^{0}_{+1}e$ is the symbol for a positron.
c. ${}^{0}_{0}\gamma$ is the symbol for gamma radiation.

16.3 **a.** Since the element is potassium, the element symbol is K and the atomic number is 19. The potassium isotopes with mass number 39, 40, and 41 will have the atomic symbol ${}^{39}_{19}K$, ${}^{40}_{19}K$, and ${}^{41}_{19}K$, respectively.
b. Each isotope has 19 protons and 19 electrons, but they differ in the number of neutrons present. Potassium-39 has $39 - 19 = 20$ neutrons, potassium-40 has $40 - 19 = 21$ neutrons, and potassium-41 has $41 - 19 = 22$ neutrons.

16.5

Medical Use	Atomic Symbol	Mass Number	Number of Protons	Number of Neutrons
Heart imaging	${}^{201}_{81}\text{Tl}$	201	81	120
Radiation therapy	${}^{60}_{27}\text{Co}$	60	27	33
Abdominal scan	${}^{67}_{31}\text{Ga}$	67	31	36
Hyperthyroidism	${}^{131}_{53}\text{I}$	131	53	78
Leukemia treatment	${}^{32}_{15}\text{P}$	32	15	17

16.7 **a.** Since the element is copper, the element symbol is Cu and the atomic number is 29. The copper isotope with mass number 64 will have the atomic symbol ${}^{64}_{29}\text{Cu}$.
b. Since the element is selenium, the element symbol is Se and the atomic number is 34. The selenium isotope with mass number 75 will have the atomic symbol ${}^{75}_{34}\text{Se}$.
c. Since the element is sodium, the element symbol is Na and the atomic number is 11. The sodium isotope with mass number 24 will have the atomic symbol ${}^{24}_{11}\text{Na}$.
d. Since the element is nitrogen, the element symbol is N and the atomic number is 7. The nitrogen isotope with mass number 15 will have the atomic symbol ${}^{15}_{7}\text{N}$.

16.9 **a.** beta particle (β, ${}^{0}_{-1}e$)
b. alpha particle (α, ${}^{4}_{2}\text{He}$)
c. neutron (n, ${}^{1}_{0}n$)
d. argon-38 (${}^{38}_{18}\text{Ar}$)
e. carbon-14 (${}^{14}_{6}\text{C}$)

16.11 **a.** 1. Alpha particles do not penetrate the skin.
b. 3. Gamma radiation requires shielding protection that includes lead or thick concrete.
c. 1. Alpha particles can be very harmful if ingested.

16.13 The mass number of the radioactive atom is reduced by four when an alpha particle (${}^{4}_{2}\text{He}$) is emitted. The unknown product will have an atomic number that is two less than the atomic number of the radioactive atom.

a. ${}^{208}_{84}\text{Po} \longrightarrow {}^{204}_{82}\text{Pb} + {}^{4}_{2}\text{He}$
b. ${}^{232}_{90}\text{Th} \longrightarrow {}^{228}_{88}\text{Ra} + {}^{4}_{2}\text{He}$
c. ${}^{251}_{102}\text{No} \longrightarrow {}^{247}_{100}\text{Fm} + {}^{4}_{2}\text{He}$
d. ${}^{220}_{86}\text{Rn} \longrightarrow {}^{216}_{84}\text{Po} + {}^{4}_{2}\text{He}$

16.15 The mass number of the radioactive atom is not changed when a beta particle (${}^{0}_{-1}e$) is emitted. The unknown product will have an atomic number that is one greater than the atomic number of the radioactive atom.

a. ${}^{25}_{11}\text{Na} \longrightarrow {}^{25}_{12}\text{Mg} + {}^{0}_{-1}e$
b. ${}^{20}_{8}\text{O} \longrightarrow {}^{20}_{9}\text{F} + {}^{0}_{-1}e$
c. ${}^{92}_{38}\text{Sr} \longrightarrow {}^{92}_{39}\text{Y} + {}^{0}_{-1}e$
d. ${}^{60}_{26}\text{Fe} \longrightarrow {}^{60}_{27}\text{Co} + {}^{0}_{-1}e$

16.17 The mass number of the radioactive atom is not changed when a positron (${}^{0}_{+1}e$) is emitted. The unknown product will have an atomic number that is one less than the atomic number of the radioactive atom.

a. ${}^{26}_{14}\text{Si} \longrightarrow {}^{26}_{13}\text{Al} + {}^{0}_{+1}e$
b. ${}^{54}_{27}\text{Co} \longrightarrow {}^{54}_{26}\text{Fe} + {}^{0}_{+1}e$
c. ${}^{77}_{37}\text{Rb} \longrightarrow {}^{77}_{36}\text{Kr} + {}^{0}_{+1}e$
d. ${}^{93}_{45}\text{Rh} \longrightarrow {}^{93}_{44}\text{Ru} + {}^{0}_{+1}e$

16.19 Balance the mass numbers and the atomic numbers in each nuclear equation.

a. ${}^{28}_{13}\text{Al} \longrightarrow {}^{28}_{14}\text{Si} + {}^{0}_{-1}e$ $\quad ? = {}^{28}_{14}\text{Si}$ beta decay
b. ${}^{180m}_{73}\text{Ta} \longrightarrow {}^{180}_{73}\text{Ta} + {}^{0}_{0}\gamma$ $\quad ? = {}^{0}_{0}\gamma$ gamma emission
c. ${}^{66}_{29}\text{Cu} \longrightarrow {}^{66}_{30}\text{Zn} + {}^{0}_{-1}e$ $\quad ? = {}^{0}_{-1}e$ beta decay
d. ${}^{238}_{92}\text{U} \longrightarrow {}^{234}_{90}\text{Th} + {}^{4}_{2}\text{He}$ $\quad ? = {}^{238}_{92}\text{U}$ alpha decay
e. ${}^{188}_{80}\text{Hg} \longrightarrow {}^{188}_{79}\text{Au} + {}^{0}_{+1}e$ $\quad ? = {}^{188}_{79}\text{Au}$ positron emission

16.21 Balance the mass numbers and the atomic numbers in each nuclear equation.

a. ${}^{1}_{0}n + {}^{9}_{4}\text{Be} \longrightarrow {}^{10}_{4}\text{Be}$ $\quad ? = {}^{10}_{4}\text{Be}$

b. ${}^{1}_{0}n + {}^{131}_{52}\text{Te} \longrightarrow {}^{132}_{53}\text{I} + {}^{0}_{-1}e$ $\quad ? = {}^{132}_{53}\text{I}$

c. ${}^{1}_{0}n + {}^{27}_{13}\text{Al} \longrightarrow {}^{24}_{11}\text{Na} + {}^{4}_{2}\text{He}$ $\quad ? = {}^{27}_{13}\text{Al}$

d. ${}^{4}_{2}\text{He} + {}^{27}_{13}\text{Al} \longrightarrow {}^{30}_{15}\text{P} + {}^{1}_{0}n$ $\quad ? = {}^{30}_{15}\text{P}$

16.23 **a.** 2. Absorbed dose can be measured in rad.

b. 3. Biological damage can be measured in mrem.

c. 1. Activity can be measured in mCi.

d. 2. Absorbed dose can be measured in Gy.

16.25 $8\ \cancel{\text{mGy}} \times \dfrac{1\ \cancel{\text{Gy}}}{1000\ \cancel{\text{mGy}}} \times \dfrac{100\ \text{rad}}{1\ \cancel{\text{Gy}}} = 0.8\ \text{rad (1 SF)}$

Thus, a technician exposed to a 5 rad dose of radiation received more radiation than one exposed to 8 mGy (0.8 rad) of radiation.

16.27 **a.** $70.0\ \cancel{\text{kg body mass}} \times \dfrac{4.20\ \mu\text{Ci}}{1\ \cancel{\text{kg body mass}}} = 294\ \mu\text{Ci (3 SFs)}$

b. from 50 rad of gamma radiation (we use a biological damage factor of 1, so 1 rad of gamma radiation = 1 rem):

$$50\ \cancel{\text{rad}} \times \frac{1\ \text{Gy}}{100\ \cancel{\text{rad}}} = 0.5\ \text{Gy (1 SF)}$$

16.29 **a.** **2.** two half-lives:

$$34\ \cancel{\text{days}} \times \frac{1\ \text{half-life}}{17\ \cancel{\text{days}}} = 2.0\ \text{half-lives}$$

b. **1.** one half-life:

$$20\ \cancel{\text{min}} \times \frac{1\ \text{half-life}}{20\ \cancel{\text{min}}} = 1\ \text{half-life}$$

c. **3.** three half-lives:

$$21\ \cancel{\text{h}} \times \frac{1\ \text{half-life}}{7\ \cancel{\text{h}}} = 3\ \text{half-lives}$$

16.31 **a.** After one half-life, one-half of the sample would be radioactive:

$$80.0\ \text{mg of}\ {}^{99m}_{43}\text{Tc} \xrightarrow{\text{1 half-life}} 40.0\ \text{mg of}\ {}^{99m}_{43}\text{Tc (3 SFs)}$$

b. After two half-lives, one-fourth of the sample would still be radioactive:

$$80.0\ \text{mg of}\ {}^{99m}_{43}\text{Tc} \xrightarrow{\text{1 half-life}} 40.0\ \text{mg of}\ {}^{99m}_{43}\text{Tc} \xrightarrow{\text{2 half-lives}} 20.0\ \text{mg of}\ {}^{99m}_{43}\text{Tc (3 SFs)}$$

c. $18\ \cancel{\text{h}} \times \dfrac{1\ \text{half-life}}{6.0\ \cancel{\text{h}}} = 3.0\ \text{half-lives}$

$$80.0\ \text{mg of}\ {}^{99m}_{43}\text{Tc} \xrightarrow{\text{1 half-life}} 40.0\ \text{mg of}\ {}^{99m}_{43}\text{Tc} \xrightarrow{\text{2 half-lives}}$$
$$20.0\ \text{mg of}\ {}^{99m}_{43}\text{Tc} \xrightarrow{\text{3 half-lives}} 10.0\ \text{mg of}\ {}^{99m}_{43}\text{Tc (3 SFs)}$$

d. $24\ \cancel{\text{h}} \times \dfrac{1\ \text{half-life}}{6.0\ \cancel{\text{h}}} = 4.0\ \text{half-lives}$

$$80.0\ \text{mg of}\ {}^{99m}_{43}\text{Tc} \xrightarrow{\text{1 half-life}} 40.0\ \text{mg of}\ {}^{99m}_{43}\text{Tc} \xrightarrow{\text{2 half-lives}} 20.0\ \text{mg of}\ {}^{99m}_{43}\text{Tc} \xrightarrow{\text{3 half-lives}}$$
$$10.0\ \text{mg of}\ {}^{99m}_{43}\text{Tc} \xrightarrow{\text{4 half-lives}} 5.00\ \text{mg of}\ {}^{99m}_{43}\text{Tc (3 SFs)}$$

16.33 The radiation level in a radioactive sample is cut in half with each half-life; the half-life of Sr-85 is 65 days.

a. For the radiation level to drop to one-fourth of its original level, $\frac{1}{4} = \frac{1}{2} \times \frac{1}{2}$ or two half-lives

$$2 \cancel{\text{half-lives}} \times \frac{65 \text{ days}}{1 \cancel{\text{half-life}}} = 130 \text{ days (2 SFs)}$$

b. For the radiation level to drop to one-eighth of its original level, $\frac{1}{8} = \frac{1}{2} \times \frac{1}{2} \times \frac{1}{2}$ or three half-lives

$$3 \cancel{\text{half-lives}} \times \frac{65 \text{ days}}{1 \cancel{\text{half-life}}} = 195 \text{ days (2 SFs)}$$

16.35 a. Since the elements calcium and phosphorus are part of bone, any calcium or phosphorus atom, regardless of isotope, will be carried to and become part of the bony structures of the body. Once there, the radiation emitted by any radioactive isotope can be used to diagnose or treat bone diseases.

b. Strontium (Sr) acts much like calcium (Ca) because both are Group 2A (2) elements. The body will accumulate radioactive strontium in bones in the same way that it incorporates calcium. Radioactive strontium is harmful to children because the radiation it produces causes more damage in cells that are dividing rapidly.

16.37 $4.0 \cancel{\text{mL solution}} \times \frac{45 \ \mu\text{Ci}}{1 \cancel{\text{mL solution}}} = 180 \ \mu\text{Ci}$ of selenium-75 (2 SFs)

16.39 Nuclear fission is the splitting of a large atom into smaller fragments with a simultaneous release of large amounts of energy.

16.41 ${}^{1}_{0}n + {}^{235}_{92}\text{U} \longrightarrow {}^{131}_{50}\text{Sn} + {}^{103}_{42}\text{Mo} + 2{}^{1}_{0}n + \text{energy} \quad ? = {}^{103}_{42}\text{Mo}$

16.43 a. Neutrons bombard a nucleus in the fission process.

b. The nuclear process that occurs in the Sun is fusion.

c. Fission is the process in which a large nucleus splits into smaller nuclei.

d. Fusion is the process in which small nuclei combine to form larger nuclei.

16.45

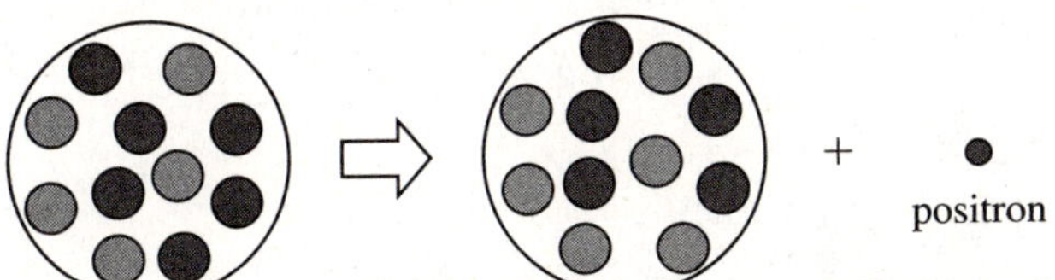

16.47

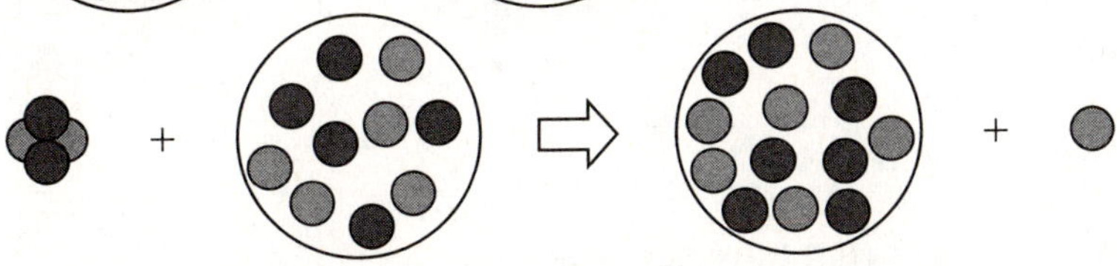

16.49 Half of a radioactive sample decays with each half-life:

$6.4 \ \mu\text{Ci of } {}^{14}_{6}\text{C} \xrightarrow{\text{1 half-life}} 3.2 \ \mu\text{Ci of } {}^{14}_{6}\text{C} \xrightarrow{\text{2 half-lives}} 1.6 \ \mu\text{Ci of } {}^{14}_{6}\text{C} \xrightarrow{\text{3 half-lives}}$
$0.80 \ \mu\text{Ci of } {}^{14}_{6}\text{C}$

$\therefore$ the activity of carbon-14 drops to 0.80 μCi in three half-lives or 3×5730 years, which makes the age of the painting 17 200 years.

16.51 a. ${}^{25}_{11}\text{Na}$ has 11 protons and $25 - 11 = 14$ neutrons.

b. ${}^{61}_{28}\text{Ni}$ has 28 protons and $61 - 28 = 33$ neutrons.

c. ${}^{84}_{37}\text{Rb}$ has 37 protons and $84 - 37 = 47$ neutrons.

d. ${}^{110}_{47}\text{Ag}$ has 47 protons and $110 - 47 = 63$ neutrons.

16.53 **a.** gamma emission
b. positron emission
c. alpha decay

16.55 **a.** ${}^{225}_{90}\text{Th} \longrightarrow {}^{221}_{88}\text{Ra} + {}^{4}_{2}\text{He}$ **b.** ${}^{210}_{83}\text{Bi} \longrightarrow {}^{206}_{81}\text{Tl} + {}^{4}_{2}\text{He}$
c. ${}^{137}_{55}\text{Cs} \longrightarrow {}^{137}_{56}\text{Ba} + {}^{0}_{-1}e$ **d.** ${}^{126}_{50}\text{Sn} \longrightarrow {}^{126}_{51}\text{Sb} + {}^{0}_{-1}e$
e. ${}^{18}_{9}\text{F} \longrightarrow {}^{18}_{8}\text{O} + {}^{0}_{+1}e$

16.57 **a.** ${}^{4}_{2}\text{He} + {}^{14}_{7}\text{N} \longrightarrow {}^{17}_{8}\text{O} + {}^{1}_{1}\text{H}$ $? = {}^{17}_{8}\text{O}$
b. ${}^{4}_{2}\text{He} + {}^{27}_{13}\text{Al} \longrightarrow {}^{30}_{14}\text{Si} + {}^{1}_{1}\text{H}$ $? = {}^{1}_{1}\text{H}$
c. ${}^{1}_{0}n + {}^{235}_{92}\text{U} \longrightarrow {}^{90}_{38}\text{Sr} + 3{}^{1}_{0}n + {}^{143}_{54}\text{Xe}$ $? = {}^{143}_{54}\text{Xe}$
d. ${}^{23\text{m}}_{12}\text{Mg} \longrightarrow {}^{23}_{12}\text{Mg} + {}^{0}_{0}\gamma$ $? = {}^{23}_{12}\text{Mg}$

16.59 **a.** ${}^{16}_{8}\text{O} + {}^{16}_{8}\text{O} \longrightarrow {}^{28}_{14}\text{Si} + {}^{4}_{2}\text{He}$ **b.** ${}^{18}_{8}\text{O} + {}^{249}_{98}\text{Cf} \longrightarrow {}^{263}_{106}\text{Sg} + 4{}^{1}_{0}n$
c. ${}^{222}_{86}\text{Rn} \longrightarrow {}^{218}_{84}\text{Po} + {}^{4}_{2}\text{He}$ **d.** ${}^{80}_{38}\text{Sr} \longrightarrow {}^{80}_{37}\text{Rb} + {}^{0}_{+1}e$

16.61 $120\ \cancel{\text{nCi}} \times \dfrac{1 \times 10^{-9}\ \cancel{\text{Ci}}}{1\ \cancel{\text{nCi}}} \times \dfrac{3.7 \times 10^{10}\ \text{Bq}}{1\ \cancel{\text{Ci}}} = 4400\ \text{Bq} = 4.4 \times 10^{3}\ \text{Bq}\ (2\ \text{SFs})$

16.63 Half of a radioactive sample decays with each half-life:
$1.2\ \text{mg of } {}^{32}_{15}\text{P} \xrightarrow{\text{1 half-life}} 0.60\ \text{mg of } {}^{32}_{15}\text{P} \xrightarrow{\text{2 half-lives}} 0.30\ \text{mg of } {}^{32}_{15}\text{P}$
∴ two half-lives must have elapsed during this time (28.6 days), yielding the half-life for phosphorus-32:

$$\frac{28.6\ \text{days}}{2\ \text{half-lives}} = 14.3\ \text{days/half-life}\ (3\ \text{SFs})$$

16.65 **a.** ${}^{47}_{20}\text{Ca} \rightarrow {}^{47}_{21}\text{Sc} + {}^{0}_{-1}e$
b. First, calculate the number of half-lives that have elapsed:

$$18\ \cancel{\text{days}} \times \frac{1\ \text{half-life}}{4.5\ \cancel{\text{days}}} = 4.0\ \text{half-lives}$$

Now we can calculate the number of milligrams of calcium-47 that remain:
$16\ \text{mg of } {}^{47}_{20}\text{Ca} \xrightarrow{\text{1 half-life}} 8.0\ \text{mg of } {}^{47}_{20}\text{Ca} \xrightarrow{\text{2 half-lives}} 4.0\ \text{mg of } {}^{47}_{20}\text{Ca} \xrightarrow{\text{3 half-lives}}$
$2.0\ \text{mg of } {}^{47}_{20}\text{Ca} \xrightarrow{\text{4 half-lives}} 1.0\ \text{mg of } {}^{47}_{20}\text{Ca}$

c. Half of a radioactive sample decays with each half-life:
$4.8\ \text{mg of } {}^{47}_{20}\text{Ca} \xrightarrow{\text{1 half-life}} 2.4\ \text{mg of } {}^{47}_{20}\text{Ca} \xrightarrow{\text{2 half-lives}} 1.2\ \text{mg of } {}^{47}_{20}\text{Ca}$
∴ two half-lives have elapsed.

$$2\ \cancel{\text{half-lives}} \times \frac{4.5\ \text{days}}{1\ \cancel{\text{half-life}}} = 9.0\ \text{days}\ (2\ \text{SFs})$$

16.67 First, calculate the number of half-lives that have elapsed:

$$24\ \cancel{\text{h}} \times \frac{1\ \text{half-life}}{6.0\ \cancel{\text{h}}} = 4.0\ \text{half-lives}$$

Now we can calculate the number of milligrams of technicium-99m that remain:
$120\ \text{mg of } {}^{99\text{m}}_{43}\text{Tc} \xrightarrow{\text{1 half-life}} 60.\ \text{mg of } {}^{99\text{m}}_{43}\text{Tc} \xrightarrow{\text{2 half-lives}} 30.\ \text{mg of } {}^{99\text{m}}_{43}\text{Tc} \xrightarrow{\text{3 half-lives}}$
$15\ \text{mg of } {}^{99\text{m}}_{43}\text{Tc} \xrightarrow{\text{4 half-lives}} 7.5\ \text{mg of } {}^{99\text{m}}_{43}\text{Tc}\ (2\ \text{SFs})$

16.69 In the fission process, an atom splits into smaller nuclei with a simultaneous release of large amounts of energy. In fusion, two (or more) small nuclei combine (fuse) to form a larger nucleus, with a simultaneous release of large amounts of energy.

16.71 Fusion occurs naturally in the Sun and other stars.

16.73 a. ${}^{180}_{80}\text{Hg} \longrightarrow {}^{176}_{78}\text{Pt} + {}^{4}_{2}\text{He}$

b. ${}^{198}_{79}\text{Au} \longrightarrow {}^{198}_{80}\text{Hg} + {}^{0}_{-1}e$

c. ${}^{82}_{37}\text{Rb} \longrightarrow {}^{82}_{36}\text{Kr} + {}^{0}_{+1}e$

16.75 ${}^{1}_{0}n + {}^{238}_{92}\text{U} \longrightarrow {}^{239}_{93}\text{Np} + {}^{0}_{-1}e$

16.77 First, calculate the number of half-lives that have elapsed since the technician was exposed:

$$36\,\cancel{\text{h}} \times \frac{1\text{ half-life}}{12\,\cancel{\text{h}}} = 3.0\text{ half-lives}$$

Because the activity of a radioactive sample is cut in half with each half-life, the activity must have been double its present value before each half-life. For 3.0 half-lives, we need to double the value three times:

$$2.0\ \mu\text{Ci of }{}^{42}_{19}\text{K} \xleftarrow{\text{1 half-life}} 4.0\ \mu\text{Ci of }{}^{42}_{19}\text{K} \xleftarrow{\text{2 half-lives}} 8.0\ \mu\text{Ci of }{}^{42}_{19}\text{K} \xleftarrow{\text{3 half-lives}} 16\ \mu\text{Ci of }{}^{42}_{19}\text{K}$$

∴ the activity of the sample when the technician was exposed was 16 μCi. (2 SFs)

16.79 First, calculate the number of half-lives that have elapsed:

$$27\,\cancel{\text{d}} \times \frac{1\text{ half-life}}{4.5\,\cancel{\text{d}}} = 6.0\text{ half-lives}$$

Because the activity of a radioactive sample is cut in half with each half-life, the activity must have been double its present value before each half-life. For 6.0 half-lives, we need to double the value six times:

$$1.0\ \mu\text{Ci of }{}^{47}_{20}\text{Ca} \xleftarrow{\text{1 half-life}} 2.0\ \mu\text{Ci of }{}^{47}_{20}\text{Ca} \xleftarrow{\text{2 half-lives}} 4.0\ \mu\text{Ci of }{}^{47}_{20}\text{Ca} \xleftarrow{\text{3 half-lives}} 8.0\ \mu\text{Ci of }{}^{47}_{20}\text{Ca}$$
$$\xleftarrow{\text{4 half-lives}} 16\ \mu\text{Ci of }{}^{47}_{20}\text{Ca} \xleftarrow{\text{5 half-lives}} 32\ \mu\text{Ci of }{}^{47}_{20}\text{Ca} \xleftarrow{\text{6 half-lives}} 64\ \mu\text{Ci of }{}^{47}_{20}\text{Ca}$$

∴ the initial activity of the sample was 64 μCi (2 SFs).

16.81 Half of a radioactive sample decays with each half-life:

$$64\ \mu\text{Ci of }{}^{201}_{81}\text{Tl} \xrightarrow{\text{1 half-life}} 32\ \mu\text{Ci of }{}^{201}_{81}\text{Tl} \xrightarrow{\text{2 half-lives}} 16\ \mu\text{Ci of }{}^{201}_{81}\text{Tl} \xrightarrow{\text{3 half-lives}}$$
$$8.0\ \mu\text{Ci of }{}^{201}_{81}\text{Tl} \xrightarrow{\text{4 half-lives}} 4.0\ \mu\text{Ci of }{}^{201}_{81}\text{Tl}$$

∴ four half-lives must have elapsed during this time (12 days), yielding the half-life for thallium-201:

$$\frac{12\text{ days}}{4\text{ half-lives}} = 3.0\text{ days/half-life (2 SFs)}$$

16.83 ${}^{48}_{20}\text{Ca} + {}^{244}_{94}\text{Pu} \rightarrow {}^{292}_{114}\text{Fl}$

Answers to Combining Ideas from Chapters 15 and 16

CI.29 a. $C_2O_4^{2-}(aq) \rightarrow 2CO_2(g)$

Balance charge with e^-: $C_2O_4^{2-}(aq) \rightarrow 2CO_2(g) + 2e^-$

b. $MnO_4^-(aq) \rightarrow Mn^{2+}(aq)$

Balance O with H_2O: $MnO_4^-(aq) \rightarrow Mn^{2+}(aq) + 4H_2O(l)$

Balance H with H^+: $8H^+(aq) + MnO_4^-(aq) \rightarrow Mn^{2+}(aq) + 4H_2O(l)$

Balance charge with e^-: $5e^- + 8H^+(aq) + MnO_4^-(aq) \rightarrow Mn^{2+}(aq) + 4H_2O(l)$

balanced reduction half-reaction

c. $2 \times [5e^- + 8H^+(aq) + MnO_4^-(aq) \rightarrow Mn^{2+}(aq) + 4H_2O(l)]$

$5 \times [C_2O_4^{2-}(aq) \rightarrow 2CO_2(g) + 2e^-]$

Overall: $16H^+(aq) + 2MnO_4^-(aq) + 5C_2O_4^{2-}(aq) \rightarrow 2Mn^{2+}(aq) + 10CO_2(g) + 8H_2O(l)$

d. In the titration equation, 1 mol of MnO_4^- reacts with 1 mol of $C_2O_4^{2-}$.

$$0.758 \text{ g } \cancel{Na_2C_2O_4} \times \frac{1 \text{ mol } \cancel{Na_2C_2O_4}}{134.00 \text{ g } \cancel{Na_2C_2O_4}} \times \frac{1 \text{ mol } \cancel{C_2O_4^{2-}}}{1 \text{ mol } \cancel{Na_2C_2O_4}} \times \frac{2 \text{ mol } \cancel{MnO_4^-}}{5 \text{ mol } \cancel{C_2O_4^{2-}}} \times \frac{1 \text{ mol } KMnO_4}{1 \text{ mol } \cancel{MnO_4^-}}$$

$= 0.002\,26$ mol of $KMnO_4$

$$24.6 \text{ } \cancel{\text{mL } KMnO_4 \text{ solution}} \times \frac{1 \text{ L solution}}{1000 \text{ } \cancel{\text{mL solution}}} = 0.0246 \text{ L of } KMnO_4 \text{ solution}$$

$$\text{molarity of } KMnO_4 = \frac{\text{moles of solute}}{\text{liters of solution}} = \frac{0.002\,26 \text{ mol } KMnO_4}{0.0246 \text{ L solution}}$$

$= 0.0919$ M $KMnO_4$ solution (3 SFs)

CI.31 a.

$$0.121 \text{ } \cancel{\text{g Mg}} \times \frac{1 \text{ } \cancel{\text{mol Mg}}}{24.31 \text{ } \cancel{\text{g Mg}}} \times \frac{1 \text{ mol } H_2}{1 \text{ } \cancel{\text{mol Mg}}} = 0.004\,98 \text{ mol of } H_2 \text{ (smaller amount of product)}$$

$$50.0 \text{ } \cancel{\text{mL solution}} \times \frac{1 \text{ } \cancel{\text{L}}}{1000 \text{ } \cancel{\text{mL}}} \times \frac{1.00 \text{ } \cancel{\text{mol HCl}}}{1 \text{ } \cancel{\text{L solution}}} \times \frac{1 \text{ mol } H_2}{2 \text{ } \cancel{\text{mol HCl}}} = 0.0250 \text{ mol of } H_2$$

∴ Mg is the limiting reactant.

b. $T = 33\,°C + 273 = 306$ K; $P = 750.$ mmHg; $n = 0.004\,98$ mol of H_2 (from **a**)

$$V = \frac{nRT}{P} = \frac{(0.004\,98 \text{ } \cancel{\text{mol}})\left(\dfrac{62.4 \text{ } \cancel{\text{L}} \cdot \cancel{\text{mmHg}}}{\cancel{\text{mol}} \cdot \cancel{\text{K}}}\right)(306 \text{ } \cancel{\text{K}})}{750. \text{ } \cancel{\text{mmHg}}} \times \frac{1000 \text{ mL}}{1 \text{ } \cancel{\text{L}}} = 127 \text{ mL of } H_2 \text{ (3 SFs)}$$

c. $$50.0 \text{ } \cancel{\text{mL solution}} \times \frac{1.00 \text{ g}}{1 \text{ } \cancel{\text{mL}}} = 50.0 \text{ g of solution}$$

$\Delta T = T_{final} - T_{initial} = 33\,°C - 22\,°C = 11\,°C$

$$50.0 \text{ } \cancel{\text{g solution}} \times \frac{4.184 \text{ J}}{\cancel{\text{g}}\,\cancel{°C}} \times 11 \text{ } \cancel{°C} = 2.3 \times 10^3 \text{ J (2 SFs)}$$

d. $$\frac{2.3 \times 10^3 \text{ J}}{0.121 \text{ g Mg}} = 1.9 \times 10^4 \text{ J/g of Mg (2 SFs)}$$

$$\frac{2.3 \times 10^3 \text{ } \cancel{\text{J}}}{0.121 \text{ } \cancel{\text{g Mg}}} \times \frac{1 \text{ kJ}}{1000 \text{ } \cancel{\text{J}}} \times \frac{24.31 \text{ } \cancel{\text{g Mg}}}{1 \text{ mol Mg}} = 460 \text{ kJ/mol of Mg (2 SFs)}$$

CI.33 a.

Isotope	Number of Protons	Number of Neutrons	Number of Electrons
${}^{27}_{14}Si$	14	13	14
${}^{28}_{14}Si$	14	14	14
${}^{29}_{14}Si$	14	15	14
${}^{30}_{14}Si$	14	16	14
${}^{31}_{14}Si$	14	17	14

b. Electron configuration of Si: $1s^22s^22p^63s^23p^2$
Abbreviated electron configuration: $[Ne]3s^23p^2$

c.

$${}^{28}_{14}Si \quad 27.977 \times \frac{92.23}{100} = 25.80 \text{ amu}$$

$${}^{29}_{14}Si \quad 28.977 \times \frac{4.683}{100} = 1.357 \text{ amu}$$

$${}^{30}_{14}Si \quad 29.974 \times \frac{3.087}{100} = 0.9253 \text{ amu}$$

Atomic mass of Si = 28.08 amu

d. ${}^{27}_{14}Si \rightarrow {}^{27}_{13}Al + {}^{0}_{+1}e$

${}^{31}_{14}Si \rightarrow {}^{31}_{15}P + {}^{0}_{-1}e$

e. 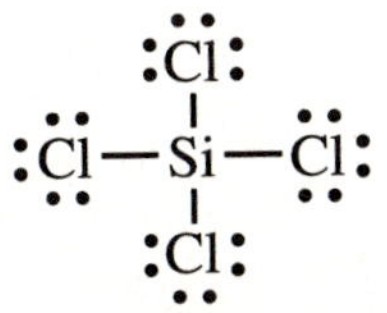

The central atom Si has four electron groups bonded to four chlorine atoms; $SiCl_4$ has a tetrahedral shape.

f. Half of a radioactive sample decays with each half-life:

$16\ \mu Ci$ of ${}^{31}_{14}Si \xrightarrow{\text{1 half-life}} 8.0\ \mu Ci$ of ${}^{31}_{14}Si \xrightarrow{\text{2 half-lives}} 4.0\ \mu Ci$ of ${}^{31}_{14}Si \xrightarrow{\text{3 half-lives}} 2.0\ \mu Ci$ of ${}^{31}_{14}Si$

Therefore, 3 half-lives have passed.

$$3\ \cancel{\text{half-lives}} \times \frac{2.6 \text{ h}}{1\ \cancel{\text{half-life}}} = 7.8 \text{ h (2 SFs)}$$

CI.35 a. ${}^{238}_{92}U \rightarrow {}^{234}_{90}Th + {}^{4}_{2}He$ $\quad ? = {}^{4}_{2}He$

b. ${}^{234}_{90}Th \rightarrow {}^{234}_{91}Pa + {}^{0}_{-1}e$ $\quad ? = {}^{234}_{91}Pa$

c. ${}^{226}_{88}Ra \rightarrow {}^{222}_{86}Rn + {}^{4}_{2}He$ $\quad ? = {}^{226}_{88}Ra$

17

Organic Chemistry

Learning Goals

- Write the IUPAC names and draw the condensed structural formulas for alkanes with substituents.
- Write the IUPAC names and formulas for alkenes and alkynes; draw the condensed structural formulas for the monomers that form a polymer.
- Describe the bonding in benzene; name aromatic compounds and draw their condensed structural formulas.
- Write the IUPAC and common names for alcohols, phenols, and ethers; draw the condensed structural formulas when given their names.
- Write the IUPAC and common names for aldehydes and ketones; draw the condensed structural formulas when given their names.
- Write the IUPAC and common names for carboxylic acids and esters; draw the condensed structural formulas when given their names.
- Write the common names for amines and the IUPAC and common names for amides; draw the condensed structural formulas when given their names.

Chapter Outline

Chapter Opener: Firefighter

17.1 Alkanes and Naming Substituents

17.2 Alkenes, Alkynes, and Polymers

17.3 Aromatic Compounds

Chemistry Link to the Environment: Some Common Aromatic Compounds

17.4 Alcohols, Phenols, and Ethers

Chemistry Link to Health: Some Important Alcohols and Phenols

17.5 Aldehydes and Ketones

Chemistry Link to the Environment: Vanilla

17.6 Carboxylic Acids and Esters

Chemistry Link to Health: Carboxylic Acids in Metabolism

17.7 Amines and Amides

Chemistry Link to the Environment: Alkaloids: Amines in Plants

Answers and Solutions to Text Problems

17.1 **a.** Pentane has a carbon chain of five carbon atoms.
b. Heptane has a carbon chain of seven carbon atoms.

17.3 Two structures are isomers if they have the same molecular formula, but different arrangements of atoms.

a. The condensed structural formulas represent the same molecule; the only difference is due to rotation of the structure. Each has a $CH_3—$ group attached to the middle carbon in a three-carbon chain.

b. The molecular formula of both these condensed structural formulas is C_6H_{14}. However, they represent structural isomers because the C atoms are bonded in a different order; they have different arrangements. In the first, there is a $CH_3—$ group attached to carbon 3 of a five-carbon chain, and in the other, there are $CH_3—$ groups attached to carbon 2 and carbon 3 of a four-carbon chain.

c. The molecular formula of both these skeletal formulas is C_8H_{18}. However, they represent structural isomers because the C atoms are bonded in a different order; they have different arrangements. In the first, there is a $CH_3—$ group attached to carbon 4 of a seven-carbon chain, and in the other, there is a $CH_3—$ group attached to carbon 2 of a seven-carbon chain.

17.5 **a.** 2-fluorobutane
b. 2,2-dimethylpropane
c. 2-chloro-3-methylpentane

17.7 Draw the main chain with the number of carbon atoms indicated by the alkane name. For example, butane has a main chain of four carbon atoms, and hexane has a main chain of six carbon atoms. Attach substituents on the carbon atoms indicated. For example, in 3-methylpentane, a $CH_3—$ group is bonded to carbon 3 of a five-carbon chain.

a. 2-methylbutane 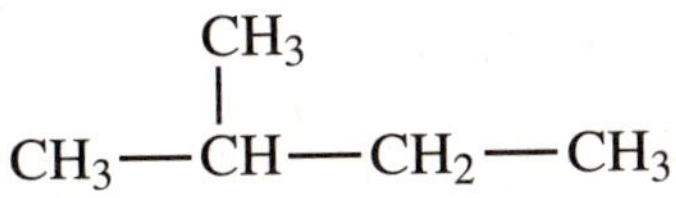

b. 3,3-dichloropentane 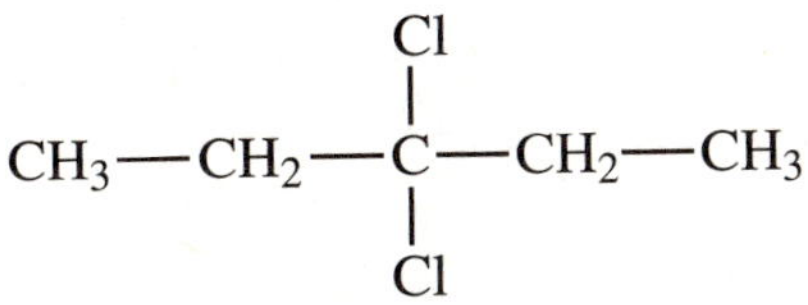

c. 2,3,5-trimethylhexane 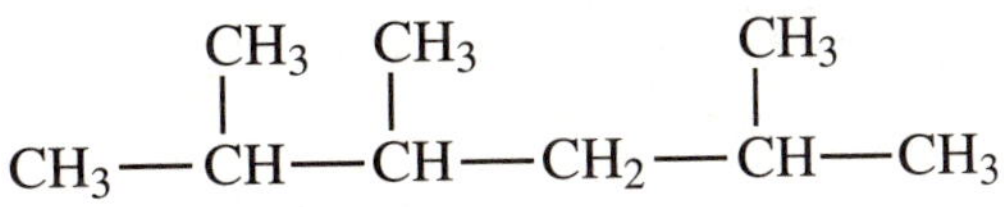

17.9 **a.** 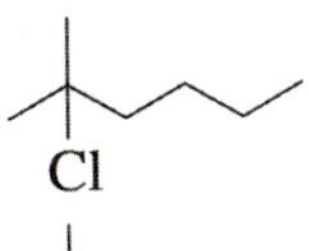

b. 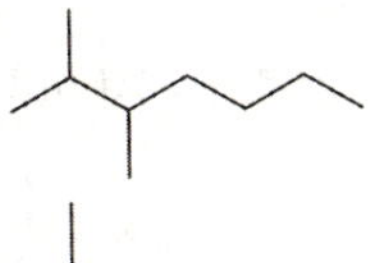

c. 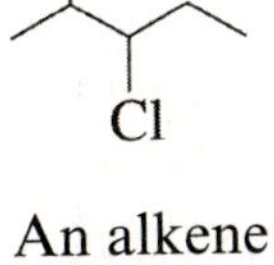

17.11 **a.** An alkene has a carbon–carbon double bond.
b. An alkyne has a carbon–carbon triple bond.
c. An alkene has a carbon–carbon double bond.

17.13 **a.** The two-carbon compound with a double bond is named ethene.
b. This is a three-carbon alkene with a methyl substituent. The name is 2-methylpropene.
c. The five-carbon compound with a triple bond between carbon 2 and carbon 3 is named 2-pentyne.

17.15 a. Propene is the three-carbon alkene. $H_2C{=}CH{-}CH_3$

b. 1-Hexyne is the six-carbon compound with a triple bond between carbon 1 and carbon 2. $HC{\equiv}C{-}CH_2{-}CH_2{-}CH_2{-}CH_3$

c. 2-Methyl-1-butene has a four-carbon chain with a double bond between carbon 1 and carbon 2 and a methyl group attached to carbon 2.

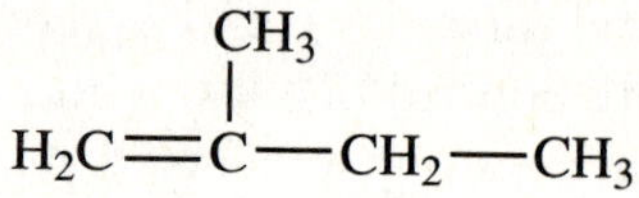

17.17 A polymer is a long-chain molecule consisting of small units (called monomers) that are repeated many times.

17.19 Polypropylene is a polymer of the monomer propylene (propene).

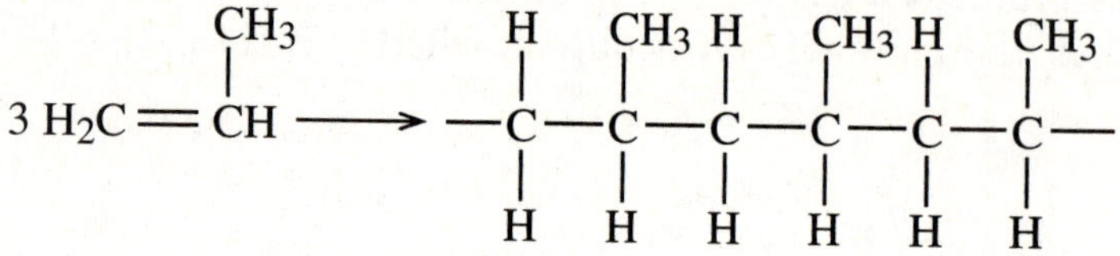

17.21 The six-carbon ring with alternating single and double bonds is benzene.

a. 2-chlorotoluene

b. ethylbenzene

c. 1,3,5-trichlorobenzene

17.23 a.

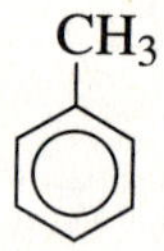

b.

c.

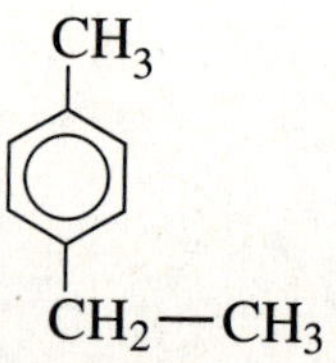

17.25 a. This compound has a two-carbon chain (ethane). The final *e* is dropped, and *ol* added to indicate an alcohol. The IUPAC name is ethanol.

b. This compound has a four-carbon chain with a hydroxyl group attached to carbon 2. The IUPAC name is 2-butanol.

c. This compound is the six-carbon benzene ring with a hydroxyl group attached. The IUPAC name is phenol.

17.27 a. 1-Propanol has a three-carbon chain with a hydroxyl group attached to carbon 1. $CH_3{-}CH_2{-}CH_2{-}OH$

b. Methyl alcohol has a hydroxyl group attached to a one-carbon alkane. $CH_3{-}OH$

c. 3-Pentanol has a five-carbon chain with a hydroxyl group attached to carbon 3.

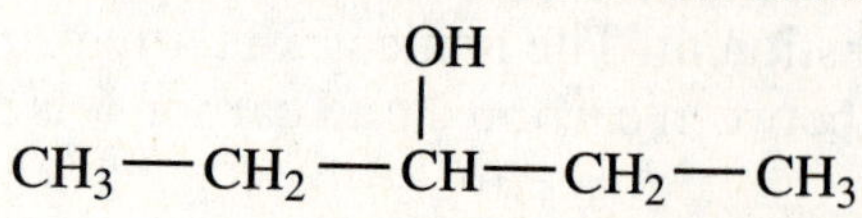

d. 2-Methyl-2-butanol has a four-carbon chain with a methyl and hydroxyl group attached to carbon 2.

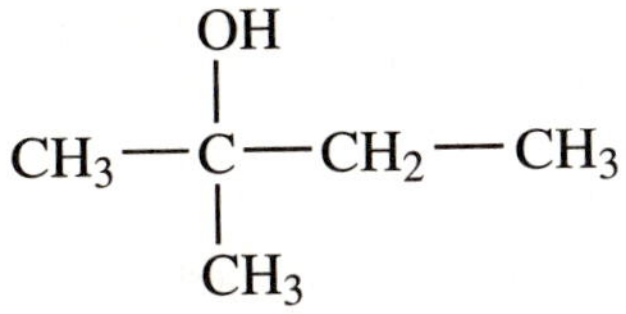

17.29 a. ethyl methyl ether **b.** dipropyl ether **c.** methyl propyl ether

17.31 a. acetaldehyde **b.** methyl propyl ketone **c.** formaldehyde

17.33 a. propanal **b.** 2-methyl-3-pentanone **c.** benzaldehyde

17.35 a. Acetaldehyde is the common name of the aldehyde with two carbons.

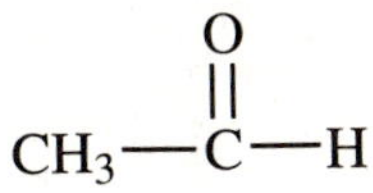

b. 2-Pentanone has a ketone group on carbon 2 of a five-carbon chain.

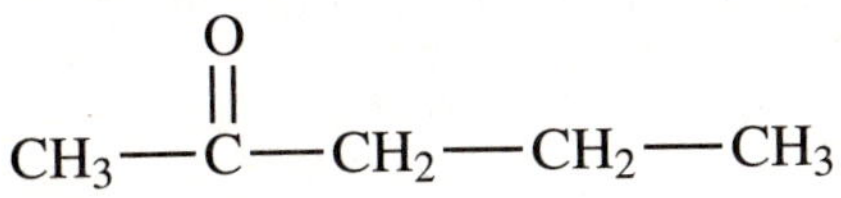

c. Butyl methyl ketone has a four-carbon group and a one-carbon group on either side of the carbonyl carbon.

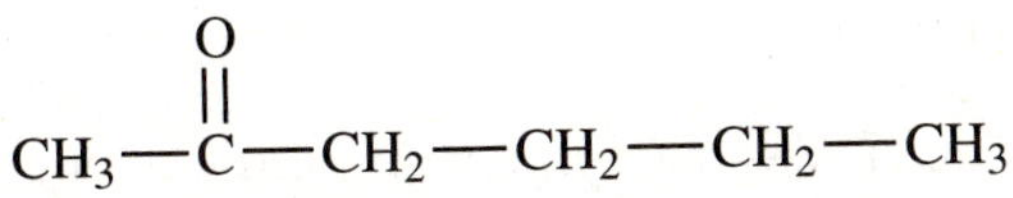

17.37 a. Ethanoic acid (acetic acid) is the carboxylic acid with two carbons.

b. Propanoic acid (propionic acid) is the three-carbon carboxylic acid.

c. 3-Methylhexanoic acid is a six-carbon carboxylic acid with a methyl group on carbon 3 of the chain.

d. 3,4-Dibromobenzoic acid is an aromatic carboxylic acid with bromine atoms on carbon 3 and carbon 4 of the ring.

17.39 a. Propionic acid is the common name of the three-carbon carboxylic acid.

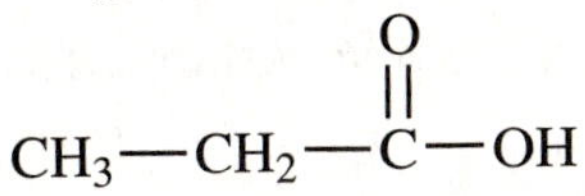

b. Benzoic acid is a carboxylic acid group attached to a benzene ring.

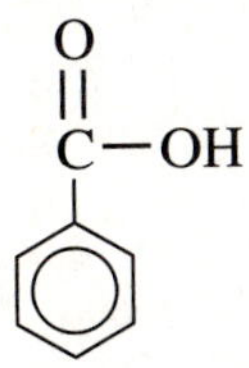

c. 2-Chloroethanoic acid is a carboxylic acid that has a two-carbon chain with a —Cl on carbon 2.

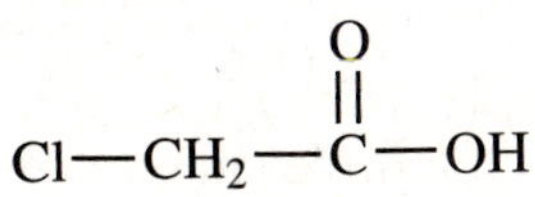

d. 3-Hydroxypropanoic acid is a carboxylic acid that has a three-carbon chain with a hydroxyl group on carbon 3.

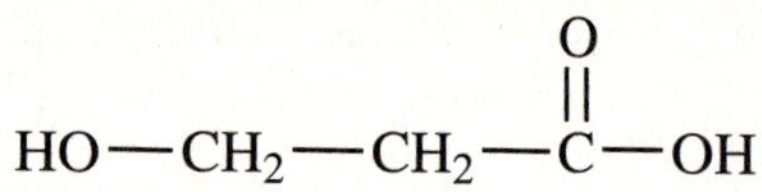

17.41 a. The ester has a one-carbon part from the alcohol methanol and a two-carbon part from the carboxylic acid acetic (ethanoic) acid.

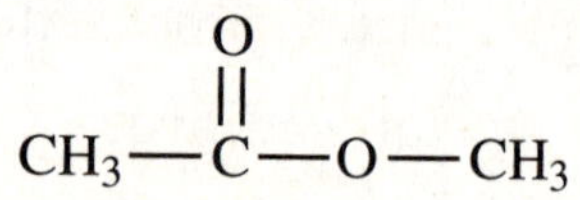

b. The ester has a one-carbon part from the alcohol methanol and a five-carbon part from the carboxylic acid pentanoic acid.

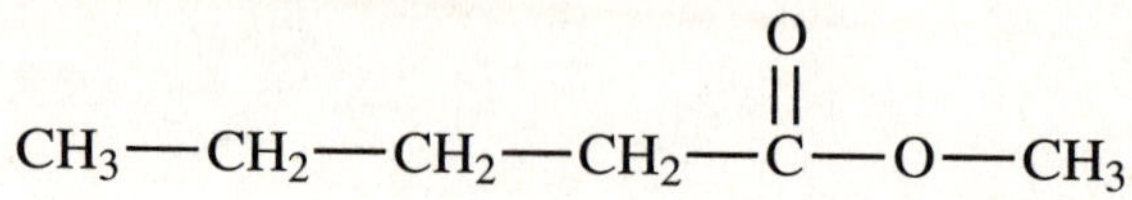

17.43 A carboxylic acid and an alcohol react to give an ester with the elimination of water.

a. 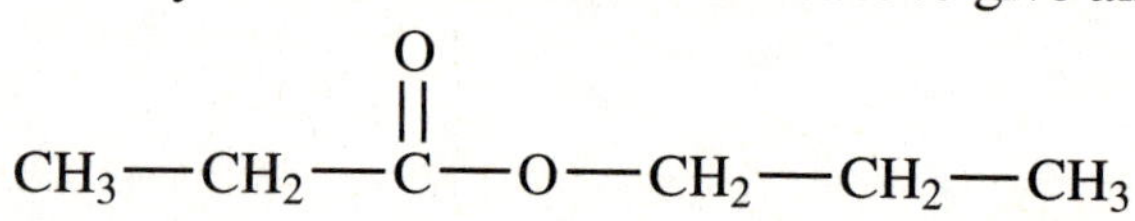

b. 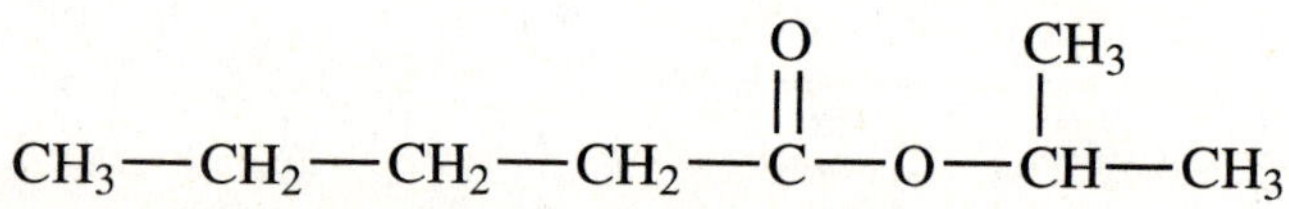

17.45 a. The name of this ester is methyl methanoate. The carbonyl portion of the ester contains one carbon; the name is derived from methanoic acid. The alkyl portion has one carbon, which is methyl.

b. The name of this ester is methyl ethanoate. The carbonyl portion of the ester contains two carbons; the name is derived from ethanoic acid. The alkyl portion has one carbon, which is methyl.

c. The name of this ester is ethyl propanoate. The carbonyl portion of the ester contains three carbons; the name is derived from propanoic acid. The alkyl portion has two carbons, which is ethyl.

17.47 a. Acetic acid is the two-carbon carboxylic acid. Methanol gives a one-carbon alkyl group.

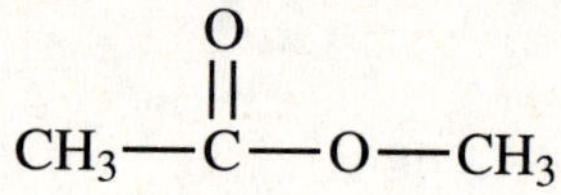

b. Formic acid is the one-carbon carboxylic acid bonded to the four-carbon 1-butanol.

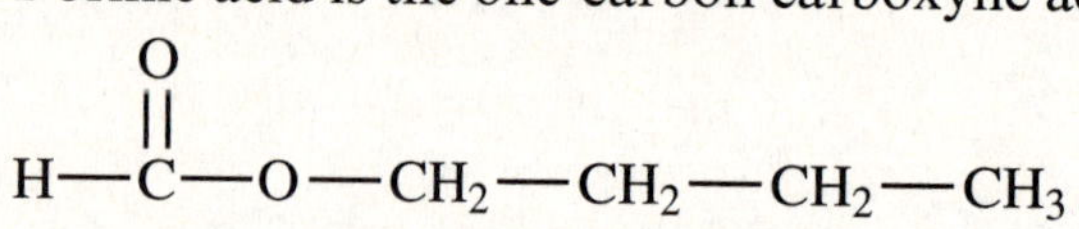

c. Pentanoic acid is the five-carbon carboxylic acid bonded to the two-carbon ethanol.

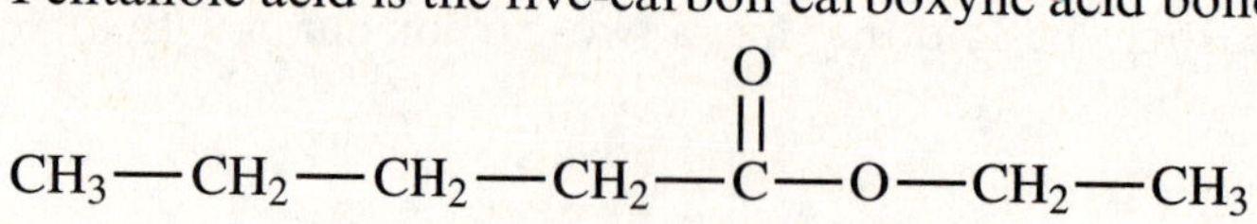

d. Propanoic acid is the three-carbon carboxylic acid bonded to the three-carbon 1-propanol.

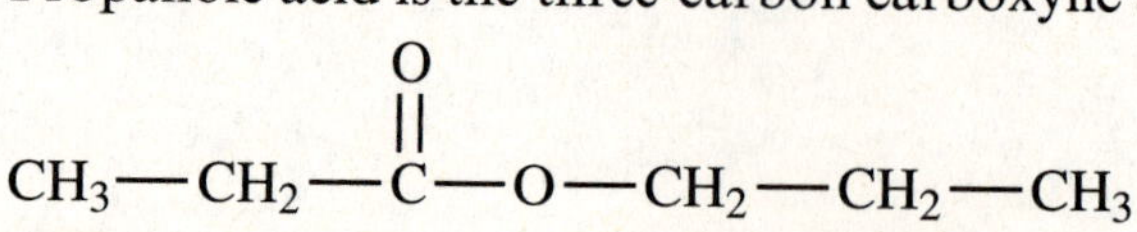

17.49 The common name of an amine consists of naming the alkyl groups bonded to the nitrogen atom in alphabetical order followed by *amine*.

a. A methyl group attached to $-NH_2$ is methylamine.

b. A methyl and a propyl group attached to the nitrogen atom would be named methylpropylamine.

c. One methyl and two ethyl groups attached to the nitrogen atom is named diethylmethylamine.

17.51 Carboxylic acids react with amines to form amides with the elimination of water.

a. 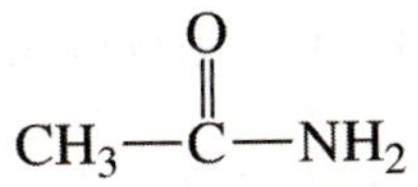

b. 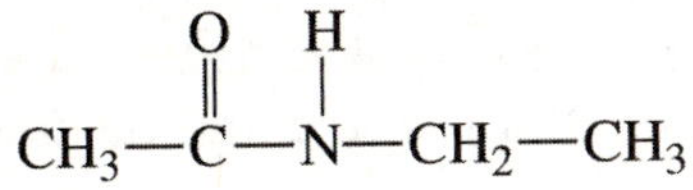

c. 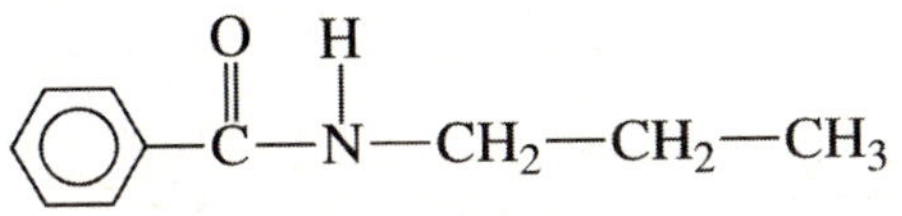

17.53 a. Ethanamide (acetamide) has a two-carbon carbonyl portion bonded to an amino group.

b. 2-Chlorobutanamide (2-chlorobutyramide) is a four-carbon amide with a $-Cl$ on carbon 2.

c. Methanamide (formamide) has only the carbonyl carbon bonded to the amino group.

17.55 a. Propionamide is the common name of the amide of propanoic acid, which has three carbon atoms.

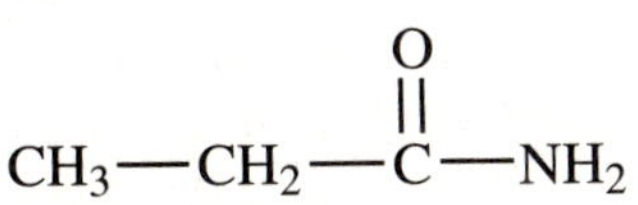

b. 2-Methylpentanamide is an amide of the five-carbon pentanoic acid with a methyl group attached to carbon 2.

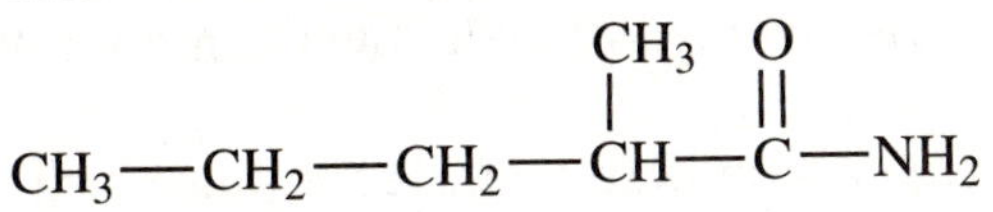

c. Methanamide is the simplest of the amides, with only one carbon atom.

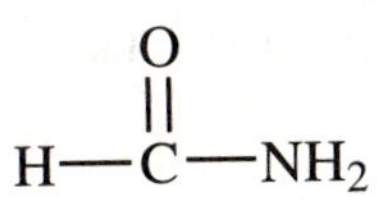

17.57 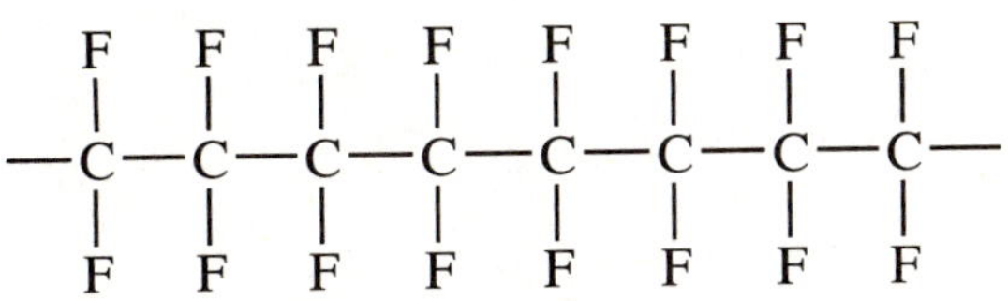

17.59 a. 3,7-dimethyl-6-octenal

b. The *en* signifies that a double bond is present.

c. The *al* signifies that an aldehyde is present.

17.61 a. 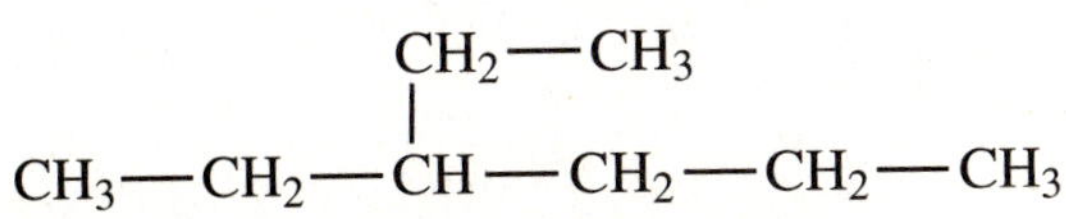

b. $CH_3-CH=CH-CH_2-CH_3$

c. $CH_3-C\equiv C-CH_2-CH_2-CH_3$

17.63 a. This compound contains a four-carbon chain with two methyl groups attached to carbon 2. The IUPAC name is 2,2-dimethylbutane.

b. This compound has a four-carbon chain with a triple bond between carbon 1 and carbon 2. The IUPAC name is 1-butyne.

c. This compound contains a five-carbon chain with a double bond between carbon 2 and carbon 3. The IUPAC name is 2-pentene.

17.65 a. This compound has a methyl group attached to the aromatic ring of benzene and is known as methylbenzene or toluene.

b. A benzene ring with a —Cl attached to carbon 1 and carbon 2 of the ring is named 1,2-dichlorobenzene.

c. A benzene ring with an ethyl group on carbon 1 and a methyl group on carbon 4 of the ring is named 1-ethyl-4-methylbenzene; if the name toluene is used as the parent compound, the ethyl group is considered to be on carbon 4 of the ring and the name would be 4-ethyltoluene.

17.67 a.

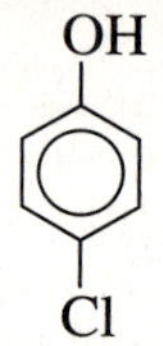

b.

$$CH_3-\underset{}{\overset{\displaystyle CH_3}{\overset{|}{C}H}}-\overset{\displaystyle OH}{\overset{|}{C}H}-CH_2-CH_3$$

c.

$$CH_3-CH_2-\overset{\displaystyle O}{\overset{\|}{C}}-CH_2-CH_3$$

17.69 a. With a hydroxyl group on carbon 3 and a —Cl on carbon 4 of the aromatic aldehyde benzaldehyde, the compound will be named 4-chloro-3-hydroxybenzaldehyde.

b. With a —Cl attached to carbon 3 of the three-carbon aldehyde, the compound will be named 3-chloropropanal.

c. With a —Cl attached to carbon 2 of the five-carbon ketone, the compound will be named 2-chloro-3-pentanone.

17.71 a. 4-Chlorobenzaldehyde is a benzene with an aldehyde group and a chlorine atom on carbon 4 of the ring.

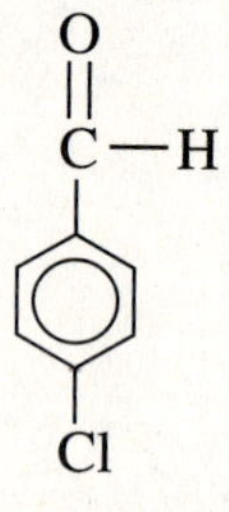

b. 3-Chloropropionaldehyde is a three-carbon aldehyde with a —Cl located two carbons from the carbonyl group.

$$Cl-CH_2-CH_2-\overset{\displaystyle O}{\overset{\|}{C}}-H$$

c. Ethyl methyl ketone (butanone) is a four-carbon ketone.

$$CH_3-\overset{\displaystyle O}{\overset{\|}{C}}-CH_2-CH_3$$

17.73 a. With a methyl group attached to carbon 3 of the four-carbon carboxylic acid, the compound is named 3-methylbutanoic acid.

b. The name of this ester is methyl benzoate. The carbonyl portion of the ester is derived from the aromatic benzoic acid. The alkyl portion has one carbon, which is methyl.

c. The name of this ester is ethyl propanoate. The carbonyl portion of the ester contains three carbons; the name is derived from propanoic acid. The alkyl portion has two carbons, which is ethyl.

17.75 a. $CH_3—CH_2—NH_2$

b. $CH_3—NH—CH_3$

c. 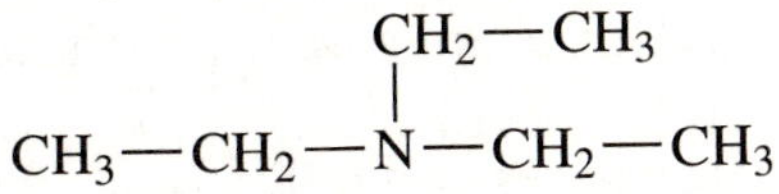

17.77 There are eight alcohols (isomers) that have the molecular formula $C_5H_{12}O$.

$CH_3—CH_2—CH_2—CH_2—CH_2—OH$ 1-pentanol

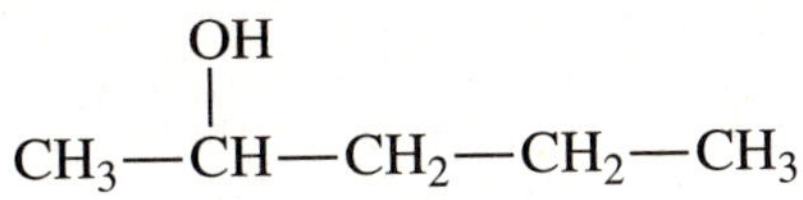 2-pentanol

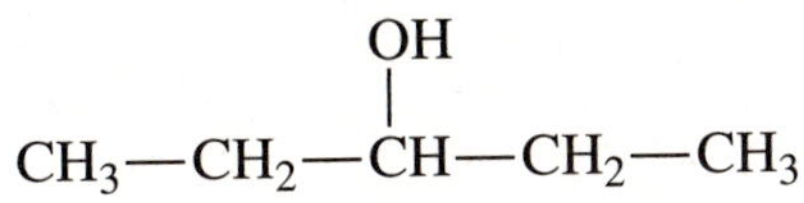 3-pentanol

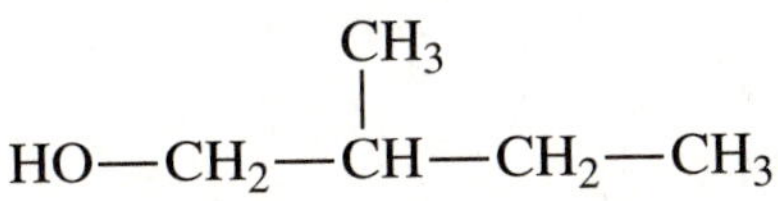 2-methyl-1-butanol

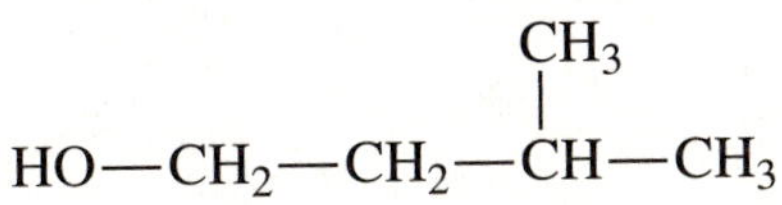 3-methyl-1-butanol

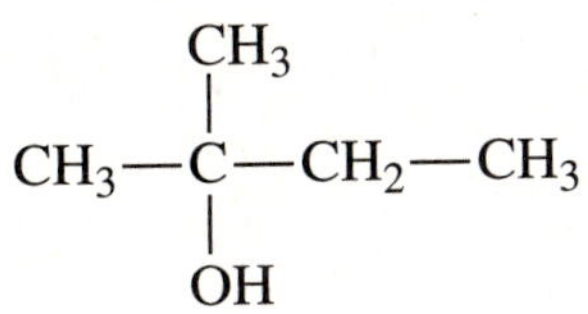 2-methyl-2-butanol

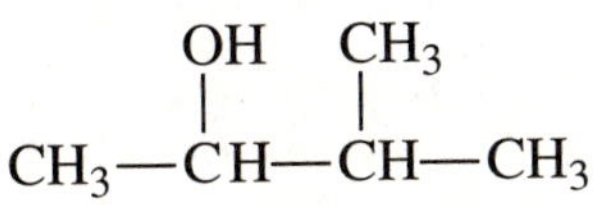 3-methyl-2-butanol

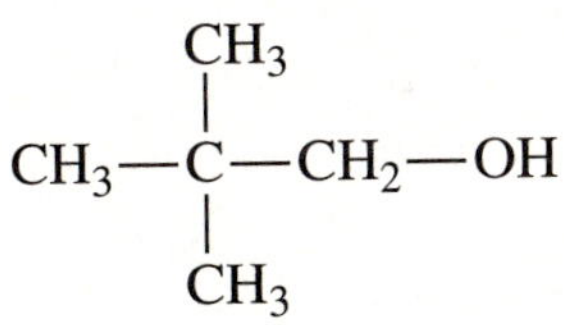 2,2-dimethyl-1-propanol

17.79

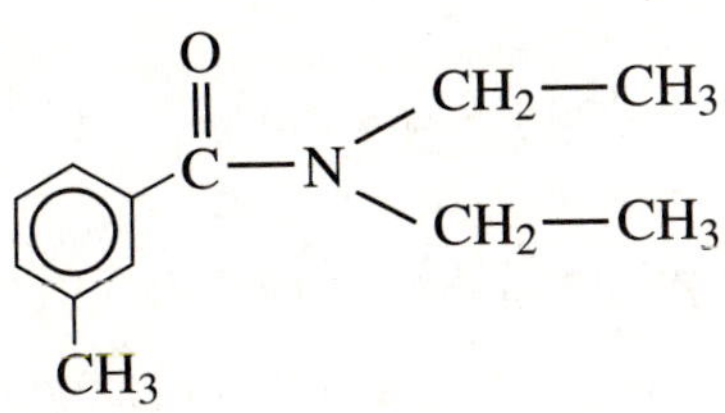

18
Biochemistry

Learning Goals

- Classify a carbohydrate as an aldose or ketose; draw the open-chain and Haworth structures for glucose, galactose, and fructose.
- Describe the monosaccharide units and linkages in disaccharides and polysaccharides.
- Describe some properties of lipids.
- Describe protein functions, and draw structures for amino acids and dipeptides.
- Identify the levels of structure of a protein.
- Describe the role of an enzyme in an enzyme-catalyzed reaction.
- Describe the structure of the nucleic acids in DNA and RNA.
- Describe the synthesis of protein from mRNA.

Chapter Outline

Chapter Opener: Forensic Toxicologist

18.1 Carbohydrates

Chemistry Link to Health: Hyperglycemia and Hypoglycemia

18.2 Disaccharides and Polysaccharides

Chemistry Link to Health: How Sweet Is My Sweetener?

18.3 Lipids

Chemistry Link to Health: Olestra: A Fat Substitute

Chemistry Link to Health: Trans Fatty Acids and Hydrogenation

18.4 Proteins

Chemistry Link to Health: Essential Amino Acids

18.5 Protein Structure

18.6 Proteins as Enzymes

18.7 Nucleic Acids

18.8 Protein Synthesis

Answers and Solutions to Text Problems

18.1 Hydroxyl (—OH) groups are found in all monosaccharides along with a carbonyl (C=O) group on the first or second carbon of the chain.

18.3 A ketopentose has five carbon atoms, several hydroxyl groups, and a ketone group.

18.5 **a.** This six-carbon monosaccharide has a carbonyl group on carbon 2; it is a ketohexose.
b. This five-carbon monosaccharide has a carbonyl group on carbon 1; it is an aldopentose.

18.7 In the open-chain structure of D-galactose, the hydroxyl group on carbon 4 extends to the left. In the open-chain structure of D-glucose, this hydroxyl goes to the right.

18.9 In the ring portion of the Haworth structure of D-glucose, there are five carbon atoms and an oxygen atom.

18.11 a. This is the α form because the —OH on carbon 2 is down.
b. This is the α form because the —OH on carbon 1 is down.

18.13 a. When this disaccharide is hydrolyzed, galactose and glucose are produced. The glycosidic bond is a β-1,4 bond since the ether bond is drawn up from carbon 1 of the galactose unit, which is on the left in the drawing, to carbon 4 of the glucose on the right. β-Lactose is the name of this disaccharide since the —OH group on carbon 1 of the glucose unit is drawn up.
b. When this disaccharide is hydrolyzed, two molecules of glucose are produced. The glycosidic bond is an α-1,4 bond since the ether bond is drawn down from carbon 1 of the glucose unit on the left to carbon 4 of the glucose on the right. α-Maltose is the name of this disaccharide since the —OH group on the rightmost glucose unit is drawn down.

18.15 a. Isomaltose is a disaccharide.
b. Isomaltose contains two molecules of α-glucose.
c. The glycosidic link in isomaltose is an α-1,6-glycosidic bond.
d. The downward position of the —OH group on carbon 1 of the second glucose makes it α-isomaltose.

18.17 a. Another name for ordinary table sugar is sucrose.
b. Lactose is the disaccharide found in milk and milk products.
c. Maltose is also called malt sugar.
d. Lactose contains the monosaccharides galactose and glucose.

18.19 a. Cellulose is a polysaccharide that is not digestible by humans.
b. Amylose and amylopectin (in starch) are the storage forms of carbohydrates in plants.
c. Amylose is the polysaccharide which contains only α-1,4-glycosidic bonds.
d. Glycogen is the most highly branched polysaccharide.

18.21 a. Lauric acid contains only carbon–carbon single bonds; it is saturated.
b. Linolenic acid has three carbon–carbon double bonds; it is polyunsaturated.
c. Stearic acid contains only carbon–carbon single bonds; it is saturated.

18.23 Glyceryl trimyristate (trimyristin) has three myristic acids (14-carbon saturated fatty acid) forming ester bonds with glycerol.

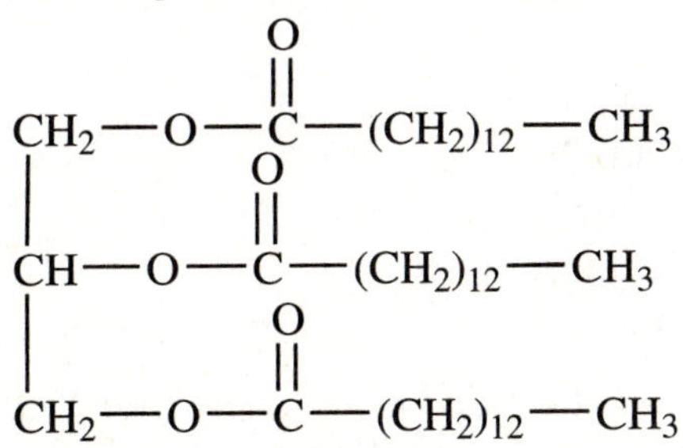

18.25 Safflower oil contains fatty acids with two or three carbon–carbon double bonds; olive oil contains a large amount of oleic acid, which has only one (monounsaturated) carbon–carbon double bond.

18.27 a. Partial hydrogenation means that some of the double bonds in the unsaturated fatty acids have been converted to single bonds by the addition of hydrogen.
b. Since the margarine has more saturated fatty acids than the original vegetable oil, the fatty acid chains pack together more tightly and remain solid at higher temperatures.

18.29

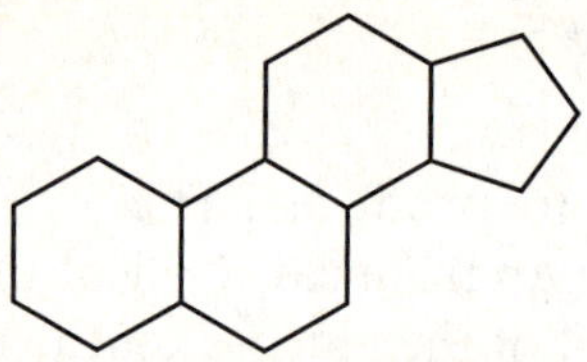

18.31 All amino acids contain a carboxylate ($—COO^-$) group and an ammonium ($—NH_3^+$) group on the alpha carbon.

18.33 a.

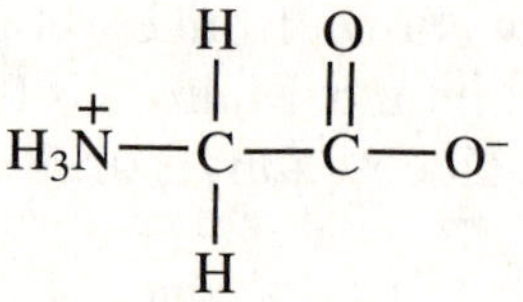

b.

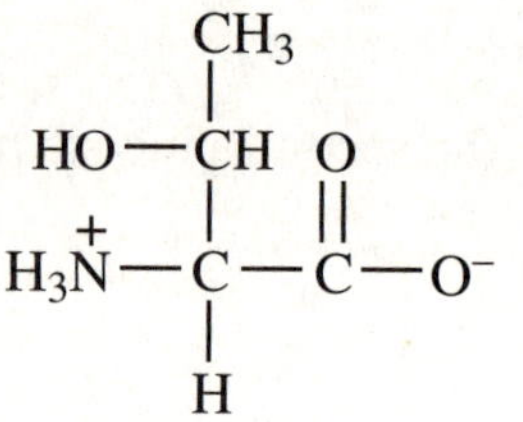

c.

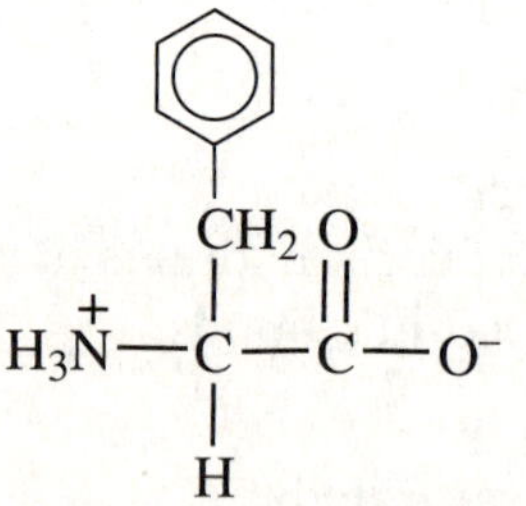

18.35 a. Glycine, which has only a —H as the R group, is hydrophobic (nonpolar).

b. Threonine has an R group that contains the polar —OH group; threonine is hydrophilic (polar, neutral).

c. Phenylalanine has an R group with a nonpolar benzene ring; phenylalanine is hydrophobic (nonpolar).

18.37 Amino acids have both three-letter and one-letter abbreviations.

a. Ala is the three-letter abbreviation for the amino acid alanine.

b. V is the one-letter abbreviation for the amino acid valine.

c. Lys is the three-letter abbreviation for the amino acid lysine.

d. C is the one-letter abbreviation for the amino acid cysteine.

18.39 In a peptide, the amino acids are joined by peptide bonds (amide bonds). The first amino acid has a free ammonium group and the last one has a free carboxylate group that ionize in biological environments.

a.

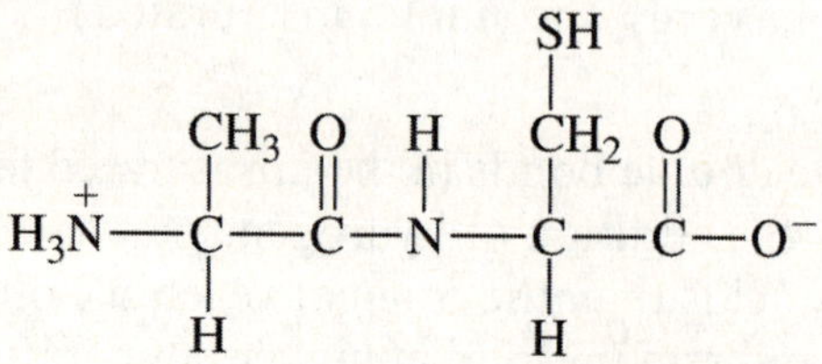

Ala-Cys, AC

b. 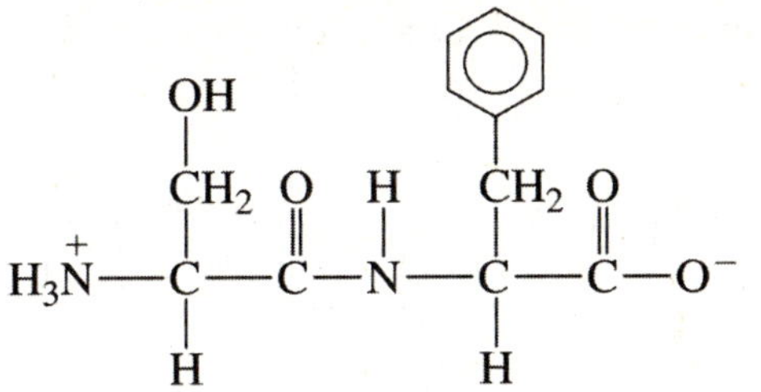

Ser-Phe, SF

c. 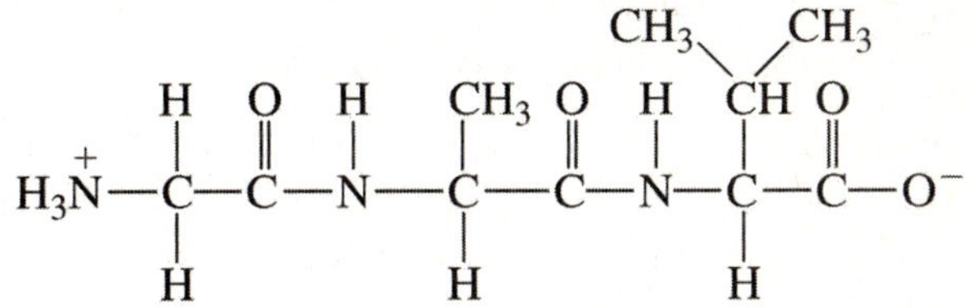

Gly-Ala-Val, GAV

18.41 The possible primary structures of a tripeptide containing one valine and two serines are Val-Ser-Ser, Ser-Val-Ser, and Ser-Ser-Val.

18.43 In the α helix, hydrogen bonds form between the oxygen atom in the carbonyl group and hydrogen in the amide group in the next turn of the helix of the polypeptide chain. In a β-pleated sheet, side-by-side hydrogen bonds occur between parallel peptides or across sections of a long polypeptide chain.

18.45 **a.** The two cysteine residues have —SH groups, which react to form a disulfide bond.
b. Serine has a polar —OH group that can form a hydrogen bond with the carboxyl group of aspartic acid.
c. Two leucine residues have nonpolar hydrocarbon R groups; they would have a hydrophobic interaction.

18.47 **a.** Cysteine, with its —SH group, can form disulfide cross-links.
b. Leucine and valine are likely to be found on the inside of the protein structure because they have nonpolar side groups and are hydrophobic.
c. The cysteine and aspartic acid would be on the outside of the protein because they have polar R groups.
d. The order of the amino acids (the primary structure) provides the side chains (R groups), whose interactions determine the tertiary structure of the protein.

18.49 **a.** An enzyme (1) has a tertiary structure that recognizes the substrate.
b. The substrate (3) has a structure that fits the active site of the enzyme.
c. The combination of an enzyme and its substrate forms the enzyme–substrate complex (2).

18.51 **a.** The equation for an enzyme–catalyzed reaction is:
$E + S \rightleftarrows ES \longrightarrow EP \longrightarrow E + P$
E = enzyme, S = substrate, ES = enzyme–substrate complex,
EP = enzyme–product complex, P = products
b. The active site is a region or pocket within the tertiary structure of an enzyme that accepts the substrate, aligns the substrate for reaction, and catalyzes the reaction.

18.53 DNA contains two purines, adenine (A) and guanine (G), and two pyrimidines, cytosine (C) and thymine (T). RNA contains the same bases, except thymine (T) is replaced by the pyrimidine uracil (U).
a. Thymine is present in DNA.
b. Cytosine is present in both DNA and RNA.

18.55 The two DNA strands are held together by hydrogen bonds between the complementary bases in each strand.

18.57 **a.** Since T pairs with A, if one strand of DNA has the sequence —AAAAAA—, the second DNA strand would be —TTTTTT—.
b. Since C pairs with G, if one strand of DNA has the sequence —GGGGGG—, the second DNA strand would be —CCCCCC—.
c. Since T pairs with A, and C pairs with G, if one strand of DNA has the sequence —AGTCCAGGT—, the second DNA strand would be —TCAGGTCCA—.

d. Since T pairs with A and C pairs with G, if one strand of DNA has the sequence —CTGTATACGTT—, the second DNA strand would be —GACATATGCAA—..

18.59 The two DNA strands separate (in a way that is similar to the unzipping of a zipper) to allow each of the bases to pair with its complementary base (A binds with T, C binds with G), which produces two exact copies of the original DNA.

18.61 The three types of RNA are messenger RNA (mRNA), ribosomal RNA (rRNA), and transfer RNA (tRNA).

18.63 In transcription, the information contained in DNA is transferred to mRNA molecules. Transcription begins when the section of a DNA to be copied unwinds, and each base in the DNA template strand is paired with its complementary base to form the messenger RNA molecule: G with C, C with G, T with A, and A with U (uracil).

18.65 The strand of mRNA would have the following sequence: —GGCUUCCAAGUG—.

18.67 a. The codon CUG in mRNA codes for the amino acid leucine (Leu).
b. The codon UCC in mRNA codes for the amino acid serine (Ser).
c. The codon GGU in mRNA codes for the amino acid glycine (Gly).
d. The codon AGG in mRNA codes for the amino acid arginine (Arg).

18.69 a. Melezitose is a trisaccharide.
b. Melezitose contains two glucose molecules and a fructose molecule.

18.71

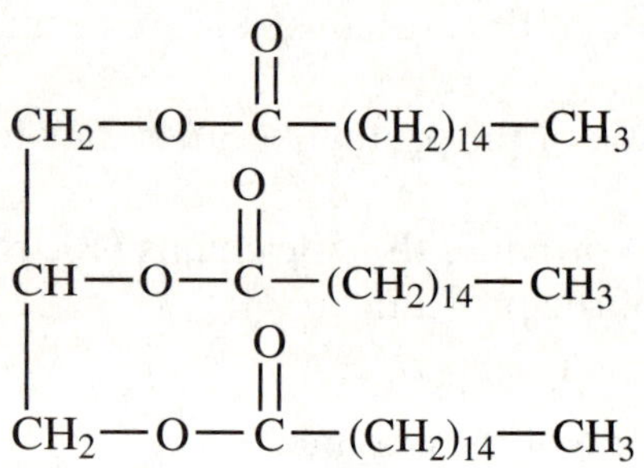

18.73 a.

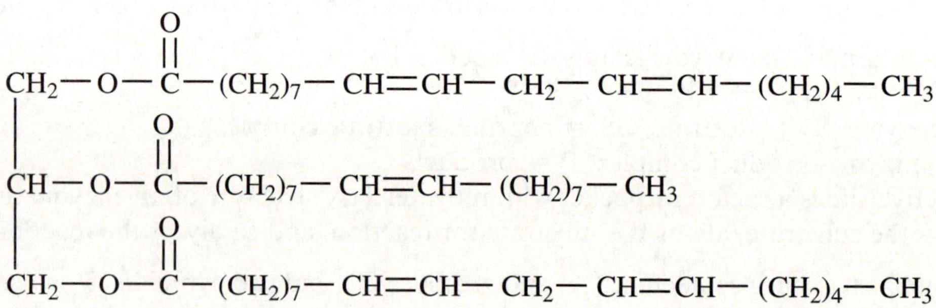

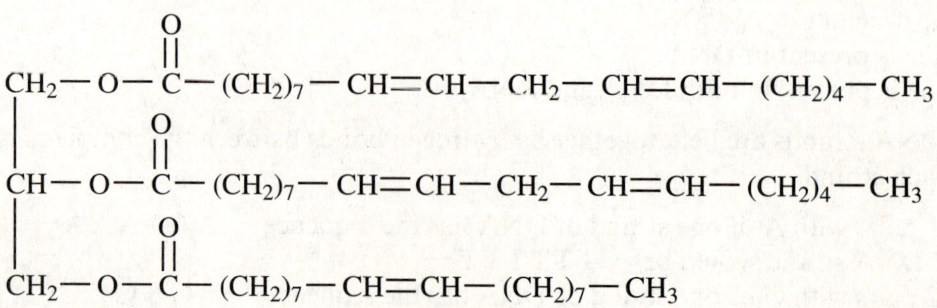

b.

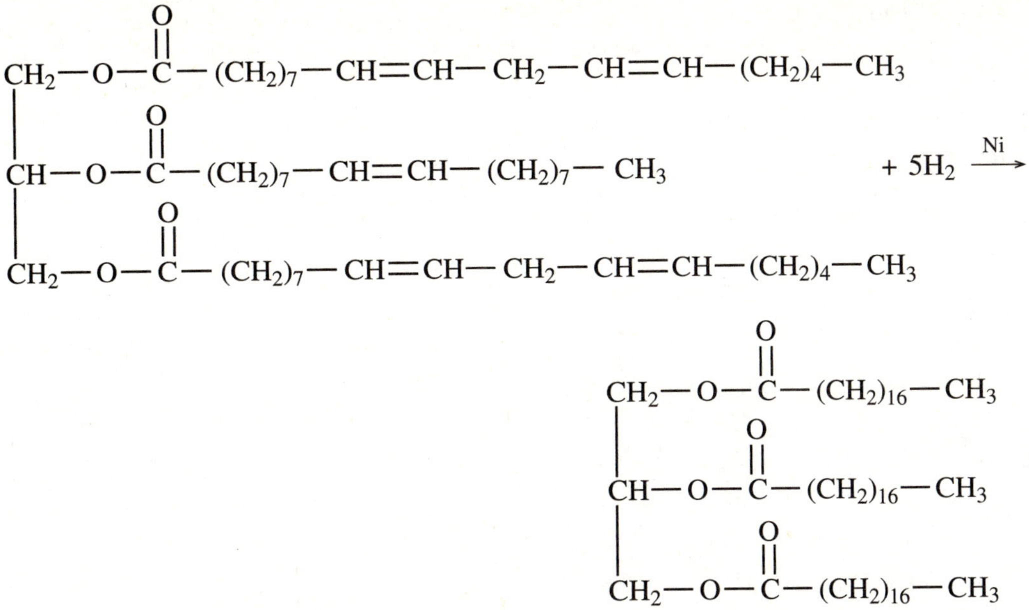

18.75 **a.** Yes, rice and garbanzo beans would provide all the essential amino acids.

b. No, a combination of lima beans and cornmeal would be lacking the essential amino acid tryptophan.

c. No, a salad of garbanzo beans and and lima beans would be lacking the essential amino acid tryptophan.

18.77 **a.** The polar R groups of asparagine and serine would interact by hydrogen bond.

b. The R groups of aspartic acid and lysine would form a salt bridge (ionic bond).

18.79 **a.**

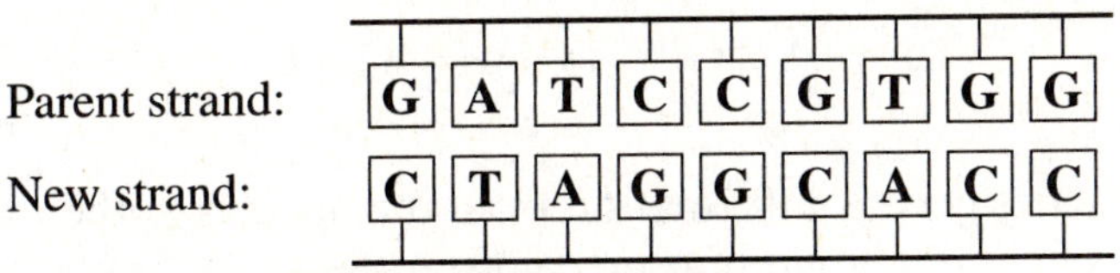

b.

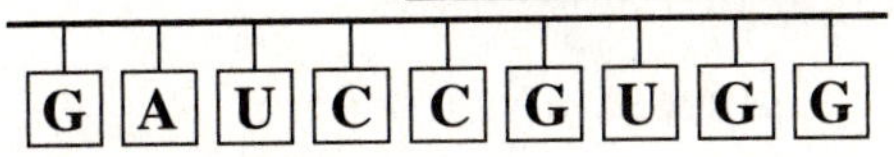

c. -Asp-Pro-Trp-

18.81 Glucose and galactose differ only at carbon 4: in the open-chain structure of glucose the —OH group on carbon 4 is drawn on the right side, and in the open-chain structure of galactose, it is drawn on the left side.

18.83

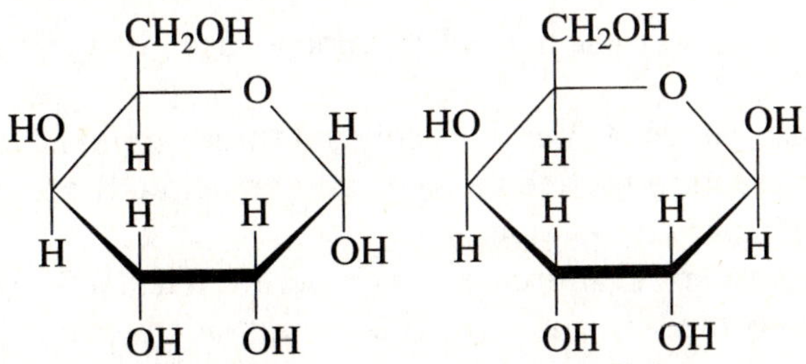

α-D-Gulose β-D-Gulose

18.85

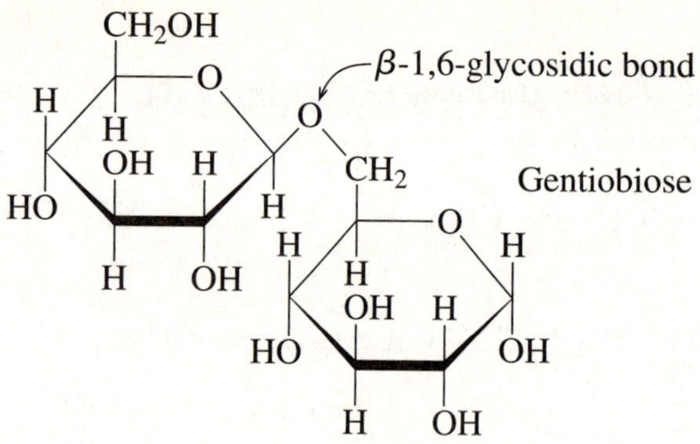

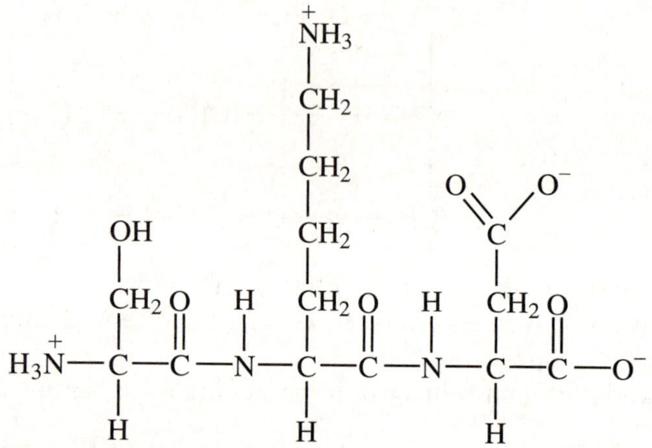

18.87

18.89 **a.** thymine and deoxyribose
b. adenine and ribose
c. cytosine and ribose
d. guanine and deoxyribose

18.91 **a.** —CTGAATCCG—
b. —ACGTTTGATCGA—
c. —TAGCTAGCTAGC—

18.93 **a.** Messenger RNA (mRNA) carries genetic information from the nucleus to the ribosomes.
b. Messenger RNA (mRNA) acts as a template for protein synthesis.

18.95 Raffinose contains the monosaccharides galactose, glucose, and fructose.

18.97 **a.** The secondary structure of a protein depends on hydrogen bonds to form a helix or a pleated sheet. The tertiary structure is determined by the interactions of R groups such as disulfide bonds and salt bridges and determines the three-dimensional structure of the protein.
b. Nonessential amino acids can be synthesized by the body, but essential amino acids must be supplied by the diet.
c. Polar amino acids have hydrophilic R groups, while nonpolar amino acids have hydrophobic R groups.
d. Dipeptides contain two amino acids, whereas tripeptides contain three amino acids.
e. An ionic bond (salt bridge) is an interaction between a basic and an acidic R group; a disulfide bond links the —SH groups of two cysteines.
f. The α helix is the secondary shape like a spiral staircase or corkscrew. The β-pleated sheet is a secondary structure that is formed by many peptide chains side by side.
g. The tertiary structure of a protein is its three-dimensional structure. In the quaternary structure, two or more peptide subunits interact to form a biologically active protein.

18.99 **a.** The mRNA sequence would be: —CGA AAA GUU UUU—
b. From the table of mRNA codons, the amino acids would be -Arg-Lys-Val-Phe-.

18.101 Using the genetic code, the codons indicate the following amino acid sequence: START-Tyr-Gly-Gly-Phe-Leu-STOP

Answers to Combining Ideas from Chapters 17 and 18

CI.37 a. The —OH group in BHT is bonded to a carbon atom in an aromatic ring, which means that BHT has a phenol functional group.

b. BHT is referred to as an "antioxidant" because it, rather than the food, reacts with oxygen in the food container, thus preventing or retarding spoilage of food.

c. $15\ \cancel{\text{oz cereal}} \times \dfrac{1\ \cancel{\text{lb}}}{16\ \cancel{\text{oz}}} \times \dfrac{1\ \cancel{\text{kg}}}{2.205\ \cancel{\text{lb}}} \times \dfrac{50.\ \text{mg BHT}}{1\ \cancel{\text{kg cereal}}} = 21\ \text{mg of BHT (2 SFs)}$

CI.39 a. acetylene $HC{\equiv}CH$

$2C_2H_2(g) + 5O_2(g) \xrightarrow{\Delta} 4CO_2(g) + 2H_2O(g) + \text{energy}$

b. $8.5\ \cancel{\text{L } C_2H_2\ (STP)} \times \dfrac{1\ \cancel{\text{mol } C_2H_2}}{22.4\ \cancel{\text{L } C_2H_2\ (STP)}} \times \dfrac{5\ \cancel{\text{mol } O_2}}{2\ \cancel{\text{mol } C_2H_2}} \times \dfrac{32.00\ \text{g } O_2}{1\ \cancel{\text{mol } O_2}} = 30.\ \text{g of } O_2\ \text{(2 SFs)}$

c. $30.0\ \cancel{\text{g } C_2H_2} \times \dfrac{1\ \cancel{\text{mol } C_2H_2}}{26.04\ \cancel{\text{g } C_2H_2}} \times \dfrac{4\ \cancel{\text{mol } CO_2}}{2\ \cancel{\text{mol } C_2H_2}} \times \dfrac{22.4\ \text{L } CO_2\ (STP)}{1\ \cancel{\text{mol } CO_2}}$

$= 51.6\ \text{L of } CO_2\ \text{(STP) (3 SFs)}$

CI.41 a. Adding NaOH will saponify the glyceryl tristearate (fat), breaking it up into fatty acid salts and glycerol, which are soluble and will wash down the drain.

b.

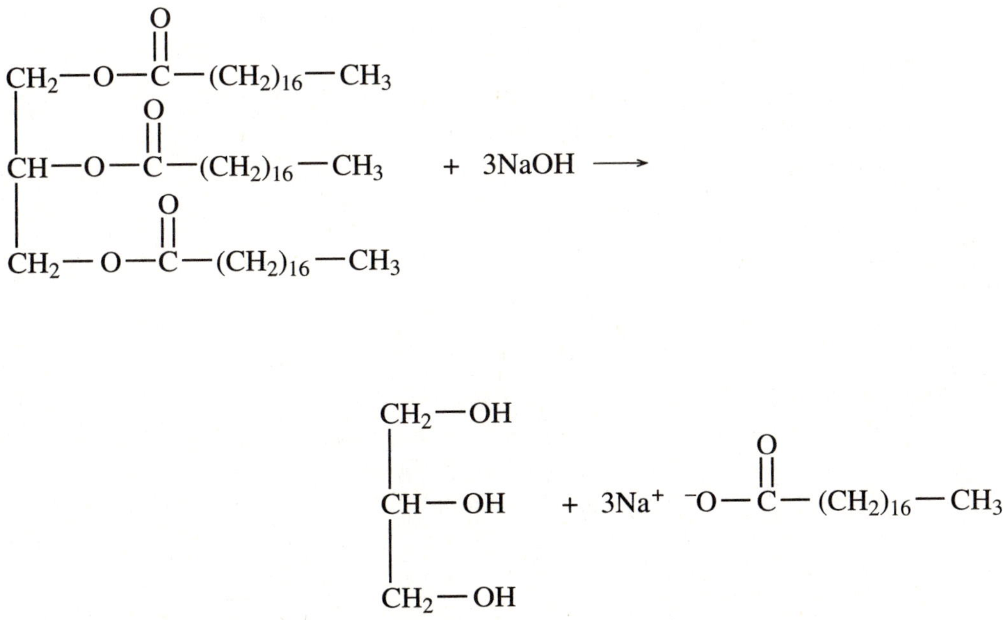

Glycerol | Salts of stearic acid

c. Tristearin $= C_{57}H_{110}O_6$

molar mass of tripalmitolein

$(C_{57}H_{110}O_6) = 57(12.01\ \text{g}) + 110(1.008\ \text{g}) + 6(16.00\ \text{g}) = 891.5\ \text{g}$

$10.0\ \cancel{\text{g tristearin}} \times \dfrac{1\ \cancel{\text{mol tristearin}}}{891.5\ \cancel{\text{g tristearin}}} \times \dfrac{3\ \cancel{\text{mol NaOH}}}{1\ \cancel{\text{mol tristearin}}} \times \dfrac{1000\ \text{mL solution}}{0.500\ \cancel{\text{mol NaOH}}}$

$= 67.3\ \text{mL of NaOH solution (3 SFs)}$

CI.43 a. 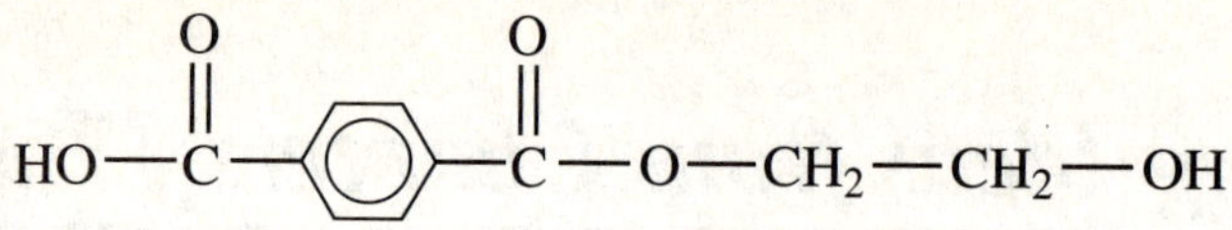

b. 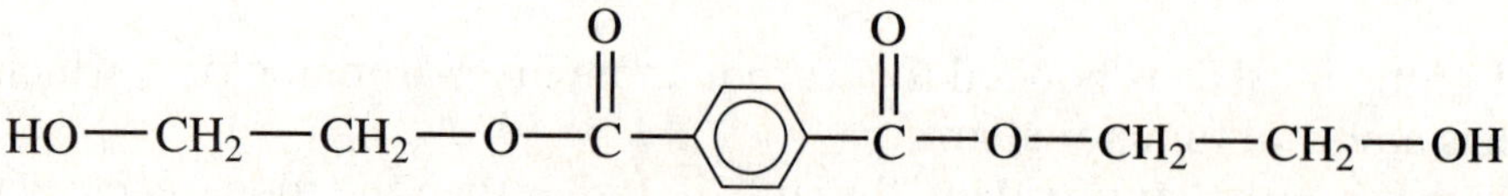

c. $2.4 \times 10^9 \text{ lb PETE} \times \dfrac{1 \text{ kg PETE}}{2.205 \text{ lb PETE}} = 1.1 \times 10^9 \text{ kg of PETE (2 SFs)}$

d. $1.1 \times 10^9 \text{ kg PETE} \times \dfrac{1000 \text{ g PETE}}{1 \text{ kg PETE}} \times \dfrac{1 \text{ mL PETE}}{1.38 \text{ g PETE}} \times \dfrac{1 \text{ L PETE}}{1000 \text{ mL PETE}}$

$= 8.0 \times 10^8 \text{ L of PETE (2 SFs)}$

e. $8.0 \times 10^8 \text{ L PETE} \times \dfrac{1 \text{ landfill}}{2.7 \times 10^7 \text{ L PETE}} = 30. \text{ landfills (2 SFs)}$